Bayesian Network

edited by
Dr. Ahmed Rebai

Bayesian Network
Edited by Dr. Ahmed Rebai

Republished by InTech
Janeza Trdine 9, 51000 Rijeka, Croatia

Publishing Process Manager Iva Lipovic

Technical Editor Goran Bajac

Cover Designer Martina Sirotic

Additional hard copies can be obtained from orders@intechopen.com

Bayesian Network, Edited by Dr. Ahmed Rebai
p. cm.
ISBN 978-953-307-124-4

For EU product safety concerns:
IN TECH d.o.o., Prolaz Marije Krucifikse Kozulić 3, 51000 Rijeka, Croatia,
info@intechopen.com or visit our website at intechopen.com.

Meet the editor

Prof. Ahmed Rebai received his engineer degree in Statistical Genetics in 1991 from the Institut National Agronomique de Paris-Grignon, France and his PhD from the same university in 1995 on developing methods and computational tools for genetic mapping. He is full-professor of Bioinformatics and laboratory director at the Centre of Biotechnology of Sfax (CBS) since 2008. Dr. Rebai published more than 200 journal articles, edited one book and filed three patents. According of Scopus database, Dr Rebai has 2364 citations in scientific publications and his Hirsch index is 28. Dr Rebai was awarded the National Medal (4th category) of The Tunisian Republic in Science and Education in 2006. His recent work in evolutionary biology includes the conservation of charge clusters in eukaryotic and prokaryotic proteomes and their functional annotation.

Contents

Preface

Bayesian networks are graphical models that represent the probabilistic relationships among a large number of variables and perform probabilistic inference with those variables. They constitute a formal framework for the representation and communication of decisions resulting from reasoning under uncertainty. Bayesian networks, which were named after Thomas Bayes (1702-1761), one of the founders of the probability theory, have emerged from several mathematical researches made in the 1980s, and particularly from works on belief networks, causal networks and influence diagrams.

Bayesian networks were first known in the 1990s as Probabilistic Expert Systems, inspired by the seminal book of Judea Pearl (1988), who was a pioneer of the probabilistic approach to artificial intelligence and is referred to as the founder of Bayesian networks. Bayesian networks are thus at least 22 years old and during the last two decades a lot of work has beendone on learning and inference with Bayesian networks. The last ten years particularly saw a massive increase in the application of BN to real-world problems, including diagnosis, forecasting, manufacturing control, information retrieval, prediction and even planning. Almost all scientific and technical fields have seen the successful use of BN as a tool for modelling the complex relationships among a large number of variables and for doing inference. The most recent applications have been in information and communications technologies, biomedicine, genomics and bioinformatics.

The first decade of this new millennium saw the emergence of excellent algorithms for learning Bayesian networks from data and for doing inference in Bayesian networks and influence diagrams. According to Google Scholar, the number of research papers and technical reports on Bayesian networks is over fifty thousand and at least seven specific books on Bayesian networks were published in 2009.

Despite this abundance of literature, there is still a need for specialized books that present original contributions both in methodology and applications of Bayesian networks. This book emphasizes these two aspects and is intended for users (current or potential) of Bayesian networks in both academic institutions (researchers, teachers, students) and industry (engineers, analysts, etc.) who want to stay up to date with Bayesian network algorithms and technologies and their use in building probabilistic expert systems and modelling complex systems.

The book is organized in two major parts. The first part, extending from chapter 1 to 10, mainly deals with theory and algorithms for learning and inference in Bayesian networks. The second part, composed of all subsequent chapters, gives selected applications of Bayesian networks in several fields, including fault diagnosis, information technology, telecommunication networks, traffic flow, building design and biology.

The book chapters are original manuscripts written by experienced researchers that have made significant contributions to the field of Bayesian networks. Although all chapters are self- contained, the reader should be familiar with texts written in mathematical and statistical language to gain full benefit from the book. I am convinced that this book will be a very useful tool for everyone who is concerned with modelling systems containing causality with inherent uncertainty and I hope that readers will find not only technical aspects for using and implementing Bayesian networks to solve their problem, but also new ideas on how their current research and work can benefit from one of the major tools of the 21st century.

Editor

Dr. Ahmed Rebai
Unit of Bioinformatics and Biostatistics,
Centre of Biotechnology of Sfax

1

Learning parameters and structure of Bayesian networks using an Implicit framework

Hanen Ben Hassen*, Lobna Bouchaala*,
Afif Masmoudi** and Ahmed Rebai*
**Unit of Bioinformatics and Biostatistics, Centre of Biotechnology of Sfax, Tunisia*
*** Laboratory of Probability and Statistics, Faculty of Science of Sfax, Tunisia*

1. Introduction

A large amount of work has been done in the last ten years on learning parameters and structure in Bayesian networks (BNs) (see for example Neapolitan, 2005). Within the classical Bayesian framework, learning parameters in BNs is based on priors; a prior distribution of the parameters (prior conditional probabilities) is chosen and a posterior distribution is then derived given the data and priors, using different estimations procedures (for example Maximum a posteriori (MAP) or Maximum likelihood (ML),...). The Achille's heal of the Bayesian framework resides in the choice of priors. Defenders of the Bayesian approach argue that using priors is, in contrary, the strength of this approach because it is an intuitive way to take into account the available or experts knowledge on the problem. On the other side, contradictors of the Bayesian paradigm have claimed that the choice of a prior is meaningless and unjustified in the absence of prior knowledge and that different choices of priors may not lead to the same estimators. In this context, the choice of priors for learning parameters in BNs has remained problematic and a controversial issue, although some studies have claimed that the sensitivity to priors is weak when the learning database is large.
Another important issue in parameter learning in BNs is that the learning datasets are seldom complete and one have to deal with missing observations. Inference with missing data is an old problem in statistics and several solutions have been proposed in the last three decades starting from the pioneering work of (Dempster et al., 1977). These authors proposed a famous algorithm that iterates, until convergence towards stationary point, between two steps, one called Expectation or E-step in which the expected values of the missing data are inferred from the current model parameter configuration and the other, called Maximization or M-step, in which we look for and find the parameter values that maximize a probability function (e.g. likelihood). This algorithm, known as the Expectation-Maximization (or EM) algorithm has become a routine technique for parameters estimation in statistical models with missing data in a wide range of applications. Lauritzen, (1995) described how to apply the EM algorithm to learn parameters for known structure BNs using either Maximum-Likelihood (ML) or maximum a posteriori (MAP) estimates (so called EM-MAP) (McLachlan et al., 1997).
Learning structure (graphical structure of conditional dependencies) in BNs is a much more complicated problem that can be formally presented in classical statistics as a model selection problem. In fact, it was shown that learning structure from data is an NP-hard problem

(Chickering et al., 2004) and that the number of structures for a given number of nodes is super-exponential (Robinson, 1977), making the exploration of the space of all possible structures practically infeasible. Structure learning in BNs has been the subject of active research in the last five years, boosted by the application to high-throughput data in biology, and different heuristics have been proposed. Two major classes of methods can be distinguished; those based on optimizing a score function (finding the structure that maximizes the joint probabilities of the network or some function of it) and those based on correlations (see Leray, (2006) for a review).
Hassairi et al., (2005) have proposed a new inference framework in statistical models that they named "Implicit inference". Implicit inference can be shortly defined as "Bayesian inference without priors" which seems like a nonsense at first sight. In fact, Implicit inference derives a special kind of posterior distribution (called Implicit distribution) that corresponds to an improper choice of the prior distribution (see details below). We recently applied this new Implicit inference framework to learning parameters in BNs with complete (Ben Hassen et al., 2008) and incomplete data (Ben Hassen et al., 2009). In this last work, a novel algorithm, similar to EM (that was called I-EM) was proposed and was shown to have better convergence properties compared to it. For structure learning in BNs, we also proposed a new score function (Implicit score) and implemented it within well known algorithms (Bouchaala et al., 2010).
In this chapter, we give a thorough presentation of the Implicit method applied to parameters and structure learning in BNs and discuss its advantages and caveats. An example application is given to illustrate the use of our method.

2. Inference with the Implicit Method

2.1 A quick tour in the Implicit world

The basic idea of the Bayesian theory is to consider any unknown parameter θ as a random variable and to determine its posterior (conditional) distribution given data and an assumed prior distribution (see for example Robert, 1994). The choice of a prior is generally based on the preliminary knowledge of the problem.
Recently, Hassairi et al., (2005) introduced the concept of Implicit distribution which can be described as a kind of posterior distribution of a parameter given data. To explain the principle of Implicit distribution let us consider a family of probability distributions $\{p(x/\theta),\ \theta \in \Theta\}$ parameterized by an unknown parameter θ in a set Θ, where x is the observed data.
The Implicit distribution $p(\theta/x)$ is calculated by multiplying the likelihood function $p(x/\theta)$ by a counting measure σ if Θ is a countable set and by a Lebesgue measure σ if Θ is an open set (σ depends only on the topological structure of Θ) and then dividing by a norming constant $c(x) = \int_\Theta p(x/\theta)\sigma(d\theta)$. Therefore the Implicit distribution is given by the following formula $p(\theta/x) = (c(x))^{-1}p(x/\theta)\sigma(\theta)$ and plays the role of a posterior distribution of θ given x in the Bayesian method, corresponding to a particular improper prior which depends only on the topology of Θ (without any statistical assumption). The Implicit distribution, which exists for most (but not all) statistical models, can be used for the estimation of the parameter θ following a Bayesian methodology. In fact, the Implicit estimator $\widehat{\theta}$ of θ corresponds to the mean (first moment) of the Implicit distribution.

2.2 A simple example: Implicit estimation in binomial distribution case

To illustrate how the Implicit method proceeds let us consider a simple example. Let $X = (N_1, N_2)$ be a random variable following a binomial distribution with unknown parameters $N = N_1 + N_2$ and $\theta = (\theta_1, \theta_2)$. We first estimate N by the Implicit method after that we use the estimate $\widehat{N}$ to estimate θ. After some calculations, we obtain

$$P(N/X) = \frac{P(X/N)}{C(X)} = C_N^{\check{N}_1}\theta_1^{N-\check{N}_1}(1-\theta_1)^{\check{N}_1+1},$$

where $\check{N}_1 = N - N_1 = \sum_{i=2}^{r} N_i$.

So, the Implicit distribution of N given $X = (N_1, ..., N_r)$ is a Pascal distribution with parameters $1-\theta_1$ and $\check{N}_1 + 1$. Suppose that θ_1 is known, the Implicit estimator $\widehat{N}$ of N is the mean of the Pascal distribution:

$$\widehat{N} = E(N/X) = \sum_{N \geq 0} N C_N^{\check{N}_1}\theta_1^{N-\check{N}_1}(1-\theta_1)^{\check{N}_1+1}.$$

Let N_{ob} be the number of observations and take

$$\theta_{k_0} = \max\{\frac{N_k}{N_{ob}}; \frac{N_k}{N_{ob}} \leq \frac{1}{r-1} \text{ and } 1 \leq k \leq r\}.$$

After some calculations, we have

$$\widehat{N} = \frac{(\check{N}_{k_0}+1)}{1-\theta_{k_0}} = N_{ob} + \frac{N_{k_0}}{\check{N}_{k_0}},$$

where $\check{N}_{k_0} = N_{ob} - N_{k_0}$

Consequently, the probability of the next observation to be in state x^k given a dataset D is obtained by

$$\widehat{\theta}_k = P(X_{N_{ob}+1} = x^k/D) = \frac{N_k+1}{\widehat{N}+r}, 1 \leq k \leq r \text{ and } k \neq k_0 \tag{2.1}$$

and $\widehat{\theta}_{k_0} = 1 - \sum_{i \neq k_0} \widehat{\theta}_i$

other examples and selected applications of Implicit distributions can be found in the original paper (Hassairi et al., 2005).

2.3 Implicit inference with Bayesian Networks

Formally, a Bayesian network is defined as a set of variables $X = \{X_1, ..., X_n\}$ with :(1) a network structure S that encodes a set of conditional dependencies between variables in X, and (2) a set P of local probability distributions associated with each variable. Together, these components define the joint probability distribution of X.

The network structure S is a directed acyclic graph (DAG). The nodes in S correspond to the variables in X_i. Each X_i denotes both the variable and its corresponding node, and $Pa(X_i)$ the parents of node X_i in S as well as the variables corresponding to those parents. The lack of possible arcs in S encode conditional independencies. In particular, given structure S, the joint probability distribution for X is given by the product of all specified conditional probabilities:

$$P(X_1, ..., X_n) = \prod_{i=1}^{n} P(X_i / Pa(X_i)) \tag{3.1}$$

a factorization that is known as *the local Markov property* and states that each node is independent of its non descendant given the parent nodes. For a given BN the probabilities will thus depend only on the structure of the parameters set.

3. Learning parameters from complete data

In this section we consider the learning of parameters in BNs with discrete variable, that is for every node i the associated random variable X_i takes r_i states :

$$node\ 1 \rightarrow X_1 \in \{x_1^1, ..., x_1^{r_1}\}$$

$$node\ 2 \rightarrow X_2 \in \{x_2^1, ..., x_2^{r_2}\}$$

$$\vdots$$

$$node\ i \rightarrow X_i \in \{x_i^1, ..., x_i^{r_i}\}$$

$$\vdots$$

$$node\ n \rightarrow X_n \in \{x_n^1, ..., x_n^{r_n}\}.$$

Let D be a dataset and let N_{ijk} be a number of observations in D for which the node i is in state k and its parents are in state j that is $X_i = x_i^k$ and $Pa(X_i) = x_i^j$. Note that, since each node might have two or more parents, state j corresponds to a combination of states of the parents. For example if a node has three parents, each having three states, then there are 27 states of the parents and j takes values from 1 to 27.
The distribution of X_i is multinomial with parameters N_{ij} and $\theta_{ij} = (\theta_{ij2}, ..., \theta_{ijr_i})$, where $N_{ij} = \sum_{k=1}^{r_i} N_{ijk}$ and $\theta_{ijk} = P(X_i = x_i^k / Pa(X_i) = x^j)$; $k = 1, ..., r_i$ and $\sum_{k=1}^{r_i} \theta_{ijk} = 1$

$$P(X_i = (N_{ij1}, ..., N_{ijr_i}) / Pa(X_i) = x^j) = N_{ij}! \prod_{k=1}^{r_i} \frac{\theta_{ijk}^{N_{ijk}}}{N_{ijk}!}.$$

Then N_{ij} and θ_{ij} are unknown parameters that will be estimated by the Implicit method. Given a network S, consider for node i, N_{ijob} is the observed number of occurrences of the node i and its parents are in the state j.
Let $\theta_{ijk(0)} = \frac{N_{ijk(0)}}{N_{ijob}} = \max\{\frac{N_{ijk}}{N_{ijob}};\ \frac{N_{ijk}}{N_{ijob}} \leq \frac{1}{r_i - 1}$ and $1 \leq k \leq r_i\}$.
The application of the Implicit method gives the following estimation of N_{ij} and θ_{ij}:

$$\widehat{N}_{ij} = N_{ijob} + \frac{N_{ijk(0)}}{\overset{\vee}{N}_{ijk(0)}}; \tag{3.2}$$

where $\overset{\vee}{N}_{ijk(0)} = N_{ijob} - N_{ijk(0)}$ and

$$\widehat{\theta}_{ijk} = \frac{N_{ijk}+1}{\widehat{N}_{ij}+r_i} \quad \text{if } k \neq k(0) \tag{3.3}$$

and

$$\widehat{\theta}_{ijk(0)} = 1 - \sum_{k \neq k(0)} \widehat{\theta}_{ijk}$$

4. Learning parameters from incomplete data

Consider a dataset D with missing data, we compute the Implicit distribution $P(\theta/D)$ and use the distributions in turn to compute expectation of parameters of interest. Let X be a random variable that follows a multinomial distribution with parameters N and $\theta = (\theta_1, ..., \theta_r)$ such that $Y = (N_1, ..., N_r) \subset X$ and $Z = (N_1^*, ..., N_r^*) \subset X$ denote the observed and unobserved variables, respectively. So, $X = (N_1 + N_1^*, ..., N_r + N_r^*)$
and $P(\theta/Y) = \sum_Z P(Z/Y)P(\theta/Y, Z)$

To estimate the parameters θ_{ijk} of the network, with incomplete dataset, we propose a new iterative algorithm named Implicit EM (or in short I-EM) algorithm. Consider a node i with parents in the state j and a dataset D which contains $N_{ij}^{(0)}$ observed and unobserved values in such state. Let $N_{ijob}^{(0)}$ the observed values in D, so $N_{ij}^{(0)} > N_{ijob}^{(0)}$ and $N_{ij}^{(0)} - N_{ijob}^{(0)}$ represents the number of unobserved states.
So, the initial conditions for a node i are:
$N_{ij}^{(0)}$ is the number of observed and unobserved states.
$\theta_{ijk}^{(0)}$ is the observed frequency of the node i in the state k given its parents in the state j. Then,
$N_{ijk}^{(0)} = N_{ij}^{(0)}\theta_{ijk}^{(0)}$ is the number of observed occurrences of the node i in the state k and its parents in the state j.
$N_{ijob}^{(0)} = \sum_{k=1}^{r_i} N_{ijk}^{(0)}$
The I-EM algorithm is iterative and involves three steps; the first step consists in getting the maximum of the conditional frequencies, the second step estimates the number of observations from the first step and the third computes the other conditional probabilities. Formally, the algorithm iterates through the following steps, until convergence:
(1) Choose the maximum frequency $k(0)$
(2) Estimate the number of observations $N_{ij}^{(1)}$
(3) Compute the conditional probabilities $\theta_{ijk}^{(1)}$
with the
stop condition being:

Compute the sum of estimated occurrences $\sum_{k=1}^{r_i} N_{ijk}^{(t)}$

if $\sum_{k=1}^{r_i} N_{ijk}^{(t)} > N_{ij}^{(0)}$ then stop, otherwise continue steps (1) to (3).

The philosophy of our algorithm is to virtually fill the missing data for all nodes until all missing cells in the database are completed. A detailed description and a formal proof of convergence of the I-EM algorithm is given in (Ben Hassen et al., 2009).

5. Learning Bayesian Network Structure

Learning Bayesian Network structure from database is an NP-hard problem and several algorithms have been developed to obtain a sub-optimal structure from a database. Most of the widely used methods are score metric-based methods. By these methods a scoring metric is defined and computed for each candidate structure and a search strategy (algorithm) is used to explore the space of possible, alternative structures and identify the one (or those) having the highest score.

5.1 Score metrics

A scoring criteria for a DAG is a function that assigns a value to each DAG based on the data. Cooper and Hersovits (1992) proposed a score based on a Bayesian approach with Dirichlet priors(known as BD: Bayesian Dirichlet). Starting from a prior distribution on the possible structure $P(B)$, the objective is to express the posterior probability of all possible structures ($P(B|D)$ or simply $P(B, D)$) conditional on a dataset D:

$$S_{BD}(B,D) = P(B,D) = \int_{\Theta} P(D|\Theta,B)P(\Theta|B)P(B)d\Theta = P(B)\int_{\Theta} P(D|\Theta,B)P(\Theta|B)d\Theta$$

The BD score is analitycally expressed as:

$$S_{BD}(B,D) = P(B)\prod_{i=1}^{n}\prod_{j=1}^{q_i}\frac{(r_i-1)!}{(N_{ij}+r_i-1)!}\prod_{k=1}^{r_i}N_{ijk}! \tag{5.2}$$

The BIC (Bayesian Information Criteria) score metric was proposed by Schwartz (1978) and is defined as:

$$S_{BIC} = logL(D|\theta^{MV},B) - \frac{1}{2}Dim(B)logN \tag{5.3}$$

where θ^{MV} is the maximum likelihood estimate of the parameters, B is the BN structure and $Dim(B)$ is the dimension of the network defined by : $Dim(B) = \sum_{i=1}^{n} Dim(X_i, B)$ and $Dim(B) = (r_i - 1)q_i$

Another common score in structure learning is the Mutual Information (MI). The Mutual Information between two random variables X and Y, denoted by $I(X,Y)$ is defined by Chow and Liu (1968):

$$I(X,Y) = H(X) - H(X|Y) \tag{5.4}$$

Where $H(X)$ is the entropy of random variables X defined as:
$H(X) = -\sum_{i=1}^{r_x} P(X = x_i)log(P(X = x_i))$
and
$H(X|Y) = -\sum_{i=1}^{r_x}\sum_{j=1}^{r_y} P(X = x_i/Y = y_j)log(P(X = x_i|Y = y_j))$ where r_x and r_y are the number of discrete states for variables X and Y, respectively.

5.2 Algorithms for structure learning

One of the most used algorithms is the K2 algorithm (Cooper and Herskovits (1992). This algorithm proceeds as follows: we assume an initial ordering of the nodes to reduce computational complexity and assume that the potential parent set of node X_i can include only those nodes that precede it in the input ordering.
Chow et al., (1968) proposed a method derived from the Maximum Weight Spaning Tree (MWST). This method associates a weight to each potential edges $X_i - X_j$ of the tree. This weight may be the MI(equation 5.4), or the local variation of the score proposed by (Heckerman et al., 1994). Given the weight matrix, we can use the Kruskal algorithm (Kruskal 1956) to obtain a directed tree by choosing a root and then browsing the tree by an in-depth search.
The GS (Greedy Search) algorithm takes an initial graph, then associates a score for each neighborhood. The graph with the highest score in this neighborhood is then chosen as the starting graph for the next iteration.

5.3 The Implicit Score (IS)

The Implicit Score(IS) have the same derivation as the the BD score in which the Implicit estimators of the paremeters (see equations 3.2 and 3.3) are used rather than Bayesian estimators (Bouchaala et al., 2010). The expression of the Implicit score (IS) is thus obtained by substituting in equation 5.2 N_{ijk} by $\widehat{N}_{ijk}\widehat{\theta}_{ijk}$ and N_{ij} by $\widehat{N}_{ij}$:

$$S_{IS}(B,D) = P(B) \prod_{i=1}^{n} \prod_{j=1}^{q_i} \frac{(r_i - 1)!}{(\widehat{N}_{ij} + r_i - 1)!} \prod_{k=1}^{r_i} \widehat{N}_{ij}\widehat{\theta}_{ijk}! \tag{5.5}$$

.

We implemented this score within K2, MWST and GS algorithms for network structure learning. Performance of IS was evaluated on a benchmark database (ASIA network (lauritzen and Spiegelhater, 1988) in comparison to other score metrics, namely BIC, BD and MI.
The experiments were carried out on different datasets randomly selected from the ASIA database (20,000 data points). The dataset size was varied from 100 to 1000 (in order to test robusteness to small databases)and 20 replicates were performed for each database size. The performance of each score was evaluated by four criteria : the average (over the replicates) numbers of missings edges, additional edges, reversed edges and correct edges (relative to the true structure inferred from the whole database).
Table 1 below shows that the Implicit score yields improved performance over other scores when used with the MWST and GS algorithm, and have similar performance when implemented within K2 algorithm.

6. Application to real data: thyroid cancer prognosis

To illustrate how the Implicit method proceed, we consider an example on thyroid cancer. The dataset comprises data on 92 thyroid cancer patients described in Rebai et al., (2009a,b). We considered only five nodes with two states each:
Therapeutic response (TR): no response (1)/complete remission (2)
Metastasis (MET) yes (1)/no (2).
Thyroglobulin level (TG) low: $\leq$ 30 ng/mL (1); high: $>$ 30 ng/mL (2)).
The genotype of a single nucleotide polymorphism within the HER2 gene (HER2): genotype AA(1); genotype AG (2)(here genotype GG was totally absent).
The genotype of a single nucleotide polymorphism within the estrogen receptor gene (ER): genotype AA and AG(1); genotype GG (2) (note here that genotypes AA and AG were merged

MWST Algorithm	IS	BIC	MI	Best Result
Correct Arc	4,39	2,62	2,71	8
Reversed Arc	1,93	3,08	3,08	0
Missing Arc	1,68	2,3	2,21	0
Extra Arc	0,68	1,32	1,22	0

(A)

K2 Algorithm	IS	BIC	BD	Best Result
Correct Arc	4,66	4,7	4,88	8
Reversed Arc	1,59	1,69	1,71	0
Missing Arc	1,75	1,61	1,41	0
Extra Arc	1,51	1,34	1,85	0

(B)

GS Algorithm	BIC-BIC	MI-BD	IS-BIC	IS-BD	Best Result
Correct Arc	4,18	4,08	5,28	5,42	8
Reversed Arc	1,92	2,34	0,82	0,92	0
Missing Arc	1,9	1,58	1,9	1,66	0
Extra Arc	0,88	1,82	0,62	1,26	0

(C)

Table 1. Comparative Analysis of the Implicit score (IS) with BD, BIC and MI scores implemented within (A) MWST algorithm, (B)K2 algorithm and (C) GS algorithm.

together because A is a risk allele). These two polymorphisms were included due to their highly significant association, inferred by bivariate and multivariate statistical tests, with the three other variables (see Rebai et al., 2009b for more details on the data).
The structure obtained by the K2 algorithm with the Implicit score is given in figure 1. Note that the same structure was obtained by the BD score.

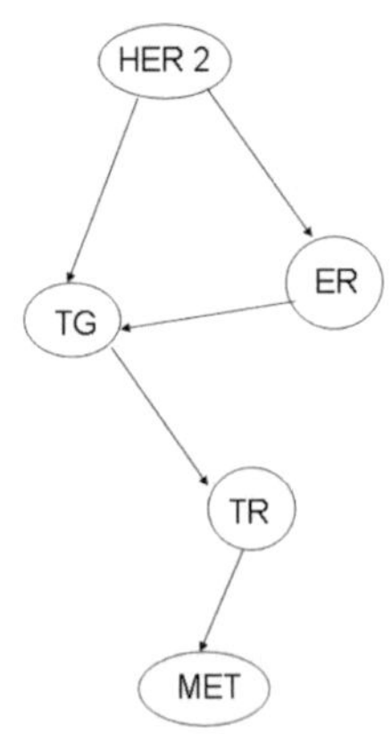

Fig. 1. The structure obtained by the K2 algorithm with the Implicit score

Using this structure we estimated the parameters by the Implicit approach. For parameter notations, nodes are denoted as: (1)ER, (2)HER2, (3)TG, (4)TR and (5)MET. Parameter t_{ijk} corresponds to the node i in state k and its parents in state j. According to the structure in figure1, one node (HER2) has no parents, three nodes have one parent and one node has two parents (TG). Consequently we have two parameters for HER2, four for ER, TR and MET and eight for TG.

parameter	estimated value	parameter	estimated value
t111	0.20608440	t112	0.79391560
t121	0.50137741	t122	0.49862259
t211	0.72054405	t212	0.27945595
t311	0.47058824	t312	0.52941176
t321	0.53736875	t322	0.46263125
t331	0.43298969	t332	0.56701031
t341	0.85161290	t342	0.14838710
t411	0.21479714	t412	0.78520286
t421	0.94267026	t422	0.05732974
t511	0.07434944	t512	0.92565056
t521	0.92560895	t522	0.07439105

Table 2. Parameters Estimates from a complete dataset of 94 thyroid cancer patients based on structure in Fig1.

If we look at the TR node and particularly the probability of the occurrence of a positive response to therapy (t412) we see that it is high (almost 80 %) when the parent (TG) is at state 1, that is for patients with low TG levels while it is small (about 6 %) for patients with high TG levels (t422). This confirm the high prognostic value of TG level, well recognized by clinicians. Another expected result is that the probability of having metastasis is very high (92 %) when the patient does not respond to therapy (t512). However, an original result is that the probability of having a high TG levels is small (about 15 %) when the patient carries non-risk genotypes at the two single nucleotide polymoprhisms (t342)compared to corresponding probabilities to carriers of a risk genotype for at least one SNP (50 % on average). This means that the two SNP can be used as early prognostic factors that predict the increase in TG levels, which might be of help for therapeutic adjustment (preventive treatment,..).
In order to test the robustness of the Implicit method in parameter learning, we introduced 5 % missing data by randomly deleting 5 % of the data for each node. Table 3 gives the parameters estimates and shows that the change in parameters estimates is slight except for the node without parents (HER2). This property of Implicit estimators has already been reported in Ben Hassen et al., (2009) and is expected because nodes without parents are expected to be more sensitive to missing data.

7. Conclusion

In this chapter, we described the Implicit method, a new framework for learning structure and probabilities in Bayesian networks. We showed how our method proceeds with complete and incomplete data. The use of the Implicit method was illustrated on a real and original dataset of thyroid cancer.
The Implicit method is a new approach that can be seen as a prior-free Bayesian approach. It has the advantages of Bayesian methods without their drawbacks. In fact, the choice of prior information in Bayesian approaches has always been problematic and has been advanced by many critics to be the major weakness of such methods. Implicit method avoids the problem of priors and leads to estimators and algorithms that are easier to derive and to implement.
We showed here and in our previous work that the Implicit score when implemented within

parameter	estimated value	parameter	estimated value
t111	0.2173913	t112	0.7826087
t121	0.3333333	t122	0.6666667
t211	0.7437071	t212	0.2562929
t311	0.4615385	t312	0.5384615
t321	0.5381062	t322	0.4618938
t331	0.4545455	t332	0.5454545
t341	0.9166667	t342	0.08333333
t411	0.2005571	t412	0.7994429
t421	0.9589041	t422	0.04109589
t511	0.0787401	t512	0.9212598
t521	0.948718	t522	0.05128205

Table 3. Table of estimated parameters for a 5 % rate of missing data for thyroid cancer patients

traditional algorithms for structure learning (and particularly the MWST algorithm) leads to better results and seems to be more robust when the database is of relatively small size. This might be a very useful property for applications in medical prognosis or diagnosis of rare diseases, where the number of patients has been a limiting factor to the use of Bayseian networks for modeling the complex relationship between several predicting factors, such as clinical, molecular, biochemical and genetical factors.
The easy implementation of the Implicit algorithm for parameters learning in Bayseian networks with missing data and its performance compared to the EM algorithm and particularly its faster convergence, is one of the reasons that can lead to its adoption for many applications in computational biology and genomics (see Needham et al., 2007).
In its current version, the Implicit method can only handle Bayesian networks with discrete variables. This of course encloses a wide range of applications, but the generalization to networks with continuous or mixed variables is our next challenge and will be addressed in the near future.

Acknowledgments

This work was supported by the Ministry of Higher Education and Scientific Research, Tunisia.

8. References

Ben Hassen, H., Masmoudi, A. and Rebai, A., 2008. Causal inference in Biomolecular Pathways using a Bayesian network approach and an Implicit method. *J. Theoret. Biol.* 4, 717-724.

Ben Hassen, H., Masmoudi, A. and Rebai, A., 2009. Inference in signal transduction pathways using EM algorithm and an Implicit Algorithm : Incomplete data case. *J. comp. Biol.* 16, 1227-1240.

Bouchaala, L., Masmoudi, A., Gargouri, F and Rebai, A., 2010. Improving algorithms for structure learning in Bayesian Networks using a new Implicit score. *Expert. Syst. Appl.*. 37, 5470-5475.

Chickering, D., Heckerman, D. and Meek, C., 2004. Large-Sample Learning of Bayesian Networks is NP-Hard. *J. Mach. Learn. Res.* 5, 1287-1330.

Cooper, G.F. and Herskovits, E., 1992. A Bayesian method for the induction of probabilistic networks from data. *Mach. Learn.* 9, 309-347.

Chow , C. K., Liu, C. N., 1968. Approximating discrete probability distributions with dependence trees. *IEEE T. Knowl. Data. En.*.20, 1-13.

Chrisman, L., 1996. A road map to research on Bayesian networks and other decomposable probabilistic models, *Technical Report, School of Computer Science, CMU, Pittsburgh, PA.*

Cooper, G., Hersovits,E. 1992. Abayesian method for the induction of probabilistic networks from data, *Mach Learn.* 9, 309-347.

Dempster, A.P., Laird, N.M. and Rubin, D.B., 1977. Maximum likelihood from incomplete data via the EM algorithm (with discussion). *J. Roy. Statist. Soc. B.* 39, 1-38.

Hassairi, A., Masmoudi, A. and Kokonendji, C., 2005. Implicit distributions and estimation. *Comm. Statist. Theory Methods.* 34, 245-252.

Heckerman, D., Geiger, D. and Chickering, M., 1994. Learning Bayesian Networks : The combination of knowledge and statistical data, *Proceedings of the* 10th *conference on uncertainty in artificial intelligence.* San francisco, CA, USA : Morgan Kaufmann Publishers. 293-301.

Krause, P., 1996. Learning probabilistic networks. *Technical Report, Philips Research Laboratories, UK.*

Kruskal, J., 1956. On the shortest spaning subtree of a graph and traveling salesman problem. *Proceedings of the American Mathematical Society.* 7, 48-50.

Leray, P., 2006. Habilitation a diriger les recherches: *Automatique et Traitement du Signal Réseaux bayésiens : apprentissage et modélisation de systèmes complexes.* Université de Rouen, UFR des Sciences.

Linggi, B. and Carpenter G., 2006. ErbB receptors: new insights on mechanisms and biology. *Trends. Cell. Biol.* 16, 649-656.

Lauritzen, S.L., 1995. "The EM algorithm for graphical association models with missing data." *Comput. Statist. Data Anal.* 19, 191-201.

Lauritzen, S. and Spiegelhater, D., 1988. Local computation with probabilities on graphical structures and their application on expert systems. J. Roy. Stat. Soc. 50, 157-224.

Mamitsuka, H., 2005. Essential Latent Knowledge for Protein-Protein Interactions: Analysis by Unsupervised Learning Approach *IEEE/ACM Trans. Comput. Biol. Bioinform.* 2, 119-130.

McLachlan, G.J. and Krishnan, T., 1997. *The EM algorithm and extensions.* Wiley, New York.

Mukhopadhyay, N., 2006. Some Comments on Hassairi et al.'s "Implicit Distributions and Estimation". *Comm. Statist. Theory Methods.* 35, 293-297.

Nakayama, T., Asai, S., Sato, N. and Soma, M., 2006. Genotype and haplotype association study of the STRK1 region on 5q12 among Japanese: a case-control study. *Stroke.* 37, 69-76.

Needham, C.J., Bradford, J.R., Bulpitt, A.J., Westhead, D.R., 2007. A Primer on Learning in Bayesian Networks for Computational Biology. *PLoS Comput. Biol.* 3, 8, e129.

Normanno, N., De Luca, A., Bianco, C., Strizzi, L., Mancino, M., Maiello, M.R., Carotenuto, A., De Feo, G., Caponigro, F. and Salomon, D.S., 2006. Epidermal growth factor receptor (EGFR) signaling in cancer. *Gene.* 366, 2-16.

Rebai M, Kallel I, Hamza F, Charfeddine S, Kaffel R, Guermazi F and Rebai A. 2009a. Association of EGFR and HER2 Polymorphisms with Risk and Clinical Features of Thyroid Cancer. *Genet. Test. Mol. Biomarkers,* 13: 779-784.

Rebai M., Kallel I, Charfeddine S., Hamza F, Guermazi F., Rebaï A. 2009b. Association of polymorphisms in oestrogen and thyroid hormone receptors with thyroid cancer risk. *J Rec. Signal Transd.*, 29: 113-118.

Robinson, R.W,1977. Counting unlabeled acyclic digraphs. *Comb. Mathem.*, 622, 28-43

Robert, C.P., 1994. *The Bayesian Choice: A decision-Theoretic Motivation.* Springer-Verlag, New-York.

Schwarz G., 1978. Estimating the dimension of a model. *Ann. Statist.* 6, 461-464.

2

Design of evolutionary methods applied to the learning of Bayesian network structures

Thierry BROUARD, Alain DELAPLACE, Muhammad Muzzamil LUQMAN, Hubert CARDOT and Jean-Yves RAMEL
University François Rabelais
France

1. Introduction

Bayesian networks (BN) are a family of probabilistic graphical models representing a joint distribution for a set of random variables. Conditional dependencies between these variables are symbolized by a Directed Acyclic Graph (DAG). Two classical approaches are often encountered when automatically determining an appropriate graphical structure from a database of cases. The first one consists in the detection of (in)dependencies between the variables (Cheng et al., 2002; Spirtes et al., 2001). The second one uses a scoring metric (Chickering, 2002a). But neither the first nor the second are really satisfactory. The first one uses statistical tests which are not reliable enough when in presence of small datasets. If numerous variables are required, it is the computing time that highly increases. Even if score-based methods require relatively less computation, their disadvantage lies in that the searcher is often confronted with the presence of many local optima within the search space of candidate DAGs. Finally, in the case of the automatic determination of the appropriate graphical structure of a BN, it was shown that the search space is huge (Robinson, 1976) and that is a NP-hard problem (Chickering et al., 1994) for a scoring approach.

In this field of research, evolutionary methods such as Genetic Algorithms (GA) (De Jong, 2006) have already been used in various forms (Acid & de Campos, 2003; Larrañaga et al., 1996; Muruzábal & Cotta, 2004; Van Dijk, Thierens & Van Der Gaag, 2003; Wong et al., 1999; 2002). Among these works, two lines of research are interesting. The first idea is to effectively reduce the search space using the notion of equivalence class (Pearl, 1988). In (Van Dijk, Thierens & Van Der Gaag, 2003) for example the authors have tried to implement a genetic algorithm over the partial directed acyclic graph space in hope to benefit from the resulting non-redundancy, without noticeable effect. Our idea is to take advantage both from the (relative) simplicity of the DAG space in terms of manipulation and fitness calculation and the unicity of the equivalence classes' representations.

One major difficulty when tackling the problem of structure learning with scoring methods – evolutionary methods included – is to avoid the premature convergence of the population to a local optimum. When using a genetic algorithm, local optima avoidance is often ensured by preserving some genetic diversity. However, the latter often leads to slow convergence and difficulties in tuning the GA parameters.

To overcome these problems, we designed a general genetic algorithm based upon dedicated operators: mutation, crossover but also a mutual information-driven repair operator which ensures the closeness of the previous. Various strategies were then tested in order to find a balance between speed of convergence and avoidance of local optima. We focus particularly onto two of these: a new adaptive scheme to the mutation rate on one hand and sequential niching techniques on the other.

The remaining of the chapter is structured as follows: in the second section we will define the problem, ended by a brief state of the art. In the third section, we will show how an evolutionary approach is well suited to this kind of problem. After briefly recalling the theory of genetic algorithms, we will describe the representation of a Bayesian network adapted to genetic algorithms and all the needed operators necessary to take in account the inherent constraints to Bayesian networks. In the fourth section the various strategies will then be developed: adaptive scheme to the mutation rate on one hand and niching techniques on the other hand. The fifth section will describe the test protocol and the results obtained compared to other classical algorithms. A study of the behavior of the used strategies will also be given. And finally, the sixth section will present an application of these algorithms in the field of graphic symbol recognition.

2. Problem settings and related work

2.1 Settings

A probabilistic graphical model can represent a whole of conditional relations within a field $X = \{X_1, X_2, \ldots, X_n\}$ of random variables having each one their own field of definition. Bayesian networks belong to a specific branch of the family of the probabilistic graphical models and appear as a directed acryclic graph (DAG) symbolizing the various dependences existing between the variables represented. An example of such a model is given Fig. 1.

A Bayesian network is denoted $B = \{G, \theta\}$. Here, $G = \{X, E\}$ is a directed acyclic graph whose set of vertices X represents a set of random variables and its set of arcs E represents the dependencies between these variables. The set of parameters θ holds the conditional probabilities for each vertices, depending on the values taken by its parents in G. The probability $k = \{P(X_k|Pa(X_k))\}$, where $Pa(X_k)$ are the parents of variable X_k in G. If X_k has no parents, then $Pa(X_k) = \emptyset$.

The main convenience of Bayesian networks is that, given the representation of conditional independences by its structure and the set θ of local conditional distributions, we can write the global joint probability distribution as:

$$P(X_1, \ldots, X_n) = \prod_{k=1}^{n} P(X_k|Pa(X_k)) \tag{1}$$

2.2 Field of applications of Bayesian networks

Bayesian networks are encountered in various applications like filtering junk e-mail (Sahami et al., 1998), assistance for blind people (Lacey & MacNamara, 2000), meteorology (Cano et al., 2004), traffic accident reconstruction (Davis, 2003), image analysis for tactical computer-aided decision (Fennell & Wishner, 1998), market research (Jaronski et al., 2001), user assistance in

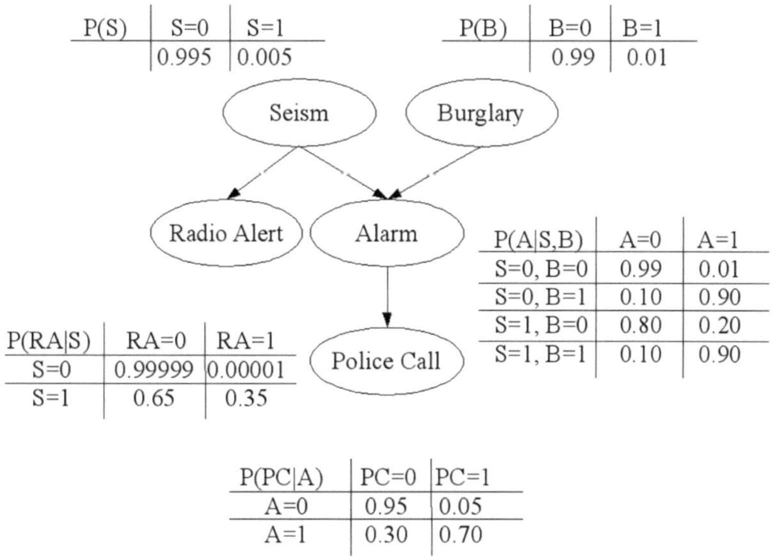

Fig. 1. Example of a Bayesian network.

software use (Horvitz et al., 1998), fraud detection (Ezawa & Schuermann, 1995), human-machine interaction enhancement (Allanach et al., 2004).

The growing interest, since the mid-nineties, that has been shown by the industry for Bayesian models is growing particularly through the widespread process of interaction between man and machine to accelerate decisions. Moreover, it should be emphasized their ability, in combination with Bayesian statistical methods (i.e. taking into account prior probability distribution model) to combine the knowledge derived from the observed domain with a prior knowledge of that domain. This knowledge, subjective, is frequently the product of the advice of a human expert on the subject. This property is valuable when it is known that in the practical application, data acquisition is not only costly in resources and in time, but, unfortunately, often leads to a small knowledge database.

2.3 Training the structure of a Bayesian network

Learning Bayesian network can be broken up into two phases. As a first step, the network structure is determined, either by an expert, either automatically from observations made over the studied domain (most often). Finally, the set of parameters θ is defined here too by an expert or by means of an algorithm.

The problem of learning structure can be compared to the exploration of the data, i.e. the extraction of knowledge (in our case, network topology) from a database (Krause, 1999). It is not always possible for experts to determine the structure of a Bayesian network. In some cases, the determination of the model can therefore be a problem to resolve. Thus, in (Yu et al., 2002) learning the structure of a Bayesian network can be used to identify the most obvious relationships between different genetic regulators in order to guide subsequent experiments.

The structure is then only a part of the solution to the problem but itself a solution.

Learning the structure of a Bayesian network may need to take into account the nature of the data provided for learning (or just the nature of the modeled domain): continuous variables – variables can take their values in a continuous space (Cobb & Shenoy, 2006; Lauritzen & Wermuth, 1989; Lerner et al., 2001) –, incomplete databases (Heckerman, 1995; Lauritzen, 1995). We assume in this work that the variables modeled take their values in a discrete set, they are fully observed, there is no latent variable i.e. there is no model in the field of non-observable variable that is the parent of two or more observed variables.

The methods used for learning the structure of a Bayesian network can be divided into two main groups:

1. Discovery of independence relationships: these methods consist in the testing procedures on allowing conditional independence to find a structure;
2. Exploration and evaluation: these methods use a score to evaluate the ability of the graph to recreate conditional independence within the model. A search algorithm will build a solution based on the value of the score and will make it evolve iteratively.

Without being exhaustive, belonging to the statistical test-based methods it should be noted first the algorithm PC, changing the algorithm SGS (Spirtes et al., 2001). In this approach, considering a graph $G = \{X, E, \theta\})$, two vertices X_i and X_j from X and a subset of vertices $S_{X_i,X_j} \in X/\{X_i, X_j\}$, the vertices X_i and X_j are connected by an arc in G if there is no S_{X_i,X_j} such as $(Xi \perp Xj | S_{X_i,X_j})$ where $\perp$ denotes the relation of conditional independence. Based on an undirected and fully connected graph, the detection of independence allows us to remove the corresponding arcs until the obtention the skeleton of the expected DAG. Then follow two distinct phases: i) detection and determination of the V-structures[1] of the graph and ii) orientation of the remaining arcs. The algorithm returns a directed graph belonging to the Markov's equivalence class of the sought model. The orientation of the arcs, except those of V-structures detected, does not necessarily correspond to the real causality of this model. In parallel to the algorithm PC, another algorithm, called IC (Inductive Causation) has been developed by the team of Judea Pearl (Pearl & Verma, 1991). This algorithm is similar to the algorithm PC, but starts with an empty structure and links couples of variables as soon as a conditional dependency is detected (in the sense that there is no identified subset conditioning S_{X_i,X_j} such as $(Xi \perp Xj | S_{X_i,X_j})$). The common disadvantage to the two algorithms is the numerous tests required to detect conditional independences. Finally, the algorithm BNPC – Bayes Net Power Constructor – (Cheng et al., 2002) uses a quantitative analysis of mutual information between the variables in the studied field to build a structure G. Tests of conditional independence are equivalent to determine a threshold for mutual information (conditional or not) between couples of involved variables. In the latter case, a work (Chickering & Meek, 2003) comes to question the reliability of BNPC.

Many algorithms, by conducting casual research, are quite similar. These algorithms propose a gradual construction of the structure returned. However, we noticed some remaining shortcomings. In the presence of an insufficient number of cases describing the observed domain, the statistical tests of independence are not reliable enough. The number of tests to be independently carried out to cover all the variables is huge. An alternative is the

[1] We call V-structure, or convergence, a triplet (x, y, z) such as y depends on x and $z (x \rightarrow y \leftarrow z)$.

use of a measure for evaluating the quality of a structure knowing the training database in combination with a heuristic exploring a space of options.

Scoring methods use a score to evaluate the consistency of the current structure with the probability distribution that generated the data. Thus, in (Cooper & Herskovits, 1992) a formulation was proposed, under certain conditions, to compute the Bayesian score, (denoted BD and corresponds in fact to the marginal likelihood we are trying to maximize through the determination of a structure G). In (Heckerman, 1995) a variant of Bayesian score based on an assumption of equivalency of likelihood is presented. BDe, the resulting score, has the advantage of preventing a particular configuration of a variable X_i and of its parents $Pa(X_i)$ from being regarded as impossible. A variant, BDeu, initializes the prior probability distributions of parameters according to a uniform law. In (Kayaalp & Cooper, 2002) authors have shown that under certain conditions, this algorithm was able to detect arcs corresponding to low-weighted conditional dependencies. AIC, the Akaike Information Criterion (Akaike, 1970) tries to avoid the learning problems related to likelihood alone. When penalizing the complexity of the structures evaluated, the AIC criterion focuses the simplest model being the most expressive of extracted knowledge from the base D. AIC is not consistent with the dimension of the model, with the result that other alternatives have emerged, for example CAIC – Consistent AIC – (Bozdogan, 1987). If the size of the database is very small, it is generally preferable to use AICC – Akaike Information Corrected Criterion – (Hurvich & Tsai, 1989). The MDL criterion (Rissanen, 1978; Suzuki, 1996) incorporates a penalizing scheme for the structures which are too complex. It takes into account the complexity of the model and the complexity of encoding data related to this model. Finally, the BIC criterion (Bayesian Information Criterion), proposed in (Schwartz, 1978), is similar to the AIC criterion. Properties such as equivalence, breakdown-ability of the score and consistency are introduced. Due to its tendency to return the simplest models (Bouckaert, 1994), BIC is a metric evaluation as widely used as the BDeu score.

To efficiently go through the huge space of solutions, algorithms use heuristics. We can found in the literature deterministic ones like K2 (Cooper & Herskovits, 1992), GES (Chickering, 2002b), KES (Nielsen et al., 2003) or stochastic ones like an application of Monte Carlo Markov Chains methods (Madigan & York, 1995) for example. We particularly notice evolutionary methods applied to the training of a Bayesian network structure. Initial work is presented in (Etxeberria et al., 1997; Larrañaga et al., 1996). In this work, the structure is build using a genetic algorithm and with or without the knowledge of a topologically correct order on the variables of the network. In (Larrañaga et al., 1996) an evolutionary algorithm is used to conduct research over all topologic orders and then the K2 algorithm is used to train the model. Cotta and Muruzábal (Cotta & Muruzábal, 2002) emphasize the use of phenotypic operators instead of genotypic ones. The first one takes into account the expression of the individual's allele while the latter uses a purely random selection. In (Wong et al., 1999), structures are learned using the MDL criterion. Their algorithm, named MDLEP, does not require a crossover operator but is based on a succession of mutation operators. An advanced version of MDLEP named HEP (Hybrid Evolutionary Programming) was proposed (Wong et al., 2002). Based on a hybrid technique, it limits the search space by determining in advance a network skeleton by conducting a series of low-order tests of independence: if X and Y are independent variables, the arcs $X \rightarrow Y$ and $X \leftarrow Y$ can not be added by the mutation operator. The algorithm forbids the creation of a cycle during and after the mutation. In

(Van Dijk & Thierens, 2004; Van Dijk, Thierens & Van Der Gaag, 2003; Van Dijk, Van Der Gaag & Thierens, 2003) a similar method was proposed. The chromosome contains all the arcs of the network, and three alleles are defined: none, $X \rightarrow Y$ and $X \rightarrow Y$. The algorithm acts as Wong's one (Wong et al., 2002) but only recombination and repair are used to make the individuals evolve. The results presented in (Van Dijk & Thierens, 2004) are slightly better than these obtained by HEP. A search, directly done in the equivalence graph space, is presented in (Muruzábal & Cotta, 2004; 2007). Another approach, where the algorithm works in the limited partially directed acyclic graph is reported in (Acid & de Campos, 2003). These are a special form of PDAG where many of these could fit the same equivalence class. Finally, approaches such as Estimation of Distribution Algorithms (EDA) are applied in (Mühlenbein & PaaB, 1996). In (Blanco et al., 2003), the authors have implemented two approaches (UMDA and PBIL) to search structures over the PDAG space. These algorithms were applied to the distribution of arcs in the adjacency matrix of the expected structure. The results appear to support the approach PBIL. In (Romero et al., 2004), two approaches (UMDA and MIMIC) have been applied to the topological orders space. Individuals (i.e. topological orders candidates) are themselves evaluated with the Bayesian scoring.

3. Genetic algorithm design

Genetic algorithms are a family of computational models inspired by Darwin's theory of Evolution. Genetic algorithms encode potential solutions to a problem in a chromosome-like data structure, exploring and exploiting the search space using dedicated operators. Their actual form is mainly issued from the work of J.Holland (Holland, 1992) in which we can find the general scheme of a genetic algorithm (see Algorithm. 1) called canonical GA. Throughout the years, different strategies and operators have been developed in order to perform an efficient search over the considered space of individuals: selection, mutation and crossing operators, etc.

Algorithm 1 Holland's canonical genetic algorithm (Holland, 1992)

/ Initialization */*
$t \leftarrow 0$
Randomly and uniformly generate an initial population P_0 of λ individuals and evaluate them using a fitness function f
/ Evolution */*
repeat
  Select P_t for the reproduction
  Build new individuals by application of the crossing operator on the beforehand selected individuals
  Apply a mutation operator to the new individuals: individuals obtained are affected to the new population P_{t+1}
  / Evaluation */*
  Evaluate the individuals of P_{t+1} using f
  $t \leftarrow t+1$;
  / Stop */*
until a definite criterion is met

Applied to the search for Bayesian networks structures, genetic algorithm pose two problems:

1. The constraint on the absence of circuits in the structures creates a strong link between the different genes and alleles of a person, regardless of the chosen representation. Ideally, operators should reflect this property.
2. Often, a heuristic searching over the space of solutions (genetic algorithm, greedy algorithm and so on.) finds itself trapped in a local optimum. This makes it difficult to find a balance between a technique able to avoid this problem, with the risk of overlooking many quality solutions, and a more careful exploration with a good chance to compute only a locally-optimal solution.

If the first item involves essentially the design of a thoughtful and evolutionary approach to the problem, the second point characterizes an issue relating to the multimodal optimization. For this kind of problem, there is a particular methodology: the niching.

We now proceed to a description of a genetic algorithm adapted to find a good structure for a Bayesian network.

3.1 Representation

As our search is performed over the space of directed acyclic graphs, each invidual is represented by an adjacency matrix. Denoting with N the number of variables in the domain, an individual is thus described by an $N \times N$ binary matrix Adj_{ij} where one of its coefficients a_{ij} is equal to 1 if an oriented arc going from X_i to X_j in G exists.

Whereas the traditional genetic algorithm considers chromosomes defined by a binary alphabet, we chose to model the Bayesian network structure by a chain of N genes (where N is the number of variables in the network). Each gene represents one row of the adjacency matrix, that's to say each gene corresponds to the set of parents of one variable. Although this non-binary encoding is unusual in the domain of structure learning, it is not an uncommon practice among genetic algorithms. In fact, this approach turns out to be especially practical for the manipulation and evaluation of candidate solutions.

3.2 Fitness Function

We chose to use the Bayesian Information Criterion (BIC) score as the fitness function for our algorithm:

$$S_{BIC}(B,D) = log\left(L(D|B,\theta^{MAP})\right) - \frac{1}{2} \times dim(B) \times log(N) \tag{2}$$

where D represents the training data, θ^{MAP} the MAP-estimated parameters, and $dim()$ is the dimension function defined by Eq. 3:

$$dim(B) = \sum_{i=1}^{n}(r_i - 1) \times \prod_{X_k \in Pa(X_k)} r_k \tag{3}$$

where r_i is the number of possible values for X_i. The fitness function $f(individual)$ can be written as in Eq. 4:

$$f(individual) = \sum_{k=1}^{n} f_k(X_k, Pa(X_k)) \tag{4}$$

where f_k is the local BIC score computed over the family of variable X_k.

The genetic algorithm takes advantage of the breakdown of the evaluation function and evaluates new individuals from their inception, through crossing, mutation or repair. The impact of any change – on local – an individual's genome shall be immediately passed on to the phenotype of it through the computing of the local score. The direct consequence is that the evaluation phase of the generated population took actually place for each individual – depending on the changes made – as a result of changes endured by him.

3.3 Setting up the population

We choose to initialize the population of structures by the various trees (depending on the chosen root vertex) returned by the MWST algorithm. Although these n trees are Markov-equivalent, the initialization can generate individuals with relevant characteristics. Moreover, since early generations, the combined action of the crossover and the mutation operators provides various and good quality individuals in order to significantly improve the convergence time. We use the undirected tree returned by the algorithm: each individual of the population is initialized by a tree directed from a randomly-chosen root. This mechanism introduces some diversity in the population.

3.4 Selection of the individuals

We use a rank selection where each one of the λ individuals in the population is selected with a probability equal to:

$$P_{select}(individual) = 2 \times \frac{\lambda + 1 - rank(individual)}{\lambda \times (\lambda + 1)} \tag{5}$$

This strategy allows promote individuals which best suit the problem while leaving the weakest one the opportunity to participate to the evolution process. If the major drawback of this method is to require a systematic classification of individuals in advance, the cost is negligible. Other common strategies have been evaluated without success: the roulette wheel (prematured convergence), the tournament (the selection pressure remained too strong) and the fitness scaling (Forrest, 1985; Kreinovich et al., 1993). The latter aims to allow in the first instance to prevent the phenomenon of predominance of "super individuals" in the early generations while ensuring when the population converges, that the mid-quality individuals did not hamper the reproduction of the best ones.

3.5 Repair operator

In order to preserve the closeness of our operators over the space of directed acyclic graphs, we need to design a repair operator to convert those invalid graphs (typically, cyclic directed graphs) into valid directed acyclic graphs. When one cycle is detected within a graph, the operator suppresses the one arc in the cycle bearing the weakest mutual information. The mutual information between two variables is defined as in (Chow & Liu, 1968):

$$W(X_A, X_B) = \sum_{X_A, X_B} \frac{N_{ab}}{N} \times log\left(\frac{N_{ab} \times N}{N_a \times N_b}\right) \tag{6}$$

Where the mutual information $W(X_A, X_B)$ between two variables X_A and X_B is calculated according to the number of times N_{ab} that $X_A = a$ and $X_B = b$, N_a the number of times $X_A = a$ and so on. The mutual information is computed once for a given database. It may happen that an individual has several circuits, as a result of a mutation that generated and/or inverted several arcs. In this case, the repair is iteratively performed, starting with deleting the shortest circuit until the entire circuit has been deleted.

3.6 Crossover Operator

A first attempt was to create a one-point crossover operator. At least, the operator used has been developed from the model of (Vekaria & Clack, 1998). This operator is used to generate two individuals with the particularity of defining the crossing point as a function of the quality of the individual. The form taken by the criterion (BIC and, in general, by any decomposable score) makes it possible to assign a local score to the set $\{X_i, Pa(X_i)\}$. Using these different local scores we can therefore choose to generate an individual which received the best elements of his ancestors. This operation is shown Fig. 2. This generation can be performed only if a DAG is produced (the operator is closed). In our experiments, P_{cross}, the probability that an individual is crossed with another is set to 0.8.

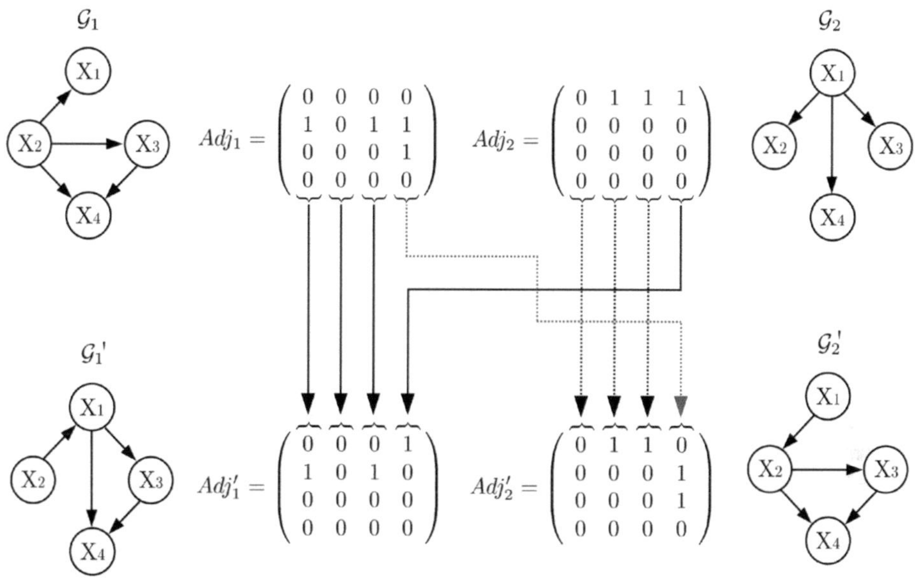

Fig. 2. The crossover operator and the transformation it performs over two DAGs

3.7 Mutation operator

Each node of one individual has a P_{mute} probability of losing or gaining one parent or to see one of its incoming arcs reverted (ie. reversing the relationship with one parent).

3.8 Other Parameters

The five best individuals from the previous population are automatically transferred to the next one. The rest of the population at $t + 1$ is composed of the $S - 5$ best children where S is the size of the population.

4. Strategies

Now, after describing our basic GA, we will present how it can be improved by i) a specific adaptive mutation scheme and ii) an exploration strategy: the niching.

The many parameters of a GA are usually fixed by the user and, unfortunately, usually lead to sub-optimal choices. As the amount of tests required to evaluate all the conceivable sets of parameters will be eventually exponential, a natural approach consists in letting the different parameters evolve along with the algorithm. (Eiben et al., 1999) defines a terminology for self-adaptiveness which can be resumed as follows:

- Deterministic Parameter Control: the parameters are modified by a deterministic rule.
- Adaptive Parameter Control: consists in modifying the parameters using feedback from the search.
- Self-adaptive Parameter Control: parameters are encoded in the individuals and evolve along.

We now present three techniques. The first one, an adaptive parameter control, aims at managing the mutation rate. The second one, an evolutionary method tries to avoid local optima using a penalizing scheme. Finaly, the third one, another evolutionary method, makes many populations evolve granting sometimes a few individuals to go from one population to another.

4.1 Self-adaptive scheme of the mutation rate

As for the mutation rate, the usual approach consists in starting with a high mutation rate and reducing it as the population converges. Indeed, as the population clusters near one optimum, high mutation rates tend to be degrading. In this case, a self-adaptive strategy would naturally decrease the mutation rate of individuals so that they would be more likely to undergo the minor changes required to reach the optimum.

Other strategies have been proposed which allow the individual mutation rates to either increase or decrease, such as in (Thierens, 2002). There, the mutation step of one individual induces three differently rated mutations: greater, equal and smaller than the individual's actual rate. The issued individual and its mutation rate are chosen accordingly to the qualitative results of the three mutations. Unfortunately, as the mutation process is the most costly operation in our algorithm, we obviously cannot choose such a strategy. Therefore, we designed the following adaptive policy.
We propose to conduct the search over the space of solutions by taking into account information on the quality of later search. Our goal is to define a probability distribution which drives the choice of the mutation operation. This distribution should reflect the performance of the mutation operations being applied over the individuals during the previous iterations of the search.

Let us define $P(i, j, op_{mute})$ the probability that the coefficient a_{ij} of the adjacency matrix is modified by the mutation operation op_{mute}. The mutation decays according to the choice of i, j and op_{mute}. We can simplify the density of probability by conditioning a subset of $\{i, j, op_{mute}\}$ by its complementary. This latter being activated according to a static distribution of probability. After studying all the possible combination, we have chosen to design a process to control

$P(i|op_{mute}, j)$. This one influences the choice of the source vertex knowing the destination vertex and for a given mutation operation. So the mutation operator can be rewritten such as shown by Algorithm 2.

Algorithm 2 The mutation operator scheme

for $j = 1$ to n **do**
  if $Pa(X_j)$ mute with a probability P_{mute} **then**
    choose a mutation operation among these allowed on $Pa(X_j)$
    apply $op_{mute}(i, j)$ with the probability $P(i|op_{mute}, j)$
  end if
end for

Assuming that the selection probability of $Pa(X_j)$ is uniformly distributed and equals a given P_{mute}, Eq. 7 must be verified:

$$\begin{cases} \sum_{op_{mute}} \delta_{op_{mute}}^{(i,j)} P(i|op_{mute}, j) = 1 \\ \delta_{op_{mute}}^{(i,j)} = \begin{cases} 1 \text{ if } op_{mute}(i,j) \text{ is allowed} \\ 0 \text{ else} \end{cases} \end{cases} \tag{7}$$

The diversity of the individuals lay down to compute $P(i|op_{mute}, j)$ for each allowed op_{mute} and for each individual X_j. We introduce a set of coefficients denoted $\zeta(i, j, op_{mute}(i, j))$ where $1 \leq i, j \leq n$ and $i \neq j$ to control $P(i|op_{mute}, j)$. So we define:

$$P(i|op_{mute}, j) = \frac{\zeta(i, j, op_{mute}(i, j))}{\sum \delta_{op_{mute}}^{(i,j)} \zeta(i, j, op_{mute}(i, j))} \tag{8}$$

During the initialization and without any prior knowledge, $\zeta(i, j, op_{mute}(i, j))$ follows an uniform distribution:

$$\zeta(i, j, op_{mute}(i, j)) = \frac{1}{n-1} \begin{cases} \forall 1 \leq i, j \leq n \\ \forall \, op_{mute} \end{cases} \tag{9}$$

Finally, to avoid the predominance of a given op_{mute} (probability set to 1) and a total lack of a given op_{mute} (probability set to 0) we add a constraint given by Eq. 10:

$$0.01 \leq \zeta(i, j, op_{mute}(i, j)) \leq 0.9 \begin{cases} \forall 1 \leq i, j \leq n \\ \forall \, op_{mute} \end{cases} \tag{10}$$

Now, to modify $\zeta(i, j, op_{mute}(i, j))$ we must take in account the quality of the mutations and either their frequencies. After each evolution phase, the $\zeta(i, j, op_{mute}(i, j))$ associated to the op_{mute} applied at least one time are reestimated. This compute is made according to a parameter γ which quantifies the modification range of $\zeta(i, j, op_{mute}(i, j))$ and depends on ω which is computed as the number of successful applications of op_{mute} minus the number of detrimental ones in the current population. Eq. 11 gives the computation. In this relation, if we set γ =0 the algorithm acts as the basic genetic algorithm previoulsy defined.

$$\zeta(i, j, op_{mute}(i, j)) = \begin{cases} min\,(\zeta(i, j, op_{mute}(i, j)) \times (1-\gamma)^{\omega}, 0.9) \text{ if } \omega > 0 \\ max\,(\zeta(i, j, op_{mute}(i, j)) \times (1-\gamma)^{\omega}, 0.01) \text{ else} \end{cases} \tag{11}$$

The regular update $\zeta(i, j, op_{mute}(i, j))$ leads to standardize the $P(i|op_{mute}, j)$ values and avoids a prematured convergence of the algorithm as seen in (Glickman & Sycara, 2000) in which the mutation probability is strictly decreasing. Our approach is different from an EDA one: we drive the evolution by influencing the mutation operator when an EDA makes the best individuals features probability distribution evolve until then generated.

4.2 Niching

Niching methods appear to be a valuable choice for learning the structure of a Bayesian network because they are well-adapted to multi-modal optimization problem. Two kind of niching techniques could be encountered: spatial ones and temporal ones. They all have in common the definition of a distance which is used to define the niches. In (Mahfoud, 1995), it seemed to be expressed a global consensus about performance: spatial approach gives better results than temporal one. But the latter is easier to implement because it consists in the addition of a penalizing scheme to a given evolutionary method.

4.2.1 Sequential Niching

So we propose two algorithms. The first one is apparented to a sequential niching. It makes a similar trend to that of a classic genetic algorithm (iterated cycles evaluation, selection, crossover, mutation and replacement of individuals) except for the fact that a list of optima is maintained. Individuals matching these optima see their fitness deteriorated to discourage any inspection and maintenance of these individuals in the future.

The local optima, in the context of our method, correspond to the equivalence classes in the meaning of Markov. When at least one equivalence class has been labelled as corresponding to an optimum value of the fitness, the various individuals in the population belonging to this optimum saw the value of their fitness deteriorated to discourage any further use of these parts of the space of solutions. The determination of whether or not an individual belongs to a class of equivalence of the list occurs during the evaluation phase, after generation by crossover and mutation of the new population. The graph equivalent of each new individual is then calculated and compared with those contained in the list of optima. If a match is determined, then the individual sees his fitness penalized and set to at an arbitrary value (very low, lower than the score of the empty structure).

The equivalence classes identified by the list are determined during the course of the algorithm: if, after a predetermined number of iterations Ite_{opt}, there is no improvement of the fitness of the best individual, the algorithm retrieves the graph equivalent of the equivalence class of it and adds it to the list.

It is important to note here that the local optima are not formally banned in the population. The registered optima may well reappear in our population due to a crossover. The evaluation of these equivalence classes began, in fact until the end of a period of change. An optimum previously memorized may well reappear at the end of the crossover operation and the individual concerned undergo mutation allowing to explore the neighborhood of the optimum.

The authors of (Beasley et al., 1993) carry out an evolutionary process reset after each determination of an optimum. Our algorithm continues the evolution considering the updated list of these optima. However, by allowing the people to move in the neighborhood of the detected optima, we seek to preserve the various building blocks hitherto found, as well as reducing the number of evaluations required by multiple launches of the algorithm.

At the meeting of a stopping criterion, the genetic algorithm completes its execution thus returning the list of previously determined optima. The stopping criterion of the algorithm can also be viewed in different ways, for example:

- After a fixed number of local optima detected.
- After a fixed number of iterations (generations).

We opt for the second option. Choosing a fixed number of local optima may, in fact, appear to be a much more arbitrary choice as the number of iterations. Depending on the problem under consideration and/or data learning, the number of local optima in which the evolutionary process may vary. The algorithm returns a directed acyclic graph corresponding to the instantiation of the graph equivalent attached to the highest score in the list of optima.

An important parameter of the algorithm is, at first glance, the threshold beyond which an individual is identified as an optimum of the evaluation function. It is necessary to define a value of this parameter, which we call Ite_{opt} that is:

- Neither too small: too quickly consider an equivalence class as a local optimum slows exploring the search space by the genetic algorithm, which focuses on many local optima.
- Nor too high: loss of the benefit of the method staying too long in the same point in space research: the local optima actually impede the progress of the research.

Experience has taught us that Ite_{opt} value of between 15 and 25 iterations can get good results. The value of the required parameter Ite_{opt} seems to be fairly stable as it allows both to stay a short time around the same optimum while allowing solutions to converge around it. The value of the penalty imposed on equivalence classes is arbitrary. The only constraint is that the value is lowered when assessing the optimum detected is lower than the worst possible structure, for example: -10^{15}.

4.2.2 Sequential and spatial niching combined

The second algorithm uses the same approach as for the sequential niching combined with a technique used in parallels GAs to split the population. We use an island model approach for our distributed algorithm. This model is inspired from a model used in genetic of populations (Wright, 1964). In this model, the population is distributed to k islands. Each island can exchange individuals with others avoiding the uniformization of the genome of the individuals. The goals of all of this is to preserve (or to introduce) genetic diversity.

Some additional parameters are required to control this second algorithm. First, we denote I_{mig} the migration interval, i.e. the number of iteration of the GA between two migration phases. Then, we use R_{mig} the migration rate: the rate of individuals selected for a migration. N_{isl} is the number of islands and finally I_{size} represents the number of individuals in each island.

In order to remember the local optima encountered by the populations, we follow the next process:

- The population of each island evolves during I_{mig} iterations and then transfer $R_{mig} \times I_{size}$ individuals.

- Local optima detected in a given island are registered in a shared list. Then they can be known by all the islands.

5. Evaluation and discussion

From an experimental point of view, the training of the structure of a Bayesian network consists in:

- To have an input database containing examples of instantiation of the variables.
- To determine the conditional relationship between the variables of the model :
  - Either from statistical tests performed on several subsets of variables.
  - Either from measurements of a match between a given solution and the training database.
- To compare the learned structures to determine the respective qualities of the different algorithms used.

5.1 Tested methods

So that we can compare with existing methods, we used some of the most-used learning methods: the K2 algorithm, the greedy algorithm applied to the structures space, noted GS; the greedy algorithm applied to the graph equivalent space, noted GES; the MWST algorithm, the PC algorithm. These methods are compared to our four evolutionary algorithms learning: the simple genetic algorithm (GA); genetic algorithm combined with a strategy of sequential niching (GA-SN); the hybrid sequential-spatial genetic approach (GA-HN); the genetic algorithm with the dynamic adaptive mutation scheme GA-AM.

5.2 The Bayesian networks used

We apply the various algorithms in search of some common structures like: Insurance (Binder et al., 1997) consisting of 27 variables and 52 arcs; ALARM (Beinlich et al., 1989) consisting of 37 variables and 46 arcs. We use each of these networks to summarize:

- Four training data sets for each network, each one containing a number of databases of the same size (250, 500, 1000 & 2000 samples).
- A single and large database (20000 or 30000 samples) for each network. This one is supposed to be sufficiently representative of the conditional dependencies of the network it comes from.

All these data sets are obtained by logic probabilistic sampling (Henrion, 1988): the value of vertices with no predecessors is randomly set, according to the probability distributions of the guenine network, and then the remaining variables are sampled following the same principle, taking into account the values of the parent vertices. We use several training databases for a given network and for a given number of cases, in order to reduce any bias due to sampling error. Indeed, in the case of small databases, it is possible (and it is common) that the extracted statistics are not exactly the conditional dependencies in the guenine network. After training with small databases, the BIC score of the returned structures by the different methods are computed from the large database mentioned earlier, in order to assess qualitative measures.

5.3 Experiments

GAs: The parameters of the evolutionary algorithms are given in Table 1.
GS: This algorithm is initialized with a tree returned by the MWST method, where the root vertex is randomly chosen.
GES: This algorithm is initialized with the empty structure.
MWST: it is initialized with a root node randomly selected (it had no effect on the score of the structure obtained).
K2: This algorithm requires a topological order on the vertices of the graph. We used for this purpose two types of initialization:

- The topological order of a tree returned by the MWST algorithm (method K2-T)
- A topological order random (method K2-R)

Parameter	**Value**	**Remarks**
Population size	150	
Mutation probability	1/n	
Crossover probability	0.8	
Recombination scheme	elitist	The best solution is never lost
Stop criterion	1000 iter.	
Initialisation		See footnote[2]
Ite_{opt}	20	For GA-SN only
γ	0.5	For GA-AM only
I_{mig}	20	For GA-HN only
R_{mig}	0.1	For GA-HN only
N_{isl}	30	For GA-HN only
I_{size}	30	For GA-HN only

Table 1. Parameters used for the evolutionary algorithms.

For each instance of K2-R – i.e. for each training database considered – we are proceeding with $5 \times n$ random initialization for choosing only those returning the best BIC score.

Some of these values (crossover, mutation probability) are coming from some habits of the domain (Bäck, 1993) but especially from experiments too. The choice of the iteration number is therefore sufficient to monitor and interpret the performance of the method considered while avoiding a number of assessments distorting the comparison of results with greedy methods.

We evaluate the quality of the solutions with two criteria: the BIC score from one hand, and a graphic distance measuring the number of differences between two graphs on the other hand. The latter is defined from 4 terms: (D) the total number of different arcs between two graphs G_1 and G_2, ($\oplus$) the number of arcs existing in G_1 but not in G_2, ($\ominus$) the number of arcs existing in G_2 but not in G_1 and (inv) the number of arcs inverted in G_1 comparing to G_2. These terms are important because, when considering two graphs of the same equivalence class, some arcs could be inverted. This implies that the corresponding arcs are not oriented in the corresponding PDAG. The consequence is that G_1 and G_2 have the same BIC score but not the same graphic distance. To compare the results with we also give the score of the empty structure G_0 and the score of the reference network G_R.

5.4 Results for the INSURANCE network

Results are given Table 2 & Table 3. The evaluation is averaged over 30 databases. Table 2 shows the means and the standard deviations of the BIC scores. For a better seeing, values are all divided by 10. Values labelled by † are significantly different from the best mean score (Mann-Whitney's test).

The results in Table 2 give an advantage to evolutionary methods. While it is impossible to distinguish clearly the performance of the different evolutionary methods, it is interesting to note that these latter generally outperform algorithms like GES and GS. Only the algorithm GS has such good results as the evolutionary methods on small databases (250 and 500). We can notice too, according to a Mann-Whitney's test that, for large datasets, GA-SN & GA-AM returns a structure close to the original one. Standard deviations are not very large for the GAs, showing a relative stability of the algorithms and so, a good avoidance of local optima.

Table 3 shows the mean structural differences between the original network and these delivered by some learning algorithms. There, we can see that evolutionary methods, particularly GA-SN, return the structures which are the closest to the original one. This network was chosen because it contains numerous low-valued conditional probabilities. These are difficult to find using small databases. So even if the BIC score is rather close to the original one, graphical distances reveals some differences. First, we can see that D is rather high (the original network G_R is made with only 52 arcs, compared to D which minimum is 24.4) even if the BIC score is very close (resp. -28353 compared to -28681). Second, as expected, D decreases when the size of the learning database grows, mainly because of the (-) term. Third, GAs obtains the closest models to the original in 11 cases over 16; the 5 others are provided by GES.

5.5 Results for the ALARM network

This network contains more vertices than the INSURANCE one, but less low-valued arcs. The evaluation is averaged over 30 databases. Evolutionary algorithms obtain the best scores. But while GES provides less qualitative solutions accordingly to the BIC score, these solutions are closest to the original one if we consider the graphical distance. Here, a strategy consisting in gradually building a solution seems to produce better structures than an evolutionary search. In this case, a GA has a huge space (3×10^{237} when applying the Robinson's formula) into which one it enumerates solutions. If we increases the size of the population the results are better than these provided by GES.

5.6 Behavior of the GAs

Now look at some measures in order to evaluate the behavior of our genetic algorithms.

A repair operator was designed to avoid individuals having a cycle. Statistics computed during the tests show that the rate of individuals repaired does not seem to depend neither on the algorithm used nor on the size of the training set. It seems to be directly related to the complexity of the network. Thus, this rate is about 15% for the INSURANCE network and about 7% for the ALARM network.

The mean number of iterations before the GA found the best solution returned for the INSURANCE network is given Table 4. The data obtained for the ALARM network are the same order of magnitude. We note here that GA-HN quickly gets the best solution. This

	Insurance			
	250	**500**	**1000**	**2000**
GA	-32135 ± 290	-31200 ± 333	-29584 ± 359	-28841 ± 89†
GA-SN	-31917 ± 286	-31099 ± 282	-29766 ± 492	**-28681**±156
GA-AM	**-31826**±270	-31076 ± 151	-29635 ± 261	-28688 ± 165
GA-HN	-31958 ± 246	**-31075**±255	**-29428**±290	-28715 ± 164
GS	-32227 ± 397	-31217 ± 314	-29789 ± 225†	-28865 ± 151†
GES	-33572 ± 247†	-31952 ± 273†	-30448 ± 836†	-29255 ± 634†
K2-T	-32334 ± 489†	-31772 ± 339†	-30322 ± 337†	-29248 ± 163†
K2-R	-33002 ± 489†	-31858 ± 395†	-29866 ± 281†	-29320 ± 245†
MWST	-34045 ± 141†	-33791 ± 519†	-33744 ± 296†	-33717 ± 254†
Original	-28353			
$\mathcal{G}_0$	-45614			

Table 2. Means and standard deviations of the BIC scores (INSURANCE).

	Insurance							
	250				**500**			
	D	$\oplus$	*Inv*	$\ominus$	**D**	$\oplus$	*Inv*	$\ominus$
GA	39,6	4,4	7,2	28	34	3,1	7,6	23,3
GA-SN	**37**	3,5	7,1	26,4	35,1	3,7	7,4	24
GA-AM	37,5	4,3	6,6	26,6	33,9	3,2	7,7	23
GA-HN	38,1	3,5	7,5	27,1	**33,3**	3	7,3	23
GS	42,1	4,6	9,4	28,1	37,7	4,5	9,4	23,8
GES	39,5	3,7	7,1	28,7	35,1	3	7,1	25
K2-T	42,7	5,1	8,4	29,2	40,8	5,4	8,8	26,6
K2-R	42,4	4,8	7,2	30,4	41,8	6,5	8,8	26,6
MWST	41,7	4	7,7	30	41,3	3,5	8,3	29,5
	1000				**2000**			
	D	$\oplus$	*Inv*	$\ominus$	**D**	$\oplus$	*Inv*	$\ominus$
GA	39,6	4,4	7,2	28	27,8	4,7	8	15,1
GA-SN	30,8	3,8	7,4	19,6	**24,4**	3,4	6,7	14,3
GA-AM	31,4	4	8	19,4	27	4,3	8,4	14,3
GA-HN	**29,3**	3,6	6,5	19,2	26,6	3,6	8,6	14,4
GS	35,9	5,1	10	20,8	31,9	5,2	11,4	15,3
GES	32,4	4,1	8,1	20,2	27,5	4	8,4	15,1
K2-T	38,7	5,9	11	21,8	34,6	7,3	10,9	16,4
K2-R	39,6	8,3	8,3	23	36,1	8,5	8,5	9,1
MWST	37,7	1,7	8,3	27,7	36,3	1,2	7,9	27,2

Table 3. Mean structural differences between the original INSURANCE network and the best solutions founded by some algorithms.

makes it competitive in terms of computing time if we could detect this event.

	Insurance Net.			
	250	**500**	**1000**	**2000**
GA	364 ± 319	454 ± 295	425 ± 249	555 ± 278
GA-SN	704 ± 295	605 ± 321	694 ± 258	723 ± 234
GA-AM	398 ± 326	414 ± 277	526 ± 320	501 ± 281
GA-HN	82 ± 59	106 ± 77	166 ± 84	116 ± 27

Table 4. Mean of the necessary number of iterations to find the best structure (INSURANCE).

The averaged computing time of each algorithm is given Table 5 (for the ALARM network). We note here that GA-HN is only three times slower than GES. We note too that these computing times are rather stable when the size of the database increases.

	ALARM Net.			
	250	**500**	**1000**	**2000**
GA	3593 ± 47	3659 ± 41	3871 ± 53	4088 ± 180
GA-SN	3843 ± 58	3877 ± 44	4051 ± 59	4332 ± 78
GA-AM	3875 ± 32	4005 ± 43	4481 ± 46	4834 ± 52
GA-HN	9118 ± 269	9179 ± 285	9026 ± 236	9214 ± 244
GS	9040 ± 1866	9503 ± 1555	12283 ± 1403	16216 ± 2192
GES	3112 ± 321	2762 ± 166	$4055 \pm 3,4$	5759 ± 420
K2-T	733 ± 9	855 ± 25	1011 ± 14	1184 ± 8
K2-R	3734 ± 61	4368 ± 152	5019 ± 67	5982 ± 43
MWST	10 ± 1	10 ± 2	11 ± 1	12 ± 1

Table 5. Averaged computing times (in seconds) and standard deviations (ALARM).

6. Application

Graphics recognition deals with graphic entities in document images and is a subfield of document image analysis. These graphic entities could correspond to symbols, mathematical formulas, musical scores, silhouettes, logos etc., depending on the application domain. Documents from electronics, engineering, music, architecture and various other fields use domain-dependent graphic notations which are based on particular alphabets of symbols. These industries have a rich heritage of hand-drawn documents and because of high demands of application domains, overtime symbol recognition is becoming core goal of automatic image analysis and understanding systems. The method proposed in (Luqman et al., 2009) is a hybrid of structural and statistical pattern recognition approaches where the representational power of structural approaches is exploited and the computational efficiency of statistical classifiers is employed.

In our knowledge there are only a few methods which use Bayesian networks for graphic symbol recognition. Recently Barrat et al. (Barrat et al., 2007) have used the naive Bayes classifier in a *pure* statistical manner for graphic symbol recognition. Their system uses three shape descriptors: Generic Fourier Descriptor, Zernike descriptor & R-Signature 1D, and applies dimensionality reduction for extracting the most relevant and discriminating features to formulate a feature vector. This reduces the length of their feature vector and eventually the number of variables (nodes) in Bayesian network. The naive Bayes classifier is a powerful Bayesian classifier but it assumes a strong independence relationship among attributes given the class variable. We believe that the power of Bayesian networks is not fully explored, as instead of using predefined dependency relationships, if we find dependencies between all variable pairs from underlying data we can obtain a more powerful Bayesian network classifier. This will also help to ignore irrelevant variables and exploit the variables that are interesting for discriminating symbols in underlying symbol set.

Our method is an original adaptation of Bayesian network learning for the problem of graphic symbol recognition. For symbol representation, we use a structural signature. The signature is computed from the attributed relational graph (ARG) of symbol and is composed of geometric & topologic characteristics of the structure of symbol. We use (overlapping) fuzzy intervals for computing noise sensitive features in signature. This increases the ability of our signature to resist against irregularities (Mitra & Pal, 2005) that may be introduced in the shape of symbol by deformations & degradations. For symbol recognition, we employ a Bayesian network. This network is learned from underlying training data by using the GA-HN algorithm. A query symbol is classified by using Bayesian probabilistic inference (on encoded joint probability distribution). We have selected the features in signature very carefully to best suit them to linear graphic symbols and to restrict their number to minimum; as Bayesian network algorithms are known to perform better for a smaller number of nodes. Our structural signature makes the proposed system robust & independent of application domains and it could be used for all types of 2D linear graphic symbols.

After representing the symbols in learning set by ARG and describing them by structural signatures, we proceed to learning of a Bayesian network. The signatures are first discretized. We discretize each feature variable (of signature) separately and independently of others. The class labels are chosen intelligently in order to avoid the need of any discretization for them. The discretization of *number of nodes* and *number of arcs* achieves a comparison of similarity of symbols (instead of strict comparison of exact feature values). This discretization step also ensures that the features in signature of query symbol will look for symbols whose number of nodes and arcs lie in same intervals as that of the query symbol.

The Bayesian network is learned in two steps. First we learn the structure of the network. Despite the training algorithms are evolutionary one, they have provided stable results (for a given dataset multiple invocations always returned identical network structures). Each feature in signature becomes a node of network. The goal of structure learning stage is to find the best network structure from underlying data which contains all possible dependency relationships between all variable pairs. The structure of the learned network depicts the dependency relationships between different features in signature. Fig.3 shows one of the learned structures from our experiments. The second step is learning of parameters of network; which are conditional probability distributions $Pr(node_i|parents_i)$ associated to nodes of the network and which quantify the dependency relationships between nodes. The network parameters are obtained by maximum likelihood estimation (MLE); which is a robust parameter estimation technique and assigns the most likely parameter values to best describe a given distribution of data. We avoid null probabilities by using Dirichlet priors with MLE. The learned Bayesian network encodes joint probability distribution of the symbol signatures.

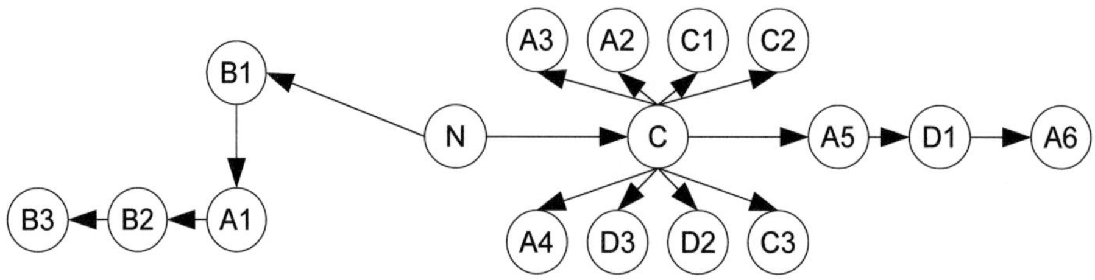

Fig. 3. Example of a Bayesian network : C = class, N = number of nodes, A1 = number of connections, A2 = number of L-junctions, A3 = number of T-junctions, A4 = number of intersections, A5 = number of parallel connections, A6 = number of successive connections, B1 (resp. B2 and B3) = number of nodes with low (resp. medium and high) density of connections, C1 (resp. C2 and C3) = number of small-length (resp. medium-length and full-length) primitives, D1 = number of small-angle (resp. medium-angle and full-angle) connections.

The conditional independence property of Bayesian networks helps us to ignore irrelevant features in structural signature for an underlying symbol set. This property states that a node is conditionally independent of its non-descendants given its immediate parents (Charniak, 1991). Conditional independence of a node in Bayesian network is fully exploited during probabilistic inference and thus helps us to ignore irrelevant features for an underlying symbol set while computing posterior probabilities for different symbol classes.

For recognizing a query symbol we use Bayesian probabilistic inference on the encoded joint probability distribution. This is achieved by using junction tree inference engine which is the most popular exact inference engine for Bayesian probabilistic inference. The inference engine propagates the evidence (signature of query symbol) in network and computes posterior probability for each symbol class. Equation 12 gives Bayes rule for our system. It states that posterior probability or probability of a symbol class c_i given a query signature *evidence* e is computed from likelihood (probability of e given c_i), prior probability of c_i and marginal likelihood (prior probability of e). The marginal likelihood $Pr(e)$ is to normalize the posterior probability; it ensures that the probabilities fall between 0 and 1.

$$Pr(c_i|e) = \frac{Pr(e,c_i)}{Pr(e)} = \frac{Pr(e|c_i) \times Pr(c_i)}{Pr(e)} \tag{12}$$

where,

$$\begin{cases} e = f_1, f_2, f_3, ..., f_{16} \\ Pr(e) = Pr(e,c_i) = \sum_{i=1}^{k} Pr(e|c_i) \times Pr(c_i) \end{cases} \tag{13}$$

The posterior probabilities are computed for all k symbol classes and the query symbol is then assigned to class which maximizes the posterior probability i.e. which has highest posterior probability for the given query symbol.

6.1 Symbols with vectorial and binary noise

The organization of four international symbol recognition contests over last decade (Aksoy et al., 2000; Dosch & Valveny, 2005; Valveny & Dosch, 2003; Valveny et al., 2007), has provided our community an important test bed for evaluation of methods over a standard dataset. These contests were organized to evaluate and test the symbol recognition methods for their scalability and robustness against binary degradation and vectorial deformations. The contests were run on pre-segmented linear symbols from architectural and electronic drawings, as these symbols are representative of a wide range of shapes (Valveny & Dosch, 2003). GREC2005 (Dosch & Valveny, 2005) & GREC2007 (Valveny et al., 2007) databases are composed of the same set of models, whereas GREC2003 (Valveny & Dosch, 2003) database is a subset of GREC2005.

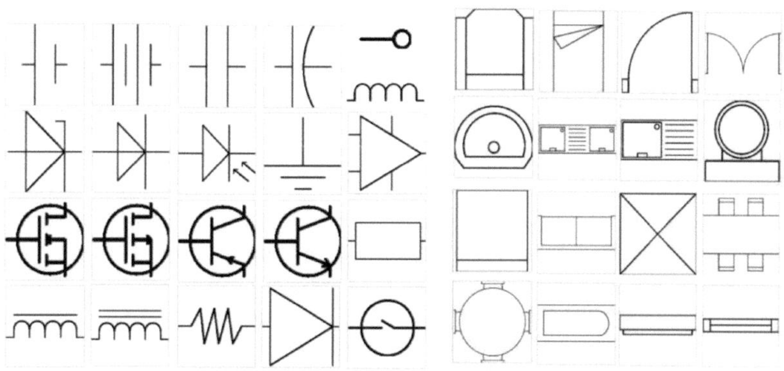

Fig. 4. Model symbols from electronic drawings and from floor plans.

We experimented with synthetically generated 2D symbols of models collected from database of GREC2005. In order to get a true picture of the performance of our proposed method on this database, we have experimented with 20, 50, 75, 100, 125 & 150 symbol classes. We generated our own learning & test sets (based on deformations & degradations of GREC2005) for our experiments. For each class the perfect symbol (the model) along with its 36 rotated and 12 scaled examples was used for learning; as the features have already been shown invariant to scaling & rotation and because of the fact that generally Bayesian network learning algorithms perform better on datasets with large number of examples. The system has been tested for its scalability on clean symbols (rotated & scaled), various levels of vectorial deformations and for binary degradations of GREC symbol recognition contest. Each test dataset was composed of 10 query symbols for each class.

Number of classes (models)	20	50	75	100	125	150
Clean symbols (rotated & scaled)	100%	100%	100%	100%	100%	99%
Hand-drawn deform. Level-1	99%	96%	93%	92%	90%	89%
Hand-drawn deform. Level-2	98%	95%	92%	90%	89%	87%
Hand-drawn deform. Level-3	95%	77%	73%	70%	69%	67%
Binary degrade	98%	96%	93%	92%	89%	89%

Table 6. Results of symbol recognition experiments.

Table 6 summarizes the experimental results. A 100% recognition rate for clean symbols illustrates the invariance of our method to rotation & scaling. Our method outperforms all GREC participants (available results from GREC2003 and GREC2005 competitions) in scalability tests and is comparable to contest participants for low levels of deformation & degradations. The recognition rates decrease with level of deformation and drop drastically for high binary degradations. This is an expected behavior and is a result of the irregularities produced in symbol signature; which is a direct outcome of the noise sensitivity of vectorization step, as also pointed out by (Llados et al., 2002). We used only clean symbols for learning and (thus) the recognition rates truely illustrate the robustness of our system against vectorial and binary noise.

6.2 Symbols with contextual noise

A second set of experimentation was performed on a synthetically generated corpus, of symbols cropped from complete documents (Delalandre et al., 2007). These experiments focused on evaluating the robustness of the proposed system against context noise i.e. the structural noise introduced in symbols when they are cropped from documents. We believe that this type of noise gets very important when we are dealing with symbols in context in complete documents and to the best of our knowledge; no results have yet been published for this type of noise. We have performed these experiments on two subsets of symbols: consisting of 16 models from floor plans and 21 models from electronic diagrams. The models are derived from GREC2005 database and are given in Fig.4. For each class the perfect symbol (model), along with its 36 rotated and 12 scaled examples was used for learning. The examples of models, for learning, were generated using ImageMagick and the test sets were generated synthetically (Delalandre et al., 2007) with different levels of context-noise in order to simulate the cropping of symbols from documents. Test symbols were randomly rotated & scaled and multiple query symbols were included for each class. The test datasets are available at (Delalandre, 2009).

Dataset	Noise	1-TOP	3-TOP
Floor plans	Level 1	84%	95%
Floor plans	Level 2	79%	90%
Floor plans	Level 3	76%	87%
Electronic diagrams	Level 1	69%	89%
Electronic diagrams	Level 2	66%	88%
Electronic diagrams	Level 3	61%	85%

Table 7. Results of symbol recognition experiments for context noise.*1*-TOP stands for the right class in given in first position and *3*-TOP stands for the right class in belonging to the first 3 answers.

Table 7 summarizes the results of experiments for context noise. We have not used any sophisticated de-noising or pretreatment and our method derives its ability to resist against context noise, directly from underlying vectorization technique, the fuzzy approach used for computing structural signature and the capabilities of Bayesian networks to cope with uncertainties. The models for electronic diagrams contain symbols consisting of complex arrangement of lines & arcs, which affects the features in structural signature as the employed vectorization technique is not able to cope with arcs & circles; as is depicted by the recognition rates for these symbols. But keeping in view the fact that we have used only clean symbols for learning and noisy symbols for testing, we believe that the results show the ability of our signature to exploit the sufficient structural details of symbols and it could be used to discriminate and recognize symbols with context noise.

7. Conclusion

We have presented three methods for learning the structure of a Bayesian network. The first one consists in the control of the probability distribution of mutation in the genetic algorithm. The second one is to incorporate a scheme penalty in the genetic algorithm so that it avoids certain areas of space research. The third method is to search through several competing populations and to allow timely exchange among these populations. We have shown experimentally that different algorithms behaved satisfactorily, in particular that they were proving to be successful on large databases. We also examined the behavior of proposed algorithms. Niching strategies are interesting, especially using the spatial one, which focuses quickly on the best solutions.

8. Acknowledgements

This work was realized using Matlab and two toolboxes dedicated to Bayesian networks manipulation: Bayesian Net Toolbox from K. P. Murphy (Murphy, 2001) and Structure Learning Package (SLP) from P. Leray & O. François (Francois & Leray, 2004).

9. References

Acid, S. & de Campos, L. M. (2003). Searching for bayesian network structures in the space of restricted acyclic partially directed graphs, *Journ. of Art. Int. Res.* **18**: 445–490.

Akaike, H. (1970). Statistical predictor identification, *Ann. Inst. Stat. Math.* **22**(1): 203–217.

Aksoy, S., Ye, M., Schauf, M., Song, M., Wang, Y., Haralick, R., Parker, J., Pivovarov, J., Royko, D., Sun, C. & Farnebäck, G. (2000). Algorithm performance contest, *Proc. of ICPR*, pp. 4870–4876.

Allanach, J., Tu, H., Singh, S., Pattipati, K. & Willett, P. (2004). Detecting, tracking and counteracting terrorist networks via hidden markov models, *Proc. of IEEE Aero. Conf.*

Bäck, T. (1993). Optimal mutation rates in genetic search, *Proc. of Int. Conf. on Genetic Algorithms*, Morgan Kaufmann, San Mateo (CA), pp. 2–8.

Barrat, S., Tabbone, S. & Nourrissier, P. (2007). A bayesian classifier for symbol recognition, *Proc. of GREC.*

Beasley, D., Bull, D. R. & Martin, R. R. (1993). A sequential niche technique for multimodal function optimization, *Evolutionary Computation* **1**(2): 101–125.

Beinlich, I. A., Suermondt, H. J., Chavez, R. M. & Cooper, G. F. (1989). The alarm monitoring system: A case study with two probabilistic inference techniques for belief networks, *Proc. of Eur. Conf. Art. Int. in Med.*, Springer Verlag, Berlin, pp. 247–256.

Binder, J., Koller, D., Russell, S. J. & Kanazawa, K. (1997). Adaptive probabilistic networks with hidden variables, *Machine Learning* **29**(2-3): 213–244.

Blanco, R., Inza, I. & Larrañaga, P. (2003). Learning bayesian networks in the space of structures by estimation of distribution algorithms., *Int. Jour. of Int. Syst.* **18**(2): 205–220.

Bouckaert, R. (1994). Properties of bayesian belief network learning algorithms, *Proc. of Uncertainty in Artificial Intelligence*, Morgan Kaufmann, San Francisco, CA, pp. 102–110.

Bozdogan, H. (1987). Model selection and akaike's information criteria (AIC): The general theory and its analytical extentions, *Psychometrika* **52**: 354–370.

Charniak, E. (1991). Bayesian networks without tears, *AI Magazine* **12**(4): 50–63.

Cheng, J., Bell, D. A. & Liu, W. (2002). Learning belief networks from data: An information theory based approach, *Artificial Intelligence* **1-2**: 43–90.

Chickering, D. (2002a). Optimal structure identification with greedy search, *Journal of Machine Learning Research* **3**: 507–554.

Chickering, D. M. (2002b). Learning equivalence classes of bayesian-network structures, *J. of Mach. Learn. Res.* **2**: 445–498.

Chickering, D. M., Geiger, D. & Heckerman, D. (1994). Learning bayesian networks is NP-hard, *Technical Report MSR-TR-94-17*, Microsoft Research.

Chickering, D. M. & Meek, C. (2003). Monotone DAG faithfulness: A bad assumption, *Technical Report MSR-TR-2003-16*, Microsoft Research.

Chow, C. & Liu, C. (1968). Approximating discrete probability distributions with dependence trees, *IEEE Trans. on Information Theory* **14(3)**(3): 462–467.

Cobb, B. R. & Shenoy, P. P. (2006). Inference in hybrid bayesian networks with mixtures of truncated exponentials, *International Jounal of Approximate Reasoning* **41**(3): 257–286.

Cooper, G. & Herskovits, E. (1992). A bayesian method for the induction of probabilistic networks from data, *Machine Learning* **9**: 309–347.

Cotta, C. & Muruzábal, J. (2002). Towards a more efficient evolutionary induction of bayesian networks., *Proc. of PPSN VII, Granada, Spain, September 7-11*, pp. 730–739.

Davis, G. (2003). Bayesian reconstruction of traffic accidents, *Law, Prob. and Risk* **2**(2): 69–89.

De Jong, K. (2006). *Evolutionary Computation: A Unified Approach*, The MIT Press.

Delalandre, M. (2009). http://mathieu.delalandre.free.fr/projects/sesyd/queries.html.

Delalandre, M., Pridmore, T., Valveny, E., Locteau, H. & Trupin, E. (2007). Building synthetic graphical documents for performance evaluation, *in* W. Liu, J. Llados & J. Ogier (eds), *Lecture Notes in Computer Science*, Vol. 5046, Springer, pp. 288–298.

Dosch, P. & Valveny, E. (2005). Report on the second symbol recognition contest, *Proc. of GREC*.

Eiben, A. E., Hinterding, R. & Michalewicz, Z. (1999). Parameter control in evolutionary algorithms, *IEEE Trans. on Evolutionary Computation* **3**(2): 124–141.

Etxeberria, R., Larrañaga, P. & Picaza, J. M. (1997). Analysis of the behaviour of genetic algorithms when learning bayesian network structure from data, *Pattern Recognition Letters* **18**(11-13): 1269–1273.

Ezawa, K. & Schuermann, T. (1995). Fraud/uncollectible debt detection using a bayesian network based learning system: A rare binary outcome with mixed data structures, *Proc. of Uncertainty in Artificial Intelligence*, Morgan Kaufmann, San Francisco (CA).

Fennell, M. T. & Wishner, R. P. (1998). Battlefield awareness via synergistic SAR and MTI exploitation, *IEEE Aerospace and Electronic Systems Magazine* **13**(2): 39–43.

Forrest, S. (1985). Documentation for prisoners dilemma and norms programs that use the genetic algorithm. University of Michigan, Ann Arbor, MI.

Francois, O. & Leray, P. (2004). BNT structure learning package: Documentation and experiments, *Technical report*, Laboratoire PSI.
URL: *http://bnt.insa-rouen.fr/programmes/BNT_StructureLearning_v1.3.pdf*

Glickman, M. & Sycara, K. (2000). Reasons for premature convergence of self-adapting mutation rates, *Proc. of Evolutionary Computation*, Vol. 1, pp. 62–69.

Heckerman, D. (1995). A tutorial on learning bayesian networks, *Technical Report MSR-TR-95-06*, Microsoft Research, Redmond, WA.

Henrion, M. (1988). Propagation of uncertainty by probabilistic logic sampling in bayes networks, *Proc. of Uncertainty in Artificial Intelligence*, Vol. 2, Morgan Kaufmann, San Francisco (CA), pp. 149–164.

Holland, J. H. (1992). *Adaptation in natural and artificial systems*, MIT Press.

Horvitz, E., Breese, J., Heckerman, D., Hovel, D. & Rommelse, K. (1998). The lumiere project: Bayesian user modeling for inferring the goals and needs of software users, *Proc. of Uncertainty in Artificial Intelligence*, Morgan Kaufmann, San Francisco (CA).

Hurvich, C. M. & Tsai, C.-L. (1989). Regression and time series model selection in small samples, *Biometrika* **76**(2): 297–307.

Jaronski, W., Bloemer, J., Vanhoof, K. & Wets, G. (2001). Use of bayesian belief networks to help understand online audience, *Proc. of the ECML/PKDD*, Freiburg, Germany.

Kayaalp, M. & Cooper, G. F. (2002). A bayesian network scoring metric that is based on globally uniform parameter priors, *Proc. of Uncertainty in Artificial Intelligence*, Morgan Kaufmann, San Francisco, CA, pp. 251–258.

Krause, P. J. (1999). Learning probabilistic networks, *Know. Eng. Rev. Arc.* **13**(4): 321–351.

Kreinovich, V., Quintana, C. & Fuentes, O. (1993). Genetic algorithms: What fitness scaling is optimal?, *Cybernetics and Systems* **24**(1): 9–26.

Lacey, G. & MacNamara, S. (2000). Context-aware shared control of a robot mobility aid for the elderly blind, *Int. Journal of Robotic Research* **19**(11): 1054–1065.

Larrañaga, P., Poza, M., Yurramendi, Y., Murga, R. & Kuijpers, C. (1996). Structure learning of bayesian networks by genetic algorithms: A performance analysis of control parameters., *IEEE Trans. on PAMI* **18**(9): 912–926.

Lauritzen, S. L. (1995). The EM algorithm for graphical association models with missing data., *Computational Statistics & Data Analysis* **19**(2): 191–201.

Lauritzen, S. L. & Wermuth, N. (1989). Graphical models for associations between variables, some of which are qualitative and some quantitative, *Annals of Statistics* **17**(1): 31–57.

Lerner, U., Segal, E. & Koller, D. (2001). Exact inference in networks with discrete children of continuous parents, *Proc. of UAI*, Morgan Kaufmann, San Francisco, CA, pp. 319–332.

Llados, J., Valveny, E., Sańchez, G. & Marti, E. (2002). Symbol recognition: Current advances and perspectives, *Lecture Notes in Computer Science*, Vol. 2390, Springer, pp. 104–128.

Luqman, M. M., Brouard, T. & Ramel, J.-Y. (2009). Graphic symbol recognition using graph based signature and bayesian network classifier, *International Conference on Document Analysis and Recognition*, IEEE Comp. Soc., Los Alamitos, CA, USA, pp. 1325–1329.

Madigan, D. & York, J. (1995). Bayesian graphical models for discrete data, *Int. Stat. Rev.* **63**(2): 215–232.

Mahfoud, S. W. (1995). *Niching methods for genetic algorithms*, PhD thesis, University of Illinois at Urbana-Champaign, Urbana, IL, USA. IlliGAL Report 95001.

Mitra, S. & Pal, S. (2005). Fuzzy sets in pattern recognition and machine intelligence, *Fuzzy Sets and Systems* **156**(3): 381–386.

Mühlenbein, H. & PaaB, G. (1996). From recombination of genes to the estimation of distributions, *Proc. of PPSN*, Vol. 1411, pp. 178–187.

Murphy, K. (2001). The bayes net toolbox for matlab, *Comp. Sci. and Stat.* **33**: 331–350.

Muruzábal, J. & Cotta, C. (2004). A primer on the evolution of equivalence classes of bayesian-network structures, *Proc. of PPSN*, Birmingham, UK, pp. 612–621.

Muruzábal, J. & Cotta, C. (2007). A study on the evolution of bayesian network graph structures, *Studies in Fuzziness and Soft Computing* **213**: 193–214,.

Nielsen, J. D., Kocka, T. & Peña, J. M. (2003). On local optima in learning bayesian networks, *Proc. of Uncertainty in Art. Int.*, Morgan Kaufmann, San Francisco, CA, pp. 435–442.

Pearl, J. (1988). *Probabilistic Reasoning in Intelligent Systems: Networks of Plausible Inference*, 1st edn, Morgan Kaufmann, San Francisco (CA).

Pearl, J. & Verma, T. S. (1991). A theory of inferred causation, *in* J. F. Allen, R. Fikes & E. Sandewall (eds), *Proc. of Princ. of Know. Repr. and Reas.*, Morgan Kaufmann, San Mateo, California, pp. 441–452.

Rissanen, J. (1978). Modelling by shortest data description., *Automatica* **14**: 465–471.

Robinson, R. (1976). Counting unlabeled acyclic digraphs, *Proc. of Combinatorial Mathematics V*, Royal Melbourne Institute of Technology, Am. Math. Soc., Australia, pp. 28–43.

Romero, T., Larrañaga, P. & Sierra, B. (2004). Learning bayesian networks in the space of orderings with estimation of distribution algorithms, *Int. Jour. of Pat. Rec. and Art. Int.* **18**(4): 607–625.

Sahami, M., Dumais, S., Heckerman, D. & Horvitz, E. (1998). A bayesian approach to filtering junk e-mail., *Proc. of the AAAI Work. on Text Categorization*, Madison, WI, pp. 55–62.

Schwartz, G. (1978). Estimating the dimensions of a model, *The Ann. of Stat.* **6**(2): 461–464.

Spirtes, P., Glymour, C. & Scheines, R. (2001). *Causation, Prediction and Search*, 2nd edn, MIT Press.

Suzuki, J. (1996). Learning bayesian belief networks based on the minimum description length principle: An efficient algorithm using the B & B technique, *Proc. of Int. Conf. on Machine Learning*, pp. 462–470.

Thierens, D. (2002). Adaptive mutation rate control schemes in genetic algorithms, *Technical Report UU-CS-2002-056*, Inst. of Information and Computing Sciences, Utrecht Univ.

Valveny, E. & Dosch, P. (2003). Symbol recognition contest: A synthesis, *Proc. of GREC*, pp. 368–386.

Valveny, E., Dosch, P., Fornes, A. & Escalera, S. (2007). Report on the third contest on symbol recognition, *Proc. of GREC*, pp. 321–328.

Van Dijk, S. & Thierens, D. (2004). On the use of a non-redundant encoding for learning bayesian networks from data with a GA, *Proc. of PPSN*, pp. 141–150.

Van Dijk, S., Thierens, D. & Van Der Gaag, L. (2003). Building a ga from design principles for learning bayesian networks, *Proc. of Genetic and Evol. Comp. Conf.*, pp. 886–897.

Van Dijk, S., Van Der Gaag, L. C. & Thierens, D. (2003). A skeleton-based approach to learning bayesian networks from data., *Proc. of Princ. and Prac. of Knowl. Disc. in Databases*, Cavtat-Dubrovnik, Croatia, pp. 132–143.

Vekaria, K. & Clack, C. (1998). Selective crossover in genetic algorithms: An empirical study, *Proc. of PPSN, Amsterdam, The Netherlands, September 27-30, 1998*, pp. 438–447.

Wong, M., Lam, W. & Leung, K. S. (1999). Using evolutionary programming and minimum description length principle for data mining of bayesian networks, *IEEE Trans. on PAMI* **21**(2): 174–178.

Wong, M., Lee, S. Y. & Leung, K. S. (2002). A hybrid data mining approach to discover bayesian networks using evolutionary programming., *Proc. of the Genetic and Evol. Comp. COnf.*, pp. 214–222.

Wright, S. (1964). Stochastic processes in evolution, *in* J. Gurland (ed.), *Stochastic models in medecine and biology*, University of Wisconsin Press, Madison, WI, pp. 199–241.

Yu, J., Smith, V. A., Wang, P. P., Hartemink, A. J. & Jarvis, E. D. (2002). Using bayesian network inference algorithms to recover molecular genetic regulatory networks, *Prof. of Int. Conf. on Systems Biology (ICSB02)*.

3

Probabilistic inferences in Bayesian networks

Jianguo Ding
Jianguo.Ding@ieee.org
Interdisciplinary Centre for Security, Reliability and Trust
University of Luxembourg
L-1359 Luxembourg
LUXEMBOURG

1. Introduction

Because a Bayesian network is a complete model for the variables and their relationships, it can be used to answer probabilistic queries about them. For example, the network can be used to find out updated knowledge of the state of a subset of variables when other variables (the evidence variables) are observed. This process of computing the posterior distribution of variables given evidence is called probabilistic inference. A Bayesian network can thus be considered a mechanism for automatically applying Bayes' theorem to complex problems.
In the application of Bayesian networks, most of the work is related to probabilistic inferences. Any variable updating in any node of Bayesian networks might result in the evidence propagation across the Bayesian networks. How to examine and execute various inferences is the important task in the application of Bayesian networks.
This chapter will sum up various inference techniques in Bayesian networks and provide guidance for the algorithm calculation in probabilistic inference in Bayesian networks. Information systems are of discrete event characteristics, this chapter mainly concerns the inferences in discrete events of Bayesian networks.

2. The Semantics of Bayesian Networks

The key feature of Bayesian networks is the fact that they provide a method for decomposing a probability distribution into a set of local distributions. The independence semantics associated with the network topology specifies how to combine these local distributions to obtain the complete joint probability distribution over all the random variables represented by the nodes in the network. This has three important consequences.
Firstly, naively specifying a joint probability distribution with a table requires a number of values exponential in the number of variables. For systems in which interactions among the random variables are sparse, Bayesian networks drastically reduce the number of required values.
Secondly, efficient inference algorithms are formed in that work by transmitting information between the local distributions rather than working with the full joint distribution.
Thirdly, the separation of the qualitative representation of the influences between variables from the numeric quantification of the strength of the influences has a significant advantage for knowledge engineering. When building a Bayesian network model, one can focus first

on specifying the qualitative structure of the domain and then on quantifying the influences. When the model is built, one is guaranteed to have a complete specification of the joint probability distribution.
The most common computation performed on Bayesian networks is the determination of the posterior probability of some random variables, given the values of other variables in the network. Because of the symmetric nature of conditional probability, this computation can be used to perform both diagnosis and prediction. Other common computations are: the computation of the probability of the conjunction of a set of random variables, the computation of the most likely combination of values of the random variables in the network and the computation of the piece of evidence that has or will have the most influence on a given hypothesis. A detailed discussion of inference techniques in Bayesian networks can be found in the book by Pearl (Pearl, 2000).

- **Probabilistic semantics.** Any complete probabilistic model of a domain must, either explicitly or implicitly, represent the joint distribution which the probability of every possible event as defined by the values of all the variables. There are exponentially many such events, yet Bayesian networks achieve compactness by factoring the joint distribution into local, conditional distributions for each variable given its parents. If x_i denotes some value of the variable X_i and $\pi(x_i)$ denotes some set of values for X_i's parents $\pi(x_i)$, then $P(x_i|\pi(x_i))$ denotes this conditional distribution. For example, $P(x_4|x_2, x_3)$ is the probability of wetness given the values of sprinkler and rain. Here $P(x_4|x_2, x_3)$ is the brief of $P(x_4|\{x_2, x_3\})$. The set parentheses are omitted for the sake of readability. We use the same expression in this thesis. The global semantics of Bayesian networks specifies that the full joint distribution is given by the product

$$P(x_1, \ldots, x_n) = \prod_i P(x_i|\pi(x_i)) \tag{1}$$

Equation 1 is also called the chain rule for Bayesian networks.

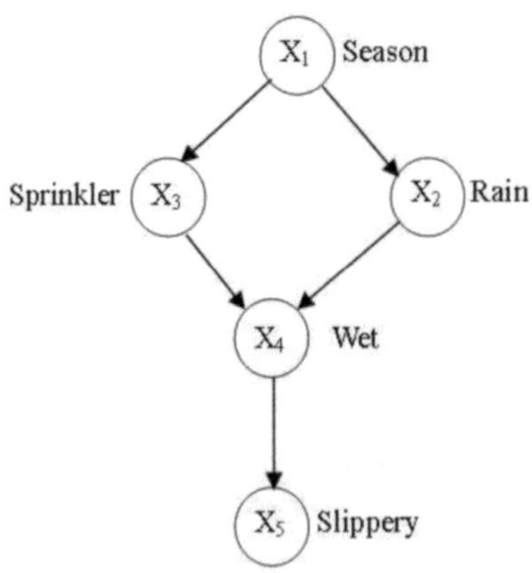

Fig. 1. Causal Influences in A Bayesian Network.

In the example Bayesian network in Figure 1, we have

$$P(x_1, x_2, x_3, x_4, x_5) = P(x_1)P(x_2|x_1)P(x_3|x_1)P(x_4|x_2, x_3)P(x_5|x_4) \tag{2}$$

Provided the number of parents of each node is bounded, it is easy to see that the number of parameters required grows only linearly with the size of the network, whereas the joint distribution itself grows exponentially. Further savings can be achieved using compact parametric representations, such as noisy-OR models, decision tress, or neural networks, for the conditional distributions (Pearl, 2000).

There are also entirely equivalent local semantics, which assert that each variable is independent of its non-descendants in the network given its parents. For example, the parents of X_4 in Figure 1 are X_2 and X_3 and they render X_4 independent of the remaining non-descendant, X_1. That is,

$$P(x_4|x_1, x_2, x_3) = P(x_4|x_2, x_3) \tag{3}$$

The collection of independence assertions formed in this way suffices to derive the global assertion in Equation 2, and vice versa. The local semantics are most useful in constructing Bayesian networks, because selecting as parents the direct causes of a given variable automatically satisfies the local conditional independence conditions. The global semantics lead directly to a variety of algorithms for reasoning.

- **Evidential reasoning.** From the product specification in Equation 2, one can express the probability of any desired proposition in terms of the conditional probabilities specified in the network. For example, the probability that the sprinkler was on, given that the pavement is slippery, is

$$\begin{aligned} & P(X_3 = on|X_5 = true) \\ &= \frac{P(X_3 = on, X_5 = true)}{P(X_5 = true)} \\ &= \frac{\sum_{x_1,x_2,x_4} P(x_1, x_2, X_3 = on, x_4, X_5 = true)}{\sum_{x_1,x_2,x_3,x_4} P(x_1, x_2, x_3, x_4, X_5 = true)} \\ &= \frac{\sum_{x_1,x_2,x_4} P(x_1)P(x_2|x_1)P(X_3 = on|x_1)P(x_4|x_2, X_3 = on)P(X_5 = true|x_4)}{\sum_{x_1,x_2,x_3,x_4} P(x_1)P(x_2|x_1)P(x_3|x_1)P(x_4|x_2, x_3)P(X_5 = true|x_4)} \end{aligned} \tag{4}$$

These expressions can often be simplified in the ways that reflect the structure of the network itself.

It is easy to show that reasoning in Bayesian networks subsumes the satisfiability problem in propositional logic and hence reasoning is NP-hard (Cooper, 1990). Monte Carlo simulation methods can be used for approximate inference (Pearl, 1987), given that estimates are gradually improved as the sampling proceeds. (Unlike join-tree methods, these methods use local message propagation on the original network structure.) Alternatively, variational methods (Jordan et al., 1998) provide bounds on the true probability.

- **Functional Bayesian networks.** The networks discussed so far are capable of supporting reasoning about evidence and about actions. Additional refinement is necessary in order to process counterfactual information. For example, the probability that "the pavement would not have been slippery had the sprinkler been *OFF*, given that the sprinkler is in fact *ON* and that the pavement is in fact slippery" cannot be computed from the information provided in Figure 1 and Equation 2. Such counterfactual probabilities require a specification in the form of functional networks, where each conditional probability $P(x_i|\pi(i))$ is replaced by a functional relationship $x_i = f_i(\pi(i), \epsilon_i)$, where ϵ_i is a stochastic (unobserved) error term. When the functions f_i and the distributions of ϵ_i are known, all counterfactual statements can be assigned unique probabilities, using evidence propagation in a structure called a "twin network". When only partial knowledge about the functional form of f_i is available, bounds can be computed on the probabilities of counterfactual sentences (Balke & Pearl, 1995) (Pearl, 2000).

- **Causal discovery.** One of the most exciting prospects in recent years has been the possibility of using Bayesian networks to discover causal structures in raw statistical data (Pearl & Verma, 1991) (Spirtes et al., 1993) (Pearl, 2000), which is a task previously considered impossible without controlled experiments. Consider, for example, the following pattern of dependencies among three events: A and B are dependent, B and C are dependent, yet A and C are independent. If you ask a person to supply an example of three such events, the example would invariably portray A and C as two independent causes and B as their common effect, namely, $A \rightarrow B \leftarrow C$. Fitting this dependence pattern with a scenario in which B is the cause and A and C are the effects is mathematically feasible but very unnatural, because it must entail fine tuning of the probabilities involved; the desired dependence pattern will be destroyed as soon as the probabilities undergo a slight change.

  Such thought experiments tell us that certain patterns of dependency, which are totally void of temporal information, are conceptually characteristic of certain causal directionalities and not others. When put together systematically, such patterns can be used to infer causal structures from raw data and to guarantee that any alternative structure compatible with the data must be less stable than the one(s) inferred; namely, slight fluctuations in parameters will render that structure incompatible with the data.

- **Plain beliefs.** In mundane decision making, beliefs are revised not by adjusting numerical probabilities but by tentatively accepting some sentences as "true for all practical purposes". Such sentences, called plain beliefs, exhibit both logical and probabilistic characters. As in classical logic, they are propositional and deductively closed; as in probability, they are subject to retraction and to varying degrees of entrenchment. Bayesian networks can be adopted to model the dynamics of plain beliefs by replacing ordinary probabilities with non-standard probabilities, that is, probabilities that are infinitesimally close to either zero or one (Goldszmidt & Pearl, 1996).

- **Models of cognition.** Bayesian networks may be viewed as normative cognitive models of propositional reasoning under uncertainty (Pearl, 2000). They handle noise and partial information by using local, distributed algorithm for inference and learning. Unlike feed forward neural networks, they facilitate local representations in which nodes correspond to propositions of interest. Recent experiments (Tenenbaum & Griffiths, 2001) suggest that they capture accurately the causal inferences made by both children and adults. Moreover, they capture patterns of reasoning that are not easily handled by any competing computational model. They appear to have many of the advantages of both the "symbolic" and the "subsymbolic" approaches to cognitive modelling.

  Two major questions arise when we postulate Bayesian networks as potential models of actual human cognition.

Firstly, does an architecture resembling that of Bayesian networks exist anywhere in the human brain? No specific work had been done to design neural plausible models that implement the required functionality, although no obvious obstacles exist.

Secondly, how could Bayesian networks, which are purely propositional in their expressive power, handle the kinds of reasoning about individuals, relations, properties, and universals that pervades human thought? One plausible answer is that Bayesian networks containing propositions relevant to the current context are constantly being assembled as needed to form a more permanent store of knowledge. For example, the network in Figure 1 may be assembled to help explain why this particular pavement is slippery right now, and to decide whether this can be prevented. The background store of knowledge includes general models of pavements, sprinklers, slipping, rain, and so on; these must be accessed and supplied with instance data to construct the specific Bayesian network structure. The store of background knowledge must utilize some representation that combines the expressive power of first-order logical languages (such as semantic networks) with the ability to handle uncertain information.

3. Reasoning Structures in Bayesian Networks

3.1 Basic reasoning structures

3.1.1 d-Separation in Bayesian Networks

d-Separation is one important property of Bayesian networks for inference. Before we define d-separation, we first look at the way that evidence is transmitted in Bayesian Networks. There are two types of evidence:

- **Hard Evidence** (instantiation) for a node A is evidence that the state of A is definitely a particular value.
- **Soft Evidence** for a node A is any evidence that enables us to update the prior probability values for the states of A.

d-Separation (Definition):
Two distinct variables X and Z in a causal network are d-separated if, for all paths between X and Z, there is an intermediate variable V (distinct from X and Z) such that either

- the connection is serial or diverging and V is instantiated or
- the connection is converging, and neither V nor any of V's descendants have received evidence.

If X and Z are not d-separated, we call them d-connected.

3.1.2 Basic structures of Bayesian Networks

Based on the definition of d-seperation, three basic structures in Bayesian networks are as follows:

1. **Serial connections**

   Consider the situation in Figure 2. X has an influence on Y, which in turn has an influence on Z. Obviously, evidence on Z will influence the certainty of Y, which then influences the certainty of Z. Similarly, evidence on Z will influence the certainty on X through Y. On the other hand, if the state of Y is known, then the channel is blocked, and X and Z become independent. We say that X and Z are d-separated given Y, and when the state of a variable is known, we say that it is instantiated (hard evidence).

We conclude that evidence may be transmitted through a serial connection unless the state of the variable in the connection is known.

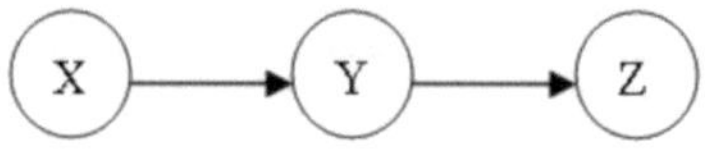

Fig. 2. Serial Connection. When Y is Instantiated, it blocks the communication between X and Z.

2. **Diverging connections**
   The situation in Figure 3 is called a diverging connection. Influence can pass between all the children of X unless the state of X is known. We say that $Y_1, Y_2, \ldots, Y_n$ are d-separated given X.
   Evidence may be transmitted through a diverging connection unless it is instantiated.

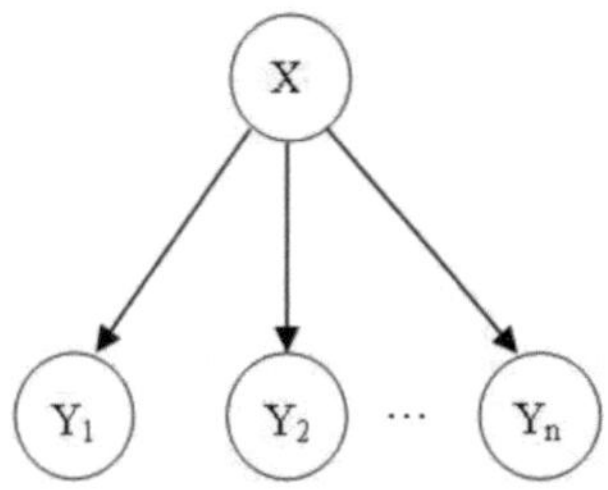

Fig. 3. Diverging Connection. If X is instantiated, it blocks the communication between its children.

3. **Converging connections**

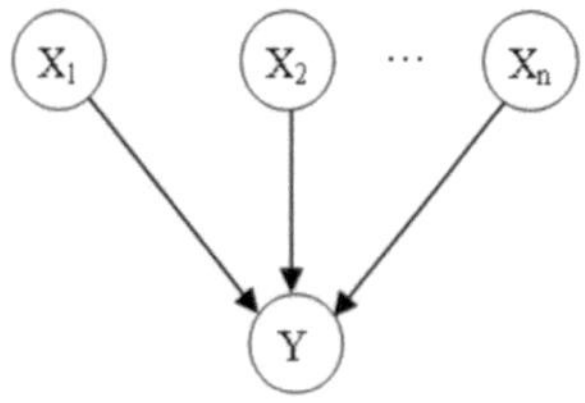

Fig. 4. Converging Connection. If Y changes certainty, it opens for the communication between its parents.

A description of the situation in Figure 4 requires a little more care. If nothing is known about Y except what may be inferred from knowledge of its parents $X_1, \ldots, X_n$, then the parents are independent: evidence on one of the possible causes of an event does not tell us anything about other possible causes. However, if anything is known about the consequences, then information on one possible cause may tell us something about the other causes.

This is the explaining away effect illustrated in Figure 1. X_4 (pavement is wet) has occurred, and X_3 (the sprinkler is on) as well as X_2 (it's raining) may cause X_4. If we then get the information that X_2 has occurred, the certainty of X_3 will decrease.

Likewise, if we get the information that X_2 has not occurred, then the certainty of X_3 will increase.

The three preceding cases cover all ways in which evidence may be transmitted through a variable.

4. Classification of Inferences in Bayesian Networks

In Bayesian networks, 4 popular inferences are identified as:

1. Forward Inference

   Forward inferences is also called predictive inference (from causes to effects). The inference reasons from new information about causes to new beliefs about effects, following the directions of the network arcs. For example, in Figure 2, $X \rightarrow Y \rightarrow Z$ is a forward inference.

2. Backward Inference

   Backward inferences is also called diagnostic inference (from effects to causes). The inference reasons from symptoms to cause, Note that this reasoning occurs in the opposite direction to the network arcs. In Figure 2 , $Z \rightarrow Y$ is a backward inference. In Figure 3 , $Y_i \rightarrow X (i \in [1, n])$ is a backward inference.

3. Intercausal Inference

   Intercausal inferences is also called **explaining away** (between parallel variables). The inference reasons about the mutual causes (effects) of a common effect (cause). For example, in Figure 4, if the Y is instantiated, X_i and $X_j (i, j \in [1, n])$ are dependent. The reasoning $X_i \leftrightarrow X_j (i, j \in [1, n])$ is an intercausal inference. In Figure 3, if X is not instantiated, Y_i and $Y_j (i, j \in [1, n])$ are dependent. The reasoning $Y_i \leftrightarrow Y_j (i, j \in [1, n])$ is an intercausal inference.

4. Mixed inference

   Mixed inferences is also called combined inference. In complex Bayesian networks, the reasoning does not fit neatly into one of the types described above. Some inferences are a combination of several types of reasoning.

4.1 Inference in Bayesian Networks

4.1.1 inference in basic models

- in Serial Connections

  - the **forward inference** executes with the evidence forward propagation. For example, in Figure 5, consider the inference $X \rightarrow Y \rightarrow Z$. [1]

    If Y is instantiated, X and Z are independent, then we have following example:

    $P(Z|XY) = P(Z|Y);$

    $P(Z^+|Y^+) = 0.95;$

    $P(Z^-|Y^+) = 0.05;$

    $P(Z^+|Y^-) = 0.01;$

[1] Note: In this chapter, $P(X^+)$ is the abbreviation of $P(X = true)$, $P(X^-)$ is the abbreviation of $P(|X = false)$. For simple expression, we use $P(Y|X)$ to denote $P(Y = true|X = true)$ by default. But in express $P(Y^+|X)$, X denotes both situations $X = true$ and $X = false$.

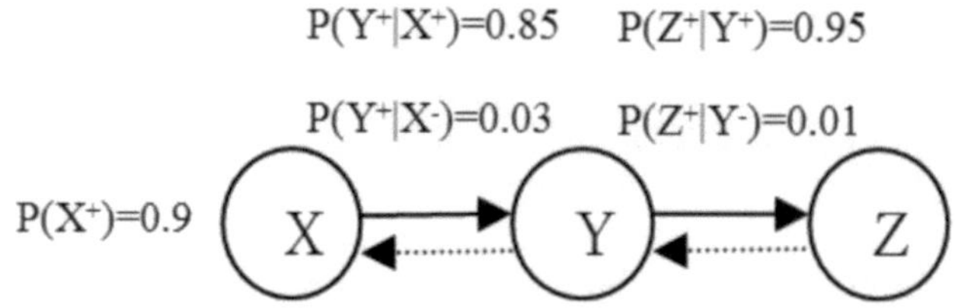

Fig. 5. Inference in Serial Connection

$P(Z^-|Y^-) = 0.99$;
if Y is not instantiated, X and Z are dependent, then
$P(Z^+|X^+Y) = P(Z^+|Y^+)P(Y^+|X^+) + P(Z^+|Y^-)P(Y^-|X^+)$
$= 0.95 * 0.85 + 0.01 * 0.15 = 0.8075 + 0.0015 = 0.809$;
$P(Z^-|X^-Y) = P(Z^-|Y^+)P(Y^+|X^-) + P(Z^-|Y^-)P(Y^-|X^-)$
$= 0.05 * 0.03 + 0.99 * 0.97 = 0.0015 + 0.9603 = 0.9618$.

- the **backward inference** executes the evidence backward propagation. For example, in Figure 5, consider the inference $Z \rightarrow Y \rightarrow X$.
  1. If Y is instantiated ($P(Y^+) = 1$ or $P(Y^-) = 1$), X and Z are independent, then

$$P(X|YZ) = P(X|Y) = \frac{P(X)P(Y|X)}{P(Y)} \tag{5}$$

$P(X^+|Y^+Z) = P(X^+|Y^+) = \frac{P(X^+)P(Y^+|X^+)}{P(Y^+)} = \frac{09*0.85}{1} = 0.765$;
$P(X^+|Y^-Z) = P(X^+|Y^-) = \frac{P(X^+)P(Y^-|X^+)}{P(Y^-)} = \frac{09*0.15}{1} = 0.135$.

  2. If Y is not instantiated, X and Z are dependent (See the dashed lines in Figure 5). Suppose $P(Z^+) = 1$ then
$P(X^+|YZ^+) = \frac{P(X^+YZ^+)}{P(YZ^+)} = \frac{P(X^+YZ^+)}{\sum_X P(XYZ^+)}$;

$P(X^+YZ^+) = P(X^+Y^+Z^+) + P(X^+Y^-Z^+) = 0.9 * 0.85 * 0.95 + 0.9 * 0.15 * 0.05 = 0.72675 + 0.00675 = 0.7335$;

$\sum_X P(XYZ^+) = P(X^+Y^+Z^+) + P(X^+Y^-Z^+) + P(X^-Y^+Z^+) + P(X^-Y^-Z^+)$
$= 0.9 * 0.85 * 0.95 + 0.9 * 0.15 * 0.99 + 0.1 * 0.03 * 0.95 + 0.1 * 0.97 * 0.01$
$= 0.72675 + 0.13365 + 0.00285 + 0.00097 = 0.86422$;

$P(X^+|YZ^+) = \frac{P(X^+YZ^+)}{\sum_X P(XYZ^+)} = \frac{0.7335}{0.86422} = 0.8487$.

In serial connections, there is no intercausal inference.

- in Diverging Connections
  - the **forward inference** executes with the evidence forward propagation. For example, in Figure 6, consider the inference $Y \rightarrow X$ and $Y \rightarrow Z$, the goals are easy to obtain by nature.

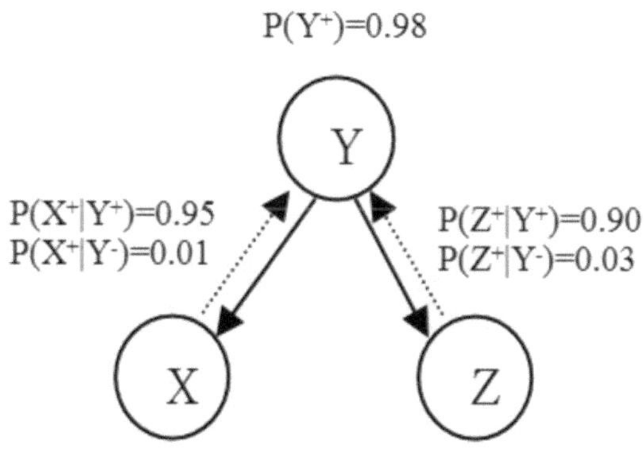

Fig. 6. Inference in Diverging Connection

- the **backward inference** executes with the evidence backward propagation, see the dashed line in Figure 6, consider the inference $(XZ) \rightarrow Y$, X and Z are instantiated by assumption, suppose $P(X^+ = 1)$, $P(Z^+ = 1)$. Then,

$$P(Y^+|X^+Z^+) = \frac{P(Y^+X^+Z^+)}{P(X^+Z^+)} = \frac{P(Y^+)P(X^+|Y^+)P(Z^+|Y^+)}{P(X^+Z^+)}$$
$$= \frac{0.98 * 0.95 * 0.90}{1} = 0.8379 \qquad (6)$$

- the intercausal inference executes between effects with a common cause. In Figure 6, if Y is not instantiated, there exists intercausal inference in diverging connections. Consider the inference $X \rightarrow Z$,
$P(X^+|YZ^+) = \frac{P(X^+YZ^+)}{P(YZ^+)} = \frac{P(X^+Y^+Z^+)+P(X^+Y^-Z^+)}{P(Y^+Z^+)+P(Y^-Z^+)}$;
$= \frac{0.98*0.95*0.90+0.02*0.01*0.03}{0.98*0.90+0.02*0.03} = 0.94936.$

- in Converging Connections,
  - the **forward inference** executes with the evidence forward propagation. For example, in Figure 7, consider the inference $(XZ) \rightarrow Y$, $P(Y|XZ)$ is easy to obtain by the definition of Bayesian Network in by nature.
  - the **backward inference** executes with the evidence backward propagation. For example, in Figure 7, consider the inference $Y \rightarrow (XZ)$.
  $P(Y) = \sum_{XZ} P(XYZ) = \sum_{XZ}(P(Y|XZ)P(XZ))$,

  $P(XZ|Y) = \frac{P(Y|XZ)P(XZ)}{P(Y)} = \frac{P(Y|XZ)P(X)P(Z)}{\sum_{XZ}(P(Y|XZ)P(XZ))}$.

  Finally,
  $P(X|Y) = \sum_Z P(XZ|Y)$,
  $P(Z|Y) = \sum_X P(XZ|Y)$.
  - the **intercausal inference** executes between causes with a common effect, and the intermediate node is instantiated, then $P(Y^+) = 1$ or $P(Y^-) = 1$. In Figure 7, consider the inference $X \rightarrow Z$, suppose $P(Y^+) = 1$,

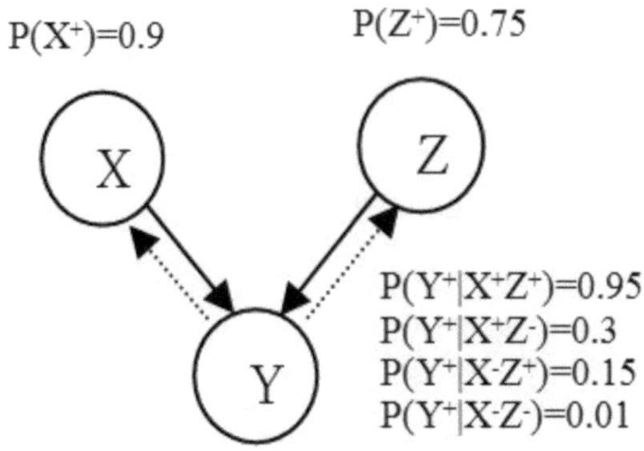

Fig. 7. Inference in Converging Connection

$$P(Z^+|X^+Y^+) = \frac{P(Z^+X^+Y^+)}{P(X^+Y^+)} = \frac{P(Z^+X^+Y^+)}{\sum_Z P(X^+Y^+Z)};$$

$$P(Z^+X^+Y^+) = P(X^+)P(Z^+)P(Y^+|X^+Z^+);$$

$$\sum_Z P(X^+YZ) = P(X^+Y^+Z^+) + P(X^+Y^+Z^-);$$

$$P(Z^+|X^+Y^+) = \frac{P(Z^+X^+Y^+)}{\sum_Z P(X^+Y^+Z)} = \frac{P(X^+)P(Z^+)P(Y^+|X^+Z^+)}{P(X^+Y^+Z^+)+P(X^+Y^+Z^-)}.$$

4.1.2 inference in complex model

For complex models in Bayesian networks, there are single-connected networks, multiple-connected, or event looped networks. It is possible to use some methods, such as Triangulated Graphs, Clustering and Join Trees (Bertele & Brioschi, 1972) (Finn & Thomas, 2007) (Golumbic, 1980), etc., to simplify them into a polytree. Once a polytree is obtained, the inference can be executed by the following approaches.

Polytrees have at most one path between any pair of nodes; hence they are also referred to as singly-connected networks.

Suppose X is the query node, and there is some set of evident nodes $E, X \notin E$. The posterior probability (belief) is denoted as $\mathbb{B}(X) = P(X|E)$, see Figure 8.

E can be splitted into 2 parts: E^+ and E^-. E^- is the part consisting of assignments to variables in the subtree rooted at X, E^+ is the rest of it.

$\pi_X(E^+) = P(X|E^+)$

$\lambda_X(E^-) = P(E^-|X)$

$$\mathbb{B}(X) = P(X|E) = P(X|E^+E^-) = \frac{P(E^-|XE^+)P(X|E^+)}{P(E^-|E^+)} = \frac{P(E^-|X)P(X|E^+)}{P(E^-|E^+)} = \alpha\pi_X(E^+)\lambda_X(E^-) \tag{7}$$

α is a constant independent of X.

where

$$\lambda_X(E^-) = \begin{cases} 1 & \textit{if evidence is } X = x_i \\ 0 & \textit{if evidence is for another } x_j \end{cases} \tag{8}$$

$$\pi_X(E^+) = \sum_{u_1,...,u_m} P(X|u_1,...,u_m) \prod_i \pi_X(u_i) \tag{9}$$

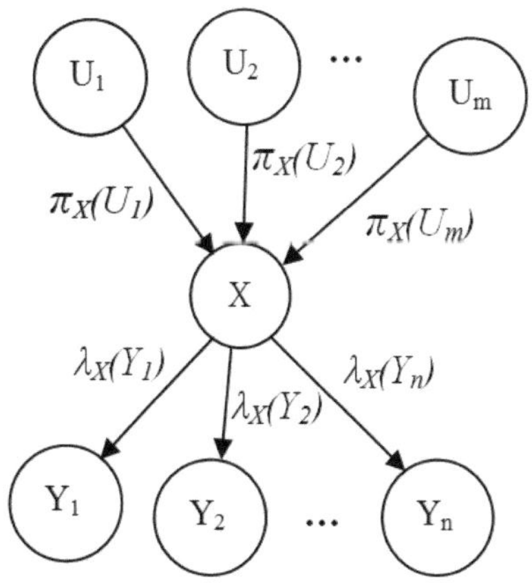

Fig. 8. Evidence Propagation in Polytree

1. Forward inference in Polytree
   Node X sends π messages to its children.

$$\pi_X(U) = \{ \begin{array}{ll} 1 & if\ x_i \in X\ is\ entered \\ 0 & if\ evidentce\ is\ for\ another\ value\ x_j \\ \sum_{u_1,\ldots u_m} P(X|u_1,\ldots u_m) \prod_i \pi_X(u_i) & otherwise \end{array} \tag{10}$$

2. Backward inference in Polytree Node X sends new λ messages to its parents.

$$\lambda_X(Y) = \prod_{y_j \in Y} [\sum_j P(y_j|X)\lambda_X(y_j)] \tag{11}$$

4.2 Related Algorithms for Probabilistic Inference

Various types of inference algorithms exist for Bayesian networks (Lauritzen & Spiegelhalter, 1988) (Pearl, 1988) (Pearl, 2000) (Neal, 1993). Each class offers different properties and works better on different classes of problems, but it is very unlikely that a single algorithm can solve all possible problem instances effectively. Every resolution is always based on a particular requirement. It is true that almost all computational problems and probabilistic inference using general Bayesian networks have been shown to be NP-hard by Cooper (Cooper, 1990). In the early 1980's, Pearl published an efficient message propagation inference algorithm for polytrees (Kim & Pearl, 1983) (Peal, 1986). The algorithm is exact, and has polynomial complexity in the number of nodes, but works only for singly connected networks. Pearl also presented an exact inference algorithm for multiple connected networks called loop cutset conditioning algorithm (Peal, 1986). The loop cutset conditioning algorithm changes the connectivity of a network and renders it singly connected by instantiating a selected subset of nodes referred to as a loop cutset. The resulting single connected network is solved by the polytree algorithm, and then the results of each instantiation are weighted by their prior probabilities. The complexity of this algorithm results from the number of different instantiations that must be considered. This implies that the complexity grows exponentially with the size of the loop cutest being $O(d^c)$, where d is the number of values that the random variables can take, and c is the size of the loop cutset. It is thus important to minimize the size of the

loop cutset for a multiple connected network. Unfortunately, the loop cutset minimization problem is NP-hard. A straightforward application of Pearl's algorithm to an acyclic digraph comprising one or more loops invariably leads to insuperable problems (Koch & Westphall, 2001) (Neal, 1993).
Another popular exact Bayesian network inference algorithm is Lauritzen and Spiegelhalter's clique-tree propagation algorithm (Lauritzen & Spiegelhalter, 1988). It is also called a "clustering" algorithm. It first transforms a multiple connected network into a clique tree by clustering the triangulated moral graph of the underlying undirected graph and then performs message propagation over the clique tree. The clique propagation algorithm works efficiently for sparse networks, but still can be extremely slow for dense networks. Its complexity is exponential in the size of the largest clique of the transformed undirected graph.
In general, the existent exact Bayesian network inference algorithms share the property of run time exponentiality in the size of the largest clique of the triangulated moral graph, which is also called the induced width of the graph (Lauritzen & Spiegelhalter, 1988).

5. Conclusion

This chapter summarizes the popular inferences methods in Bayesian networks. The results demonstrates that the evidence can propagated across the Bayesian networks by any links, whatever it is forward or backward or intercausal style. The belief updating of Bayesian networks can be obtained by various available inference techniques. Theoretically, exact inferences in Bayesian networks is feasible and manageable. However, the computing and inference is NP-hard. That means, in applications, in complex huge Bayesian networks, the computing and inferences should be dealt with strategically and make them tractable. Simplifying the Bayesian networks in structures, pruning unrelated nodes, merging computing, and approximate approaches might be helpful in the inferences of large scale Bayeisan networks.

Acknowledgment

This work is supported by ERCIM (the European Research Consortium for Informatics and Mathematics).

6. References

A. Balke and J. Pearl. Counterfactuals and policy analysis in structural models. Proceedings of the Eleventh Conference on Uncertainty in Artificial Intelligence, pages 11-18, 1995. Morgan Kaufmann.

Bertele, U. and Brioschi, F. (1972). Nonserial Dynamic Programming. Academic Press, London, ISBN-13: 978-0120934508.

G. Cooper. Computational complexity of probabilistic inference using Bayesian belief networks. Artificial Intelligence, 42:393-405, 1990.

Finn V. Jensen and Thomas D. Nielsen (2007). Bayesian Networks and Decision Graphs. Springer, ISBN-13:978-0-387-68281-5.

Golumbic, M. C. (1980). Algorithmic Graph Theory and Perfect Graphs. Academic Press, London,ISBN-13: 978-0122892608.

M. Goldszmidt and J. Pearl. Qualitative Probabilities for Default Reasoning, Belief Revision, and Causal Modeling. Artificial Intelligence, 84(1-2): 57-112, July 1996.

M. I. Jordan, Z. Ghahramani, T. S. Jaakkola and L. K. Saul. An Introduction to Variational Methods for Graphical Models. M. I. Jordan (Ed.), Learning in Graphical Models. Kluwer, Dordrecht, The Netherlands, 1998.

Jin H. Kim and Judea Pearl. A computational model for combined causal and diagnostic reasoning in inference systems. In Proceedings of the Eighth International Joint Conference on Artificial Intelligence (IJCAI-83), pages 190-193, 1983. Morgan Kaufmann.

F. L. Koch, and C. B. Westphall. Decentralized Network Management Using Distributed Artificial Intelligence. Journal of Network and systems management, Vol. 9, No. 4, December 2001.

S. L. Lauritzen and D. J. Spiegelhalter. Local Computations with Probabilities on Graphical Structures and Their Application to Expert Systems. Journal of the Royal Statistical Society, Series B 50:157-224, 1988.

R. M. Neal, Probabilistic Inference Using Markov Chain Monte Carlo methods, Tech. Rep. CRG-TR93-1, University of Toronto, Department of Computer Science, 1993.

J. Pearl. A constraint-propagation approach to probabilistic reasoning, Uncertainty in Artificial Intelligence. North-Holland, Amsterdam, pages 357-369, 1986.

J. Pearl. Evidential Reasoning Using Stochastic Simulation of Causal Models. Artificial Intelligence, 32:247-257, 1987.

J. Pearl. Probabilistic Reasoning in Intelligent Systems: Networks of Plausible Inference. Morgan Kaufmann, San Mateo, CA, 1988.

J. Pearl. Causality: Models, Reasoning, and Inference. Cambridge, England: Cambridge University Press. New York, NY, 2000.

J. Pearl and T. Verma. A theory of inferred causation. J. A. Allen, R. Fikes and E. Sandewall (Eds.), Principles of Knowledge Representation and Reasoning. Proceedings of the Second International Conference, pages 441-452. Morgan Kaufmann, San Mateo, CA, 1991.

P. Spirtes, C. Glymour and R. Scheines. Causation, Prediction, and Search. Springer-Verlag, New York, 1993.

J. B. Tenenbaum and T. L. Griffiths. Structure learning in human causal induction. Advances in Neural Information Processing Systems, volume 13, Denver, Colorado, 2001. MIT Press.

4

Advanced algorithms of bayesian network learning and inference from inconsistent prior knowledge and sparse data with applications in computational biology and computer vision

Rui Chang
University of California, San Diego
USA

1. Introduction

Bayesian networks are a popular class of graphical probabilistic models for researches and applications in the field of Artificial Intelligence. Bayesian network are built on Bayes' theorem (16) and allow to represent a joint probability distribution over a set of variables in the network. In Bayesian probabilistic inference, the joint distribution over the set of variables in a Bayesian network can be used to calculate the probabilities of any configuration of these variables given fixed values of another set of variables, called observations or evidence. Bayesian networks have been widely used for efficient probabilistic inference and data mining in many fields, such as computational biology and computer vision (17; 18).

Before we can generate useful prediction and reasoning by Bayesian networks, it is required to construct these network models from any resources. Over decades, enormous algorithms have been proposed to construct (we use construct and model interchangeably in this chapter) these Bayesian networks. These methods can be roughly classified into two categories: i) top-down modeling methods and ii) reverse-engineering methods. Top-down modeling methods seek for direct solutions to Bayesian network structure and parameter assignments from any prior knowledge resources and domain experts. Currently, this class of methods usually recruits both probability elicitation procedures from domain experts (23) and quantitative knowledge engineering process to disclose the Bayesian network structure and parameters. The advantages of this type of methods are the direct assignment of the parameters and structures from domain knowledge and experts without computational complications. However, in most domains, these methods encounter practical obstacles due to the actual availability of quantitative information and to the limitation of an expert knowledge. In contrast, reverse-engineering approaches utilize machine learning algorithms to train (learn) Bayesian network structure and parameters from a collection of past observations. This process belongs to unsupervised learning in machine learning theory. The advantage of this class approaches is that, a training machine can automatically determine a best Bayesian network model with structure and parameters which optimally fits to the training data under the judgments of an object function or scoring function. (in stead of manually evaluation in top-down methods). This score function is often the posterior probability function of a Bayesian network structure and parameters given the training data. The learned single best model is called Maximum-a-Posterior (MAP) estimation which is computed from data likelihood and prior distribution. In the last twenty to ten years, reverse-engineering approaches have become mainstream researches in the field of Bayesian network modeling. Fruitful results have been achieved, especially in the efficient learning of Bayesian network structure and parameters with (in-) complete data (4; 19–21).

However, a major problem of Bayesian network learning in most existing algorithms is the demands on a large amount of training samples to achieve good generalization performance. The generalization performance of a learned Bayesian network largely depends on the amount of training dataset and the quality of the prior provided to the learning process. Specially, if training data is scarce, it becomes crucial to use various forms of prior knowledge to improve the accuracy of learned models and avoid overfitting. Moreover, although the maximum a posteriori estimation, i.e., the selection of a single best Bayesian network model from the data by learning, is useful for the case of large data sets, independence assumptions among the network variables often make this single model vulnerable to overfitting. In realistic problems, the data basis is often very sparse and hardly sufficient to select one adequate model, i.e., there is considerable model uncertainty. In fact, selecting one single Bayesian model can then lead to strongly biased inference results. Therefore, it is preferable to adopt full Bayesian approaches, such as model averaging, to incorporate these model uncertainties.

2. Overview

2.1 Advanced Bayesian Network Modeling and Inference from Consistent and Inconsistent Prior Knowledge

As the first part of our methodology, we propose novel methods to make use of prior qualitative knowledge in a domain to construct Bayesian networks and generate quantitative probability predictions from these models. These algorithms stem from the observations that in many domains, enormous amounts of priori qualitative knowledge have been accumulated by original studies. This type of knowledge is often represented in terms of qualitative relational statements between two or more entities. For example, in biomedical domain, such a statement can be *smoking increases the risk of lung cancer*. In this statement, two entities are *smoking* and *lung cancer* and these two entities are connected to each other through a directed and functional relation: *increase*. The semantics encoded in this statement is: smoking positively influences lung cancer so that the probability and risk of lung cancer is increased under the condition of smoking. In genomics research, a common knowledge about biological molecular interactions would be a transcript factor binds to a gene and up-regulate this gene's expression level in a cell. In computer vision, qualitative statement can be among action units. For instance, "cheek raiser" tends to happen with "lip corner puller", when smiling. In this statement, cheek raiser increases the occurrence probability of lip corner puller. Similar qualitative statements can be found in many other domains, such as economy, politics, science and engineering indicating that our proposed methods have great promises in these fields. In fact, these inequality constraints have been proposed and used in qualitative probabilistic inference process, such as qualitative probabilistic network (25). However, due to the lacks of quantitative measurements in these qualitative knowledge and constraints, they have been ignored in the quantitative modeling of Bayesian networks.

In our top-down Bayesian inference method, we designed a knowledge model which captures the entities and their relationships in the statement. Various qualitative relations are mapped into mathematically meaningful constraints and inequalities over the Bayesian network structure and parameter space. Due to their qualitativeness, these constraints eventually define a prior distribution in the model space, i.e. model uncertainty. These constraints reduce the set of all possible Bayesian models to those which are consistent with the set of statements considered. This class of consistent models is used to perform full Bayesian inference which can be approximated by Monte Carlo methods, i.e. the quantitative inference and reasoning can be calculated in each of the Bayesian model and these quantitative results are averaged and weighted by the model posterior probability. This is even analytically tractable for smaller networks and statement sets.

Notably, qualitative knowledge are often inconsistent, i.e. there may exist contradicting qualitative statements on entities and/or their relations which eventually affect the model uncertainty in the constructed Bayesian network model space. Therefore, it is imperative to develop methods for reconciling inconsistent qualitative knowledge and for modeling Bayesian networks and performing quantitative prediction. To this end, we further propose a novel framework for performing quantitative Bayesian inference with model averaging based on the inconsistent qualitative statements as a coherent extension of framework of quantitative Bayesian inference based on a set of consistent hypotheses introduced above (33). Our method interprets the qualitative statements by a vector of knowledge features whose structure can be represented by a hierarchical Bayesian network. The prior probability for each qualitative knowledge component is calculated as the joint probability distribution over the features and can be decomposed into the production of the conditional probabilities of the knowledge features. These knowledge components define multiple Bayesian model classes in the hyperspace. Within each class, a set of constraints on the ground Bayesian model space can be generated. Therefore, the distribution of the ground model space can be decomposed into a set of weighted distributions determined by each model class. This framework is used to perform full Bayesian inference which can be approximated by Monte Carlo methods, but is analytically tractable for smaller networks and statement sets.

2.2 Related Works

In discrete model, qualitative causal knowledge have been utilized for abstract probabilistic graphical models, i.e. qualitative probabilistic network (QPN) (6) and reasoning algorithms in QPN have been proposed (5; 9). These algorithms perform qualitative inference with sign propagation in stead of quantitative predictions and neither inconsistent hypotheses could be dealt with.

2.3 Advanced Bayesian Network Learning with Integration of Prior Knowledge and Sparse data

As the second part of the methodology section, we introduce our latest algorithm developments in learning Bayesian network models. In this method, Bayesian network learning accuracy is drastically improved by integrating generic qualitative domain knowledge with training data. We use the knowledge model designed in section 3.1 to translate the causality in qualitative domain knowledge into a set of constraints over structural and parameter space. For parameter learning, we recruit a sampling approach to recover the prior belief distribution in parameter space out of the constraints. We then propose a novel Bayesian parameter score function which integrates this informative prior as soft regulation with the quantitative data statistics. In this way, the parameter posterior distribution is combinatorial regulated by both quantitative data and prior knowledge. In the conventional Bayesian network learning algorithm, MAP estimation usually employs Dirichlet priori to further regulate the statistical counts from the data. However, as discussed above, it is often impossible to determine the correct hyperparameters of this prior distribution which may result bias in the MAP estimation. Our algorithm resolves this issue by establishing an informative prior from domain qualitative knowledge. This informative prior provides the learning machine a correctly defined model subspace to seek for global maximum. By combining each possible prior pseudo counts in this subspace with data statistical counts, we can explore multiple local maximum estimates and determine the global maximum by model selection scheme. Thus, we avoid trapping in the local maximum. This method is particular useful in accurate learning of a Bayesian network under sparse training data. These algorithms can be naturally extended to BN structural learning which is under active developments.

2.4 Related Works

Researches have proposed a number of algorithms to learn Bayesian network parameters by utilizing various forms of prior knowledge, such as dirichlet function (28; 29). In (30–32), parameter learning schemes for various graphical models incorporating parameter sharing constraints are proposed. These algorithms provide efficient solutions for parameter learning with parameter sharing constraints, i.e. parameter equality in one multinomial conditional distribution. If a parameter satisfy the constraints, it obeys the dirichlet distribution with certain normalizer. Otherwise, the prior distribution is zero. A closed form normalization solution is derived in case of parameter sharing constraints. Moreover, some simple forms of inequalities within one conditional distribution are proposed (32). In this case, no closed-form solution is possible. Though, in (30–32), constrained parameter learning problem is treated as a constraint optimization problem and efficient algorithms are developed, the forms of the constraints are limited to either parameter sharing or inequality constraints within one conditional distribution, such as $P(A|B)>P(\overline{A}|B)$. More generic and important inequality constraints, such as $P(A|B)>P(A|\overline{B})$ is not addressed by their methods.

In (35) and (37), methods are proposed to deals with the inequality constraints in parameter learning. A penalty term is designed to regulate the likelihood which is derived from the monotonic influence with form of $P(A|B)>P(A|\overline{B})$. The violation term can only penalize the likelihood when the learned local maximum violates the constraints in the sign, but it can not distinguish a set of all possible local maximums obeying the constraints. So, final solution is not necessary a global maximum. (Eq.8 in (35) and Eq.9 in (37)). This is a serious problem in case of learning with very sparse data. In this case, although ML estimation may output an estimate obeying the sign of the constraints, this ML estimation is highly probable incorrect due to the amount of data. In this case, neither (35) nor (37) could use prior statistics to correct the estimation. As stated in (37), a soft Bayesian prior which regulates the ML term is desired. A similar iterative approach with penalty function was introduced in (36). The method in (42), however, includes constraints beyond the monotonicity constraints.

In (38), an averaging scheme is proposed. This method is only feasible up to 5/6 parents. (39) proposed a similar idea to ours independently. A method which uses a soft

Bayesian prior to regulate the ML score and introduce the concept of model uncertainty in the MAP estimation. The empirical Bayes and maximum posterior estimate in (39) and $QMAP_{FBA}$,$QMAP_{FMA}$ in my paper are comparable. However, (39) indirectly translates the prior knowledge into an intractable integration which has to be approximated. The dirichlet hyperparameters is replaced by another hyperparameter (Eq.14 in (39)). Their initial idea is to assign some confidence to constraints. (Eq.7 in (39)). But it may be easier and more efficient to handle this issue in the knowledge level than score level (34). Comparatively, we work directly on the parameter space through sampling and obtain the dirichlet hyperparameters directly. Thus, we believe our method can be more efficient and feasible than their method.

3. Methods

In this section, we formally propose our top-down Bayesian network modeling algorithm, i.e. Bayesian inference with only consistent and inconsistent qualitative prior knowledge. Next, we introduce our advanced Bayesian network learning algorithm by integrating both qualitative prior knowledge and data.

3.1 Probabilistic Representation of a Qualitative Knowledge Model

Several qualitative models have been proposed in the context of Qualitative Probabilistic Networks (QPN). Qualitative knowledge models describe the process of transforming the qualitative statements into a set of probability constraints. The proposed Bayesian inference method outlined above is independent of the qualitative knowledge model, i.e. the model posterior probability is independent of the set of qualitative statements used, once the set of probabilistic inequality constraints which are translated from qualitative statements is given. Three existing qualitative models are the Wellman approach (25), the Neufeld approach (22) and the orders of magnitude approach (27). Here we follow the Wellman approach, where qualitative knowledge involves influential effects from parent nodes to child nodes which are classified according to the number of inputs from parents to child and their synergy. For the sake of simplicity, we restrict our discussion to binary-valued nodes. Logic "1" and "0" values of a node are defined as "present" and "absent" or "active" and "inactive", as synonyms, A and $\overline{A}$. For multinomial nodes, similar definitions can be applied.

3.1.1 Structural Qualitative Knowledge Model

The qualitative knowledge contained in the statements are describing two aspects of a belief network, i.e. structure and parameter. The structural knowledge of a simple network consisting node *B* and node *A* can be described with two first-order logic predicates:

$$\begin{aligned} Depend(A,B) &= 0/1 \\ Influence(A,B) &= 0/1 \end{aligned} \tag{1}$$

which describe whether A and B are dependent and whether the influence direction is from A to B; *Depend* and *Influence* are denoted by *Dp* and *I* as well as, the set of structural knowledge features is denoted by Π={Dp,I}.

3.1.2 Parameter Qualitative Knowledge Model

Under each structure feature, we extend the QPN model with two sets of dependent features, i.e. baseline qualitative knowledge features, Σ and extended qualitative knowledge features, Ψ. These two feature sets are used to describe the qualitative parameter knowledge.

3.1.2.1 Baseline Qualitative Knowledge Model

In QPN, a set of features define the basic properties of qualitative causal influences and their synergy classified by the number of inputs from parents to child which are refined in this paper and are referred to as *Baseline Qualitative Knowledge Model.* Baseline features transform qualitative statements into a primitive set of constraints on model parameter space. We discuss three cases of influences, namely single influence, joint influence and mixed joint influence. In addition, we discussed the qualitative influence derived from recurrent and/or conflicting statements. The definitions of the influences in our work are originated and refined based on the qualitative probabilistic network in (25) which enables us to translate the qualitative statements into a set of constraints in the parameter space which can be used to model the parameter distribution given the structure.

I. Single Influence

Definition 3.1 If a child node B has a parent node A and the parent imposes a isolated influence on the child, then qualitative influence between parent and child is referred to as *single influence.* Single influence can be further classified into single positive influence and single negative influence.

Definition 3.2 If presence of parent node A renders presence of child node B more likely, then the parent node is said to have a *single positive influence* on the child node. This can be represented by the inequality

$$Pr(B|A) \geq Pr(B|\overline{A}) \tag{2}$$

Definition 3.3 If presence of parent node A renders presence of child node B less likely, then parent node is said to have a *single negative influence* on child node. This can be represented by the inequality

$$Pr(B|A) \leq Pr(B|\overline{A}) \tag{3}$$

II. Joint Influence

Definition 3.4 If a child node B has more than one parent node and all parents influence the child in a joint way, then these influences between parents and child are referred to as *joint influence.* This joint influence can be either synergic (cooperative) or antagonistic (competitive) and the individual influences from the parents to the child can be either positive or negative.

Definition 3.5 If a joint influence from two or more parent nodes generates a combined influential effect larger than the single effect from each individual parent, then the joint influence is referred to as *plain synergic joint influence* or *plain synergy.*

Assume that parent nodes A and B impose positive individual influences on child node C, then the knowledge model can be defined as

$$Pr(C|A,B) \geq \left\{ \begin{array}{c} Pr(C|A,\overline{B}) \\ Pr(C|\overline{A},B) \end{array} \right\} \geq Pr(C|\overline{A},\overline{B}) \tag{4}$$

Definition 3.6 If joint influences from two or more parent nodes generate an combined influential effect larger than the sum of each single effect from an individual parent, then the joint influence is referred to as *additive synergic joint influence* or *additive synergy.*(24)

Assume in case that parent nodes A and B impose a positive individual influence on child node C, then we define

$$Pr(C|A,B) \geq Pr(C|A,\overline{B}) + Pr(C|\overline{A},B) \geq \left\{ \begin{array}{c} Pr(C|A,\overline{B}) \\ Pr(C|\overline{A},B) \end{array} \right\} \geq Pr(C|\overline{A},\overline{B}) \tag{5}$$

Similar rules can be applied to the case where A and B impose a negative individual influence on child node C. Comparing Eq. 5 with Eq. 4, we can conclude that *additive synergy* is a sufficient condition for *plain synergy* and *plain synergy* is a necessary but not sufficient condition for *additive synergy*. Therefore, if multiple parents demonstrate additive synergy, it is sufficient to judge that this influence is also plain synergy, but not vice-versa.

It is important to distinguish between plain synergy and additive synergy since they represent distinct semantic scenarios in a domain. For example, A is a protein and B is a kinase which phosphorylates protein A and produces the phosphorylated protein C. Because of the nature of this protein-protein interaction, neither B nor A alone can significantly increase the presence of C, but both together can drastically increase the presence of C which is greater than the sum of C in case of either A or B present. In this example A and B exhibit additive synergy and it is sufficiently to conclude that A and B has plain synergy as well.

Definition 3.7 If the joint influences from two or more parent nodes generate a combined influential effect less than the single effect from individual parent, then the joint influence is referred to as *antagonistic joint influence* or *antagonism*.

Assume that parent nodes A and B have independent positive single influences on child node C, the antagonistic influence of A and B can be represented by

$$Pr(C|\overline{A},\overline{B}) \leq Pr(C|A,B) \leq \left\{ \begin{array}{c} Pr(C|A,\overline{B}) \\ Pr(C|\overline{A},B) \end{array} \right\} \tag{6}$$

Similar rules can be applied to the case where A and B imposes a negative individual influence on child node C.

III. Mixed Joint Influence

In case that the joint effect on a child is formed by a mixture of positive and negative individual influences from its parents, the extraction of a probability model is not well-defined in general. Hence, we adopt the following scheme: If there are mixed influences from several parent nodes to a child node, and no additional information is given, then they are treated as independent and with equal influential strength. Assume that parent node A imposes positive single influence on child node C and parent node B imposes negative single influence on child node C, then the joint influence can be represented by

$$Pr(C|A,B) \geq Pr(C|\overline{A},B);\ Pr(C|A,\overline{B}) \geq Pr(C|\overline{A},\overline{B});$$

$$Pr(C|A,\overline{B}) \geq Pr(C|A,B);\ Pr(C|\overline{A},\overline{B}) \geq Pr(C|\overline{A},B) \tag{7}$$

Any additional structure can be brought into the CPT of the corresponding collider structure as soon as dependencies between influences are made explicit by further qualitative statements.

3.1.3 Extended Qualitative Knowledge Model

The extended qualitative knowledge model defines relative and absolute properties of probability configurations in qualitative causal influences and synergy from the baseline model. It includes the probabilistic ratio and relative difference between any number of configurations in a qualitative causal influence and the absolute probabilistic bound of any configuration in a causal influence. These extended features impose further restriction on the set of constraints generated by baseline model, therefore, restrain the uncertainty in Bayesian model space so that more accurate generalization can be achieved.

The extended qualitative knowledge features can be consistently represented by a linear inequality. In the case that node B impose single influence on node A, there are two probabilistic configurations. The linear constraints can then be written as

$$Pr(B|A) \geq, \leq R \times Pr(B|\overline{A}) + \Delta; Pr(B|A) \in [Bd_{min}, Bd_{max}]; Pr(B|\overline{A}) \in [Bd'_{min}, Bd'_{max}] \quad (8)$$

which R is *Influence Ratio*, Δ is *Influence Difference* and Bd, Bd' denote *bound*. In some cases, baseline and extended qualitative knowledge information are provided by the qualitative statements simultaneously. However, in most cases, extended knowledge features are not fully provided in the qualitative statements. In these cases, only baseline knowledge model will be used to generate constraints in model space to perform inference by model averaging. Once the qualitative knowledge is translated by the feature set $\{\Pi(Dp, I), \Lambda(\Sigma, \Psi(R, \Delta, Bd))\}$ according to Eq. 1 to Eq. 8, the distribution of ground models is defined by this knowledge. Once formulated, the Monte Carlo sampling procedure will make sure that all inequalities are satisfied for valid models.

3.1.4 Hierarchical Knowledge Model for Inconsistent Statements

The dependent qualitative knowledge feature set can be represented by a hierarchical Bayesian network (HBN) (3). Within a knowledge HBN, the structural feature Π and parameter feature Λ are two first-level composite nodes. Π can be further decomposed into two leaf nodes Dp and I. The parameter feature Λ contains two second-level composite nodes, i.e. the baseline knowledge features Σ and extended knowledge features Ψ which consists of three leaf nodes R, Δ and Bd. Thus qualitative knowledge Ω can be described as $\Omega = \{\Pi(Dp, I), \Lambda(\Sigma, \Psi(R, \Delta, Bd))\}$, where $\Sigma = (SP, SN, PlSyn, AdSyn, Ant, MxSyn)$. The hierarchical knowledge model is shown in Figure 1(a) and a tree hierarchy in Figure 1(b). The equivalent Bayesian network is shown in Figure 1(c).
Hierarchical Bayesian Networks encode conditional probability dependencies in the same way as standard Bayesian Networks. The prior probability of a qualitative knowledge Ω can be written as a joint probability of $\{\Pi, \Lambda\}$ and can be decomposed according to the dependency between each component features as follows.

$$Pr(\Omega) = Pr(\Pi)Pr(\Sigma|\Pi)Pr(\Psi|\Sigma) \quad (9)$$

where $Pr(\Psi|\Sigma) = Pr(R|\Sigma)Pr(\Delta|\Sigma)Pr(Bd|\Sigma)$, $Pr(\Pi) = Pr(Dp)Pr(I|Dp)$ and $Pr(\Sigma|\Pi) = Pr(\Sigma|I)$. The conditional probabilities of qualitative knowledge features can be calculated by counting the weighted occurrences given a set of inconsistent statements. The weight of knowledge features equals to the credibility of their knowledge sources which may be evaluated by a domain expert or determined by the source *impact factor*. If no further information on the weights is available, they are set to 1. In this case, the conditional probability of features is computed only by occurrence count. For example, we assume a set of qualitative statements, $\widetilde{S} = \{S_1, S_2, S_3\}$, about *smoking* and *lung cancer* are observed: 1) *The risk is more than 10 times greater for smokers to get lung cancer than no-smokers.* 2) *Men who smoke two packs a day increase their risk more than 25 times compared with non-smokers.* 3) *There is not significant evidence to prove that smoking directly cause lung cancer, however, clinical data suggest that lung cancer is related to smoking.* The statements can be represented by a vector of features which is shown in Figure 2. The conditional probability of the features can be calculated straightforwardly by

$$Pr(I|Dp) = (w_1 + w_2)/w_a \qquad Pr(\overline{I}|Dp) = (w_3)/w_a$$
$$Pr(r_1|\Sigma = SP) = w_1/w_b \qquad Pr(r_2|\Sigma = SP) = (w_1 + w_2)/w_b$$

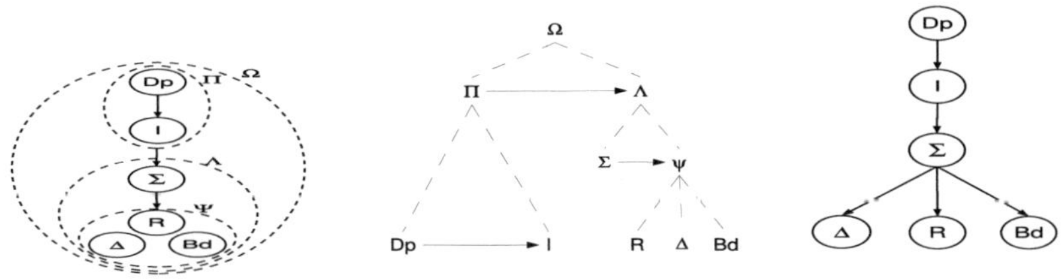

(a) HBN (b) Tree (c) BN

Fig. 1. Hierarchical Bayesian Network on Qualitative Knowledge

Stat.	Dp	I	Σ	R	Δ	B	Weight
S_1	1	1	SP	[10,∞]	null	null	w_1
S_2	1	1	SP	[25,∞]	null	null	w_2
S_3	1	0	null	null	null	null	w_3

Fig. 2. Feature-vector of Statements

where $w_a = w_1 + w_2 + w_3$, $w_b = 2w_1 + w_2$, $Pr(Dp) = 1$, $Pr(SP|I) = 1$, $r_1 = [10, 25]$ and $r_2 = [25, \infty]$. One notion is that the knowledge features $\Psi = \{R, \Delta, Bd\}$ in Figure 1(a) are continuous-valued and therefore, can be transformed to discrete attributes by dynamically defining new discrete attributes that partition the continuous feature value into a discrete set of intervals. In the above example, the continuous feature R in S_1 has value range $[10, \infty]$ and a continuous value range $[25, \infty]$ in S_2. The continuous ranges can be partitioned into two discrete intervals: $r_1 = [10, 25]$ and $r_2 = [25, \infty]$, therefore, the qualitative knowledge $\widetilde{\Omega} = \{\Omega_1, \Omega_2, \Omega_3\}$ can be transformed from $\widetilde{S} = \{S_1, S_2, S_3\}$ with discrete-valued features.

3.1.4.1 Qualitative Knowledge Integration

Once we have calculated the conditional probabilities of knowledge features, the prior probability of qualitative knowledge can be computed according to Eq. 9. Thus the inconsistent knowledge components are ready to be reconciled. The qualitative knowledge transformed from the feature vector of statements in Figure 2 can be described by $\widetilde{\Omega}$:

$$\Omega_1 = \{1, 1, SP, [10, 25], \emptyset, \emptyset\} \quad \Omega_2 = \{1, 1, SP, [25, \infty], \emptyset, \emptyset\} \quad \Omega_3 = \{1, 0, \emptyset, \emptyset, \emptyset, \emptyset\} \tag{10}$$

where Ω_k=$\{Dp_k, I_k, \Sigma_k, R_k, \Delta_k, Bd_k\}$. If the weights of statements are set to 1, the knowledge prior probability is calculated, then we have $Pr(\Omega_1)$=2/9, $Pr(\Omega_2)$=4/9 and $Pr(\Omega_3)$=1/3.

$$\begin{aligned} Pr(\Omega_1) &= Pr(Dp)Pr(I|Dp)Pr(SP|I)Pr(r_1|SP) = 2/9 \\ Pr(\Omega_2) &= Pr(Dp)Pr(I|Dp)Pr(SP|I)Pr(r_2|SP) = 4/9 \\ Pr(\Omega_3) &= Pr(Dp)Pr(\bar{I}|Dp) = 1/3 \end{aligned} \tag{11}$$

The integrated qualitative knowledge thus preserved the uncertainty from each knowledge component. Each qualitative knowledge component Ω_k defines a model class with a set of constraints on the ground model space which is generated by its features. The model class and its constraints are used for modeling Bayesian networks and performing quantitative inference.

3.2 Bayesian Inference with Consistent Qualitative Knowledge

3.2.1 Bayesian Modeling and Inference

A Bayesian model m represents the joint probability distribution of a set of variables $\mathbf{X} = X_1, X_2, ..., X_D$ (19). The model is defined by a graph structure s, which defines the structures of the conditional probabilities between variables, and a parameter vector θ, the components of which define the entries of the corresponding conditional probability tables (CPTs). Hence, a Bayesian network can be written as $m = \{s, \theta\}$. If we believe that one single model m reflects the true underlying distribution, we can perform inference based on this model. Given some observations or "evidence" E, reflected by fixed measured values of a subset of variables, $\mathbf{X}_q = E$, we wish to derive the distribution of the remaining variables $X \in \mathbf{X} \backslash \mathbf{X}_q$. It is provided by their conditional probability given the evidence in light of the model, $Pr(X|E, m)$, which can be efficiently evaluated by known methods.(26)
In contrast, the full Bayesian framework does not attempt to approximate one true underlying distribution. Instead, all available information is used in an optimal way to perform inference, without taking one single model for granted. To formalize this statement for our purposes, let us classify the set of available information into an available set of data, D, and a body of non-numeric knowledge, Ω. The a posteriori distribution of models m is then given by

$$Pr(m|D,\Omega) = \frac{Pr(D|m)\ Pr(m|\Omega)}{Pr(D,\Omega)}. \tag{12}$$

The first term in the numerator of eq. (12) is the likelihood of the data given the model, which is not directly affected by non-numeric knowledge Ω, the second term denotes the model prior, whose task is to reflect the background knowledge. We obtain

$$Pr(m|D,\Omega) = \frac{1}{Z} Pr(D|m)\ Pr(m|\Omega), \tag{13}$$

where Z is a normalization factor which will be omitted from the equations for simplicity. The first term contains the constraints of the model space by the data, and the second term the constraints imposed by the background knowledge. In the full Bayesian approach, we can perform inference by model averaging. Now, given some observation or evidence E, the (averaged) conditional distribution of the remaining variable X is performed by integrating over the models:

$$Pr(X|E,D,\Omega) = \int Pr(X|E,m) Pr(m|D,\Omega) dm = \int Pr(X|E,m) Pr(D|m) Pr(m|\Omega) dm \tag{14}$$

3.2.2 Bayesian Network Inference with Qualitative Knowledge

In this paper we consider the extreme case of no available quantitative data, $D = \emptyset$. Even in this case, it is still possible to perform proper Bayesian inference,

$$Pr(X|E,\Omega) = \int Pr(X|E,m) Pr(m|\Omega) dm. \tag{15}$$

Now the inference is based on the general background information contained in Ω alone, and the specific information provided by the measurements E. This is reflected by the fact that inference results are conditioned on both quantities in eq. (15).
In order to determine $Pr(m|\Omega)$, we need a formalism to translate a body of qualitative knowledge into an a priori distribution over Bayesian models. For this we adopt the following notation for a Bayesian model class. A Bayesian model is determined by a graph structure s and

by the parameter vector θ needed to specify the conditional probability distributions given that structure. We refer to θ as one specific CPT configuration. A Bayesian model class $\widetilde{M}$ is then given by (i) a discrete set of model structures $\widetilde{S} = \{s_1, s_2, \ldots, s_K\}$, and (ii) for each structure s_k a (eventually continuous) set of CPT configurations Θ_k. The set of member Bayesian models $m \in \widetilde{M}$ of that class is then given by $m = \{(s_k, \theta) | k \in \{1, \ldots, K\}, \theta \in \Theta_k\}$. The model distribution now reads

$$Pr(m|\Omega) = Pr(s_k, \theta|\Omega) = \frac{Pr(\theta|s_k, \Omega)Pr(s_k|\Omega)}{\sum_{a=1}^{K} \int_{\Theta_a} Pr(\theta|s_a, \Omega)d\theta Pr(s_a|\Omega)}. \tag{16}$$

In eq. (16), first the set of allowed structures is determined by means of Ω, followed by the distributions of the corresponding CPT configurations. Then, we calculate the model's posterior probability $Pr(m|\Omega)$ in eq. 16. Inference is carried out by integrating over the structure space and the structure-dependent parameter space:

$$Pr(X|E, \Omega) = \sum_{k=1}^{K} \int_{\Theta_k} Pr(X|E, s_k, \theta)Pr(s_k, \theta|\Omega)d\theta. \tag{17}$$

It is very common to express non-numeric knowledge in terms of qualitative statements about a relationship between entities. Here we assume Ω to be represented as a list of such qualitative statements. In this form, the information can be used in a convenient way to determine the model prior, eq. (16): (i) Each entity which is referenced in at least one statement throughout the list is assigned to one variable X_i. (ii) Each relationship between a pair of variables constrains the likelihood of an edge between these variables being present. (iii) The quality of that statement (e.g., "activates", "inactivates") affects the distribution over CPT entries θ given the structures. In the most general case, the statement can be used to shape the joint distribution over the class of all possible Bayesian models over the set of variables obtained from Ω.

Here we propose a simplified but easy-to-handle way for constructing the prior model distribution. We use each statement to constrain the model space to that subspace which is consistent with that statement. In other words, if a statement describes a relationship between two variables, only structures s_k which contain the corresponding edge are assigned a nonzero probability $Pr(s_k|\Omega)$. Likewise, only parameter values on that structure, which are consistent with the contents of that statement, are assigned a nonzero probability $Pr(\theta|s_k, \Omega)$. If no further information is available, the distribution is constant in the space of consistent models.

3.3 Bayesian Inference with Inconsistent Qualitative Knowledge

In this section, we propose a novel approach to make use of a set of inconsistent qualitative statements and their prior belief distribution as background knowledge for Bayesian modeling and quantitative inference.

A Bayesian model m represents the joint probability distribution of a set of variables $X = \{x_1, x_2, ..., x_N\}$ (1). The model is defined by a graph structure s and a parameter vector θ, i.e. $m = \{s, \theta\}$. In full Bayesian framework, all available information is used in an optimal way to perform inference by taking model uncertainty into account. Let us classify the set of available information into an available set of training data D and a set of inconsistent qualitative background knowledge $\widetilde{\Omega} = \{\Omega_1, \ldots, \Omega_K\}$ on a constant set of variables. The posterior distribution of models m is then given by

$$Pr(m|D,\widetilde{\Omega}) = \frac{Pr(D|m,\widetilde{\Omega})Pr(m|\widetilde{\Omega})Pr(\widetilde{\Omega})}{Pr(D,\widetilde{\Omega})} \tag{18}$$

The first term in the numerator of Eq. 18 is the likelihood of the data given the model. The second term denotes the model prior which reflects the inconsistent set of background knowledge and the last term is the prior belief of the knowledge set. Now, inference in the presence of evidence is performed by building the expectation across models:

$$Pr(X|D,E,\widetilde{\Omega}) = \int dmPr(X|E,m)Pr(D|m,\widetilde{\Omega})Pr(m|\widetilde{\Omega})Pr(\widetilde{\Omega}) \tag{19}$$

In this paper we consider the extreme case of no available quantitative data, $D = \emptyset$.

$$Pr(X|E,\widetilde{\Omega}) = \int dmPr(X|E,m)Pr(m|\widetilde{\Omega})Pr(\widetilde{\Omega}) \tag{20}$$

In this case, model prior distribution $Pr(m|\widetilde{\Omega})$ is determined soly by the inconsistent background knowledge set $\widetilde{\Omega}$. Each independent qualitative knowledge component, $\Omega_k \in \widetilde{\Omega}$, uniquely defines a model class, M_k, with a vector of features, i.e. $\widetilde{M} = \{M_1, \ldots, M_K\}$. The features are translated into a set of constraints which determine the distribution of the ground models within each model class.
First of all, the probability of a model class given the inconsistent knowledge set is written as

$$Pr(M_k|\widetilde{\Omega}) = \sum_{i=1}^{K} Pr(M_k|\Omega_i)Pr(\Omega_i|\widetilde{\Omega}) = Pr(\Omega_k) \tag{21}$$

where $\{Pr(M_k|\Omega_i) = 1, i = k\}$ and $\{Pr(M_k|\Omega_i) = 0, i \neq k\}$ since the *k-th* model class is uniquely defined by Ω_k and is independent to the other knowledge component. Secondly, the probability of a ground Bayesian model sample m in the *k-th* model class given the inconsistent knowledge set is

$$Pr(m \in M_k|\widetilde{\Omega}) = Pr(m|M_k)Pr(M_k|\widetilde{\Omega}) \tag{22}$$

Thus, the inference on X given evidence E and inconsistent knowledge set $\widetilde{\Omega}$ in Eq. 20 can be written as

$$Pr(X|E,\widetilde{\Omega}) = \sum_k \int_m dmPr(X|m,E)Pr(m|M_k)Pr(\Omega_k)$$

where $Pr(m|\widetilde{\Omega}) = \sum_k Pr(m \in M_k|\widetilde{\Omega})$ and we assume the inconsistent knowledge set to be true, i.e. $Pr(\widetilde{\Omega}) = 1$. Therefore, the inference is calculated by firstly integrating over the structure space and the structure-dependent parameter space of a ground Bayesian model from a model class according to the constraints and performing such integration iteratively over all possible model classes with the prior distribution. The integration in Eq. 23 is non-trivial to compute, however, Monte Carlo methods can be used to approximate the inference.

3.3.1 ASIA Benchmark Model

The ASIA network (10) is a popular toy belief model for testing Bayesian algorithms. The structure and parameter of actual ASIA network is shown in Figure 3.
For demonstration, we consider the inconsistent qualitative statements with regarding to single edge between *Smoking* and *Lung Cancer*, as well as the collider structure of *Lung Cancer*,

Bronchitis and *Dyspnea*. The method applies to all of the entities and their relations in the ASIA network. *1. Although nonsmokers can get lung cancer, the risk is about 10 times greater for smokers. (http://www.netdoctor.co.uk);2. The lifetime risk of developing lung cancer in smokers is approximately 10%.(http://www.chestx-ray.com/Smoke/Smoke.html);3. Men who smoke two packs a day increase their risk more than 25 times compared with non-smokers.(http://www.quit-smoking-stop.com/lung-cancer.html)4. Lifetime smoker has a lung cancer risk 20 to 30 times that of a non-smoker(http://www.cdc.gov/genomics/hugenet/ejournal/OGGSmoke.htm)5. Only 15% of smokers ultimately develop lung cancer(http://www.cdc.gov/genomics/hugenet/ejournal/OGGSmoke.htm);6. The mechanisms of cancer are not known. It is NOT possible to conclusively attribute a cause to effects whose mechanisms are not fully understood.(http://www.forces.org/evidence/evid/lung.htm);7. It is estimated that 60% of lung cancer patients have some dyspnea at the time of diagnosis rising to 90% prior to death.(http://www.lungcancer.org/health_care/focus_on_ic/ symptom/dyspnea.htm)8. Muers et al. noted that breathlessness was a complaint at presentation in 60% of 289 patients with non-small-cell lung cancer. Just prior to death nearly 90% of these patients experienced dyspnea. (2);9. At least 60% of stage 4 lung cancer victims report dyspnea.(http://www.lungdiseasefocus.com/lung-cancer/palliative-care.php);10. Significantly more patients with CLD than LC experienced breathlessness in the final year (94% CLD vs 78% LC, P < 0.001) and final week (91% CLD vs 69% LC, P < 0.001) of life. (7);11. 95% of patients with chronic bronchitis and emphysema reported Dyspnea. (8)*
Each statement is analyzed by the hierarchical knowledge model in Figure 1(a) and the extracted features are summarized in Figure 3(c). In this statement set, the first six statements represent the relation between (tobacco)smoking and lung cancer. $\{S_1, \ldots, S_5\}$ describe a *single positive (SP)* influence from smoking to lung cancer with inconsistent knowledge features of the *ratio (R)* and *bound (Bd)*. However, statement S_6 declares a contradicting knowledge suggesting that smoking is not the cause of lung cancer. $\{S_7, \ldots, S_{11}\}$ describe the synergic influence from lung cancer and bronchitis to dyspnea. Without further information, it can be represented by *plain synergy with positive individual influence*. The knowledge on the extended features in Eq. 7 of the conditional probability distribution of this collider structure is not available, however, the knowledge on the extended features of the marginalized conditional probability space are provided in these statements. For simplicity, we assume the weight of every qualitative statement equals to 1, i.e. $\{w_i = 1, i = 1, \ldots, 11\}$. Due to the parameter independency (1), we can compute the conditional probability of each local structure independently. For each local structure, we calculate the conditional probability of knowledge features by counting its occurrence frequency. For the local structure of smoking and lung cancer in the ASIA network, the prior probability of the knowledge features can be calculated as $Pr(Dp)$=5/6, $Pr(I|Dp)$=1, $Pr(\overline{I}|\overline{Dp})$=1, $Pr(SP|I)$=1, $Pr(r_1|SP)$=1/5, $Pr(r_2|SP)$=1/5, $Pr(r_3|SP)$=2/5, $Pr(r_4|SP)$=1/5, $Pr(b_1|SP)$=1/2 and $Pr(b_2|SP)$=1/2 where $r_1 = [9, 11]$, $r_2 = [20, 25]$, $r_3 = [25, 30]$ and $r_4 = [30, \infty]$; $b_1 = [9\%, 11\%]$ and $b_2 = [14\%, 16\%]$. The continuous-valued feature R and Bd are discretized into $|R| = 4$ and $|Bd| = 2$ discrete-value intervals respectively. Based on the features and their prior belief, a set of qualitative knowledge $\widetilde{\Omega} = \{\Omega_1, \ldots, \Omega_{16}\}$ is formed in Figure 3(d).

3.3.1.1 ASIA Model Monte Carlo Sampling

Given the integrated qualitative knowledge set $\widetilde{\Omega}$ with prior probabilities, we now construct the Bayesian model class and the distribution on ground model space within each class. For demonstration purposes, we assume the partial structure and its parameters, i.e. $\{\alpha, \gamma, \lambda, f\}$, to be known as in Figure 3(b). Therefore the uncertainty of ASIA model space is restricted to the uncertainty of the local structure and parameter space on *Smoking* and *Lung Can-*

cer which can be described by $Pr(m|M_k)$ and $Pr(M_k)$ defined by $\{\Omega_k|k = 1,\ldots,9\}$, i.e. $\{M_k(\Omega_k)|k = 1,\ldots,9\}$, as well as the uncertainty of the local space on *Lung Cancer*, *Bronchitis* and *Dyspnea* which can be jointly determined by three types of model class, i.e. the root-dimension model class defined by Ω_{10}, the marginal-dimension model classes of lung cancer and dyspnea defined by $\{\Omega_i|i = 11,\ldots,14\}$ and the marginal-dimension model classes of bronchitis and dyspnea defined by $\{\Omega_j|j = 15,16\}$. Thus, there are total eight possible combination of these model classes, i.e. $\{M_k(\Omega_{10},\Omega_i,\Omega_j)|k = 10,\ldots,17; i = 11,\ldots,14; j = 15,16\}$ and each combination virtually forms a complete model class which defines the set of constraints on the structure and parameter space of ground Bayesian model for the local collider structure of lung cancer, bronchitis and dyspnea. The prior probability of each combination, $Pr(M_k)$ is the product of the prior probability of its independent components, i.e.

$$Pr(M_k) = Pr(\Omega_{10})Pr(\Omega_i)Pr(\Omega_j) \tag{23}$$

For each local structure, we perform 10,000 sampling iterations. In each iteration, we select a model class M_k randomly based on the prior probability of the model class, i.e $Pr(M_k)$. In each selected model class, we randomly choose 3 samples of ground Bayesian model m, whose structure and parameter space is consistent with the class constraints $Pr(m|M_k)$ as shown in Figure 1(a). In this way, for the local structure of smoking and lung cancer, the prior bability of the model class is equivalent to its knowledge component, i.e. $Pr(M_k)$=$Pr(\Omega_k)$. We generate total N=30,000 ground model samples from model classes $\{M_k(\Omega_k)|k = 1,\ldots,9\}$ defined by Ω_k in Figure 3(d). The ground model samples are shown in Figure 4(a). For the local collider structure of lung cancer, bronchitis and dyspnea, we generate N=30,000 ground model samples from the combination of model classes defined in Eq. 23 based on $\{\Omega_k|k = 10,\ldots,16\}$ in Figure 3(d). The marginal conditional probability samples are shown in Figure 4(b) and 4(c). Without further information on lung cancer, bronchitis and dyspnea, we can set their prior probabilities to be 1/2. By taking average over the models in Figure 4(a) to 4(c), we can calculate the mean value for the conditional probability of lung cancer given smoking, i.e. $\overline{\beta_1}$=0.1255, $\overline{\beta_0}$=0.006, and of Dyspnea given lung cancer and Bronchitis, i.e. $\overline{\xi_0}$=0.2725, $\overline{\xi_1}$=0.9053, $\overline{\xi_2}$=0.5495 and $\overline{\xi_3}$=0.968. Note that since the *9th* model class defined by Ω_9 for the structure of lung cancer and smoking, i.e. $M_9(\Omega_9)$, contains no edge between the nodes, the parameter of this model class is null.

3.3.1.2 ASIA Model Inference

For each of the model sample, according to Eq. 23, we perform inferences *in silico* on the likelihood of a patient having lung cancer (Lc) given information about the patient's smoking status and clinical evidences including observation of X-ray, Dyspnea, and Bronchitis, i.e. $X_{obs} = \{Sm, Xr, Dy, Br\}$. The convergence of these prediction under a set of evidences $\widetilde{E} = \{E_1, E_2, E_3, E_4, E_5, E_6\}$ are shown in Figure 4(d). The true prediction values with parameters in Figure 3(b) under the evidence set $\widetilde{E}$ are listed below in Figure 5. The presence of bronchitis could explain away the probability of lung cancer and the presence of smoking increases the risk of getting lung cancer.

3.3.2 Breast Cancer Bone Metastasis Prediction

We apply our framework to integrate a set of inconsistent qualitative hypotheses about the molecular interactions between Smad proteins of the TGFβ signaling pathway in breast cancer bone metastasis network. From recent studies (11–15), a set of qualitative statements on

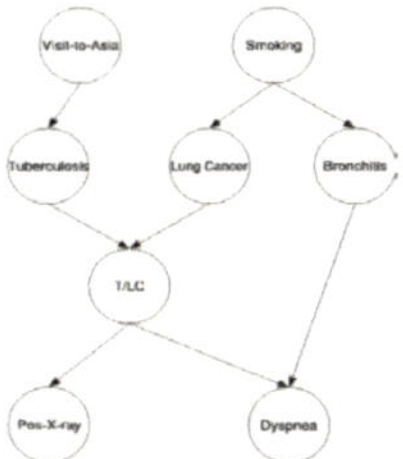

(a) ASIA Network Structure

Node Pair	CPT θ			
(Pa, Ch)	$Pr(Ch\|Pa)$		$Pr(Ch\|\overline{Pa})$	
(V,T)	$\alpha_1=0.05$		$\alpha_0=0.01$	
(S,LC)	$\beta_1=0.3$		$\beta_0=0.01$	
(S,B)	$\gamma_1=0.6$		$\gamma_0=0.3$	
$(T/LC,X)$	$\lambda_1=0.9$		$\lambda_0=0.05$	
$([Pa_1,Pa_2],Ch)$	$Pr(Ch\|Pa_1,Pa_2)$	$Pr(Ch\|Pa_1,\overline{Pa_2})$	$Pr(Ch\|\overline{Pa_1},Pa_2)$	$Pr(Ch\|\overline{Pa_1},\overline{Pa_2})$
$([T/LC,B],D)$	$\xi_3=0.9$	$\xi_2=0.7$	$\xi_1=0.8$	$\xi_0=0.1$
$([T,LC],T/LC)$	$f_3=1$	$f_2=1$	$f_1=1$	$f_0=0$

(b) ASIA Network Parameter

Stat.	Dp	I	Σ	R	Δ	Bd	Weight
S_1	1	1	SP	[9, 11]	null	null	w_1
S_2	1	1	SP	null	null	[9%, 11%]	w_2
S_3	1	1	SP	[25, ∞]	null	null	w_3
S_4	1	1	SP	[20, 30]	null	null	w_4
S_5	1	1	SP	null	null	[14%, 16%]	w_5
S_6	0	0	SP	null	null	null	w_6
S_7	1	1	SP	null	null	[60%, 90%]	w_7
S_8	1	1	SP	null	null	[60%, 90%]	w_8
S_9	1	1	SP	null	null	[60%, 100%]	w_9
$S_{10}(1)$	1	1	SP	null	null	[69%, 78%]	w_{10}
$S_{10}(2)$	1	1	SP	null	null	[91%, 94%]	w_{10}
S_{11}	1	1	SP	null	null	[94%, 96%]	w_{11}

(c) Feature-vector of Statements

Ω	Dp	I	Σ	R	Δ	Bd	$Pr(\Omega_i)$
Ω_1	1	1	SP	[9, 11]	null	[9%, 11%]	1/12
Ω_2	1	1	SP	[9, 11]	null	[14%, 16%]	1/12
Ω_3	1	1	SP	[20, 25]	null	[9%, 11%]	1/12
Ω_4	1	1	SP	[20, 25]	null	[14%, 16%]	1/12
Ω_5	1	1	SP	[25, 30]	null	[9%, 11%]	1/6
Ω_6	1	1	SP	[25, 30]	null	[14%, 16%]	1/6
Ω_7	1	1	SP	[30, ∞]	null	[9%, 11%]	1/12
Ω_8	1	1	SP	[30, ∞]	null	[14%, 16%]	1/12
Ω_9	0	0	null	null	null	null	1/6
Ω_{10}	1	1	PlSyn	null	null	null	1
Ω_{11}	1	1	SP	null	null	[60%, 69%]	3/11
Ω_{12}	1	1	SP	null	null	[69%, 78%]	4/11
Ω_{13}	1	1	SP	null	null	[78%, 90%]	3/11
Ω_{14}	1	1	SP	null	null	[90%, 100%]	1/11
Ω_{15}	1	1	SP	null	null	[91%, 94%]	1/2
Ω_{16}	1	1	SP	null	null	[94%, 96%]	1/2

(d) Integrated Qualitative Knowledge with Prior Probability

Fig. 3. ASIA Belief Network and Qualitative Statements and Knowledge in ASIA network

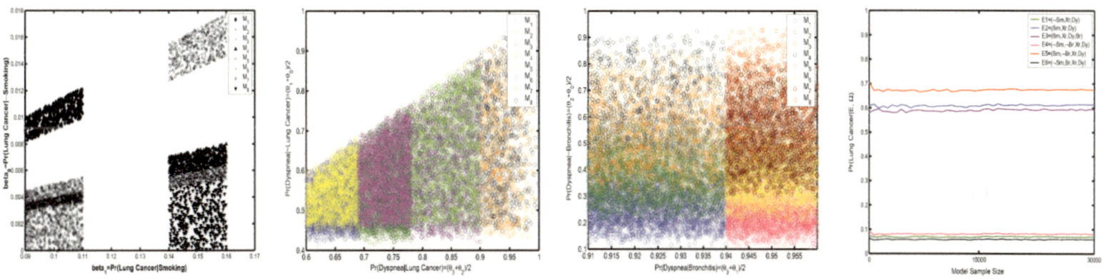

(a) Model Samples for Smoking and Lung Cancer (b) Model Samples for Lung Cancer and Dyspnea (c) Model Samples for Bronchitis and Dyspnea (d) Quantitative Bayesian Inference

Fig. 4. ASIA Model Sampling and Inference

molecular interactions in the breast cancer bone metastasis network can be extracted. A Dynamic Bayesian model can be constructed based on this set of statements as shown in Fig. 6(a) and the quantitative prediction with forward belief propagation based on a set of consistent qualitative hypotheses has been introduced in (33).

In this section, we consider the inconsistent qualitative statements with regard to the mechanism of Smad7 in blockade of the TGFβ signals. In (14), the qualitative statements can be extracted as S_1: *Smad7 directly binds to the activated type I TGF-β receptor and inhibits phosphorylation of the R-Smads.*;S_2: *Smad6 acts in a different way as Smad7. It competes with the activated Smad1 for binding to Smad4.*; In (15), the qualitative statements can be extracted as S_3: *The inhibitory activity of Smad6 and Smad7 is thought to result from an ability to interfere with receptor interaction and phosphorylation of the receptor-regulated Smads.*;S_4: *However, their inhibitory activity might also result from their ability to form a complex with receptor-activated Smads.*;Similar statements can be extracted from (13) as S_5: *I-Smads (Smad6,7) interact with type I receptors activated by type II receptors.*;S_6: *I-Smads have also been reported to compete with Co-Smad (Smad4) for formation of complexes with R-Smads (Smad2/3).*

This set of statements represent the molecular interactions between I-Smad (Smad7), R-Smad (Smad2/3) and Co-Smad (Smad4). $\{S_1, S_3, S_5\}$ report the interaction between Smad7, type I TGFβ-receptor (TβRI) and Smad2/3. $\{S_4, S_6\}$ describe the interaction between Smad7 and Smad4 to form a complex whereas S_2 provides contradicting information. Each statement is analyzed by the hierarchical knowledge model in Figure 1(a) and the extracted features are

Exp.	E_1	E_2	E_3	E_4	E_5	E_6
True	0.17	0.87	0.84	0.21	0.91	0.11
Simulation	0.07	0.61	0.59	0.08	0.67	0.06

Fig. 5. Inference Results on ASIA Network

summarized in Figure 7(a). For simplicity, we assume the weight of every qualitative statement equals to 1, i.e. $\{w_i = 1, i = 1, \ldots, 6\}$. Due to the parameter independency (1), we can compute the conditional probability of each local structure by counting the occurrence frequency of the knowledge features independently. For the local structure of Smad7, TβRI and Smad2/3, the prior probability of the knowledge features can be calculated as $Pr(Dp)$=1, $Pr(I|Dp)$=1, $Pr(\overline{I}|\overline{Dp})$=1. For the local structure of Smad7, Smad4 and phosphorylated-Smad2/3 (Smad2/3-p), $Pr(Dp)$=2/3, $Pr(\overline{Dp})$=1/3, $Pr(I|Dp)$=1, $Pr(\overline{I}|\overline{Dp})$=1. Based on the features and their prior belief, a set of qualitative knowledge $\widetilde{\Omega}$ is formed in Figure 7(b). In this experiment, the extended features of the inconsistent knowledge are not available.
We now construct the Bayesian model class and the distribution on ground model space within each class. The uncertainty of the TGFβ-Smad BCBM model space is restricted to the uncertainty of the local structure and parameter space on Smad7, TβRI and Smad4 which is defined by $\{\Omega_1,\Omega_2\}$ in Figure 7(b). The model classes can be expressed as $\{M_k(\Omega_k)|$k=1,2$\}$ and the prior probability of each model class equals to the prior probability of the knowledge, i.e. Pr(M_k)=Pr(Ω_k). We perform 10,000 sampling interactions. In each iteration, we select a model class M_k randomly based on the prior probability Pr(M_k). In each model class, we randomly generate 3 samples of the ground Bayesian model m by Monte Carlo method, whose structure and parameter space is consistent with the class constraints $Pr(m|M_k)$ as defined by Eq. 1 to Eq. 7. Therefore, we obtain N=30,000 ground models from the model classes. By taking average over the ground models, we can calculate the mean value for the conditional probability of the complex Smad4-Smad2/3-p given Smad7, Smad4 and Smad2/3-p. Note that since M_1 contains no edges between Smad7 and Smad4-Smad2/3-p, the parameter of this model class is null.
Each ground model is a Dynamic Bayesian network (DBN) which can be unrolled over time to form a series of 2TBNs (4). The prediction on the probability of bone metastasis given a set of evidences $E_i \in \{E_1, E_2, E_3\}$ in each model class, i.e. the integral in Eq. 23, can be calculated by integrating the predictions over all DBN models which is equivalent to compute firstly the mean DBN model with averaged parameters and then perform prediction on this mean DBN model (33). The simulation results and the observed bone metastasis probability in (11) are shown in Fig. 6(b) and Fig. 6(c).

3.3.3 Conclusion

In this paper, we proposed a hierarchical Bayesian model for modeling the semantics of the qualitative knowledge with a vector of features. The inconsistent knowledge components are integrated by calculating a prior distribution. The integrated qualitative knowledge set is used as prior background knowledge in modeling Bayesian networks and performing quantitative inference. We benchmarked our method with the ASIA network and applied our method to a real-world problem and simulation results suggest that our methods can reconcile the inconsistent qualitative uncertainty and produce reasonable quantitative prediction based on the inconsistent knowledge set.

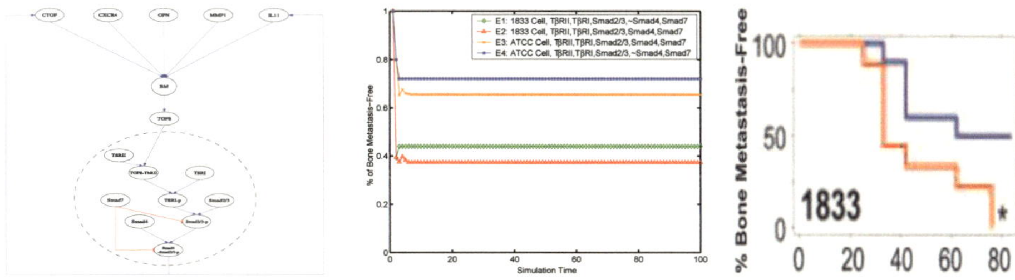

(a) Signaling Pathway (b) Simulation Results (c) Observation

Fig. 6. Integrated TGFβ-Smad BCBM Network and Prediction

Stat.	Dp	I	Σ	R	Δ	Bd	Weight
S_1	1	1	MxSyn	null	null	null	w_1
S_2	0	0	null	null	null	null	w_2
S_3	1	1	MxSyn	null	null	null	w_3
S_4	1	1	MxSyn	null	null	null	w_4
S_5	1	1	MxSyn	null	null	null	w_5
S_6	1	1	MxSyn	null	null	null	w_6

(a) Feature-vector of Statements

Ω	Dp	I	Σ	R	Δ	Bd	$Pr(\Omega_i)$
Ω_1	0	0	null	null	null	null	1/3
Ω_2	1	1	MxSyn	null	null	null	2/3
Ω_3	1	1	MxSyn	null	null	null	1

(b) Integrated Qualitative Knowledge with Prior Probability

Fig. 7. Qualitative Statements and Knowledge in TGFβ-Smad BCBM Network

3.4 Bayesian Network Learning with Informative Prior Qualitative Knowledge

We propose a framework for Bayes net parameter learning with generic prior knowledge. In this study, we use the knowledge model in section 3.1 to translate the qualitative domain knowledge into a set of inequality parameter constraints. We reconstruct the parameter priori distribution (i.e. priori pseudo counts) from these constraints. We then propose a novel Bayesian parameter score function which integrates this prior distribution with the quantitative data statistics. In this way, the parameter posterior distribution is combinatorially regulated by both quantitative data and prior knowledge.

3.4.1 Qualitative Constraints and Sampling

In general, qualitative domain knowledge can define various constraints over conditional probabilities in a BN. As described in last section, most of these constraints can be represented by a linear regression function $f(\theta_{ijk}) \leq c, \forall i, j, k$ (c is a scaler), where θ_{ijk} is the conditional probability of the state of i-th node being k, given its j-th parent configuration. In particular, one type of constraints can be derived from this function. *Cross-distribution Constraints* defines the relative relation between a pair of parameters over different conditions. If two parameters in a constraint share the same node index i and value k, but different parent configuration j, the constraint is called cross-distribution constraint. This constraints can be usually derived from causality in the qualitative knowledge.

$$\theta_{ijk} \leq, \geq \theta_{ij'k} \forall j \neq j' \tag{24}$$

Given the constraints defined by f, we can withdraw samples of parameter which are consistent with the constraints, e.g. in Eq. 24, by accept-reject sampling. Since sampling can be done at each node, it is relatively reasonable for demonstration. But node with more parent nodes, Gibbs sampling and simulated annealing can be used.

3.4.2 Qualitative Bayesian Parameter Score (QBPS)

In this study, we assume the data distribution is multinomial and prior is Dirichlet. The posterior probability of the parameter given the data in standard MAP estimation can be written as

$$logPr(\theta|G,D) = \log Pr(D|\theta,G) + \log Pr(\theta|G) - c = \log\{\alpha \prod_{i=1}^{n}\prod_{j=1}^{q_i}\prod_{k=1}^{r_i} \theta_{ijk}^{N_{ijk}+N'_{ijk}-1}\} \quad (25)$$

where θ denotes the parameters in a Bayes net and G is the network's structure. i, j, k is defined as section 3.4.1. The first term in Eq. 25 represent the data statistics which is followed by the Dirichlet prior distribution with hyperparameter N'_{ijk} (1). α is a normalizer. In standard MAP method, N'_{ijk} is usually set to a very small and equal number which results in non-informative prior.

We propose a posterior probability which employs the informative prior constraints (f) in the last section. In previous methods (35–37), f is imposed into the posterior probability as an penalty term. The MAP estimation is transformed to constrained optimization problem. However, the violation term f in these cases can only penalize the likelihood when the learned local maximum violates the constraints in the sign, but it can not distinguish a set of all possible local maximums obeying the constraints. So, final solution is not necessary a global maximum (37). Therefore, it is desired to use prior constraints (such as Eq. 24) as soft regulations to the posterior probability in Eq. 25. We name this MAP-like score function as Qualitative Bayesian Parameter Score (QBPS).

$$\log Pr(\theta|G,D,\Omega) = \log Pr(D|\theta,G) + \log Pr(\theta|G,\Omega) - c \quad (26)$$

The difference between Eq. 26 and Eq. 25 is the addition of Ω to the posterior probability in Eq. 25. The first term in Eq. 26 is the data statistics as in the standard MAP estimation. The second term $Pr(\theta|G,\Omega)$ represent the parameter's prior distribution given prior knowledge Ω. Ω can represent any forms of generic prior constraints over the parameter space, such as Eq. 24. In conventional approaches, $Pr(\theta|G)$ can be any probability function, such as Gaussian or Dirichlet distribution function with pre-defined hyperparameters. In case of multinomial data, $Pr(\theta|G)$ oftenly take the form of beta distribution due to the conjugate distribution property. Thus, the problem is to fuse the prior knowledge Ω and its associated constraints (f) over parameter space with the beta distribution $Pr(\theta|G)$ which results in the constrained beta distribution $Pr(\theta|G,\Omega)$.

In general, we can either i) fit the beta distribution into the constrained parameter space by estimating the hyperparameters of Dirichlet distribution given a vector of constrained parameter samples θ^l_{ijk} (43). These samples can be obtained based on the accept-reject sampling. In this case, we only select one local maximum prior model (one instance of hyperparameter) to substitute the uncertainty in the (priori) parameter space (all possible instances of hyperparameter) or ii) admit the model uncertainty and utilize conjugate property of beta distribution to reconstruct the (priori) parameter space distribution based on all constrained parameter samples. In this case, we have

$$Pr(\theta|\Omega,G) = \alpha \prod_{i=1}^{n}\prod_{j=1}^{q_i}\prod_{k=1}^{r_i} \theta^{l}_{ijk}{}^{M^l_{ijk}} \quad \forall l = 1, ..., L \quad (27)$$

where θ^l_{ijk} is an instance of constrained prior parameter sample and M^l_{ijk} denotes the number of 'success' cases of this instance (X_i=k, Π_i=j$|\theta^l_{ijk}$) exists in the past A (A is an arbitrary number) samples. It is equal to

$$M^l_{ijk} = A \times Pr^l(X_i = k, \Pi_i = j|\Omega) \tag{28}$$

Together, the QBPS score can be written as

$$Pr(\theta|G, D, \Omega) = \alpha \prod_{i=1}^{n} \prod_{j=1}^{q_i} \prod_{k=1}^{r_i} {\theta^l_{ijk}}^{N_{ijk}+M^l_{ijk}} \quad \forall l = 1, ..., L \tag{29}$$

where N_{ijk} is the number of occurrence in the training date for the ith node to have a value of k and for its parent to have a value of j and L is the total number of priori parameter samples from accept-reject sampling. (L is a large number) Thus, the local maximum estimation of a

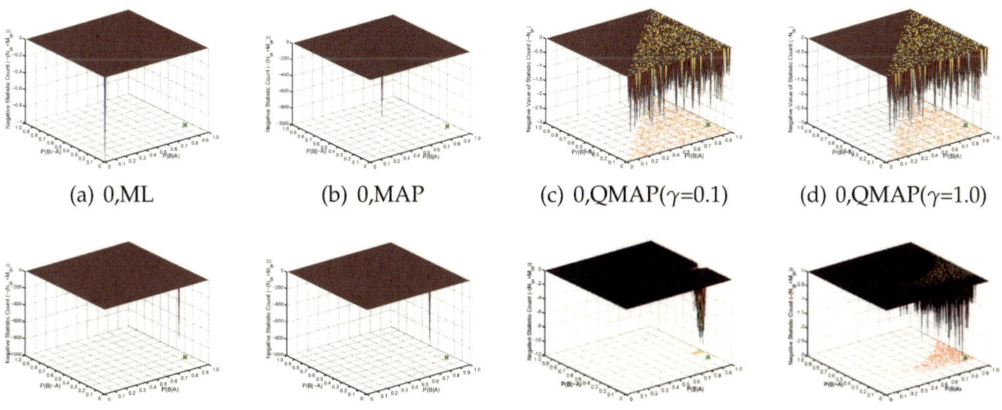

(a) 0,ML (b) 0,MAP (c) 0,QMAP(γ=0.1) (d) 0,QMAP(γ=1.0)

(e) 20,ML (f) 20,MAP (g) 20,QMAP(γ=0.1) (h) 20,QMAP(γ=1.0)

Fig. 8. Parameter Learning in Toy Network: The network contains two binary nodes. A is an activator parent of B. X,Y-axles represent conditional probability P(B|A) and P(B|$\overline{A}$) respectively; Z-axis is equal to the negative value of posterior statistical counts [-(N_{ijk}+M^l_{ijk})] in Eq. 29.

QBPS score equals to

$$\widehat{\theta}^l_{ijk} = \frac{N_{ijk} + \gamma N_0 Pr^l(X_i = k, \Pi_i = j|\Omega)}{\sum_{k=1}^{K} N_{ijk} + \gamma N_0 Pr^l(X_i = k, \Pi_i = j|\Omega)} \tag{30}$$

where N_0 is equal to the number of total data samples. Now, we further assume that A and N_0 has a ratio γ, i.e. $A = \gamma \times N_0$. From Eq. 30, we can see that ratio γ actually specified the belief-ratio between data statistics and prior knowledge statistics. If γ=0, we neglect the statistics from the prior knowledge and only trust the statistics in the data, thus, our estimation in Eq. 30 converges to ML results; If γ=+∞, we neglect the statistics in the data and only trust the prior knowledge, the results converge to the previously mentioned constraint-based probabilistic inference in (Dynamic) Bayesian inference [9,10]. If 0<γ<+∞, the QBPS score is softly regulated by both data statistics and the prior knowledge and constraints in the domain.

Since the estimation in Eq.8 is a joint effect from both inequality constraints in qualitative prior knowledge and data observation, we name it as Qualitative Maximum a Posterior (QMAP) estimation.

3.4.3 QMAP Estimation

1. QMAP Estimation with Full Bayesian Approach

As we have shown, we can reconstruct the priori parameter distribution from prior constraints. Each priori parameter sample θ^l_{ijk} together with the given structure (G) define a prior network m^l. Each priori m^l can be mapped to a posteriori. Thus, the final posterior probability of all Bayesian network models is defined over this class of prior networks m^l in terms of a set QBPS scores (Eq. 29). Our final goal is to predict future observations on variable X from the training data (D) and priori constraints Ω. Given BN structure (G), this prediction can be calculated as integration over the parameter space weighted by its posterior probability.

$$Pr(X|G,D,\Omega) = \int_\theta Pr(X|\theta,G)Pr(\theta|G,\Omega,D)d\theta \tag{31}$$

The posterior probability of the parameter given data and qualitative prior knowledge, i.e. $Pr(\theta|G,\Omega,D)$, is in-turn an integration over all possible prior models (m) in the class defined by Ω, thus, we can extend Eq. 31 as

$$Pr(X|G,D,\Omega) = \int_\theta Pr(X|\theta,G) \int_m \frac{Pr(D|\theta,G)Pr(\theta|G,m)Pr(m|\Omega)}{Pr(D)} dmd\theta \tag{32}$$

$Pr(m|\Omega)$ in Eq 32 is equal to 1 since all the valid prior models (m) are consistent with the prior constraints Ω.

The outer integration can be approximated by its local maximum if we assume the QBPS curve for each model is peaky, then we can write the inference as $Pr(X|\widehat{\theta},G)$. With full Bayesian approach, final QMAP estimation of the parameter can be optimized by integrating the set of local QBPS maximums over the prior network space, i.e. selecting the QMAP estimation which maximize the integrated QBPS score.

$$\widehat{\theta} = argmax_\theta \left\{ \int_m \frac{Pr(D|\theta,G)Pr(\theta|G,m)Pr(m|\Omega)}{Pr(D)} dm \right\} = argmax_\theta \left\{ \frac{1}{L} \sum_{l=1}^{L} \alpha \prod_{ijk} \theta_{ijk}^{N_{ijk}+M^l_{ijk}} \right\} \tag{33}$$

Note that each prior network m^l uniquely associate with a pseudo prior statistical count M^l_{ijk}. The prior network space is discrete. By taking the derivative of Eq. 33 wrt θ_{ijk}, we obtain the constrained QMAP estimation with full Bayesian approach as

$$\widehat{\theta}_{QMAP,FBA} = \frac{1}{L} \left\{ \sum_{l=1}^{L} \frac{N_{ijk} + M^l_{ijk}}{\sum_k N_{ijk} + M^l_{ijk}} \right\} \tag{34}$$

2. QMAP with Frequentist Maximization Approach

On the other hand, the final QMAP estimation can be obtained by frequentist maximum approach to select one single best estimate from the parameter posteriori space. In this way, we could pick up the maximum from a set of local maximums.

$$\widehat{\theta}_{QMAP,FMA} = argmax_{\{l\}} \left\{ \frac{N_{ijk} + M^l_{ijk}}{\sum_k N_{ijk} + M^l_{ijk}} \right\} \tag{35}$$

An example plot of posterior statistical counts in Eq. 29 is shown in Fig. 8. In case of ML learning, the M^l_{ijk} is equal to zero for all i,j,k. In case of MAP learning, we simulated a typical scenario, where the dirichlet parameters are set equally to a scalar. In this case, the dirchlet parameters tends to smooth the posterior score by adding equal amount of pseudo counts for all i,j,k. The smoothed posterior favors to the uniformly distribution in this case. By setting these prior pseudo counts to 1, conventional MAP methods try to minimize this biased smooth effect. However, the bias remains significant when the training data is relative small. In Fig. 8(g) and 8(h), we show that our proposed QMAP methods augment the posterior distribution by reconstructing the prior from the qualitative knowledge and each prior distribution sample M^l_{ijk} is combined with the data statistics to regulates posterior counts on equal opportunities. In this way, we can explore the multiple local maximums sit in the posterior space so that we ensure to select the global maximum.

3.5 Experiments

3.5.1 Experiment Design

We evaluate our proposed parameter learning methods using a realistic AU recognition data. We test our algorithm in following learning conditions: a) In extreme case, we assume there are no available training data and we use only generic qualitative domain knowledge which are derived from causality in a BN to estimate the parameter. b) In standard case, we do not employ any domain knowledge which is eventually equivalent to ML estimation. c) In an fusion case, we use both training data and generic qualitative domain knowledge to learn the parameter. We compare our results to standard ML and MAP estimation results.

3.5.2 Facial Action Unit Recognition

In this section, we apply our method to facial action unit (AU) recognition. The Facial Action Coding System (FACS) (40) is the most commonly used system for facial behavior analysis. Based on FACS, facial behaviors can be decomposed into a set of AUs, each of which is related to the contraction of a specific set of facial muscles. An automatic AU recognition system has many applications. Current AU recognition methods tend to perform AU recognition individually, ignoring their relationships with other AUs. Due to the underlying physiology and the facial anatomy, AUs often move in a coordinated and synchronized manner in order to produce a meaningful expression. To represent the dependencies among AUs, Tong et al (41) proposed to use Bayesian Network to capture the relationships among AUs. Following their work, we propose to use the same BN model to capture the relationships among the 14 most common AUs as shown in Figure 9(a), where the larger circular nodes in the model represent AUs while the smaller nodes represent their image measurements. They have demonstrated that the BN model is superior to the state of the arts AU recognition method. But to use the model, they need a large amount of training data, which is often hard to acquire. We will show that we can achieve comparable results using only a fraction of their training data. Using the model, we extract constraints based on the following rules provided by domain experts: 1. *Marginal Constraint*: In spontaneous cases, some AUs rarely occur. One example for this case is AU27, and the rule is P(AU27 = 1)$\leq$P(AU27 = 0), where 1 means presence and 0 means absence. 2. *Causality-derived Cross-distribution Constraint*: As shown in Figure 4, every link between two AU nodes has a sign provided by the domain expert. The + sign denotes positive influence,which means two AU nodes have co-occurrence relationship, while a negative sign denotes negative influence, which means the two AU nodes have mutual exclusive relationship. Considering an AU node AU_i has only one parent node AU_j, if the sign of the link is positive, we have $P(AU_i = 1|AU_j = 0) \leq P(AU_i = 1|AU_j = 1)$, e.g. $P(AU1 = 1|AU2 = 0) \leq P(AU1 = 1|AU2 = 1)$; if the sign of the link is negative, then we can get $P(AU_i = 1|AU_j = 1) \leq P(AU_i = 1|AU_j = 0)$, e.g. $P(AU6 = 1|AU27 = 1) \leq P(AU6 = 1|AU27 = 0)$. If an AU node AU_i has more than one AU parent nodes, AU^P denote all the parent nodes with positive links, and AU^N denote all the parent nodes with negative links. Then we get $P(AU_i = 1|AU^P = 0, AU^N = 1) \leq P(AU_i = 1|AU^P = 1, AU^N = 0)$, e.g. $P(AU15 = 1|AU24 = 0, AU25 = 1) \leq P(AU15 = 1|AU24 = 1, AU25 = 0)$. 3. *Range Constraint*: If an AU node AU_i has more than one parent nodes AUP , and all of them with positive influence, then $P(AU_i = 1|AU^P = 1) \geq 0.8$. If an AU node AU_i has more than one parent nodes AU^N, and all of them with negative influence, then $P(AU_i = 1|AU^N = 1) \leq 0.2$.

Please note the above constraints are due to either facial anatomy or due to certain facial patterns. They are generic enough to be applied to different databases and to different individuals.

3.5.3 Integrative Learning with domain knowledge and data

The 8000 images used in experiments are collected from Cohn and Kanades DFAT-504. In each simulation run, we randomly select 0 to 5000 samples out of 8000 samples for training and we repeat learning task for 20 times. Training data are used for learning the parameters in the AU BN (Figure 9(a)). After the learning, we select 1000 untouched samples for testing. Testing data are used to perform AU recognition through inference given learned BN. We assume the training data is complete. In the first part, we show the learning results in K-L divergence on the AU subnetwork in Figure 9(a). In the second part, we show the real classification results. We apply ML and QMAP estimation with qualitative domain knowledge defined above to learning the parameters in the AU subnetwork. The K-L divergence is shown in Figure 9(b). The x-axis and the y-axis denote training sample size and K-L divergence respectively. The K-L result is actually the mean K-L divergence which is calculated by averaging the parameter learning results over all randomly selected training samples under each specific sample size. We can see that: i) QMAP with γ=1 performs significantly better than ML estimation under every training data size. More specifically, the K-L divergence for ML estimation with 3 training sample is decreased from 2.21 to 0.24 for QMAP with γ=1. Even at 5000 training samples, the K-L divergence for ML estimation is decreased from 0.04 to close to 0 for QMAP estimation; On the other hand, we can evaluate the results by counting how many training samples are required to achieve specific desired K-L divergence level for ML, MAP and QMAP method respectively. At 3 training sample, K-L divergence for QMAP estimation is 0.24. In order to obtain equivalent or better K-L divergence level, ML estimation needs 200 samples. At 5000 training sample, K-L divergence for ML estimation is 0.04 which can be achieved by QMAP with 10 samples. These results are extremely encouraging, as using our methods with domain-specific yet generic qualitative constraints, and with a small number of manually labeled data (10), we can achieve similar learning accuracy to the ML estimation with full training dataset (5000).

The encouraging learning results of our QMAP method shed light over the usage of generic qualitative domain knowledge in learning task. Therefore, in this section, we explore an extreme case of parameter learning by ignoring all training data sample but only employing the set of qualitative constraints (same set of constraints defined above) to learn the AU subnetwork parameters. In this case, the data statistics counts in Eq. 30 is zero due to lack of training data. The parameter estimation is only determined by priori pseudo counts given the qualitative knowledge. The K-L divergence in this case is 0.0308 which is lower than K-L

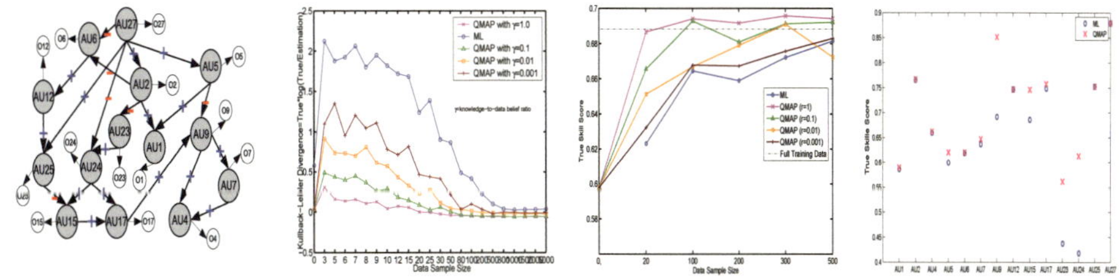

(a) AU Recognition Network (b) KL Divergence (c) Averaged AU Recognition Skill (d) AU Node Skill with 200 samples

Fig. 9. Comparison of AU recognition network parameter learning results from ML and QMAP respectively. a) AU Recognition Network with AU nodes and measurement nodes; b) K-L divergence measurement of parameter learning in AU network based on training dataset with various sample size. Comparison of AU recognition skill using the BN learned from ML and QMAP respectively. We compare QMAP to standard ML skills. c) AU Recognition Network; d) AU Recognition skill score at 200 training samples on AU nodes;

divergence of ML learning with full training data (5000 training samples). Meanwhile, this K-L divergence level corresponds to that of QMAP learning with γ=1 at 25 data samples.

3.5.4 Classification

In this section, we want to study the performance of the proposed learning methods by using such learned BN model for AU classification. For AU classification, we need feed the BN model with AU measurements computed from Gabor Wavelet jets. Given the AU measurements, we want to infer the true states of each AU using the model parameters learnt with our method. Specifically, we want to study the AU recognition performance under different amount of training data including the extreme case of using no training data at all, and compare the classification results with those in (36). We perform classification based on the learned AU network from ML and our proposed QMAP approach in section 3.5.3). For demonstration, we select the learned AU network parameter under training dataset with representative sample size: 0, 20, 100, 200, 300 and 500. After learning, we randomly select 1000 untouched data samples for classification test. Figure 9(c) shows the AU recognition results. The x-axis represent the training data size for learning AU network parameters (in case of 0 training size, no training data but only qualitative prior knowledge is used for AU network parameter estimation) and y-axis denotes the true skill score (the difference between true positive rate and false positive) respectively. The true skill is calculated by averaging all AU nodes' skill score. We can see from Figure 9(c), the true skill score for QMAP with various belief-ratio (γ) is significantly better than the skill score for ML estimation under nearly all training data sample size except for QMAP with γ=0.01. In particular, even at sparse training data (20 samples), the average true skill score for all AU nodes increases from 0.6229 for ML estimation to 0.6866 for QMAP with γ=1, to 0.6655 for QMAP with γ=0.1, to 0.6512 for QMAP with γ=0.01 and to 0.6322 for QMAP with γ=0.001; At 100 training samples, true skill score further enhances from 0.6644 for ML estimation to 0.6940 for QMAP with γ=1, to 0.6928 for QMAP with γ=0.1, to 0.6668 for QMAP with γ=0.01 and 0.6677 for QMAP with γ=0.001. While training sample size grows to 200, 300, and 500 samples, the true skill score from QMAP with γ=1.0 is equal to 0.6916, 0.6957 and 0.6942 respectively and tends to converge. In the same case, ML estimation shows consistently lower classification ability than QMAP. Please note that, using full training dataset (7000 samples for training and 1000 samples for testing), true skill score for ML estimation converge at 0.6883 (shown as the black dashed line in Figure. 9(c)). From the above results, we can conclude that i) our proposed QMAP estimation by integrating domain-specific yet very generic qualitative prior constraints with quantitative training data significantly improves the AU recognition results comparing to ML estimation at all sample size spanning from sparse data to rich data. This observation is particularly true with γ=1; ii) Our proposed QMAP estimations (with different γ) needs much fewer training samples for AU network to achieve equivalent and even better AU recognition results than ML estimation. iii) Comparing the true skill score of QMAP estimation to the score of ML estimation with full training dataset, we can see that, with a much smaller number of manually labeled data (around 35 samples) ,QMAP with γ=1 can already achieve much better AU recognition results than ML estimation with full training dataset (7000 samples). While decreasing the weight on prior knowledge to γ=0.1, QMAP requires from 80 to 250 training samples to achieve better AU classification results than ML estimation with full training dataset. When γ reduces to 0.01, QMAP needs around 300 samples to outperform ML estimation with full training dataset. This number keeps increasing while γ reduces. When γ=0.001, the true skill score of QMAP tends to converge with ML estimation. Therefore, in practice, we shall put a larger weight on qualitative prior knowledge as long as our knowledge are valid in a domain. The above observation is also consistent with our K-L measurements in Figure 9(b). In summary, we demonstrate that by our approach, qualitative prior constraints can be integrated into standard BN parameter learning to achieve significantly improved prediction results. Next, we want to compare our results with a well developed method in AU recognition (36). To this end, we compare the true skill score of our QMAP at 200 training samples to the skill score of Constrained-ML (CML) estimation (Figure4(b) in (36)) at 300 training samples. The true skill of each AU node of our QMAP is plot with optimized γ is shown in 9(d). Firstly, we can see that our QMAP approach significantly improves the true skill on AU node number 5, 9, 15, 23 and 24. Slightly improve the skill on AU node 1, 7, 17. The rest skill is equivalent to ML estimation. Comparatively, our method boost the skills on those AU nodes (6, 23, 12, 25, 17, 24, 9, 4) whose skill score is worse than ML estimation in (36).

4. References

[1] D. Heckerman.(1996). A tutorial on learning with bayesian networks, Microsoft Research, USA, Tech. Rep.

[2] D. J. Dudgeon and M. Lertzman.(1998). Dyspnea in the advanced cancer patient, J. of Pain and Symptom Management, vol. 16(4), pp. 212-219.

[3] E. Gyftodimos and P. Flach.(2002). Hierarchical bayesian networks: A probabilistic reasoning model for structured domains, Proceedings of the ICML-2002 Workshop on Development of Representations, pp. 23-30.

[4] K. Murphy.(2002). Dynamic bayesian networks:representation, inference and learning, Ph.D. dissertation, University of California, Berkeley, USA.

[5] M. J. Druzdzel and M. Henrion.(1993). Efficient reasoning in qualitative probabilistic networks, Proceedings of the Eleventh National Conference on Artificial Intelligence. Washington DC: AAAI Press, 1993, pp. 548-553.

[6] M. P. Wellman.(1990). Fundamental concepts of qualitative probabilistic networks, Artificial Intelligence, vol. 44, pp. 257-303.

[7] P. Edmonds, S. Karlsen, S. Khan, and J. Addington-Hall.(2001). A comparison of the palliative care needs of patients dying from chronic respiratory diseases and lung cancer, Palliative Medicine, vol. 15(4), pp. 287-295.
[8] R. Kinsman, R. Yaroush, E. Fernandez, J. Dirks, M. Schocket, and J. Fukuhara.(1983). Symptoms and experiences in chronic bronchitis and emphysema, Chest, vol. 83, pp. 755-761.
[9] S. Renooij, L. C. van der Gaag, and S. Parsons.(2002). Context-specific sign-propagation in qualitative probabilistic networks, Artificial Intelligence, vol. 140, pp. 207-230.
[10] S. L. Lauritzen and D. J. Spiegelhalter.(1998). Local computations with probabilities on graphical structures and their application to expert systems, J. Royal Statistics Society B, vol. 50(2), pp. 157-194.
[11] Y. Kang, W. He, S. Tulley, G. P. Gupta, I. Serganova, C. R. Chen, K. Manova-Todorova, R. Blasberg, W. L. Gerald, and J. Massague.(2005). Breast cancer bone metastasis mediated by the smad tumor suppressor pathway, Proceedings of the National Academy of Sciences of the USA.
[12] Y. Kang, P. M. Siegel, W. Shu, M. Drobnjak, S. M. Kakonen, C. Cordón-Cardo, T. A. Guise, and J. Massagué.(2003). A multigenic program mediating breast cancer metastasis to bone, Cell, vol. 3, no. 6, pp. 537-549.
[13] K. Miyazono.(2000). Positive and negative regulation of TGF-β signaling, Journal of Cell Science, vol. 113, no. 7, pp. 1101-1109.
[14] Y. Shi and J. Massagué.(2003). Mechanisms of TGF-β signaling from cell membrane to the nucleus, Cell, vol. 113.
[15] Y. Zhang and R. Derynck.(1999). Regulaiton of smad signalling by protein associations and signalling crosstalk, Trends in Cell Biology, vol. 9, no. 7, pp. 274-279.
[16] Thomas Bayes, *An essay towards solving a Problem in the Doctrine of Chances*, Philosophical Transactions of the Royal Society of London,1763.
[17] S.L. Lauritzen and D.J. Spiegelhalter, *Local Computations with Probabilities on Graphical Structures and Their Application to Expert Systems*, Journal of the Royal Statistical Society, 1988
[18] Judea Pearl,*Probabilistic Reasoning in Intelligent Systems: Networks of Plausible Inference*,Morgan Kaufmann Publishers, Inc., San Mateo, CA, USA,1988
[19] David Heckerman, *A Tutorial on Learning with Bayesian Networks*,1996
[20] David Heckerman, *Learning Bayesian Networks: The Combination of Knowledge and Statistical Data*, KDD Workshop, 1994
[21] Nir Friedman and Moises Goldszmidt, *Learning Bayesian networks with local structure*, Learning in graphical models,1999
[22] Eric Neufeld, *A probabilistic commonsense reasoner*, International Journal of Intelligent Systems, 1990
[23] M.J. Druzdzel and L.C. van der Gaag, *Elicitation of probabilities for belief networks: combining qualitative and quantitative information*, Proceedings of the Eleventh Conference on Uncertainty in Artificial Intelligence, 1995
[24] Silja Retive Approaches to Quantifying Probabilistic Networks,Ph.D Thesis Universiteit Utrecht,2001
[25] Michael P. Wellman, *Fundamental Concepts of Qualitative Probabilistic Networks*, Artificial Intelligence, 1990
[26] M. Dejori and M. Stetter, *Identifying interventional and pathogenic mechanisms by generative inverse modeling of gene expression profiles*, J. Comput. Biology,1135-1148,2004

[27] Jesús Cerquides and Ramon López de Màntaras, *Knowledge Discovery With Qualitative Influences and Synergies*,Proceedings of the Second European Symposium on Principles of Data Mining and Knowledge Discovery,1998
[28] D. Heckerman. Learning Bayesian Networks: The Combination of Knowledge and Statistical Data. *Proc. KDD Workshop*,
[29] D. Geiger and D. Heckerman. A characterization of the dirichlet distribution through global and local parameter independence. *The Annals of Statistics*, 25:1344Ű1369, 1997.
[30] R. S. Niculescu. Exploiting parameter domain knowledge for learning in bayesian networks. *Technical Report CMU-TR-05-147*, Carnegie Mellon University, 2005.
[31] R. S. Niculescu, T. Mitchell, and R. B. Rao. Parameter related domain knowledge for learning in graphical models. *In Proceedings of SIAM Data Mining conference*, 2005.
[32] R. S. Niculescu, T. Mitchell, and R. B. Rao. Bayesian Network Learning with Parameter Constraints. *Journal of Machine Learning Research*, 7:1357Ű1383, 2006.
[33] R. Chang, M. Stetter, and W. Brauer. Quantitative Inference by Qualitative Semantic Knowledge Mining with Bayesian Model Averaging. *IEEE Transaction on Knowledge and Data Engineering*, Vol. 20, No. 12, December, 2008.
[34] R. Chang, W. Brauer, and M. Stetter. Modeling semantics of inconsistent qualitative knowledge for quantitative Bayesian network inference. *Neural Networks*, 21(2-3): 182-192, 2008.
[35] F. Wittig and A. Jameson, Exploiting Qualitative Knowledge in the Learning of Conditional Probabilities of Bayesian Networks. *The 16th Conference on Uncertainty in Artificial Intelligence*, USA, 2000.
[36] Yan Tong and Qiang Ji, Learning Bayesian Networks with Qualitative Constraints, IEEE Conference on Computer Vision and Pattern Recognition (CVPR), 2008.
[37] E. Altendorf, A. C. Restificar and T. G. Dietterich: Learning from Sparse Data by Exploiting Monotonicity Constraints. *The 21st Conference on Uncertainty in Artificial Intelligence*, USA, 2005: 18-26.
[38] Linda van der Gaag, B. Hans and Ad Feelders, Monotonicity in Bayesian Networks. *The 20th Conference on Uncertainty in Artificial Intelligence*, USA, 2004.
[39] Y. Mao and G. Lebanon, Domain Knowledge Uncertainty and Probabilistic Parameter Constraints. *The 25th Conference on Uncertainty in Artificial Intelligence*, USA, 2009.
[40] P. Ekman and W. Friesen. Facial Action Coding System: A Technique for the Measurement of Facial Movement. Consulting Psychologists Press, 1978.
[41] Yan Tong, Wenhui Liao, Zheng Xue and Qiang Ji, A Unified Probabilistic Framework for Spontaneous Facial Activity Modeling and Understanding, IEEE Conference on Computer Vision and Pattern Recognition (CVPR), 2007.
[42] Wenhui Liao and Qiang Ji, Learning Bayesian Network Parameters Under Incomplete Data with Qualitative Domain Knowledge, Pattern Recognition, Volume 42 , Issue 11, Pages 3046-3056, 2009.
[43] T. P. Minka, Estimating a dirichlet distribution, 2003. [Online].

5

Group decision making using Bayesian network inference with qualitative expert knowledge

Wichian Premchaiswadi and Nipat Jongsawat
Graduate School of Information Technology in Business, Siam University
Thailand

1. Introduction

In group decision making, different experts often think about the same problem in quite different ways. They frequently have different opinions for decision making about the same situation. Using a Bayesian network structure for optimizing problems, different experts who work as a group for projects may have different solutions for indentifying the causal relationships among variables in the BN model and quantifying graphical models with numerical probabilities. For example, expert-1 may state that "making a decision in situation A causes situation B and making a decision in situation B causes situation C". But expert-2 may state that "making a decision in situation B causes situation A and making a decision in situation A causes situation C". Even in a simple case of decision making, the expert knowledge obtained from different experts is quite different. It is typically not possible to avoid contradictions among different expert's solutions in group decision making.

In this article, we propose a practical framework and a methodology for transforming expert knowledge or final group decision making statements into a set of qualitative statements and probability inequality constraints for inference in a Bayesian Network. First, we need to identify a set of alternatives on which the experts have opinions and then consider the problem of constructing a group preference ranking. If such a group preference ranking can be created, then one could utilize the alternative at the top of the ranked list the alternative preferred by the group. Second, after we obtain the most preferred alternative or statement such as "A causes B and then B causes C" from the group decision making, we propose a formal method to transform knowledge, represented by a set of qualitative statements, into an a priori distribution for Bayesian probabilistic models. The mathematical equation for Bayesian inference is derived based on knowledge obtained from the final group decision statements. The set of model parameters, consistent with the statements, and the distribution of models in the structure-dependent parameter space are presented. We also propose a simplified method for constructing the "a priori" model distribution. Each statement obtained from the experts is used to constrain the model space to the subspace which is consistent with the statement provided. Finally, we present qualitative knowledge models and then show a complete formalism of how to translate a set of qualitative statements into probability inequality constraints. Several cases of Bayesian influence are classified and the probability inequality constraints presented in each case are described.

This article is organized as follows: Section 2 presents more detail about the background of Bayesian networks and some perspectives of qualitative causal relationships in the Bayesian approach. Section 3 addresses the method of constructing a group preference ranking and group decision making from the individual preferences obtained from the experts performing group work. Section 4 addresses the methods to transform a final solution or expert knowledge into an "a priori" distribution for Bayesian probabilistic models in more detail. Section 5 describes the method used to translate a set of qualitative statements into probability inequality constraints and presents different cases of influences in BN model. Section 6 presents a conclusion and discusses some perspectives and ideas for future work.

2. Background

This section is intended to describe the background of Bayesian networks and some perspectives of qualitative causal relationships in the Bayesian approach. Bayesian networks (also called belief networks, Bayesian belief networks, causal probabilistic networks, or causal networks) are acyclic directed graphs in which nodes represent random variables and arcs represent direct probabilistic dependencies among the nodes (Pearl, 1988). Bayesian networks are a popular class of graphical probabilistic models for research and application in the field of artificial intelligence. They are motivated by Bayes' theorem (Bayes, 1763) and are used to represent a joint probability distribution over a set of variables. This joint probability distribution can be used to calculate the probabilities for any configuration of the variables. In Bayesian inference, the conditional probabilities for the values of a set of unconstrained variables are calculated given fixed values of another set of variables, which are called observations or evidence. Bayesian models have been widely used for efficient probabilistic inference and reasoning (Pearl, 1988: Lauritzen & Spiegelhalter, 1988) and numerous algorithms for learning the Bayesian network structure and parameters from data have been proposed (Heckerman, 1994: Heckerman, 1996: Friedman & Goldszmidt, 1999). The causal structure and the numerical parameters of a Bayesian network can be obtained using two distinct approaches (Cheng, et al., 2001: Nipat & Wichian, 2009). First, they can be obtained from an expert. Second, they can also be learned from a dataset or data residing in a database. The structure of a Bayesian network is simply a representation of independencies in the data and the numerical values are a representation of the joint probability distributions that can be inferred from the data (Singh & Valtorta, 1995: Spirtes & Meek, 1995). In practice, some combination of these two approaches is typically used. For example, the causal structure of a model is acquired from an expert, while the numerical parameters of the model are learned from the data in a database.

For realistic problems, the database is often very sparse and hardly sufficient to select one adequate model. This is considered as model uncertainty. Selecting one single model can lead to strongly biased inference results. On the other hand, in science and industry, there is an enormous amount of qualitative knowledge available. This knowledge is often represented in terms of qualitative causal relationships between two or more entities. For example, in the statement: "smoking increases the risk of lung cancer," the two entities: smoking and lung cancer are related to each other. Moreover, the smoking entity positively influences the lung cancer entity since lung cancer risk is increased in the case of smoking. It is therefore desirable to make use of this body of evidence in probability inference modeling.

3. Group Preference Ranking and Group Decision

In this section, we present the first step which is identifying the group solution for a BN model of the proposed framework (see Fig. 1). We describe several methods for experts to make decisions for identifying the relationship between variables in a Bayesian network model and arriving at a final BN solution representing the group.

The general case is one in which we have a group of experts and a set of alternatives, for example "A activates B and B activates C", "B activates A and A activates C", and "C activates A and A activates B", on which the experts have opinions. We assume that each expert has a preference ranking on the set of alternatives. That is, using these preferences, each expert can order the alternatives in a list such that if alternative A activates B, and B activates C are in the list, then the experts have an agreement with that alternative. A set of individual preference rankings, one for each expert in the group, is called a group preference schedule. One goal of the first portion of our proposed practical framework is to consider the problem of constructing a group preference ranking from the individual preferences (that is, from the group preference schedule). If such a group preference ranking can be created, then one could call the alternative at the top of the group list the alternative selected by the group of experts. However, such a group ranking may not be possible, and moreover, even if it is possible, the alternative at the top of the list may not be one that would win the majority selection in an election among all options. Thus the second goal of our work in the first step is to consider other possible ways of picking the most preferred choice, especially if none of the alternatives would receive a majority selection in an election among all alternatives. We will identify the properties of the decision process that corresponds to our ideas about the characteristics such decision processes should have.

Example 1. Suppose that we have a group of three experts, labeled expert-1, expert-2, and expert-3, and a set of three variables, labeled A, B, and C. For this example, assume the individual preference rankings are as follows:

Expert-1: A→B→C ; Expert-2: B→C→A ; Expert-3: C→A→B

Using pairwise comparisons and a simple-majority rule, we see that both expert-1 and expert-3 agree that "A causes B", and therefore, because the vote is 2 to 1, the group should agree with "A causes B". Therefore, on the basis of this information, we would propose that the group preference ranking should be "A causes B and then B causes C; (A→B→C)". However, both expert-2 and expert-3 agree with "C causes A", and therefore the group should agree with "C causes A". We conclude that the proposed group preference ranking in this example is not transitive: The experts agree with A→B→C→A. This cyclic, or intransitive, behavior is normally considered unacceptable for a preference ranking. We conclude that even in this simple situation, the majority rule decision process can lead to unacceptable preference rankings. The intransitive phenomena do occur when the number of variables and alternatives increase. That is for many groups and sets of preferences, the group preferences determined by the pairwise majority rule voting are intransitive. What are some ways to cope with the results of this example?

Let's consider again the simple situation of three experts and three alternatives. Then each expert has 6 different preference rankings-that is, 6 ways in which the 3 alternatives can be listed: 3 choices for the alternative listed first, 2 choices for the alternative listed second, and 1 choice for the alternative listed last. Because there are 6 experts, there are 6 x 6 x 6 = 216

different preference schedules for the group. The likelihood of intransitive group preferences depends on how the experts select their individual preference rankings. For instance, if we know that two of the experts have the same preference ranking, then that preference ranking will be the preference ranking for the group, and intransitivity will not occur. As another example, if two experts have alternative Z as the top choice, then intransitive group preferences will never occur. However, intransitive group preferences can still occur if experts select their individual preferences at random. This situation is more complicated but it is not considered in this article because we assume that the experts use their own experience to make their own decisions. They will not make a decision at random.
In light of this discussion about the difficulties encountered with simple-majority voting, we look for other ways to achieve our primary goal of finding ways for groups to make decisions. We introduce the concept of sequential voting or selection: a sequence of votes where at each vote, a choice is to be made between two alternatives. In any situation with an odd number of experts, this process always yields a result, and this result can be taken as a most preferred alternative. However, as we show in an example below, this method also has problems.
Example 2. Suppose that the relationship between variables is to be identified by first considering, for example, A and B and then considering the impact on the last variable.
Expert-1 considers A and B first and states that B causes A and then A causes C: B→A→C.
Expert-2 considers A and C first and state that A causes C and then C causes B: A→C→B.
Expert-3 considers B and C first and state that C causes B and then B causes A: C→B→A.
The results in this example show that we are in the unfortunate situation of having a group preference that depends on the sequence in which the selections were taken.
We have illustrated some of the problems with simple-majority rule and sequential selecting decision processes. We turn next to another approach to the problem: assigning points to a pair of variables of each order on the basis of their relative rankings and defining a group preference ranking by adding the points assigned to each alternative by all experts.
Example 3. We will illustrate the technique by considering five experts and four variables (See Table 1). Each expert makes a series of decisions at each order-level. For example, expert-1 makes a decision that "making a decision in situation A causes situation C" in a first order level, C causes B in a second order level, and B causes D in a third order level. Each expert assigns 3 point to the first order level, 2 point to the second order level, and so on. For a specific alternative, add the points assigned by all experts. The alternative with the most points is the most preferred, the alternative with the second largest number of points is the second most preferred, and so on. This method is known as the Borda count group decision process (María & Jose, 2007). We observe that this decision process has an implicit relative strength of preferences. The relative strengths of all preferences are the same.

Order	Expert-1	Expert-2	Expert-3	Expert-4	Expert-5	Points
1	A→C	D→A	B→A	C→B	A→C	3
2	C→B	A→C	A→C	B→D	C→D	2
3	B→D	C→B	C→D	D→A	D→B	1

Table 1. A group of five experts and four alternatives

The group preference ranking is obtained by adding the points assigned to each alternative (A→C: 10 points, C→B: 6 points, D→A: 4 points, B→D: 3 points, B→A: 3 points, C→D: 3 points, D→B: 1 points).

We conclude that the group preference ranking is A causes C, C causes B, and B causes D. The alternative D→A has 4 points but it is not included because A is a parent node in the first order level so that D cannot cause A.

By considering a few examples, we have identified shortcomings of some common decision processes in group decision making. With the last technique, problems are still possible to occur when two alternatives at the same level have the same score. However, this section proposes several techniques in the decision process to produce a group preference ranking and a final group solution.

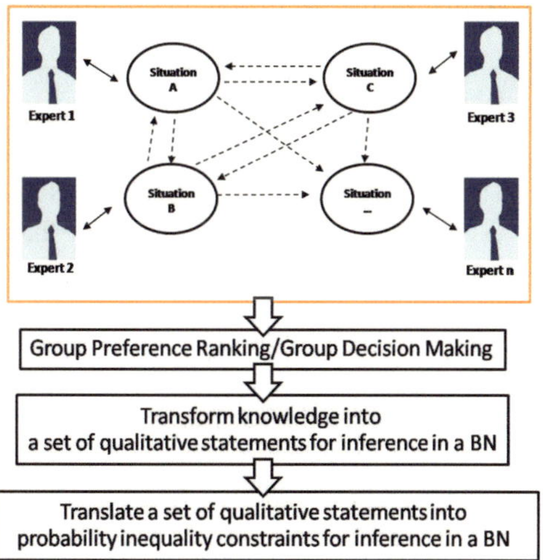

Fig. 1. A practical framework

4. Methods

In this section, we describe a methodology to use qualitative expert knowledge obtained from the previous step for inferencing in a Bayesian network. We proceed from the decision-making assumptions and the general equation for Bayesian inference based on final group decision making statements obtained from the experts to a detailed method to transform knowledge, represented by a set of qualitative statements, into an a priori distribution for Bayesian probabilistic models.

For simplicity, let's consider a simple case of decision making in which the body of expert knowledge ω consists of a single statement ω = "making a decision in situation A causes situation B". We know that there are 2 random events or variables A and B, which we assume are binary, and we need to consider the set of all possible Bayesian models on A and B. The set of possible model structures are described in the following categories: 1) S_1: A and B have no causal relationship between them, 2) S_2: A and B have some causal relationship between them but the direction of influence cannot be identified, 3) S_3: A causes B, and 4) S_4: B causes A. "making a decision in situation A causes situation B" directly states a causal influence of A on B. We use the statement "A activates B" to constrain the space of structures: $P(S_3 \mid \omega) = 1$; $P(S_n \mid \omega)=0$, n=1,2,4). The ω is represented as a qualitative statement described by the expert, A causes B. The graph structure (S_3) encodes the probability distribution

$$P(A,B) = P(B \mid A)P(A) \tag{1}$$

No further information on P(A) is available; however, P(B | A) can be further constrained. The corresponding Conditional Probability Table (CPT) is shown in Table 2.

A	P(B=1) | A
0	θ_0
1	θ_1

Table 2. Conditional probability table

The values of the conditional probabilities from the components of the parameter vector $\theta = (\theta_0, \theta_1)$ of the model class with structure S_3. θ_0 is the probability of B is active when A is not active. θ_1 is the probability of B is active when A is active. From the statement, we now can infer that the probability of B is active when A is active is higher than the same probability with A inactive. The P(B | A) when P(A) is available is higher than the P(B | A) when P(A) is not available. The inequality relationship is obtained as follows:

$$P(B=1 \mid A=1) \geq P(B=1 \mid A=0),\ \theta_1 \geq \theta_0 \tag{2}$$

Hence, the set of model parameters consistent with that statement is given by

$$\Theta_3 = \{(\theta_0, \theta_1) \mid 0 \leq \theta_0 \leq 1 \land \theta_0 \leq \theta_1 \leq 1\} \tag{3}$$

and the distribution of models in the structure-dependent parameter space becomes

$$P(\theta \mid S_3, \omega) = \begin{cases} 1, \theta \in \Theta_3 \\ 0, \text{else} \end{cases} \tag{4}$$

A Bayesian model m represents the joint probability distribution of a set of variables $X = X_1, X_2, X_3, \ldots, X_D$. The model is defined by a graph structure, which determines the structures of the conditional probabilities between variables, and a parameter vector θ, the components of which define the entries of the corresponding conditional probability tables (CPTs). Hence, a Bayesian network can be written as $m = \{s, \theta\}$. Given some observations or evidence E, reflected by fixed measured values of a subset of variables, the conditional probability given the evidence in light of the model is described as P(X | E, m).

The full Bayesian network model does not attempt to approximate the true underlying distribution. Instead, all available information is used in an optimal way to perform inference, without taking one single model for granted. To formalize this statement for our purposes, let us classify the set of available information into an available set of data D and a body of nonnumeric expert knowledge ω. The probability distribution of model m is given by

$$P(m|D, \omega) = \frac{P(D|m)P(m|\omega)}{P(D, \omega)} \tag{5}$$

The first parameter value D of $P(D, \omega)$ is the likelihood of the data given the model, which is not directly affected by nonnumeric expert knowledge ω, the second parameter value ω denotes the model a priori, whose task is to reflect the background knowledge. For simplicity, the numerator $P(D, \omega)$ of $P(m \mid D, \omega)$ will be omitted from the equation (5). The term $P(D \mid m)$ contains the constraints of the model space by the data, and the term $P(m \mid \omega)$ contains the constraints imposed by the expert knowledge. Hence, given some observation or evidence E, the conditional distribution of the remaining variable X is performed by integrating over the models.

$$P(X \mid E, D, \omega) = \int P(X \mid E, m)P(m \mid D, \omega)dm \quad (6)$$
$$= \int P(X \mid E, m)P(D \mid m)P(m \mid \omega)dm \quad (7)$$

In this article, we consider the case of no available quantitative data; D is assigned a null value. The term D and $P(D \mid m)$ will be omitted from equation (6) and (7). Even in this case, it is still possible to perform a proper Bayesian inference.

$$P(X \mid E, \omega) = \int P(X \mid E, m)\, P(m \mid \omega)dm \quad (8)$$

Now, the inference is based on the general information (contained in ω) obtained from experts, and the specific information provided by the measurement E. In order to determine $P(m \mid \omega)$, we need a formalism to translate the qualitative expert knowledge into an a priori distribution over Bayesian models. The following notations are adopted for a Bayesian model class. A Bayesian model is determined by a graph structures and by the parameter vector θ needed to specify the conditional probability distributions given that structure. The parameter vector θ is referred to by one specific CPT configuration. A Bayesian model class is then given by 1) a discrete set of model structures $S = \{s_1, s_2, s_3, \ldots, s_K\}$ and for each structure s_k, a set of CPT configurations Θ_k. The set of member Bayesian models $m \in M$ of that class is then given by $m = \{(s_k, \theta) \mid k \in \{1, \ldots, K\}, \theta \in \Theta_k\}$. The model distribution is shown in (9).

$$P(m|\omega) = P(s_k, \theta|\omega) = \frac{P(\theta|s_k, \omega)P(s_k|\omega)}{\sum_{a=1}^{K} \int_{\Theta_a} P(\theta|s_a, \omega)d\theta\, P(s_a|\omega)} \quad (9)$$

In (9), the set of allowed structures is determined by means of ω, followed by the distributions of the corresponding CPT configurations. The model's a posterior probability $P(m \mid \omega)$ is calculated as shown in (9). Inference is carried out by integrating over the structure space and the structure-dependent parameter space.

$$P(X|E, \omega) = \sum_{k=1}^{K} \int_{\Theta_3} P(X|E, s_k, \theta)P(s_k, \theta|\omega)d\theta \quad (10)$$

It is common to express nonnumeric expert knowledge in terms of qualitative statements about a relationship between entities. The ω is represented as a list of such qualitative statements. The following information is essential to determine the model a priori (10): First,

each entity which is referenced in at least one statement throughout the listed is assigned to one variable X_i. Second, each relationship between a pair of variables constrains the likelihood of an edge between these variables being presented. Last, the quality of the statement such as activates or inactivates affects the distribution over CPT entries θ given the structure. The statement can be used to shape the joint distribution over the class of all possible Bayesian models over the set of variables obtained from ω in the general case.
We propose a simplified method for constructing the a priori model distribution. Each statement is used to constrain the model space to that subspace which is consistent with that statement. In other words, if a statement describes a relationship between two variables, only structures s_k which contain the corresponding edge are assigned a nonzero probability $P(s_k | \omega)$. Likewise, only parameter values on that structure, which are consistent with the contents of that statement, are assigned a nonzero probability $P(\theta | s_k, \omega)$. If no further information is available, the distribution remains constant in the space of consistent models.
Having derived the Bayesian model class (s_3, Θ_3) consistent with the statement, we can now perform inference by using an equation (10). Under the condition of A is set to active (E = {A = 1}), let us ask what is the probability of having B active. We can determine this by integrating over all models with nonzero probability and averaging their respective inferences, which can be done analytically in this simple case.

$$\begin{aligned} P(B = 1|E, \omega) &= \sum_{k=1}^{K} P(s_k, \omega) \int_{\Theta 3} P(B|A, s_k, \theta)P(\theta| s_k, \omega)d\theta \\ &= \omega \int_{\Theta 3} P(B=1|A=1, \theta)d\theta \\ &= \omega \int_0^1 \int_{\theta_0}^1 \theta_1 d\theta_1\, d\theta_0 = 2/3 \end{aligned} \tag{11}$$

where ω = 2 is the normalizing factor in the parameter space of θ = (θ_0 , θ_1) such that

$$\omega \int_0^1 \int_{\theta_0}^1 d\theta_1 d\theta_0 = 1 \tag{12}$$

It is worth noting that, as long as simple inequalities are considered as statements, the problem remains analytically tractable even in higher dimensions. In general, integration during Bayesian inference can become intractable using analytical methods.

5. Probabilistic Representation of a Qualitative Expert Knowledge Model

The model from the previous section is derived to provide a full formalism of how to translate a set of qualitative statements into probability inequality constraints. Several qualitative models have been proposed in the context of qualitative probabilistic networks. Qualitative knowledge models describe the process of transforming qualitative statements into a set of probability constraints. The proposed Bayesian inference method outlined above is independent of the qualitative knowledge model. The model's a posterior probability is independent of the set of qualitative statements used, once the set of probabilistic inequality constraints which are translated from qualitative statements is determined. Three existing qualitative models are the Wellman approach (Wellman, 1990) the Neufeld approach (Neufeld, 1990), and the orders of magnitude approach (Cerquides &

Lopez, 1998). In this article, we utilize the Wellman approach, where qualitative expert knowledge involves influential effects from parent variables to child variables which are classified according to the number of inputs from parent to child and their interaction. For reasons of simplicity, binary-valued variables are used in our examples. The values of a variable or node defined as "present" and "absent" or "active" and "inactive" are represented as logical values "1" and "0" (as synonyms A and $\overline{A}$). For multinomial variables, similar definitions can be applied.

Qualitative influences with directions can be defined based on the number of influences imposed from parent to child. There are three cases of influences, namely, single influence, joint influence, and mixed joint influence. In addition, there are recurrent statements and conflicting statements. The first issue can be solved by using a Dynamic Bayesian Network (DBN) (Murphy, 2002: Premchaiswadi & Jongsawat, 2010) and the second issue can be solved by adopting a voting scheme. The definitions of influence presented in this article are refined based on the QPN in (Wellman, 1990). They are used to translate the qualitative expert statements into a set of constraints in the parameter space which can be used to model the parameter distribution given the structure. For a more general understanding of the explanation in this section, we assume that we obtained a set of final group decision making statements, transformed them into a set of qualitative statements, and explained those using different case studies in each criterion of probability inequality constraints for inference in a Bayesian Network. The BN model of each case study in each criterion is shown in Fig. 2.

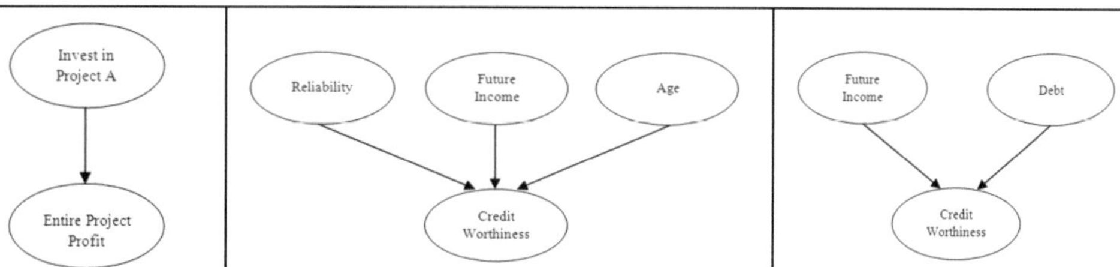

Fig. 2A. Example of single positive and negative influence.

Fig. 2B. Example of plain synergy influence. Reliability, future income, and age synergically influence credit worthiness.

Fig. 2C. Example of mixed joint influence. Debt and future income influence on credit worthiness.

Fig. 2. The BN of each case study in each criterion

5.1 Single Influence

In the statement, "investing in project A increases the profit of the entire project in such good economic situations," investing in project A is the parent node which has a single positive influence on child node the profit of the entire project.

$$P(\text{Entire Project Profit}|\text{Invest A}) \geq P(\text{Entire Project Profit}|\overline{\text{Invest A}})$$

In another statement, "investing in project A reduces the profit of the entire project in such a severe economic crisis," investing in project A is the parent node which imposes a single negative influence on child node the profit of the entire project.

$$P(\text{Entire Project Profit}|\text{Invest A}) \leq P(\text{Entire Project Profit}|\overline{\text{Invest A}})$$

The graphical representation of the above qualitative statements from an expert is shown in Fig. 2A.

5.2 Joint Influence

Let us consider credit worthiness of individual causes. Several risk factors have been identified for credit worthiness. According to the Thai credit bureau report, the three most prominent risk factors are reliability, future income, and age. The chance of getting credit worthiness increases as an individual gets higher future income, age, and reliability. This knowledge about credit worthiness factors can be encoded by a qualitative causality model. According to the statements, the main risk factors that influence credit worthiness by positive synergy as shown in Fig. 2B.

The joint influence of these three factors together is more significant than individual influences from any of these factors alone. We can represent this synergy by the inequalities

$$P(CW|R, FI, A) \geq \begin{Bmatrix} P(CW|R, \overline{FI}, \overline{A}) \\ P(CW|\overline{R}, FI, \overline{A}) \\ P(CW|\overline{R}, \overline{FI}, A) \end{Bmatrix} \qquad P(CW|R, FI, A) \geq \begin{Bmatrix} P(CW|R, FI, \overline{A}) \\ P(CW|R, \overline{FI}, A) \\ P(CW|\overline{R}, FI, A) \end{Bmatrix}$$

and

$$P(CW|R, FI, A) \geq P(CW|\overline{R}, \overline{FI}, \overline{A})$$

If we assume these risk factors pair wise symmetric, we can further derive the following inequalities:

$$\begin{matrix} P(CW|R, FI, \overline{A}) \\ P(CW|R, \overline{FI}, A) \\ P(CW|\overline{R}, FI, A) \end{matrix} \geq \begin{Bmatrix} P(CW|R, \overline{FI}, \overline{A}) \\ P(CW|\overline{R}, FI, \overline{A}) \\ P(CW|\overline{R}, \overline{FI}, A) \end{Bmatrix}$$

where CW, R, FI, and A stands for Credit Worthiness, Reliability, Future Income, and Age. Note that often but not always, the combined influence refers to the sum of independent influences from each parent node to each child node. Assume that parent nodes R and FI impose negative individual influence on child node CW, then the knowledge model can be defined as

$$P(\overline{CW}|R, FI) \geq \begin{Bmatrix} P(\overline{CW}|R, \overline{FI}) \\ \\ P(\overline{CW}|\overline{R}, FI) \end{Bmatrix} \geq P(\overline{CW}|\overline{R}, \overline{FI})$$

5.2 Mixed Joint Influence

Generally, the extraction of a probability model is not well defined if the joint affect on a child is formed by a mixture of positive and negative individual influences from its parents.

Therefore, we adopted the following scheme: If there are mixed influences from several parent nodes on a child node, and no additional information is given, then these are treated as independent and with equal influential strength.
For example, future income imposes a positive single influence on credit worthiness and debt imposes a negative single influence on credit worthiness, then the joint influence can be represented by

$$P(CW|FI, D) \geq P(CW|\overline{FI}, D),$$
$$P(CW|FI, \overline{D}) \geq P(CW|\overline{FI}, \overline{D}),$$
$$P(CW|FI, \overline{D}) \geq P(CW|FI, D),$$
$$P(CW|\overline{FI}, \overline{D}) \geq P(CW|\overline{FI}, D).$$

A credit worthiness case study for a mixed joint influence is shown in Fig. 2C.
Once formulated, we can use a Monte Carlo sampling procedure to make sure that all inequalities are satisfied for valid models. Any additional structure can be brought into the CPT of the corresponding structure as soon as the dependencies between influences are made explicit by further qualitative statements.

6. Conclusion and Future Work

In this paper, we presented several techniques in the decision process to produce a group preference ranking and a final group solution. After that we established mathematical equations for Bayesian inference based on a final group solution obtained from experts. We also described in detail a method to transform knowledge, represented by a set of qualitative statements, into an "a priori" distribution for Bayesian probabilistic models. The set of model parameters consistent with the statements and the distribution of models in the structure-dependent parameter space were presented. A simplified method for constructing the "a priori" model distribution was proposed. Each statement was used to constrain the model space to a subspace which is consistent with the statements. Next, we provided a full formalism of how to translate a set of qualitative statements into probability inequality constraints. Several cases of Bayesian influence were classified and the probability inequality constraints presented in each case are described.
For future research, we intend to construct multiple objective decision-making methods and its applications based on the concepts proposed in this article. We will apply the concepts to a specific case study using a set of group decision making statements and report the simulation results.

7. References

Bayes T. (1763). An Essay Towards Solving a Problem in the Doctrine of Chances. *Philosophical Trans*, Royal Soc. of London.

Cerquides, J. & Lopez de Mantaras, R. (1998). Knowledgge Discovery with Qualitative Influences and Synergies. *In Proceedings of the Second European Symposium*, Principle of Data Mining and Knowledge Discovery (PKDD).

Friedman, N. & Goldszmidt, M. (1999). Learning Bayesian Networks with Local Structure. Learning in Graphical Models.

Heckerman, D.; Geiger, D. & Chickering, D. M. (1995). Learning Bayesian Networks: The Combination of Knowledge and Statistical Data. *Machine Learning*, Vol. 20, p.197-243.

Heckerman, D. (1996). A Tutorial on Learning with Bayesian Networks. *Technical Report MSR-TR-95-06*, Microsoft, http://research.microsoft.com/research/pubs/view.aspx?msr_tr_id=MSR-TR-95-06.

Jongsawat, N. & Premchaisawadi, W. (2009). A SMILE Web-based Interface for Learning the Causal Structure and Performing a Diagnosis of a Bayesian Network. *In Proceedings of the 2009 IEEE International Conference on Systems, Man, and Cybernetics*, San Antonio, TX, USA, p.382-388.

Jongsawat, N. & Premchaisawadi, W. (2010). Bayesian Network Inference with Qualitative Expert Knowledge for Decision Support Systems. *In Proceedings of the 11th ACIS International Conference on Software Engineering, Artificial Intelligence, Networking and Parallel/Distributed Computing*, The University of Greenwich, London, United Kingdom.

Lauritzen, S. L. & Spiegelhalter, D. J. (1988). Local Computations with Probabilities on Graphical Structures and Their Application to Expert Systems, *Journal of the Royal Statistical Society (B)*, Vol. 50, p.157-224.

María, T. E. & Jose María, M. (2007). Aggregation of Individual Preference Structures in Ahp-Group Decision Making. *Journal of Group Decision and Negotiation, Springer Netherlands*, Vol. 16, No. 4, July, p.287-301.

Murphy, K. (2002). Dynamic Bayesian Networks: Representation, Inference, and Learning. *PhD dissertation*, University of California, Berkeley.

Neufeld, E. (1990). A Probabilistic Commonsense Reasoner. *International Journal of Intelligent Systems*, Vol. 5, p.565-594.

Pearl, J. (1988). Probabilistic Reasoning in Intelligent Systems. *Networks of Plausible Inference*, San Mateo, CA, Morgan Kaufmann Publishers.

Singh, M. & Valtorta, M. (1995). Construction of Bayesian Network Structures from Data: A Brief Survey and an Efficient Algorithm, *International Journal of Approximate Reasoning*, Vol. 12, p.111-131.

Spirtes, P. & Meek, C. (1995). Learning Bayesian Networks with Discrete Variables from Data. *In Proceedings of the First International Conference on Knowledge Discovery and Data Mining*, Montreal, QU. Morgan Kaufmann.

Wellman, M. P. (1990). Fundamental Concepts of Qualitative Probabilistic Networks. *Artificial Intelligence*, Vol. 40, pp.257-303.

6

Forming object concept using Bayesian network

Tomoaki Nakamura and Takayuki Nagai
The University of Electro-Communications
Japan

1. Introduction

As the recent developments in humanoid robotics, there is growing interest in object recognition and learning, since they are essential tasks for robots to work in our surrounding environments. Most frameworks for recognition and learning are based only on visual features. It seems that those are insufficient for 'understanding' of objects, since each object has its own intended use leading to the function, which is the key to object concept (Landau et al., 1998; Stark et al., 1996). Of course, appearance is deeply related with functions, since many objects have certain forms resulting from their functions. This fact is especially-pronounced in hand tools. Thus the visual learning and recognition of hand tools may succeed to some extent. However, such classification does not give any information on their functions. The important point is not classification in its own right but rather inference of the function through the classification. We believe that must be the basis of 'understanding', which we call object concept. Therefore objects must be learned, i.e. categorized, and recognized through their functions.

In this chapter objects (hand tools) are modeled as the relationship between appearance and functions. The proposed approach uses the model, which relates appearance and functions, for learning and recognizing objects.

The appearance is defined as a visual feature of the object, while the function is defined as certain changes in work objects caused by a tool. Each function is represented by a feature vector which quantifies the changes in the work object. Then the function is abstracted from these feature vectors using the Bayesian learning approach (Attias, 1999). All information can be obtained by observing the scene, in which a man uses the hand tool. For the model of object concept, Bayesian Network is utilized. The conditional probability tables, which are parameters of the model, are estimated by applying EM algorithm to the observed visual features and function information. This process can be seen as the learning of objects based on their functions. Since the function and appearance are stochastically connected in the model, inference of unseen object's function is possible as well as recognizing its category.

Related works are roughly classified into three categories. One of these is an attempt to recognize objects through their functions (Rivlin et al., 1995; Stark et al., 1996; Woods et al., 1995). Although those works share the same idea with us, the authors do not consider the learning process of object function. Thus the function of each object must be defined and programmed manually. Secondly, unsupervised visual categorization of objects has been studied extensively (Fergus et al., 2003; Sivic et al., 2005). However, function is not taken into consideration. Thirdly, there has been research on object recognition through human action (Kojima et al., 2004). The authors relate object recognition with human action, which represents how to use it, rather than the object function itself. In (Ogura et al., 2005), authors have reported the

model for robot tool use. However, they do not consider categorization and the robot can not cope with unknown objects. The proposed framework differs from those works in important ways. The key point of the proposed approach is learning of the relationship between appearance and function. This approach may lead to a computational model for the affordance (Gibbson, 1979).
This chaper is organized as follows: the following section discusses an object concept model based on the Bayesian Network. Then, the details of the model such as object appearance and the function model are described in section **3**. Section **4** shows some experimental results to validate the proposed framework and this chapter is conclued in section **5**.

2. Forming Object Concept

2.1 Bayesian Network for Object Concept

To 'understand' objects a novel framework, which differs from conventional matching-based 'recognition' approach, is required. Here we define 'understanding' of an object as inference of its function. For example, to understand 'scissors' is to infer their function, that is, cutting the work objects. Here is the problem to be considered, that is, what is the definition of the function? Especially by almost all hand tools, the work object undergoes some physical change. For example, scissors change shape and number of the work object, and pens can change surface brightness. These various changes in a scene can be observed as a feature vector, which results in our definition of function. A detail description of the function will be given in the following section.
The schematic diagram of the above discussion is shown in Fig.1 (a). Then Fig.1(a) can be rewritten using graphical model as in Fig.1(b). It should be noted that the following relationship is used to rewrite Fig.1(a) to Fig.1(b).

$$P(I)P(O|I)P(X_V|O)P(F|O) = P(O)P(I|O)P(X_V|O)P(F|O). \tag{1}$$

Thus the problem considered in this chapter results in the parameter estimation and inference using the graphical model in Fig.1(b). Of course the model is too simple to explain all aspects of object understanding. In fact, more complex factors such as usage of the tool etc. are important and should be taken into account. This is an issue in the future and now we focus our discussion on the implementation of the system based on the model in Fig.1(b).
The Bayesian Network in Fig.1(b) has four nodes; one of these is unobservable object concept O and the other nodes are observable object(scene) ID I, visual feature X_V and function F. To be precise F is not observable. In Fig.1(c), details of the node F is illustrated. In the figure X_F and Z_F represent observable feature vector and 'abstract function', which is abstracted from feature vectors using Bayesian learning approach, respectively.

2.2 Learning Algorithm

From Fig.1(c), the joint probability of I, X_V, X_F O and Z_F can be written as

$$P(I, X_V, X_F, O, Z_F) = P(O)P(I|O)P(X_V|O)P(Z_F|O)P(X_F|Z_F). \tag{2}$$

The parameters in the above equation $P(O)$, $P(I|O)$ $P(X_V|O)$ and $P(Z_F|O)$ are estimated using the EM algorithm, as the model contains unobserved latent variable. It should be noted that $P(X_F|Z_F)$ is given by the abstract function model(Gaussian Mixture Model) as we describe later. Let the parameters be $\boldsymbol{\theta}$, the problem is a maximization of the following equation:

$$L(D) = \log \sum_{Z_F} \sum_{O} P(I, X_V, X_F, O, Z_F|\boldsymbol{\theta}). \tag{3}$$

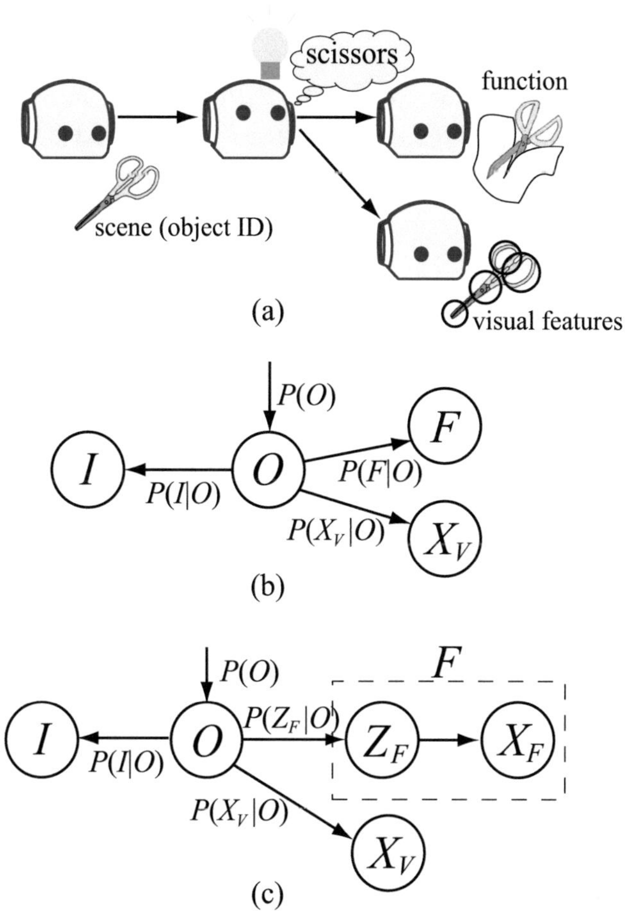

Fig. 1. A model of object concept. (a)Schematic diagram. (b)Graphical model representation of (a). (c)Details of the node F in (b).

By applying Jensen's inequality, we obtain

$$L(D) = \log \sum_{Z_F} \sum_{O} q(O, Z_F | I, X_V, X_F, \hat{\boldsymbol{\theta}}) \frac{P(I, X_V, X_F, O, Z_F | \boldsymbol{\theta})}{q(O, Z_F | I, X_V, X_F, \hat{\boldsymbol{\theta}})}$$
$$\geq F(q, \boldsymbol{\theta}) = \sum_{Z_F} \sum_{O} q(O, Z_F | I, X_V, X_F, \hat{\boldsymbol{\theta}}) \log \frac{P(I, X_V, X_F, O, Z_F | \boldsymbol{\theta})}{q(O, Z_F | I, X_V, X_F, \hat{\boldsymbol{\theta}})}. \quad (4)$$

Then the lower limit $F(q, \boldsymbol{\theta})$ is maximized iteratively with respect to q and $\boldsymbol{\theta}$ one after the other. The maximization with respect to q is to compute

$$q(O, Z_F | I, X_V, X_F, \hat{\boldsymbol{\theta}}) = \frac{P(O)P(I|O)P(X_V|O)P(Z_F|O)P(X_F|Z_F)}{\sum_{Z_F} \sum_O P(O)P(I|O)P(X_V|O)P(Z_F|O)P(X_F|Z_F)}. \quad (5)$$

On the other hand the maximization with respect to $\boldsymbol{\theta}$ is equivalent to maximize the Q-function;

$$Q(\boldsymbol{\theta}) = \langle P(I, X_V, X_F, Z_F, O|\boldsymbol{\theta})\rangle_{q(O,Z_F|I,X_V,X_F,\hat{\boldsymbol{\theta}})}. \tag{6}$$

The parameter $\boldsymbol{\theta}$ can be updated by solving $\partial Q(\boldsymbol{\theta})/\partial\boldsymbol{\theta} = 0$. The EM algorithm alternates the following two steps starting from initial values and converges to a local minimum.

[E-step] Compute Equation 5.

[M-step]

$$P(O) \propto \sum_I \sum_j \sum_{Z_F} \{n(I, X_{Vj})q(Z_F, O|I, X_{Vj}, X_F, \hat{\boldsymbol{\theta}})\}, \tag{7}$$

$$P(I|O) \propto \sum_j \sum_{Z_F} \{n(I, X_{Vj})q(Z_F, O|I, X_{Vj}, X_F, \hat{\boldsymbol{\theta}})\}, \tag{8}$$

$$P(X_{Vj}|O) \propto \sum_I \sum_{Z_F} \{n(I, X_{Vj})q(Z_F, O|I, X_{Vj}, X_F, \hat{\boldsymbol{\theta}})\}, \tag{9}$$

$$P(Z_F|O) \propto \sum_j \sum_I \{n(I, X_{Vj})q(Z_F, O|I, X_{Vj}, X_F, \hat{\boldsymbol{\theta}})\}, \tag{10}$$

where X_{Vj} represents the j-th dimension of X_V and $n(I, X_{Vj})$ denotes how many times $\{I, X_{Vj}\}$ occurred in the observations. It should be noted that $P(X_V|O)$ can be written as $P(X_V|O) = \prod_j P(X_{Vj}|O)$.

2.3 Inference

An object (category) can be recognized from observed visual information and function using the learned model as

$$\underset{O}{\operatorname{argmax}}\, P(O|X_V, X_F, I) = \underset{O}{\operatorname{argmax}} \frac{P(O)P(I|O)P(X_V|O)\sum_{Z_F}\{P(Z_F|O)P(X_F|Z_F)\}}{\sum_O [P(O)P(I|O)P(X_V|O)\sum_{Z_F}\{P(Z_F|O)P(X_F|Z_F)\}]}. \tag{11}$$

It is worth noting that Equation 11 is for the known object I. In order to apply Equation 11 to the unseen object $\hat{I}$, $P(\hat{I}|O)$ and $P(O)$ must be recalculated using the EM-algorithm described in the foregoing section. At this time $P(X_V|O)$ and $P(Z_F|O)$ are fixed. This idea is called fold-in heuristics described in (Hofmann, 2001).
It is possible to infer the unseen object's function only from the observed visual information. Inversely, typical appearance of the object that has a specific function can be derived. Inference of object function can be carried out by

$$\underset{Z_F}{\operatorname{argmax}}\, P(Z_F|X_V, \hat{I}) = \underset{Z_F}{\operatorname{argmax}} \frac{\sum_O P(O)P(\hat{I}|O)P(X_V|O)P(Z_F|O)}{\sum_O P(O)P(\hat{I}|O)P(X_V|O)}. \tag{12}$$

The fold-in heuristics should be applied to the calculation of $P(\hat{I}|O)$ and $P(O)$.

3. Visual Information and Functions

3.1 Object appearance (X_V)

There are two different attributes of object parts. One is functional parts and the other is non-functional ones. The clipper blade and scissors handle are examples of functional parts, which

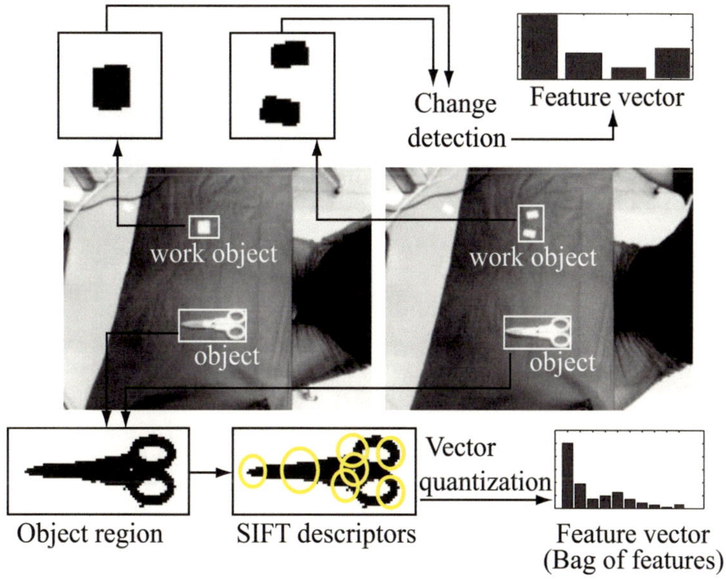

Fig. 2. The image processing for the object appearance and functions.

are requisite for scissors. The relative location of these parts is also important. On the other hand, non-functional parts are not directly linked to the object function. The object shape reflects these two types of parts. Therefore, only functional parts should be extracted to capture the relationship between appearances and functions correctly. We use SIFT descriptors (Lowe, 2004) in order to extract parts of the object and then the object appearance is represented by bag of features model.

The lower part of Fig.2 illustrates the processing for computing visual information. At first the object region is extracted from images as shown in the figure. The SIFT descriptors are computed in the object region. These computed descriptors are vector quantized using the pre-defined code book and frequency count is taken for the bag of features representation.

In Fig.3, some examples of the actual SIFT key points and histograms. Each histogram given in the figure is only a part of whole 500 dimensional information. One can see the similarites between within class objects.

3.2 Feature Extraction for Functions (X_F)

As we mentioned earlier, the function of a tool is defined as the pattern of certain changes in its work object. It is very important to select changes to be observed, since it directly affects the ability of the system to discover object functions.

Here four features are computed considering properties of general hand tools.

(1)Color change on the surface of the work object; this change can be captured by computing the correlation coefficient between color histograms of the work object before and after manipulation ($\boldsymbol{C}_{before}$ and $\boldsymbol{C}_{after}$).

$$x_C = Cr(\boldsymbol{C}_{after}, \boldsymbol{C}_{before}), \quad (13)$$

where $Cr(\boldsymbol{a}, \boldsymbol{b}) = (\boldsymbol{a} \cdot \boldsymbol{b})/|\boldsymbol{a}||\boldsymbol{b}|$.

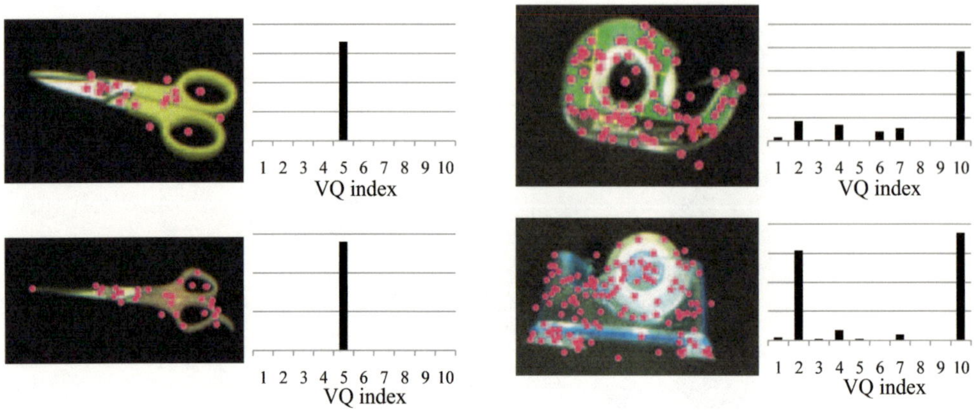

Fig. 3. Examples of SIFT keypoints and histograms.

(2)Contour change of the work object; to capture this change the correlation coefficient between Fourier descriptors of the work object before and after manipulation is computed ($\boldsymbol{F}_{before}$ and $\boldsymbol{F}_{after}$).

$$x_F = Cr(\boldsymbol{F}_{after}, \boldsymbol{F}_{before}). \tag{14}$$

(3)Barycentric position change of the work object; the relative distance between barycentric positions of a work object before and after manipulation ($\boldsymbol{G}_{before}$ and $\boldsymbol{G}_{after}$) is computed.

$$x_G = |\boldsymbol{G}_{after} - \boldsymbol{G}_{before}|. \tag{15}$$

(4)Change in number of the work object; this can be detected by counting the connected components relevant to a work object.

$$x_N = N_{after} - N_{before}, \tag{16}$$

where N_{before} and N_{after} represent the numbers of connected components of the work object before and after manipulation, respectively.
Then, the feature vector can be written as $\boldsymbol{x} = (x_N, x_G, x_C, x_F)$. The upper part of Fig.2 illustrates an example of the feature extraction. As shown in the figure, above four features are extracted from images before and after manipulation.

3.3 Bayesian Learning of Functions

3.3.1 The graphical model for GMM

The object functions are modeled by Gaussian Mixture Model (GMM) as in Fig.4 using the feature vectors described above. This modeling process corresponds to abstraction of object functions. Figure 5 shows 3D-plot of features that motivates us to use GMM. Three clusters, which represent different functions, can clearly be seen in the figure. The Variational Bayes (VB) framework (Attias, 1999) is used for the parameter estimation, since the number of abstract functions can be estimated as an optimal model structure.
Figure 4 illustrates the graphical model for the GMM. This model corresponds to F, which is denoted by the dashed box, in the whole model Fig.1(c). In Fig.4, $\boldsymbol{x}_n (n = 1, \cdots, N)$ represents the observable change vectors of the work objects and N is the number of training samples.

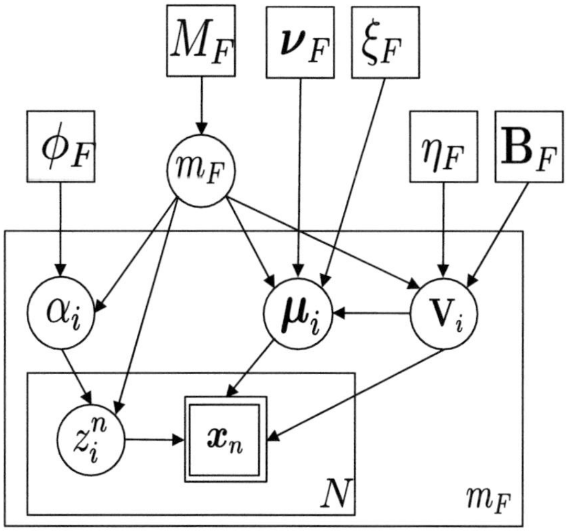

Fig. 4. The detailed graphical model for functions.

m_F, $\boldsymbol{\mu}_i$, $\boldsymbol{V}_i$ and α_i denote the number of functions, i-th component of m_F sets of mean vectors, precision matrices and mixture ratios, respectively. All of these parameters have their own prior distributions. The prior distribution of the multinomial distribution $\boldsymbol{\alpha} = \{\alpha_1, \cdots, \alpha_{m_F}\}$ is the Dirichlet distribution with the degrees of freedom ϕ_F. The prior distribution of the mean vector $\boldsymbol{\mu}_i$ is the Gaussian distribution with the mean vector $\boldsymbol{\nu}_F$ and the precision matrix $\xi_F \boldsymbol{V}_i$. $\boldsymbol{V}_i$ has the following Wishart distribution, which is parameterized by η_F and $\boldsymbol{B}_F$, as the prior distribution:

$$\mathcal{W}(\boldsymbol{V}_i|\eta_F, \boldsymbol{B}_F) \propto |\boldsymbol{V}_i|^{1/2(\eta_F - d - 1)} \exp(-\frac{1}{2}\mathrm{Tr}(\boldsymbol{V}_i \boldsymbol{B}_F)), \quad (17)$$

where $\mathcal{W}()$ represents the Wishart distribution. The model structure m_F also has the uniform distribution M_F as its prior distribution. z_i^n denotes a latent variable, which represents the functions.

In the variational Bayesian approach, the following marginal likelihood of the observations $\boldsymbol{D} = \{\boldsymbol{x}_1, \cdots, \boldsymbol{x}_N\}$ is considered:

$$\mathcal{L}(\boldsymbol{D}) = \log P(\boldsymbol{D}) = \log \sum_{m_F} \sum_{\boldsymbol{Z}_F} \int_\theta P(\boldsymbol{D}, \boldsymbol{Z}_F, \boldsymbol{\theta}, m_F) d\boldsymbol{\theta}, \quad (18)$$

where $\boldsymbol{Z}_F = \{z_i^n\}_{n=1,i=1}^{N,m_F}$ and $\boldsymbol{\theta}$ represent latent variables and a set of model parameters, respectively. Now the variational posterior q is introduced to make the problem tractable.

$$q(\boldsymbol{Z}_F, \boldsymbol{\theta}, m_F) = q(m_F) q(\boldsymbol{Z}_F | m_F) \prod_k^K q(\boldsymbol{\theta}_k | m_F), \quad (19)$$

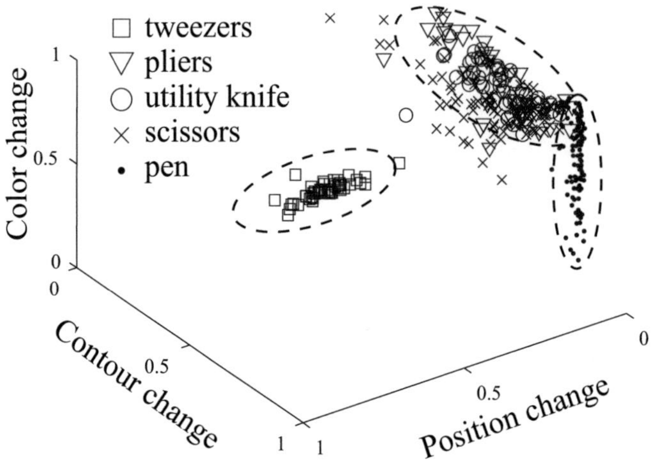

Fig. 5. 3D-plot of feature vectors of object functions.

where $\boldsymbol{\theta}$ is assumed to be decomposed into k independent parameters $\boldsymbol{\theta}_k (k = 1, \cdots, K)$. Then, $\mathcal{L}(\boldsymbol{D})$ can be written as follows:

$$\begin{aligned}
\mathcal{L}(\boldsymbol{D}) &= \log \sum_{m_F} \sum_{\boldsymbol{Z}_F} \int_\theta q(\boldsymbol{Z}_F, \boldsymbol{\theta}, m_F) \frac{P(\boldsymbol{D}, \boldsymbol{Z}_F, \boldsymbol{\theta}, m_F)}{q(\boldsymbol{Z}_F, \boldsymbol{\theta}, m_F)} d\boldsymbol{\theta} \\
&= \sum_{m_F} \sum_{\boldsymbol{Z}_F} \int_\theta q(\boldsymbol{Z}_F, \boldsymbol{\theta}, m_F) \log \frac{q(\boldsymbol{Z}_F, \boldsymbol{\theta}, m_F)}{P(\boldsymbol{Z}_F, \boldsymbol{\theta}, m_F | \boldsymbol{D})} d\boldsymbol{\theta} \\
&\quad + \sum_{m_F} \sum_{\boldsymbol{Z}_F} \int_\theta q(\boldsymbol{Z}_F, \boldsymbol{\theta}, m_F) \log \frac{P(\boldsymbol{D}, \boldsymbol{Z}_F, \boldsymbol{\theta}, m_F)}{q(\boldsymbol{Z}_F, \boldsymbol{\theta}, m_F)} d\boldsymbol{\theta} \\
&\equiv \mathrm{KL}(q(\boldsymbol{Z}_F, \boldsymbol{\theta}, m_F), P(\boldsymbol{Z}_F, \boldsymbol{\theta}, m_F | \boldsymbol{D})) + \mathcal{F}[q], \qquad (20)
\end{aligned}$$

where $F[q]$ and KL denote free energy and Kullback-Leibler divergence, respectively. Since $L(\boldsymbol{D})$ does not depend on q, the maximization of $F[q]$ with respect to q is equivalent to the minimization of Kullback-Leibler divergence between q and true posterior P. Therefore, variational posterior q, which maximizes $F[q]$, is the best approximation to the true posterior P. The optimum variational posterior of each parameter can be obtained by maximizing $\mathcal{F}[q]$ with respect to $\boldsymbol{\theta}_k$ using Lagrange multipliers method.

$$\begin{aligned}
\mathcal{F}[q] = \sum_{m_F} q(m_F) \Bigg\{ &\left\langle \log \frac{P(\boldsymbol{D}, \boldsymbol{Z}_F | \boldsymbol{\theta}, m_F)}{q(\boldsymbol{Z}_F | m_F)} \right\rangle_{q(\boldsymbol{Z}_F | m_F), q(\boldsymbol{\theta} | m_F)} \\
&+ \sum_{k=1}^{K} \left\langle \log \frac{P(\boldsymbol{\theta}_k | m_F)}{q(\boldsymbol{\theta}_k | m_F)} \right\rangle_{q(\boldsymbol{\theta}_k | m_F)} + \log \frac{P(m_F)}{q(m_F)} \Bigg\}. \qquad (21)
\end{aligned}$$

Derivation of $q(\boldsymbol{Z}_F | m_F)$, $q(\boldsymbol{\theta}_k | m_F)$ and $q(m_F)$ are given hereafter.

3.3.2 Derivation of $q(\mathbf{Z}_F|\mathbf{m}_F)$

$q(\mathbf{Z}_F|m_F)$ can be obtained by maxmizing $\mathcal{F}(q)$ under the constraint $\sum_{\mathbf{Z}_F} q(\mathbf{Z}_F|m_F) = 1$ using the Lagrange multiplier method. From Equation 21,

$$\mathcal{F}[q(\mathbf{Z}_F|m_F)] = \left\langle \log \frac{P(\mathbf{D}, \mathbf{Z}_F|\boldsymbol{\theta}, m_F)}{q(\mathbf{Z}_F|m_F)} \right\rangle_{q(\mathbf{Z}|m_F), q(\boldsymbol{\theta}|m_F)}. \tag{22}$$

Let λ be a Lagrange multiplier. Then the problem becomes the extreme value problem of the following functional $\mathcal{J}[q(\mathbf{Z}_F|m_F)]$,

$$\mathcal{J}[q(\mathbf{Z}_F|m_F)] = \mathcal{F}[q(\mathbf{Z}_F|m_F)] + \lambda(\sum_{\mathbf{Z}_F} q(\mathbf{Z}_F|m_F) - 1) \tag{23}$$

$$\begin{aligned}
\frac{\partial \mathcal{J}[q(\mathbf{Z}_F|m_F)]}{\partial q(\mathbf{Z}_F|m_F)} &= \frac{\partial}{\partial q(\mathbf{Z}_F|m_F)} \left[\sum_{\mathbf{Z}_F} \left\{ q(\mathbf{Z}_F|m_F) \int q(\boldsymbol{\theta}|m_F) \log \frac{P(\mathbf{D}, \mathbf{Z}_F|\boldsymbol{\theta}, m_F)}{q(\mathbf{Z}_F|m_F)} d\boldsymbol{\theta} \right\} \right. \\
&\quad \left. + \lambda(\sum_{\mathbf{Z}_F} q(\mathbf{Z}_F|m_F) - 1) \right] \\
&= \frac{\partial}{\partial q(\mathbf{Z}_F|m_F)} \int q(\boldsymbol{\theta}|m_F) \{ q(\mathbf{Z}_F|m_F) \log P(\mathbf{D}, \mathbf{Z}_F|\boldsymbol{\theta}, m_F) \\
&\quad - q(\mathbf{Z}_F|m_F) \log q(\mathbf{Z}_F|m_F) \} \, d\boldsymbol{\theta} + \lambda \\
&= \int q(\boldsymbol{\theta}|m_F) \{ \log P(\mathbf{D}, \mathbf{Z}_F|\boldsymbol{\theta}, m_F) - \log q(\mathbf{Z}_F|m_F) - 1 \} \, d\boldsymbol{\theta} + \lambda \\
&= \langle \log p(\mathbf{D}, \mathbf{Z}_F|\boldsymbol{\theta}, m_F) \rangle_{q(\boldsymbol{\theta}|m_F)} - \log q(\mathbf{Z}_F|m_F) - 1 + \lambda \\
&= 0
\end{aligned}$$

$$\Rightarrow \quad q(\mathbf{Z}_F|m_F) = \exp \left\{ \langle \log P(\mathbf{D}, \mathbf{Z}_F|\boldsymbol{\theta}, m_F) \rangle_{q(\boldsymbol{\theta}|m_F)} + \lambda - 1 \right\} \tag{24}$$

$$\begin{aligned}
\frac{\partial \mathcal{J}[q(\mathbf{Z}_F|m_F)]}{\partial \lambda} &= \sum_{\mathbf{Z}_F} q(\mathbf{Z}_F|m_F) - 1 \\
&= \sum_{\mathbf{Z}_F} \exp \left\{ \langle \log P(\mathbf{D}, \mathbf{Z}_F|\boldsymbol{\theta}, m_F) \rangle_{q(\boldsymbol{\theta}|m_F)} + \lambda - 1 \right\} - 1 = 0
\end{aligned}$$

$$\Rightarrow \quad \exp(\lambda - 1) = \frac{1}{\sum_{\mathbf{Z}_F} \exp \langle \log P(\mathbf{D}, \mathbf{Z}_F|\boldsymbol{\theta}, m_F) \rangle_{q(\boldsymbol{\theta}|m_F)}} \tag{25}$$

From Equations 24 and 25, $q(\mathbf{Z}_F|m_F)$ can be obtained by

$$\begin{aligned}
q(\mathbf{Z}_F|m_F) &= \frac{\exp \langle \log P(\mathbf{D}, \mathbf{Z}_F|\boldsymbol{\theta}, m_F) \rangle_{q(\boldsymbol{\theta}|m_F)}}{\sum_{\mathbf{Z}_F} \exp \langle \log P(\mathbf{D}, \mathbf{Z}_F|\boldsymbol{\theta}, m_F) \rangle_{q(\boldsymbol{\theta}|m_F)}} \\
&= C \exp \langle \log P(\mathbf{D}, \mathbf{Z}_F|\boldsymbol{\theta}, m_F) \rangle_{q(\boldsymbol{\theta}|m_F)},
\end{aligned} \tag{26}$$

where C represents a normalizing constant.

3.3.3 Derivation of $q(\boldsymbol{\theta}_i|m_F)$

$q(\boldsymbol{\theta}_i|m_F)$ can be also obtained through the maximization of $\mathcal{F}(q)$ under the constraint of $\sum_{\theta_i} q(\boldsymbol{\theta}_i|m_F) = 1$. From Equation 21, we can write the terms, which are dependent on $q(\boldsymbol{\theta}_i|m_F)$,

$$\mathcal{F}[q(\boldsymbol{\theta}_i|m_F)] = \left\langle \log \frac{P(\boldsymbol{D}, \boldsymbol{Z}_F|\boldsymbol{\theta}, m_F)}{q(\boldsymbol{Z}_F|m_F)} \right\rangle_{q(\boldsymbol{Z}|m_F),q(\boldsymbol{\theta}|m_F)} + \sum_j \left\langle \log \frac{P(\boldsymbol{\theta}_j|m_F)}{q(\boldsymbol{\theta}_j|m_F)} \right\rangle_{q(\boldsymbol{\theta}_j|m_F)}. \tag{27}$$

Let λ be a Lagrange multiplier. Then the problem becomes the extreme value problem of the following functional $\mathcal{J}[q(\boldsymbol{\theta}_i|m_F)]$,

$$\begin{aligned}
\mathcal{J}[q(\boldsymbol{\theta}_i|m_F)] &= \mathcal{F}[q(\boldsymbol{\theta}_i|m_F)] + \lambda(\int q(\boldsymbol{\theta}_i|m_F)d\boldsymbol{\theta}_i - 1) \\
\frac{\partial \mathcal{J}[q(\boldsymbol{\theta}_i|m_F)]}{\partial q(\boldsymbol{\theta}_i|m_F)} &= \frac{\partial}{\partial q(\boldsymbol{\theta}_i|m_F))} \Bigg[\sum_{\boldsymbol{Z}_F} \Big\{ q(\boldsymbol{Z}_F|m_F) \int q(\boldsymbol{\theta}|m_F) \log P(\boldsymbol{D}, \boldsymbol{Z}_F|\boldsymbol{\theta}, m_F) \\
&\quad - q(\boldsymbol{\theta}|m_F) \log q(\boldsymbol{Z}|m_F) d\boldsymbol{\theta} \Big\} \\
&\quad + q(\boldsymbol{\theta}_i|m_F) \log P(\boldsymbol{\theta}_i|m_F) - q(\boldsymbol{\theta}_i|m_F) \log q(\boldsymbol{\theta}_i|m_F) \Bigg] + \lambda \\
&= \langle \log P(\boldsymbol{D}, \boldsymbol{Z}_F|\boldsymbol{\theta}, m_F) \rangle_{q(\boldsymbol{Z}_F|m_F),q(\boldsymbol{\theta}_{-i}|m_F)} + \log P(\boldsymbol{\theta}_i|m_F) \\
&\quad - \log q(\boldsymbol{\theta}_i|m_F) - 1 = 0,
\end{aligned} \tag{28}$$

$$\Rightarrow \quad q(\boldsymbol{\theta}_i|m_F) = P(\boldsymbol{\theta}_i|m_F) \exp\Big(\langle \log P(\boldsymbol{D}, \boldsymbol{Z}_F|\boldsymbol{\theta}, m_F) \rangle_{q(\boldsymbol{Z}_F|m_F),q(\boldsymbol{\theta}_{-i}|m_F)} + \lambda - 1 \Big), \tag{29}$$

$$\begin{aligned}
\frac{\partial \mathcal{J}[q(\boldsymbol{\theta}_i|m_F)]}{\partial \lambda} &= \int q(\boldsymbol{\theta}_i|m_F) d\boldsymbol{\theta}_i - 1 \\
&= \int P(\boldsymbol{\theta}_i|m_F) \exp\Big(\langle \log P(\boldsymbol{D}, \boldsymbol{Z}_F|\boldsymbol{\theta}, m_F) \rangle_{q(\boldsymbol{Z}_F|m_F),q(\boldsymbol{\theta}_{-i}|m_F)} \\
&\quad + \lambda - 1 \Big) d\boldsymbol{\theta}_i - 1 = 0,
\end{aligned} \tag{30}$$

$$\Rightarrow \quad \exp(\lambda - 1) = \frac{1}{\int P(\boldsymbol{\theta}_i|m_F) \exp\Big(\langle \log P(\boldsymbol{D}, \boldsymbol{Z}_F|\boldsymbol{\theta}, m_F) \rangle_{q(\boldsymbol{Z}_F|m_F),q(\boldsymbol{\theta}_{-i}|m_F)} \Big) d\boldsymbol{\theta}_i}, \tag{31}$$

where $\boldsymbol{\theta}_{-i}$ represents all of parameters $\boldsymbol{\theta}$ except $\boldsymbol{\theta}_i$. Substituting Equation 31 into 29, we obtain the following variational posterior $q(\boldsymbol{\theta}_i|m_F)$:

$$\begin{aligned}
q(\boldsymbol{\theta}_i|m_F) &= \frac{P(\boldsymbol{\theta}_i|m_F) \exp\Big(\langle \log P(\boldsymbol{D}, \boldsymbol{Z}_F|\boldsymbol{\theta}, m_F) \rangle_{q(\boldsymbol{Z}_F|m_F),q(\boldsymbol{\theta}_{-i}|m_F)} \Big)}{\int P(\boldsymbol{\theta}_i|m_F) \exp\Big(\langle \log P(\boldsymbol{D}, \boldsymbol{Z}_F|\boldsymbol{\theta}, m_F) \rangle_{q(\boldsymbol{Z}_F|m_F),q(\boldsymbol{\theta}_{-i}|m_F)} \Big) d\boldsymbol{\theta}_i} \\
&= C_i P(\boldsymbol{\theta}_i|m_F) \exp\Big(\langle \log P(\boldsymbol{D}, \boldsymbol{Z}_F|\boldsymbol{\theta}, m_F) \rangle_{q(\boldsymbol{Z}_F|m_F),q(\boldsymbol{\theta}_{-i}|m_F)} \Big),
\end{aligned} \tag{32}$$

where C_i represents a normalizing constant.

3.3.4 Derivation of $q(m_F)$

The maximization of $F[q]$ with respect to $q(m_F)$ results in the optimum variational posterior of the model structure $q(m_F)$. Equation 21 can be rewritten as

$$\mathcal{F}[q] = \langle \mathcal{F}_{m_F} \rangle_{q(m_F)} + \left\langle \log \frac{P(m_F)}{q(m_F)} \right\rangle_{q(m_F)}, \tag{33}$$

where $\mathcal{F}_{m_F}$ represents the sum of terms, which do not contain $q(m_F)$. Since $q(\boldsymbol{Z}_F|m_F)$ and $q(\boldsymbol{\theta}|m_F)$ affect only $\mathcal{F}_{m_F}$, the maximization of $\mathcal{F}[q]$ with respect to $q(\boldsymbol{Z}_F|m_F)$ and $q(\boldsymbol{\theta}|m_F)$ is equivalent to the maximization of $\mathcal{F}_{m_F}$ with respect to $q(\boldsymbol{Z}_F|m_F)$ and $q(\boldsymbol{\theta}|m_F)$. Now let a maximum value of $\mathcal{F}_{m_F}$ with respect to $q(\boldsymbol{Z}_F|m_F)$ and $q(\boldsymbol{\theta}|m_F)$ be $\mathcal{F}^*_{m_F}$. Then, the optimum variational posterior of the model structure $q(m_F)$ can be written as

$$\mathcal{F}[q] = \langle \mathcal{F}^*_{m_F} \rangle_{q(m_F)} + \left\langle \log \frac{P(m_F)}{q(m_F)} \right\rangle_{q(m_F)}. \tag{34}$$

The above equation should be maximized with respect to $q(m_F)$ under the constraint of $\sum_m q(m_F) = 1$. A Lagrange multiplier λ is again introduced and then, it becomes the extreme value problem of the functional $\mathcal{J}[q(m_F)]$,

$$\begin{aligned}
\mathcal{J}[q(m_F)] &= \langle \mathcal{F}^*_{m_F} \rangle_{q(m_F)} + \left\langle \log \frac{P(m_F)}{q(m_F)} \right\rangle_{q(m_F)} + \lambda \left(\sum_{m_F} q(m_F) - 1 \right) \\
&= \sum_{m_F} q(m_F) \left(\mathcal{F}^*_{m_F} + \log \frac{P(m_F)}{q(m_F)} \right) + \lambda \left(\sum_{m_F} q(m_F) - 1 \right), \\
\frac{\partial \mathcal{J}[q(m_F)]}{\partial q(m_F)} &= \mathcal{F}^*_{m_F} + \log P(m_F) - \log q(m_F) - 1 + \lambda = 0, \\
& \Rightarrow q(m_F) = P(m_F) \exp\left(\mathcal{F}^*_{m_F} - 1 + \lambda \right), \qquad (35) \\
\frac{\partial \mathcal{J}[q(m_F)]}{\partial \lambda} &= \sum_{m_F} q(m_F) - 1 = 0 \\
& \Rightarrow \exp(\lambda - 1) = \frac{1}{\sum_{m_F} P(m_F) \exp\left(\mathcal{F}^*_{m_F} \right)}, \qquad (36)
\end{aligned}$$

Substituting Equation 36 into35, the following $q(m_F)$ is obtained:

$$q(m_F) = \frac{P(m_F) \exp\left(\mathcal{F}^*_{m_F} \right)}{\sum_{m_F} P(m_F) \exp\left(\mathcal{F}^*_{m_F} \right)} = C_{m_F} P(m_F) \exp\left(\mathcal{F}^*_{m_F} \right), \tag{37}$$

where C_{m_F} is a normalizing constant ensuring $\sum_{m_F} q(m_F) = 1$. The maximization of $q(m_F)$ is equivalent to the maximization of $\mathcal{F}^*_{m_F}$, since a uniform distribution $P(m_F) = M_F$ is assumed as the prior distribution of m_F. Therefore an optimum model structure m_F can be estimated through the maximization of $\mathcal{F}^*_{m_F}$ with respect to $q(\boldsymbol{Z}|m_F)$ and $q(\boldsymbol{\theta}|m_F)$.

3.3.5 Variational posteriors for GMM

Finally, we can obtain the variational posterior of latent variables for given m_F as follows:

$$q(\boldsymbol{Z}_F|m_F) = C\prod_{i=1}^{m_F}\prod_{n=1}^{N}\exp\left\{z_i^n\left(\langle\log\alpha_i\rangle_{q(\boldsymbol{\alpha}|m_F)}+\frac{1}{2}\langle\log|\boldsymbol{V}_i|\rangle_{q(\boldsymbol{V}_i|m_F)}\right.\right.$$
$$\left.\left.-\frac{1}{2}\mathrm{Tr}\left\{\langle\boldsymbol{V}_i\rangle_{q(\boldsymbol{V}_i|m_F)}\left\langle(\boldsymbol{x}_n-\boldsymbol{\mu}_i)(\boldsymbol{x}_n-\boldsymbol{\mu}_i)^T\right\rangle_{q(\mu_i|m_F)}\right\}\right)\right\}. \quad (38)$$

The variational posterior of $\boldsymbol{\alpha}$ is the following Dirichlet distribution parameterized by $\{\phi_0+\bar{N}_i\}_{i=1}^{m_F}$.

$$q(\boldsymbol{\alpha}|m_F) = \mathcal{D}\left(\{\alpha_i\}_{i=1}^{m_F}|\{\phi_0+\bar{N}_i\}_{i=1}^{m_F}\right), \quad (39)$$

where $\mathcal{D}()$ denotes Dirichlet distribution and

$$\bar{N}_i=\sum_{n=1}^{N}\bar{z}_i^n, \quad \bar{z}_i^n = \langle z_i^n\rangle_{q(z_i^n|m_F)}. \quad (40)$$

The variational posteriors of $\boldsymbol{V}_i$ is the Wishart distribution, which is parameterized by $\mathbf{B}_i$ and $\eta_F+\bar{N}_i$ as follows:

$$q(\boldsymbol{V}_i|m_F) = \mathcal{W}(\boldsymbol{V}_i|\eta_F+\bar{N}_i,\mathbf{B}_i), \quad (41)$$

where

$$\mathbf{B}_i = \mathbf{B}_F+\bar{\boldsymbol{C}}_i+\frac{\bar{N}_i\xi_F}{\bar{N}_i+\xi_F}(\bar{\boldsymbol{x}}_i-\boldsymbol{\nu}_F)(\bar{\boldsymbol{x}}_i-\boldsymbol{\nu}_F)^T, \quad (42)$$

$$\bar{\boldsymbol{x}}_i = \frac{1}{\bar{N}_i}\sum_{n=1}^{N}\bar{z}_i^n\boldsymbol{x}_n, \quad \bar{\boldsymbol{C}}_i=\sum_{n=1}^{N}\bar{z}_i^n(\boldsymbol{x}_n-\bar{\boldsymbol{x}}_i)(\boldsymbol{x}_n-\bar{\boldsymbol{x}}_i)^T. \quad (43)$$

The marginalization of $q(\boldsymbol{\mu}_i,\boldsymbol{V}_i|m_F)$ with respect to $\boldsymbol{V}_i$ yields the variational posterior of $\boldsymbol{\mu}_i$. It turns out that the variational posterior becomes the Student's t-distribution parameterized by $\bar{\boldsymbol{\mu}}_i$, $\boldsymbol{\Sigma}\mu_i$, and f_{μ_i}.

$$q(\boldsymbol{\mu}_i|m_F) = \mathcal{T}(\boldsymbol{\mu}_i|\bar{\boldsymbol{\mu}}_i,\boldsymbol{\Sigma}_{\mu_i},f_{\mu_i}), \quad (44)$$

where $\mathcal{T}()$ represents the Student's t-distribution as follows:

$$\mathcal{T}(\boldsymbol{\mu}_i|\bar{\boldsymbol{\mu}}_i,\boldsymbol{\Sigma}_{\mu_i},f_{\mu_i}) \propto \left\{1+(\boldsymbol{\mu}_i-\bar{\boldsymbol{\mu}}_i)^T(\boldsymbol{\Sigma}_{\mu_i}f_{\mu_i})^{-1}(\boldsymbol{\mu}_i-\bar{\boldsymbol{\mu}}_i)\right\}^{-\frac{d+f_{\mu_i}}{2}}, \quad (45)$$

where d represents the dimension of the input vector. Moreover each parameter can be written as

$$\bar{\boldsymbol{\mu}}_i = \frac{\bar{N}_i\bar{\boldsymbol{x}}_i+\xi_F\boldsymbol{\nu}_F}{\bar{N}_i+\xi_F}, \quad \boldsymbol{\Sigma}_{\mu_i}=\frac{\mathbf{B}_i}{(\bar{N}_i+\xi_F)f_{\mu_i}}, \quad f_{\mu_i}=\eta_F+\bar{N}_i+1-d. \quad (46)$$

The optimization process starts from an initial guess and iterates the E-step (Equation 38) and M-step (Equations 39, 41, and 44) until it converges. This variational EM-algorithm gives a local maximum of $\mathcal{F}[q]$. The optimum structure m_F, which represents the number of functions,

object category	ID	A set	B set	total
scissors	T1	7	3	10
pen	T2	8	3	11
pliers	T3	2	2	4
tweezers	T4	3	2	5
utility knife	T5	3	1	4
stapler	T6	4	1	5
tape	T7	4	2	6
colored vinyl tape	T8	2	2	4

Table 1. Number of tools in the experiment.

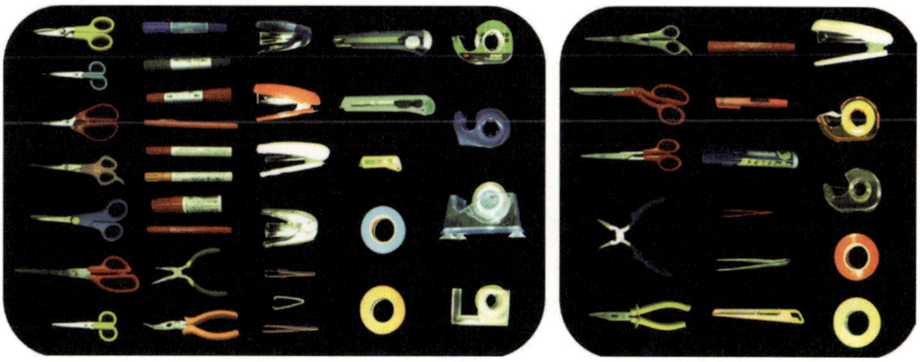

(a) A set (b) B set

Fig. 6. Hand tools used in the experiment. (a)Set A. (b)Set B.

can be also obtained by selecting m_F that maximizes $\mathcal{F}[q]$. In the later experiment, the model structure is selected in the range of $2 \leq m_F \leq 8$.

When a novel observation X_F is given to the learned model, the function Z_F^* can be estimated as follows:

$$Z_F^* = \underset{Z_F}{\text{argmax}}\, P(X_F|Z_F) = \underset{j}{\text{argmax}}\, \hat{\alpha}_j \mathcal{N}(X_F|\hat{\boldsymbol{\mu}}_j, \hat{\boldsymbol{V}}_j), \tag{47}$$

where $\mathcal{N}$ represents the Gaussian distribution. $\hat{\alpha}_j$ is the j-th component of a mode of the variational posterior $q(\boldsymbol{\alpha}|m_F)$. $\hat{\boldsymbol{\mu}}_j$ and $\hat{\boldsymbol{V}}_j$ denote modes of the variational posteriors $q(\boldsymbol{\mu}_j|m_F)$ and $q(\boldsymbol{V}_j|m_F)$, respectively. For given X_F, $P(X_F|Z_F)$ in Fig.1(c) is derived as

$$P(X_F|Z_F) = \hat{\alpha}_{Z_F} \mathcal{N}(X_F|\hat{\boldsymbol{\mu}}_{Z_F}, \hat{\boldsymbol{V}}_{Z_F}). \tag{48}$$

4. Experiments

4.1 Experimental Setup

A total of 49 objects with 8 categories, which are given in Tab.1, are employed in the experiments. These 49 objects are divided into two groups. Figure 6(a) is the set A containing a total of 33 hand tools (7 scissors, 8 pens, 2 pliers, 4 staplers, 3 tweezers, 4 tapes, 2 colored vinyl

(a)

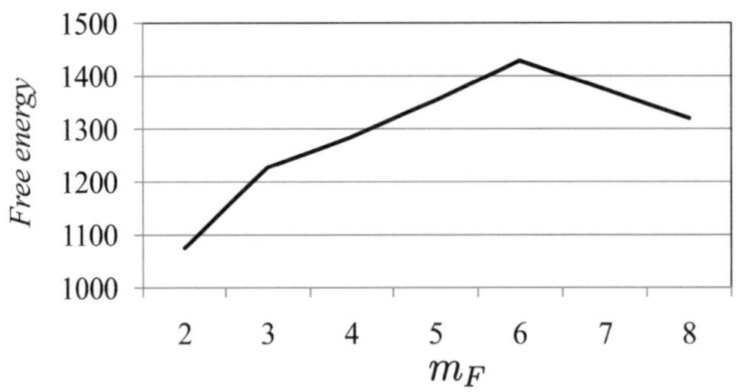

(b)

Fig. 7. (a)A snapshot of the system. (b)Structure vs. free energy.

tapes and 3 utility knives). The set B consists of a total of 16 hand tools (3 scissors, 3 pens, 2 pliers, 2 tweezers, 1 stapler, 2 tapes, 2 colored vinyl tapes and 1 utility knife), which are shown in Fig.6(b). Figure 7(a) shows the actual system setup. The camera is fixed to capture the user's hands and takes images during the manipulation. The tool and the work object are extracted based on background difference method, and then the system computes appearance and function information as we mentioned earlier. Three experiments were conducted using this system.

4.2 Finding Abstract Functions

At first the function model in Fig.4 was trained. Each of 33 hand tools (set A) used 10 times and a total of 330 feature vectors were obtained. The VB algorithm was applied to the data to estimate the parameters and optimal structure, i.e. number of abstract functions. Figure 7(b) shows free energy over the number of functions m_F. The figure implies that six functions explain the data best. In fact, we have confirmed these six functions correspond to 'cut', 'write',

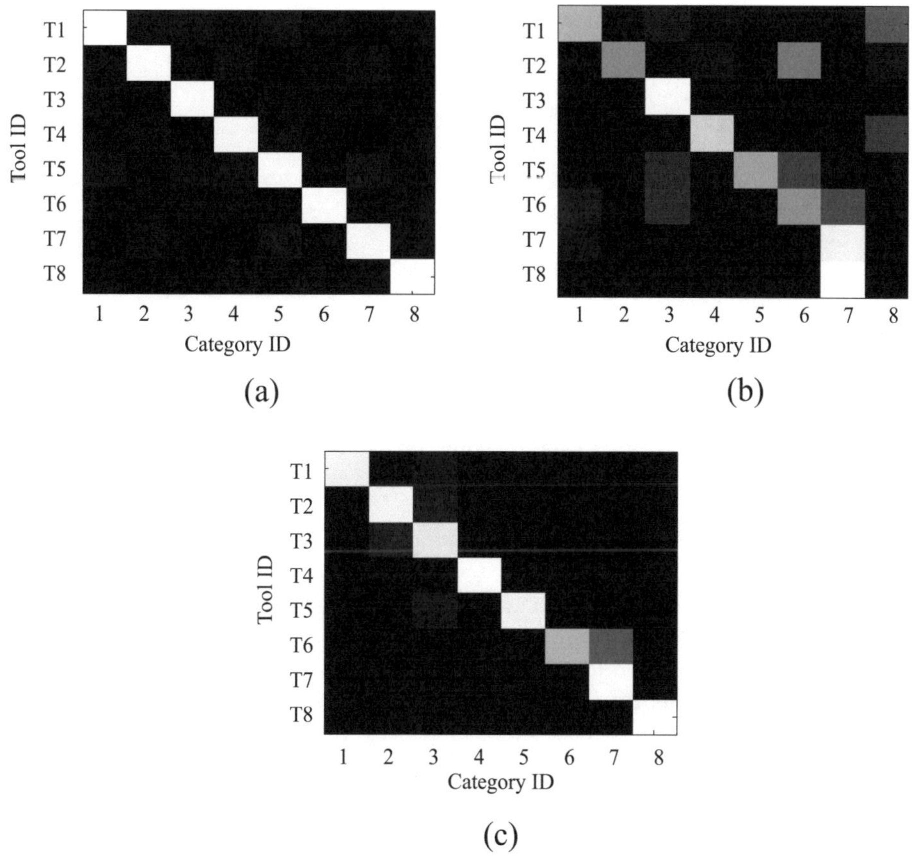

Fig. 8. (a)Correct categorization. (b)Categorization with only visual information. (c)Categorization with visual and function information.

'move'. 'deform', 'adhesion' and 'adhesion with color change'. In the following experiments, the abstract function model, which was obtained in this experiment, is used.

4.3 Results of Learning

The tools in the A set are used in the second experiment for the training of Fig.1(c). Each of 33 hand tools was used 10 times; hence the model was trained using a total of 330 data. Then Equation 11 was used to classify 330 data. The classification result is compared with ground truth to evaluate how well the objects are categorized. The result is shown in Fig.8. In these figures the horizontal and vertical axes indicate category and object indices, respectively. The white bar in the figure represents that the object is classified into the category. From the figure one can see that the system has reasonably categorized the objects by using both appearance and function information.

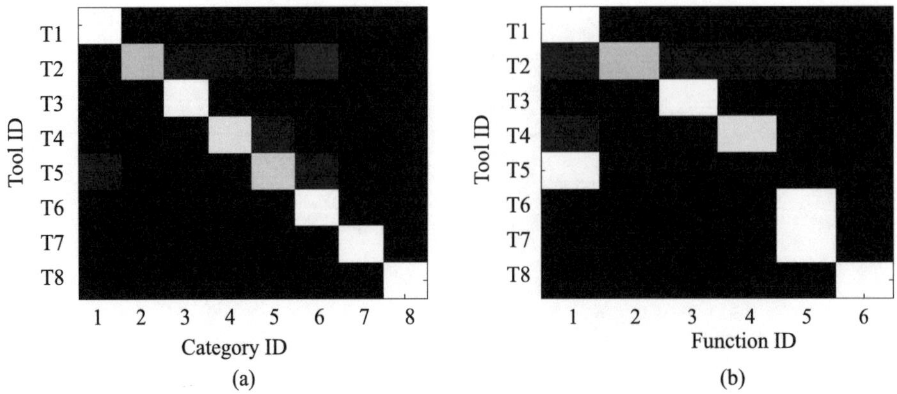

Fig. 9. Result of recognition. (a)Tool recognition from visual and function information. (b)Function recognition from only visual information.

4.4 Results of Inference

After the training in the foregoing subsection, the system observed unseen objects in the B set and recognized their categories from the observable visual information and functions. The result is given in Fig.9 (a). Then the system inferred their functions only from appearance. Equation(12) was used to identify the function. The result is given in Fig.9 (b). It can be seen that the system inferred object functions almost perfectly. In fact, inference accuracy is 95.4%.

5. Conclusions

This chapter hase discussed a novel framework for object understanding. Implementation of the proposed framework using Bayesian Network has been presented. Although the result given in this paper is preliminary one, we have shown that the system can form object concept by observing the performance by human hands. The on-line learning is left for the future works. Moreover the model should be extended so that it can represent the object usage and work objects.

6. Acknowledgements

This work was supported by Grant-in-Aid for Scientific Research (C) (20500186, 20500179) and Grant-in-Aid for Scientific Research on Innovative Areas (The study on the neural dynamics for understandig communication in terms of complex hetero systems).

7. References

Attias, H. (1999). Infering parameters and structure of latent variable models by variational bayes, *Proceedings of 15th Conference on Uncertainty in Artificial Intelligence*, pp.21–30.

Fergus, R., Perona, P., & Zisserman, A. (2003). Object class recognition by unsupervised scale-invariant learning, *Proceedings of the IEEE Conference on Computer Vision and Pattern Recognition*, vol.2, pp.264–271.

Gibbson, J. J. (1979). *The Ecological Approach to Visual Perception*, Lawrence Eribaum, Hillsdale, NJ.

Hofmann, T. (2001). Unsupervised Learning by Probabilistic Latent Semantic Analysis, *Machine Learning*, vol.42, pp.177–196.

Kojima A., Higuchi, M., Kitahashi, T., & Fukunaga, K. (2004). Toward a cooperative recognition of human behaviors and related objects, *Proceedings of Second International Workshop on Man-Machine Symbiotic Systems*, pp.195–206.

Landau, B., Smith, L., & Jones, S. (1998). Object shape, object function, and object name, *Journal of Memory and Language*, 38(ML972533):1–27.

Lowe, D. G. (2004) Distinctive image features from scale-invariant keypoints, *Int. Journal of Computer Vision*, 60(2):91–110.

Ogura, T., Okada, K., & Inaba. M. (2005). Humanoid tool operating motion generation, *Proceedings of The 23rd Annual Conf. of the Robotics Society of Japan*, 1F15, (in japanese).

Rivlin, E., Dickinson, S. J., & Rosenfeld, A. (1995). Recognition by functional parts, *Computer Vision and Image Understanding: CVIU*, 62(2), pp.164–176.

Sivic, J., Russell, B. C., Efros, A. A., Zisserman, A. & Freeman, W.T. (2005). Discovering object categories in image collections, *AI Memo*, 2005-005, pp.1–12.

Stark, L., Bowyer, K., Hoover, A., & Goldgof,D. B. (1996). Recognizing object function through reasoning about partial shape descriptions and dynamic physical properties, *Roceedings of The IEEE*, 84(11):1640–1656.

Woods, K., Cook, D., Hall, L., Bowyer, K. W., & Stark, L. (1995) Learning membership functions in a function-based object recognition system, *Journal of Artificial Intelligence Research*, 3:187–222.

7

Guiding complex design optimisation using Bayesian Networks

Artem Parakhine and John Leaney
University of Technology, Sydney, Faculty of Engineering and Information Technology
Australia

1. Introduction

In this chapter we aim to present an approach to addressing the problem of architectural optimisation of complex computer-based systems. Initially, we provide a survey of methods created with the direct aim of optimising system with respect to their non-functional qualities. An overview of commonalities and differences between these methods leads to exploration of reasons for their success in attaining their respective goals. The results of this discussion provide the information necessary to formulate the definition of the notion of *guidance* in the context of architectural design and optimisation.
Following that, we describe a new Heuristic-based Architectural Optimisation framework developed in an attempt to unify the lessons learnt from the successes and shortcomings of the surveyed optimisation approaches. Special attention is paid to the *guidance* mechanism employed by the new framework to control the progress of design optimisation. We propose to develop a suitable guidance mechanism using a Bayesian Belief Network (BBN) obtained from a combination of hybrid simulation modeling and BBN discovery algorithm.
Finally, the rest of the chapter is devoted to a description of a brief example problem. There, an attempt is made to show how the guidance mechanism can be applied to a fairly simple problem that nonetheless possesses an element of ambiguity that so often falters system design optimisation activities.

2. Complex System Design and Optimisation

The process of design optimisation is a special form of design synthesis characterised by a high level of specificity with respect to its goals. As such it relies on the following supporting elements:

measurable goals - clear and quantifiable statement of goals to enable evaluation of candidate designs.

design variability - the system design must present amenable elements. In the context of system design the change may be made at various levels: topology of component arrangement, interchangeable components, configuration within a component.

constraints - restrictions, such as budgeted cost and time, provide an essential definition of design feasibility which effectively reduces the search space to a manageable size.

Currently, all prevalent approaches to architectural system design can be separated into two groups based on their main artefacts: quality-driven and model-driven (Parakhine, 2009, p. 22).
The quality-driven design synthesis process focuses on extensive analysis of qualities and possible trade-offs. It necessitates creation of goal descriptions expressed partially in terms of choices required to attain them, which in turn contributes towards the evaluation of the candidate architectures. On the other hand, the model-driven methodology concentrates on representing the system design from the point of view of a specific domain (e.g. security) with consequent modification through application of well-defined domain-specific changes.
Although the two groups operate on different artefacts, both of them fit into the same general structure of an architectural optimisation process shown in Figure 1.

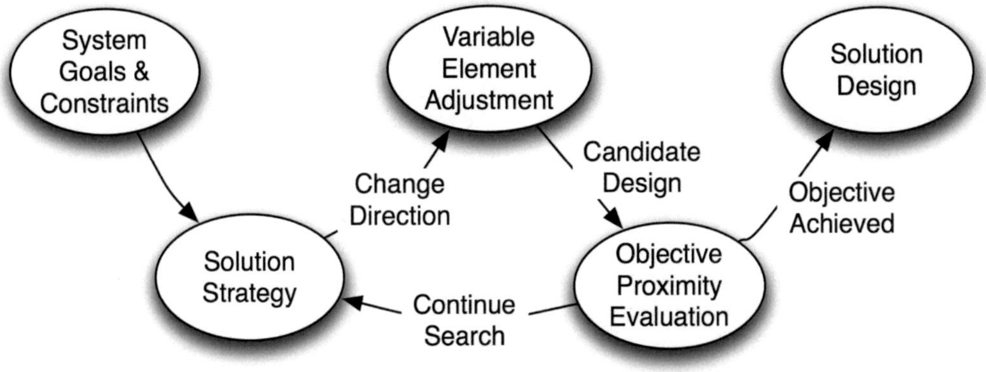

Fig. 1. A generic process of system optimisation at an architectural level.

In its simplest terms, the process of architectural optimisation can be characterised as one based on iterative derivation and evaluation aimed at satisfaction of multiple goals. The fields of computer science and system design offer a wealth of frameworks exhibiting features and capabilities of the optimisation process shown in Figure 1. Table 1 details how the requirements for facilitation of architectural optimisation have been addressed by some of the various frameworks built specifically for that purpose or built to solve general design problems in a way which makes optimisation possible.
The described optimisation approaches adopt various methods and techniques to attain their goals and produce the outcomes. Those relevant to the system design process are listed below. The presented features and techniques are grouped by the element of the optimisation process (Figure 1) whose function they were used to fulfill:

Genetic Algorithms - indicates that the framework performs evolutionary search that relies on encoding of system design in the form appropriate for breeding operators.

Meta-Heuristics - shows that the framework was created around a specific optimisation heuristic, such as simulated annealing or a tabu-search.

Function Analysis - identifies frameworks that employ a mathematical function analysis to drive the search process.

Configuration - only variations in the properties of the system components are considered.

Component - this approach to system change adopts components as variable elements.

	Goals & Constraints Focus			Solution Strategy			Design Variables			Evaluation Mechanism			End Result	
Architectural Optimisation Approach	Individual	Trade-off	Multiple	Genetic Algorithms	Meta-Heuristics	Function Analysis	Configuration	Component	Structure	Calculation	Simulation	Runtime Monitoring	Design Suggestion	Complete Design
Diaconescu, Mos, Murphy (2003)	✓				✓			✓				✓		✓
Coit & Konak (2006)	✓			✓	✓			✓	✓		✓		✓	
Bucci, Streeter, Maio (1979)	✓					✓		✓	✓	✓			✓	
Bosch & Benngston (2001)	✓					✓		✓		✓			✓	
QDAD		✓			✓			✓	✓	✓	✓		✓	
Gockhale (2004)		✓		✓			✓	✓			✓			✓
Grunske (2006)		✓		✓				✓	✓	✓	✓			✓
MDAD			✓		✓		✓	✓	✓	✓	✓			✓
Sharma & Trivedi (2005)			✓		✓		✓	✓			✓		✓	
van Gurp & Bosch (2000)			✓		✓		✓	✓	✓	✓			✓	

Table 1. Summary of approaches to architectural system optimisation.

Structure - focuses on variations in the structural arrangement of system components.

Calculation - describes a mechanism relying on a evaluation of a mathematical function.

Simulation - performs assessment of candidate architectures by constructing simulations of system operation.

Runtime Monitoring - evaluation is performed by making changes to the system and collecting metrics during its subsequent operation.

Design Suggestion - indicates that the optimisation produces a prescriptive list of design modification.

Complete Design - shows that the framework produces a system architecture specified to a relatively high level of detail.

The approaches to architectural optimisation listed in the matrix in Table 1 are arranged in the order of increasing number of goals with the lowest level containing approaches aimed at a specific non-functional system qualities such as maintainability (Bosch & Bengtsson, 2001) or reliability (Gokhale, 2004).

The first entry in the table is the Automated Quality Assurance (AQuA) framework (Diaconescu & Murphy, 2005) which attempts to automatically tailor the system to the operational conditions by varying the choice and configuration of the system components. The AQuA framework is built to continuously check the system for *performance anomalies* and apply a pre-defined *adoption decision* should such anomaly be detected. As such the framework is limited severely by the quality of the metric collection and the depth of analysis that can be performed in a timely manner during system operation so as to preempt possible reduction in performance.

Further to that, the problem of achieving specific qualities can be better addressed before the system is built. For example, an existing requirement for high reliability can be addressed by exploring variations in the choice and number of redundant components in each of the system modules (Coit & Smith, 1996). This formulation, in effect, links the single non-funcitonal property (reliability) to a single type of structural architecture change (connection of components in parallel).

However, introducing changes that may be applied in combination may render a large set of possible candidate architectures. As this approach to optimisation developed, a number of strategies such as Genetic Algorithms (Coit & Smith, 1996), integer programming (Coit, 2001) and Multiple Weighted Objective optimisation heuristic (Coit & Konak, 2006) were used to restrict and manage the pool of candidate solutions.

Bucci & Streeter considered three version of system distribution structure: centralised storage, individual storage at each node and no centralised database, or hierarchical storage where individual nodes synchronise with the centralised storage unit. At the core of this approach was a model developed to assess the cost per transaction for each of the possible three arrangements. The resultant model allowed to determine the cost-minimizing approach to structuring architecture and showed how it would affect the response time and, by extension, the overall throughput. This development has led to the creation of a more generalised model (Bucci & Maio, Sep 1982) built to handle the trade-off between performance and cost.

Another focused optimisation approach was proposed by Bosch & Bengtsson who attempted to formalise the relationship between the properties of the system design and the effort it would take to change it during maintenance cycle. The key relationship guiding the optimisation was expressed as *effort*, or a form of cost, based on the sum of productivity factors associated with each potential change.

Such focus on a single quality fails to address the situations presenting conflicting quality requirements. This led to the development of the Quality Driven Architectural Design (QDAD) family of frameworks such as Architectural Trade-off Analysis Method (Kazman et al., 1999) and Business IT Alignment Method (Chen et al., 2005). These approaches attempt to identify and manage the strong trade-off relationships existing in the design space. To achieve this, a typical QDAD method relies heavily on a collection of mature analytical models similar to the ones described above. Unfortunately, formulating, using and maintaining such combination of metrics would be a complex and arduous task as these models would have to be manually adjusted as system design changes in time and scale.

Another group of optimisation methods listed in Table 1 adopt an approach focused on design options and how they can be optimally combined to get the highest feasible level of overall system qualities.

For example the identification mechanism for "good" architectures shown by Grunske represents a culmination of the development of this type of change-centric approach (Grunske, 2003; 2006; Grunske et al., 2005). There, Grunske uses principles of graph representation, architecture refactoring and genetic algorithms to create an optimisation framework. In this framework, the mechanism of guidance relies on the genetic fitness function that is used to determine the successful mutations propagated into the next generation.

The genetic algorithm approach to multi-objective optimisation and the corollary use of fitness function is not without merit. The ways the concept of *fitness* was used by different approaches (Coit & Smith, 1996; Grunske, 2006) are combined and extended further by the detailed fitness function used by Gokhale to specifically address the trade-off between two non-functional qualities (Gokhale, 2004). Although it is recognised that the successful application of genetic algorithms relies greatly on many factors such as mutation rates and population sizes, Grunske and Gokhale show that the fitness procedures have potential to handle the complex link between architectural assessment and complex multi-objective process of architectural optimisation. Still, the problem of creating such a fitness function remains non-trivial and worsens with the increase in the number of goals.

Next, Table 1 lists the Model-Driven Architectural Design methods which are based on principles of Model-Driven Design (*MDA Guide*, 2003). Although the approaches in this group do not explicitly target optimisation, they nonetheless provide an opportunity to perform it. This is achieved by structuring the process as one comprised of models and transformations. Consequently, the optimisation can take place by applying different transformations to the underlying model. However, the separation into models and transformations ignores the possible conflicts between transformations.

Still, the static analysis of artefacts created as part of MDAD can be used to better understand the system. Additionally, it can be extended as shown by Sharma & Trivedi who used Discrete Time Markov Chain modelling to estimate the reliability, performance and security of a given system by examining the number of visits and times spent in that module calculated from a transitional probability matrix (Sharma & Trivedi, 2005). The probabilities within this matrix are affected by the architecture as different structures and behaviour would lead to the variation in time spent in each modules and the number of visits to specific modules required to finish processing. As such the matrix links compositional features of the system with he models used to estimate its individual qualities, in this case: reliability, performance and security.

The main limitation of the transitional probability matrix is that it does not provide an explanation of how design decisions affect the distributions comprising the matrix. Achieving such form of representation is, nonetheless, possible as shown by the SAABNet framework (van Gurp, 2003, p. 71). At the foundation of SAABNet is a Bayesian Belief Network or BBN which combines probabilistic and factual knowledge about the system features and understanding of its qualities into a single informational resource.

The nature of BBN notation has the power to include information from models, simulations and surveys of domain experts as well as records of architectural changes that have taken place in a given system or across multiple systems. Once assembled, the BBN can be used to determine which combination of factors is most likely to lead to a higher level of a desired quality while at the same time give the designer some idea regarding the effects the change in factors may have on other non-functional system qualities. Resultantly, the SAABNet can be employed to support both the QDAD and MDAD approaches mentioned earlier in this chapter. The SAABNet network and the corollary issues are discussed in further detail in Section 4.

3. Heuristic-based Optimisation Framework

As mentioned in the previous section, the design optimisation framework proposed to address the shortcomings of methods presented in Table 1 is focused around the iterative process borrowed from the field of control theory. The core of the framework is based on the conventional control feedback loop which exists with the intent of providing a mechanism for continuous adjustment of the system behaviour based on changes detected in the set of control variables. Leaney et al. (2004) proposed an extension of this approach into the field of design optimisation with the design process itself being the system under control. Further advances in the fields of architectural refinement (Denford et al., 2004) and heuristic encoding (Maxwell et al., 2006) spurred a development of a more detailed version of the framework; its current state shown in Figure 2.

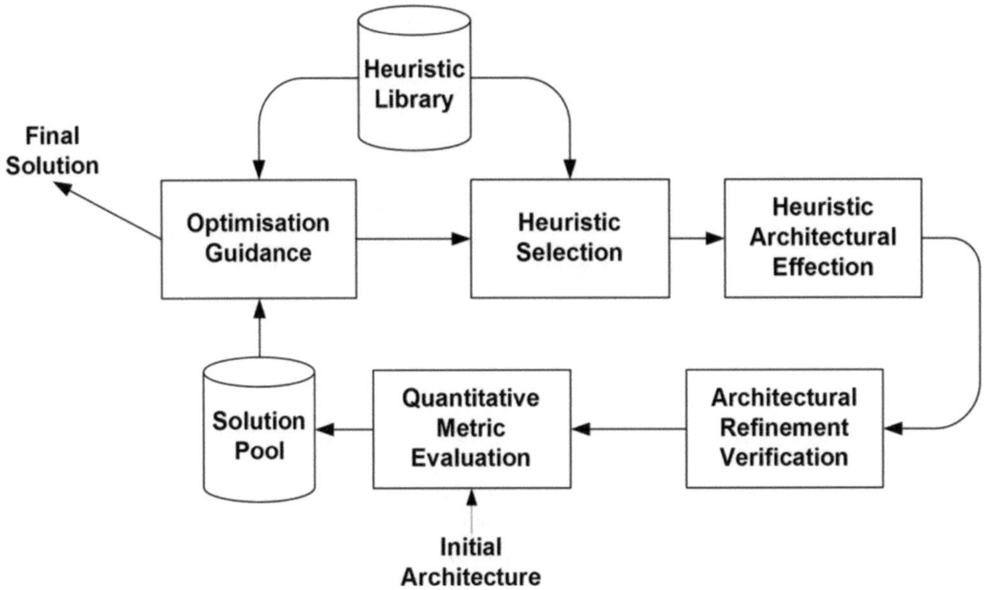

Fig. 2. General view of Heuristic-based Architecture Optimisation framework (Maxwell et al, 2005).

The Heuristic-based Architecture Optimisation (HAO) framework proposed by Maxwell et al. represents a combination of both *constructive* and *local search* methods of solution of approximation. It uses constructive heuristics to change the design of the system during optimisation while relying heavily on a guidance strategy to establish the *neighbourhood of solutions* for a current state of system design and select the appropriate change heuristic.
Specifically, it possesses the following features:

iteration The search for an optimal solution is conducted by iterating through a series of design candidates generated from the application of specific design heuristics.

memory The change heuristics, are, in essence, an abstract codification of specific transformations (Maxwell, 2007, p. 73). As such their sequential application may lead to compound effects that can provide important insights into properties of the system being designed. Hence, the HAO process records all generated solutions in the *Solution Pool* with the intent of allowing the designer to analyse and study properties of the architecture.

non-parametrised view of design Use of heuristics also allows the HAO framework to adopt a holistic view of architectural design without the need for an intermediate step of decomposition into a limited set of mutable parameters.

functional preservation The framework also ensures, through the use of *Architectural Refinement Verification* (Denford et al., 2004), that in the process of design generation none of the design's functional properties are affected in a negative way.

generality Domain-specific heuristics and quantitative metric evaluators can be used to support optimisation of systems from a multitude of domains.

decision support Finally, at its core, the HAO framework is oriented at organising and supporting the activities performed by the system designer during a search for a new design candidate

In its form, the HAO framework attempts to incorporate all the best features of the optimisation frameworks described in Section 2 whilst also exhibiting and addressing the shortcomings of those frameworks. For example, the proposed use of heuristics relates to the generality achieved by the *refactoring* approach described in (Grunske et al., 2005) and non-parametrised view of design is an extension of evolutionary chromosome-based search proposed in (Gokhale, 2004). Furthermore, the modular structure evident in the proposed framework was chosen to ensure that the modules could advance with a degree of independence. Thus, the framework can take advantages of different, possibly domain specific, libraries of heuristics as well as a range of externally defined quality evaluators.

The ultimate aim of the process is to find a solution that represents an optimal compromise on competing system qualities. To achieve a positive outcome the process requires priming, which includes obtaining a set of potential design changes encoded as heuristics (Maxwell et al., 2005), providing a baseline design, and a set of goals and constraints. Effectively, the goals and constraints must contain the following: the expression of desired non-functional qualities that should be present in the solution and the minimal levels to which these qualities must be achieved for a given solution to be considered valid.

The original design is then evaluated to establish a baseline measurements with respect to optimisation goals. Once assessed, the baseline design and its associated characteristics are placed into the *Solution Pool*, which is used establish a historical context that may be used by the designer to study the *evolvability* characteristics of the system (Rowe et al., 1998).

Consequently, the baseline design information is used by the *Optimisation Guidance* component to generate information regarding the type of change that is most likely to render positive results. This information is then applied within the *Heuristic Selection* module to find viable candidates from the options available in the *Heuristic Library*.

Following selection, the identified heuristics are applied the *Architectural Refinement Verification* is used to confirm the candidate designs preserve functional characteristics of the original architecture. Finally, the candidates are evaluated and the results are provided back to the *Optimisation Guidance* component for further analysis.

The role of *Optimisation Guidance* is to determine the best possible alternative architecture in the current solution pool and which combination of heuristics, when applied, would produce an outcome closest to the desired system form. To achieve this, the guidance mechanism uses a combination of Bayesian principles and Hybrid Simulation to form a new view of the system from the viewpoint of potential evolutionary changes directed at achieving the goals of optimisation process. The specifics of guidance composition and operations are described in the coming sections of this chapter.

4. Guidance Mechanism Theory and Application

The core operations of the *Optimisation Guidance* component are performed using a Bayesian Belief Network (BBN) representation of causal links between change heuristics and system qualities. A BBN can be viewed as a *directed acyclic graph* in which the *nodes* represent variables, or assertive propositions, and *edges* describe the presence and direction of influences which exist between connected nodes (Pearl, 1986, p. 358). The relationship where one variable *A* influences another variable *B*, denoted by the existence of a directed *arc* linking a *node A* to another *node B*, carries a nomenclature of *A* being a *parent* of *B* and, conversely, *B* being a *child* of *A*.

Each *node* in the graph can assume a number of known states according to an associated Conditional Probability Distribution (CPD). The CPD of each node defines the probabilistic features, states, of the *node* with respect to the combinations of states of its *parents*. Thus, a Bayesian Belief Network is a graph representation of a joint-probability model. This representation is obtained through the decomposition of the original model into a product of conditional probabilities for each of the comprising *nodes* (Pearl, 1986, p. 359). This definition is expressed formally in the Equation 1 below:

$$P(x_1, ..., x_n) = \prod_i P(x_i|S_i) \tag{1}$$

where:

$x_1, ..., x_n$	are	variables present in the graph
S_i	is	set of *parents* for variable x_i
$P(x_i\|S_i)$	is	conditional probability for variable x_i

The main advantage of this decomposition is that it enables the construction of a BBN to be undertaken in a modular format. Since the overall model does not have to be defined in its entirety it can be built gradually through repeated application of a simple process focusing on each individual variable x_i (Haddawy, 1999). The natural restriction of scale that results from such *localisation* means that the process can be based on human judgement to obtain qualitative relationships perceived to be significant in the context of the problem.

Generally, the process of BBN construction can be split into two phases: identification of variables and elaboration of their interrelationships. This breakdown means that in building a Bayesian network model, one can first focus on specifying the qualitative structure of the domain and then focus on quantifying the influences. When this process is finished, a complete BBN is guaranteed to be a complete specification of a joint probability distribution. This specification is subject, however, to the possible shortcomings in the awareness and judgement regarding the specifics of the domain. As a result, once completed a BBN should be verified and validated and only then can it provide a useful mechanism for probabilistic inference. This, in turn, led to the frequent use of BBNs for modelling domain knowledge and implementation of probabilistic reasoning capabilities in Decision Support Systems (Tsamardinos et al., 2006).

In recent years the concept of Bayesian Belief Networks has played a pivotal role in several attempts undertaken by researchers to represent knowledge about the non-functional qualities of systems. SAABNet or Software Architecture Assessment Belief Network, developed by van Gurp & Bosch (2000), was created with the aim of helping the designer perform qualitative assessment during the architecture design process. SAABNet utilitsed a static Bayesian Belief Network to describe the interrelationships between design qualities.

The interdependencies between various nodes described by Gurp and Bosch are defined within a framework similar to the Factor-Criteria-Metric model described by McCall (1994, p. 1086). Figure 3 shows an example of a single slice of through the SAABNet model. The slice shows all the quality factors influenced directly, or indirectly, by "implementation language".

The belief network structure at the core of SAABNet is comprised of three types of nodes:

Architectural Characteristic nodes (AC) represent various system characteristics and their interdependencies, eg. *component granularity* which in SAABNet is shown as dependent on implementation language and architectural style. The states for the nodes of this type are expressed in terms of specific measurements for features, such as "C++" and "Java" for the node named *implementation_language* or "high" and "low" for the *dynamic_binding* variable.

Quality Criteria nodes (QC) represent composing features of the external quality factors. Due to their higher level of abstraction the QC nodes are assigned states that are qualitative and measurable in the context of the system. For example *vertical_complexity* can assume state "high" and "low" and *testability* can be "good" or "bad".

Quality Factor nodes (QF) represent external quality requirements such as *reusability* and *complexity*. All of the QF nodes are given qualitative states such as "good" and "bad" that are meant to reflect the level of desired qualities that the system is perceived to possess.

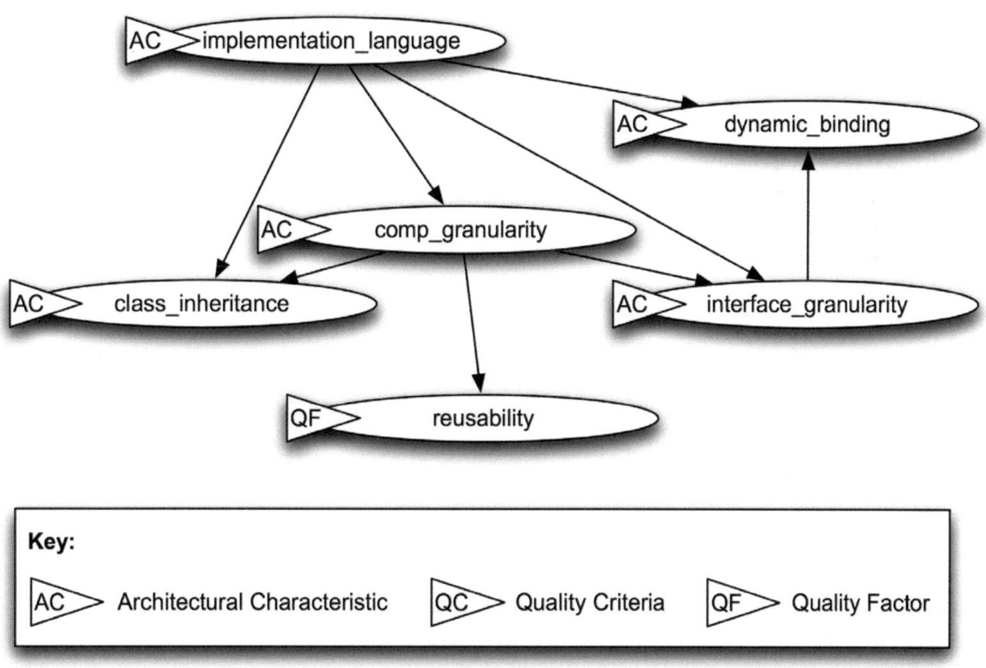

Fig. 3. An example slice thought the SAABNet described by van Gurp (2003, p. 73)

SAABNet was intended by van Gurp & Bosch to assist designers in identifying which system properties were most likely to effect their desired system qualities. In this way it would allow them to select the goals they deemed desirable and, using the properties of the BBN, identifty specific levels, or values, of system parameters needed to meet those goals. For example, should the architect be faced with the task of increasing the *configurability* and *understandability* of the system, then it is possible to use SAABNet to obtain an advisory set of measurements for architectural characteristics. In this case *interface and component granularity* could be identified as a likely characteristic to contribute towards achieving the desired qualities.

Additionally, SAABNet can provide the designer with information about variables that will need special attention during the development process. This special attention is identified, and warranted, by the variables possessing a large number of dependency relationships. Furthermore, by examining the links between various nodes in SAABNet, the designer can identify possible conflicts and trade-offs which exist in the architecture. All of these applications can provide valuable aid to a process of complex system optimisation.

Fundamentally, SAABNet represents an attempt to construct a persistent and reusable belief model of influences between architectural characteristics, their combinations, and quality requirements. Certain parallels can be drawn between SAABNet and the SAU framework proposed by Folmer & Bosch and, in their motivation and focus, the two can be said to be related with SAABNet representing an attempt to take advantage of the representative and reasoning capabilities provided by BBNs.

Naturally, the success of this approach is rooted in the assumption that the non-functional *factors* contributing to the overall quality of the system can be defined in a generally applicable manner. However, finding a general definition for *scalability*, a common quality requirement in many system designs, might omit the nuances specific to the given problem with disastrous consequences.

One way of dealing with the generic nature of qualities is to use qualitative descriptions of measurements, using terms such as 'high'/'low' and 'present'/'absent'. Although somewhat vague in their descriptive power, the chosen names for states of qualitative variables featured in the resultant BBNs are useful in a decision-making process. The contrasting binary nature of their states helps to leverage the possibly incomplete empirical view of the system qualities and the facilitate the process of quality fulfilment (van Gurp, 2003, p. 81).

4.1 System Identification

The use of belief networks with a *static structure* in the context of system analysis, and even in optimisation, has shown some promise. Notably, BBNs can facilitate exploration of the causal relationships that exist between the system structure, the decisions that led to its formulation and the qualities affected by its form. The mathematical properties of BBNs, inherited by the described networks, allow the designer to explore, albeit to a limited extent, the effects of hypothetical design changes on the qualities exhibited by the system.

However, the networks do not address the problem stemming from the variations in the strengths of existing relationships, or the formation of new ones that may take place over time in different problem domains. Furthermore, the assignment of states, nodes and actual topology within the network is uniformly manual and leads to a very high cost that must be incurred during network construction. Due to the heavy involvement of personnel in this network construction phase, the resultant networks may be affected by the failings of individuals and political issues. Nevertheless, the Bayesian approach presents a number of avenues for addressing these issues.

Specifically, it is possible to reduce the residual effects of human errors and political bias by attempting to encode the assumptions held by system stakeholders in a simulation model which is then used to produce output suitable for BBN construction. This is possible because the relationships between a system and its non-functional qualities is one of emergence. That is the features of the system interact to achieve its non-functional properties. Hence, in terms relevant to the implementation of an optimisation process, it can be said that a computer-based system design can serve as both the subject and the objective of optimisation. Its elements are used in evaluation of its fitness and present the context for choice and application of a solution strategy.

Due to these characteristics of a system architecture, the task of optimising it to achieve greater levels of system qualities relies heavily on knowing the specifics of how the qualities exhibited by the system link with the particulars of its features. As a result, the implementation of architectural optimisation guidance can be postulated as a problem comprised of two parts:

System Identification - It is necessary to recognise that any system (computer-based or otherwise) is a result of interactions of the elements composing it. The interactions result from the choice of components and their structural arrangement. Hence, to guide the optimisation process it is first necessary to identify the form for the system that takes into account the choices and interactions that lead to its exhibited qualities.

System Analysis - The guidance methodology must incorporate knowledge that would allow the designer to examine the set of features exhibited by the identified *system form*, determine the factors that have the greatest influence and elicit a strategy for changing the source system architecture.

The problem of system identification can be described as one aimed at deriving a parameter-based mathematical description of the system. This is an example of a *white-box* model. In other words, a model that is a completely transparent construct built using the well-known foundational assumptions regarding system features, which in this case refers to non-functional qualities. Such description can be as simple as a set of polynomial functions or a genetic representation which depends on presence of components or structures within the system.

This potential variety of forms gives rise to a more general system identification question:

> What is the representation of a *systems' design state* suitable for guiding the architectural optimisation of a system that is complex in its composition and interactions with its environment?

Developing the answer to this question requires a *model* that takes into account two major factors. Specifically, what a systems is and how it came into its current arrangement.

According to the definitions provided by Bunge (1979), a system can be represented as a set of non-propositional functions. Such form implies that the functions can not be evaluated to *true* or *false* given their domain alone and successful interpretation requires understanding the relationships that exist between the system and the design context. This, in turn, requires that the process of design itself must somehow be evident in the representation of the system created to answer the question stated above. The problem, however, is that system design is motivated by changes in the environment existing outside the boundaries of the system.

Seen from the optimisation guidance point of view, the design process is a series of decisions taken based on understanding the properties of all relevant artefacts, short-term and strategic goals and constraints regarding potentially applicable changes to the system. Consider a series of decisions such as this to be (Peterka, 1981, p. 9):

$$D_0, D_1, D_2, ..., D_n \tag{2}$$

Each element in the series (2) represents a designer's choice from a number of options. This choice is made under a degree of uncertainty regarding the true potential implications of the decision. At time 0, the very first decision is based on original goals and assumptions and introduces new information regarding system properties. This means that in its most general form a single decision can be decomposed into inputs u and outputs y:

$$D_t = \{u_t, y_t\} \tag{3}$$

In the above, u_t represents a set of values, for example specific choices of components or structure, imposed by the designer at time t onto the system. Whereas y_t represents the results of combining the inputs with the existing system context. It is the outputs set y_t that contains the resultant and emergent qualities of the system. These qualities, just like the outputs themselves, can only have their values influenced indirectly, through preceding design choices. However, the outputs are not restricted to qualities and can include other information such as measurements and forecasts that can serve an important role in future decisions.
Based on a combination of (2) and (3), the overall design process can be represented as a series of sets of inputs and outputs like so:

$$u_1, y_1, u_2, y_2, \ldots, u_{t-1}, y_{t-1}, u_t, y_t, \ldots \tag{4}$$

Given (4), the series of decisions between some two points in time i and j with $i < j$ can be shortened to the simpler form:

$$D_i^j = \{y_j, u_j, D_i^{j-1}\} \tag{5}$$

Within this new representation, if the value for index i is omitted then it can be assumed that D^j represents a set of all decisions from the first to the one taken at j.

$$D^t = \{D_1, D_2, \ldots, D_{t-1}, D_t\} \tag{6}$$

For a designer to make optimal decisions as part of the design process they need to decide in advance which of the available heuristics is optimal in the context of the design problem. To accomplish this, they must be able to forecast, before the input u_{t+1} is applied, what possible values the future elements in the input-output series (4) can take, for at least some of the distinct design choices available at time t. To determine this, the designer needs the conditional probability distribution:

$$p(D_{t+1}^{t+N} | D^t) \tag{7}$$

At this point the chain rule, shown below as (8), can be applied to produce the expanded equation shown in (9).

$$p(x_n, x_{n-1}, \ldots, x_1) = \prod_{k=2}^{n} p(x_k | x_{k-1}, \ldots, x_1) \cdot p(x_1) \tag{8}$$

$$p(D_{t+1}^{t+N} | D^t) = \prod_{i=t+1}^{t+N} p(D_i | D^{i-1}) \tag{9}$$

The equation (9) can be further expanded by making use of the basic relation stated as in (10) to obtain the following substitution for the product expression shown on the right side of (9):

$$p(a,b|c) = p(a|b,c)p(b|c) \tag{10}$$

$$p(D_i|D^{i-1}) = p(y_i, u_i|D^{i-1}) = p(y_i|u_i, D^{i-1}) \cdot p(u_i|D^{i-1}) \tag{11}$$

This, in turn, through substitution into (9), leads to the following expanded form:

$$p(D_{t+1}^{t+N}|D^t) = \prod_{i=t+1}^{t+N} p(y_i|u_i, D^{i-1}) \cdot p(u_i|D^{i-1}) \tag{12}$$

The factors in (12) have the following interpretation. The conditional probability distribution, as shown in (13), describes in general terms the transformation by which the input u_i is determined on the basis of the known past history of the decision process.

$$p(u_i|D^{i-1}) \tag{13}$$

The other factor, i.e. the set of conditional probability distributions, shown in (14), describe how the output y_i is a result of past history of the decision process D^{i-1} and the last choice of inputs u_i at each instance i.

$$p(y_i|u_i, D^{i-1}) \tag{14}$$

The set of conditional probability distributions shown in (14) is, in effect, the general description of the system as a compound result of the decisions which lead to its composition at point i. Hence, the requirement for the definition of a model for *systems' design state* posed earlier in this section can be fulfilled by *finding a set of conditional probability distributions (14) over a finite set of parameters for the time period required.* In effect, the *conditional probability distribution* links the elements of the proposed change to the fulfilment of design goals thus aiding the designer in devising an appropriate design strategy under uncertainty.

Importantly, the product of the conditional probability distribution (14) featured in (12) resembles closely the definition of the underlying model for Bayesian Belief Network presented in Equation 1. The formulae and nomenclature introduced in this section provide positive assessment as to the viability of the guidance approach based on use of BBNs for quality modelling. However, the analysis of existing BBN approaches discussed at the beginning of this section highlights the importance of the network construction process to the viability of BBNs as guidance tools supporting an architectural optimisation process. Consequently, the features of the BBNs relevant to their construction need to be examined.

As mentioned before, the key feature of BBNs is the *localisation* they afford, in other words, the fact that they provide a method for decomposing a probability distribution into a set of local distributions. The semantics of *independence* associated with the network topology specifies how to combine these local distributions to obtain the complete joint probability distribution (Equation 1) over all the input variables represented by the nodes in the network.

This has two important consequences:

1. When specifying a joint probability distribution as a table of values, the required a number of values increases exponentially with the number of variables. In systems in which interactions among the input variables are sparse, Bayesian networks drastically reduce the number of values required to be considered simultaneously.

2. The separation of qualitative representation of the influences between variables from the numeric quantification of the strengths of the influences has a significant advantage for knowledge engineering.

These structural and semantic features of BBNs are conducive to their construction and effectiveness, even in the presence of limited about of data. Specifically, the advantage mentioned in point 2 above is the greatest contributing factor to the simplicity of the BBN construction process. The process itself is broken down into two parts.
The first phase involves careful consideration of relevant concepts in the domain of the problem that is being modelled. In the domain of system optimisation this amounts to encoding of variations in inputs affected by the designer. These variations must also be accompanied by a recording of corresponding changes in the outputs that emerge as a result of the designer's decision. Upon completion of variable identification, the next step in BBN construction involves the assignment of dependencies between the variables. This assignment is accompanied by conditional probability information. After these two phases the resultant network is tested to verify that it matches the existing knowledge of the modelled domain.
Furthermore, the focus on domain-specific decisions, afforded by the use of Bayesian network technology, can, when applied within the context of system design optimisation, be interpreted in a dual way. The first one, as evident from work done by van Gurp, involves building one large network to represent the *system design state* in the most complete way possible. This is a complex and arduous task which incurs heavy validation and verification requirements for its outcomes. Even when this is completed, there is no guarantee that the resultant BBN structure would be able to withstand the test of time as the technological and business elements of the context evolve. For large systems, this is impracticable from an engineering standpoint.
Instead, the problem of BBN construction can be solved by adopting one of the available network discovery algorithms. These algorithms have been created to allow the structure of a Bayesian network to be learned from a sufficiently large set of observational data (Tsamardinos et al., 2003). One such approach is the Min-Max Hill-Climbing (MMHC) algorithm (Tsamardinos et al., 2006). The MMHC algorithm was chosen for implementation in the proposed guidance mechanism due to its effectiveness and simplicity of application.
The inclusion of the MMHC algorithm as part of the proposed guidance methodology introduces an additional problem concerning the availability of observational data. Within the context of a complex design problem such data may need to include specific values for variables representing quality measurements correlated with specific system and design states. Obtaining such data may not always be possible even if is the designer has access to variety of sources such as interviews, historical records, design documentation and system telemetry. To address this, the proposed methodology includes multi-paradigm simulation described in greater detail in Section 4.2.1.
However, in complex systems it is possible that different unknown dynamic relationships can lead to the same topographic configuration of the resultant Bayesian Belief Network. Hence the application of the proposed method takes place in a context where two distinct elements are present: the reality of the situation being analysed and a set of possible candidate models which can be produced from the available data set. The problem of ensuring that the identified model is the actual, or a very close, depiction of the qualitative relationships present in the system can only be handled by submitting it to the evaluation by the system's stakeholders and the designer.

4.2 Guidance Components and Model

The proposed guidance methodology relies on two supporting concepts: multi-paradigm, or hybrid simulation (Section 4.2.1), and automatic discovery of causal relationships between a system's characteristics and its qualities (Section 4.2.2). The role of the former is to provide the designer with a facility capable of modelling the current, as well as possible, structural and behavioural characteristics of the system. The latter of the two is necessary to ensure that data collected during the successive runs of the modelling and exploration phase are aggregated into a meaningful representation of causalities. In this case a BBN is used to understand the causal relationships and to identify the drivers behind the specific non-functional qualities of the system.

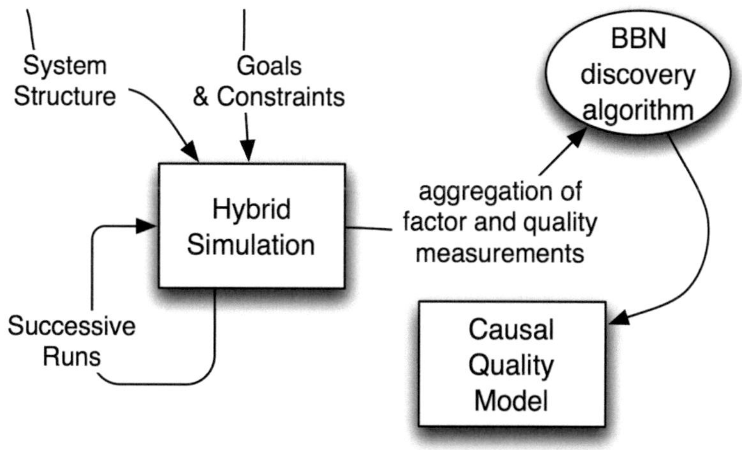

Fig. 4. The formulation process resulting in the production of guidance data.

The interaction between elements described above is depicted in Figure 4. Two elements (*Hybrid Simulation* and *BBN Discovery Algorithm*) are employed to produce a, best-effort, comprehensive understanding of the system with respect to the requirements and constraints imposed upon the designer. This interaction of composing elements is proposed as one that would help to ensure that no factor is omitted from exploration and analysis of system's design state. The exception being where the designer explicitly wishes to ignore certain factors. Within such an approach it is necessary to ensure that the simulation of the system is flexible enough to represent the ramifications of possibly very large and complex changes both on system structure and on the outside elements such as users or external processes.

4.2.1 Hybrid Simulation

The practice of exploring quality characteristics of Computer-Based Systems through construction and execution of various type of simulations is well established, as was shown in the overview of research presented in Section 2. However, the diversity of specific features that are combined to attain individual qualities in any given system demands creation of a portfolio of simulations, each targeting a single or a group of related qualities. The process of creating and maintaining such a portfolio becomes a task in itself. A task that is made difficult by issues of maintaining consistency throughout the various simulation models comprising the portfolio and creating ways to combine their outputs to construct a context describing the non-functional qualities of the system and, as a result, provide effective analysis and decision support.

Therefore, in the context of a design optimisation guidance that aims to improve a variety of possible interrelated qualities exhibited by a complex CBS, it is desirable for the designer to be able to construct a single simulation model. Doing so would address the problems of maintaining consistency across multiple models as well as difficulties in combining their outputs. However, to be successful, this single, all-encompassing model must be rigorous enough to provide a faithful representation of a complex system while remaining flexible enough to facilitate exploration of relationships between *emergent* system qualities and the effects of possible design changes.

In order to achieve this flexibility the simulation must be able to deal with concepts from three major paradigms which currently dominate the field of simulation modelling (Borshchev & Filippov, 2004):

- System Dynamics (SD) - Required to be able to represent the effects of policy introduction or modification at the highest levels of abstraction, as well as the analysis of trends and other system *properties-of-the-whole* (Forrester, 1991, p. 22).
- Discrete Events (DE) - Required to understand the issues associated with utilisation of various resources available to the system, as well as the effects of various scheduling decisions.
- Agent-Based (AB) - This paradigm allows the simulation of elements which can only be meaningfully represented as active objects with individual, purposeful behaviour (Borshchev, 2005, p. 8).

Borshchev et al. (Borshchev, 2005; Borshchev & Filippov, 2004) has proposed combining the modelling paradigms mentioned above. The proposed combination places a special emphasis on the use of Agent-Based modelling to accentuate both its flexibility and pragmatism in situations where complete information about the system may be unavailable. Additionally, the AB approach facilitates the exploration of the *emergence* of global system properties. It achieves this by examining the interactions of various system elements over time, a characteristic which can be successfully employed to support the process of causal discovery described in Section 4.2.2.

The multi-paradigm, hybrid simulation, approach (Borshchev & Filippov, 2004) possesses both abstract characteristics (system dynamics, discrete events) and pragmatic behavioural properties (agents). Combined together these approaches provide a modelling paradigm that has been shown to be flexible and powerful enough to address the simulation and optimisation needs of problems ranging from supply chain management (Almeder & Preusser, 2007) to enterprise IT cost analysis (Popkov et al., 2006).

The overall aim of the proposed optimisation guidance method is to provide a recommendation on a list of changes considered by the designer. To achieve this, it relies on a simulation model of the original system. This model has to be robust enough to allow exploration of pre-defined usage scenarios targeted, for example, at exploring the *scalability* of the system.

It is possible that multiple runs of the simulation will be required in order to obtain a better understanding of the causal relationships between various observed system factors and its overall qualities. Eventually, the framework will be able to output a large set of aggregated data combining information on the effects of variations in system factors on the system qualities measurements. This resultant set of data will serve as an input for the automatic casual discovery algorithm.

4.2.2 Bayesian Belief Network Discovery

In Figure 4 the causal discovery mechanism of the proposed design optimisation guidance method is represented by the "BBN discovery algorithm" entity. This refers to an automatic way of constructing the topology of a Bayesian Belief Network from a large set of observational data.

The BBN elucidated from the simulation output data is used as a Causal Quality Model of the system being subjected to design optimisation. In this capacity, the function of the resultant BBN with respect to the designer is somewhat similar to that of a Decision Support System. Its aim is to show the designer which factors within the system have the greatest measure of effect on the system qualities specified as acquisition concerns. Further, it aims to show how these, and other, qualities may be affected by possible design changes.

To achieve this, the BBN encapsulates the main inference relationships present in the system design with respect to the emergent qualities. These qualities, as previously discussed, may serve as the goals or the constraints of the overall optimisation process. Semantically, the BBN used to provide optimisation guidance is built with the same underlying principal node types as used by van Gurp (2003) (Section 4), which, in turn, stem from the Factor-Criteria-Metric structures defined by McCall (1994). Specifically, there are three distinct groups of nodes:

simple characteristics These nodes describe the observed characteristics of the system known to be relevant to the goals of the system optimisation.

complex characteristics Members of this group represent the knowledge of how the simple characteristics combine into more complex ones which in turn produce cumulative effect onto the system qualities.

system qualities Final layer of nodes which is created to compose the inference structure about the non-functional system characteristics pursued by the optimisation process.

It is the explicit responsibility of the designer to define which specific features of the simulation model should be processed and included as part of the data set that will be used as an input to the BBN learning algorithm. However, faced with such a task the designer may find it difficult to decide which of the features are relevant. Nevertheless, some direction can be provided in this difficult task. Specifically, the BBN that is likely to produce the best guidance should include nodes from two major groups: *principal* and *auxiliary*.

The *principal* network nodes are necessary to represent the core inference structure required to control the direction of the optimisation process. This group of nodes should include characteristics of the system that can change based on the available selection of design modification heuristics and the system qualities stated as goals of the optimisation. For example, if the designer is required to optimise scalability of a given system then the network should contain as many nodes as possible describing the issues affecting system scalability in that context.

A typical *auxiliary* network node should be determined based on the principles of a narrow quality focus. The *auxiliary* nodes are necessary primarily to ensure that the constraints of the optimisation are properly addressed within the process. Depending on the conditions imposed by the problem these constraints may be classified as *hard* or *soft*. Hard constraints ensure that the changes to the system design during optimisation do not dramatically reduce a known quality of the system. Whereas soft constraints are there to show what effects a certain course of action will have on qualities not expressly sought after as optimisation goals.

Since each node in a BBN represents a variable, which may assume a value according to the associated conditional probability distribution, it is not possible to express general relationships within the system without enumerating all the potential states every node may assume

in advance. This may make the task of creating *auxiliary* nodes difficult as it will most likely be based upon static domain expertise. However, the BBN discovery process employes a construct called a Markov Blanket (MB) (Tsamardinos et al., 2003), which helps to remove attributes that do not affect the nodes of interest and, therefore, are unnecessary.
Discovering the structure of a Bayesian network from a set of data has emerged as a major focus for research. It has been of particular interest since when source data can be represented by a time series, the edges in the graph of a Bayesian network can be used to infer likely causal relations (Spirtes et al., 2000). The actual algorithm used to *discover* the presence and orientation of links between the variables featured in the BBN at the core of the guidance framework is the Max-Min Hill-Climbing (MMHC) algorithm developed by Tsamardinos et al. (2006).
This MMHC algorithm draws upon a variety of ideas from search-and-score and local learning techniques, such as Markov Blanket discovery, to construct the skeleton of a Bayesian network corresponding to the source data and then perform a Bayesian-scoring greedy hill-climbing search to orient the edges within the graph. One of the most attractive features of the MMHC algorithm is that it has been proven to work well with the highly dimensional data sets showing that it can deal well with situations where non-functional qualities are a result of interactions between a large number of factors. Furthermore, research has shown (Tsamardinos et al., 2006, p. 30) this algorithm exhibits good scalability and accuracy when applied to learn converging sparse networks.
However, the Direct Acyclical Graph (DAG) of the BBN produced as a result of MMHC application does not represent the complete Causal Quality Model. Additional calculations must be performed to establish the probability distributions in the network. Specifically, to determine for each variable X that has a set of states S_x the probability of X being in the state $s \in S_x$ for each combination of states of its parents P_x. In order to perform this calculation an additional algorithm was developed to combine the source data from the simulation with the structural information obtained from the application of MMHC algorithm. The Section 4.3 provides a description of an example showing application of the proposed methodology to a small-scale decision scenario.

4.2.3 Guidance Methodology Synthesis

The ultimate goal of combining the elements depicted in Figure 4 is to obtain the Causal Quality Model (CQM). What the CQM effectively represents is a new, analytical view of the system that should be constructed and used in tandem with the traditionally accepted views representing static structure, processes, data and physical deployment.
Hence, the synthesis of this view should rely on the available architectural, environmental and contextual information. Specifically, to construct this view the following information is needed:

1. A set of goals and constraints: obtained from functional and non-functional requirements, business drivers, future plans and budgeting both in terms of time and money.
2. Information about system structure combined with elements deemed relevant in view of goals and constrains, these elements may include back-office workflows and customer behaviour. Simulation time can also be a very important factor and should be included as a special consideration based on the goals of optimisation.
3. Feature and metric selection must be made to identify nodes for the BBN (based on the feature and goal information), the actual structure of the DAG will be elicited using an algorithm.

4. Possible variation in characteristics of the system being optimised can also be added as part of the simulation. In this case the successive runs of simulation can be used to explore how these degrees of freedom affect the goals specified in 2.

The next section (Section 4.3) explores the issues of implementation in the context of a simple problem in order to focus on the specific of the guidance methodology and not the complexities of the optimisation problem.

4.3 Example

One of the major critiques that can be made about the examples of the BBN applications described in Section 4 is that, in both cases, the assignment of probability values for nodes affected by multiple parents is based on conglomeration of a priori knowledge held by domain experts. Although networks constructed using such information have been shown to be useful analytical tools when applied to system architectures (Trendowicz & Punter, 2003; van Gurp & Bosch, 2000), they are nonetheless exposed to the possibility of errors that can be introduced when domain experts from multiple fields are asked to quantify the combined influences effecting a specific node. The example presented in this section aims to explore the way by which the proposed architectural optimisation guidance methodology aims to address this issue in its use of Bayesian networks.
This example itself focuses on how specific knowledge about system qualities such as *Reliability* and *Performance* (shown in Table 2) can be used in combination with known details of user behaviour to reason about possible avenues for design optimisation aiming to increase the overall *Usability* of the system.

quality	P(good)	P(bad)
reliability	0.99	0.01
performance	0.86	0.14

Table 2. Likelihood values for Reliability and Performance

To this end, the process of model construction was started by creating a state machine describing an individual system user as an Agent[1]. The resultant User Agent is shown in Figure 5. This simplified version is based on data accumulated in system logs that track some aspects of behaviour exhibited by individual system users. As such this data represents only the most basic, meaningful, use-case scenario for the services offered by the system. In short, the agent represents a typical user of the search and reservation services.
Several things should be noted about the described state transitions. Firstly, transitions leading to the "Leave" state on the diagram represent instances of agents discontinuing the use of service due to poor reliability. The execution of these transitions leading to this node is controlled by a set of functions which combine the probability value associated with *Reliability* (Table 2) and the current state of the agent. Specifically, the "Leave" state can be reached from three instances of a choice function, or *branch element* node, marked by the letter 'B'. When reached, the choice functions uses probability information to determine whether the specific instance of User Agent experiences unreliable system behaviour and if, in fact, such an occurrence is observed the function immediately activates the transition to the "Leave" state. At

[1] All hybrid models used in this and other chapters of this thesis were done using AnyLogic^{TM} modelling tool (`http://www.xjtek.com/`).

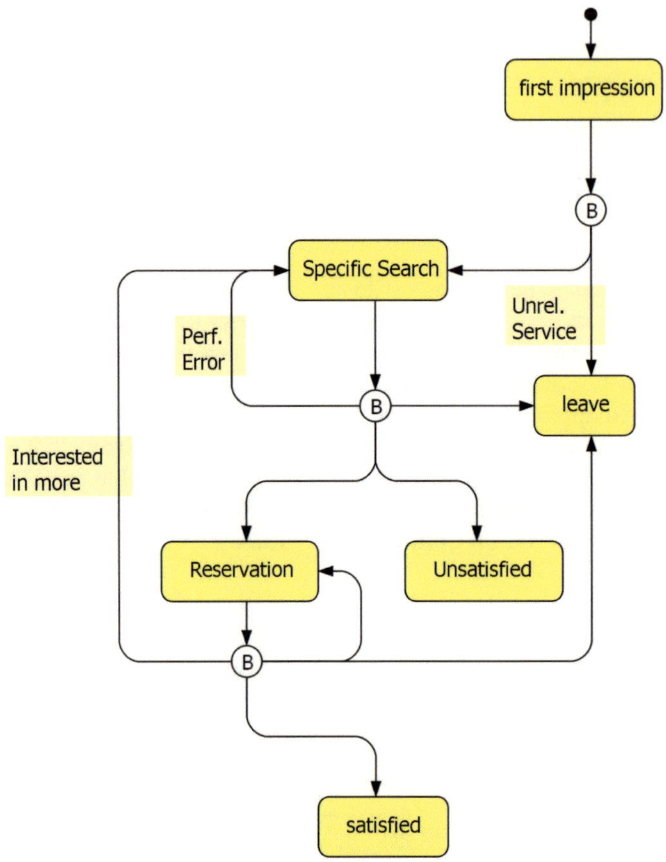

Fig. 5. Behavior Description for Useability Agent

this point the instance of the User Agent terminates operation and the system can be said to *have been unusable due to 'bad' Reliability*.

Likewise, the same choice functions also control the activation of arcs leading back to the originating node. These arcs represent retries due to poor performance. In the case of observing 'bad' *Performance* the User Agent will attempt to retry the current operation, the number of times that this will be attempted depends on the previous state of the agent. This is done with the intention of reflecting the observation that users attempting to make a reservation usually attempt more retries than the users performing searches. However, even if a given instance of the User Agent observes both 'good' *Performance* and *Reliability* it may still end up in the "Unsatisfied" state, which also exists to represent the group of system users who do not progress towards reservation.

In the context of this simple model, the measure of system *Usability* was determined by the likelihood of a User Agent reaching "Satisfied" state. To obtain observations for this measure a simulation of the behaviour exhibited by a population of 1000 User Agents was created and executed until all agents reach one of the terminating states: "Leave", "Unsatisfied" or "Satisfied". After several runs of the simulation were completed, the following numeric results were produced:

- R_t number of repeat requests due to timeouts = 222
- R_s number successful requests = 1332
- R_c number total *countable* requests = 1387
- **useability at good performance** $(1 - R_t/R_c)$ **= 0.84**
- U_f failures due to the unreliable behaviour of the system = 41
- U_t number of User agents = 1000
- **useability at good reliability** $(1 - U_f/U_t)$ **= 0.959**

A small clarification must be made to the figures above concerning *countable requests*. Due to the nature of the User agent behaviour some requests made by a specific User, even if successful, may still lead to the User abandoning the system and terminating operation by entering the "Leave" state.
The values listed above can be used to calculate the combined likelihood values, which describe the effect various states of *Reliability* and *Performance* have on *Usability* (Table 3). The values in Table 3 have been calculated by multiplying the values shown in the list of *Usability* values shown in the list above. This conditional probability information is shown in its marginalised form in the BBN depicting the structural relationships between qualities shown in Figure 6. Naturally, the specific values of 'good' for each of the qualities in question corresponds to a range of actual values for the metrics contributing to the evaluation of the corresponding quality.

reliability	good		bad	
performance	good	bad	good	bad
useable	0.80556	0.15344	0.03444	0.00659
not	0.19444	0.84656	0.96556	0.99341

Table 3. Likelihood values for Usability

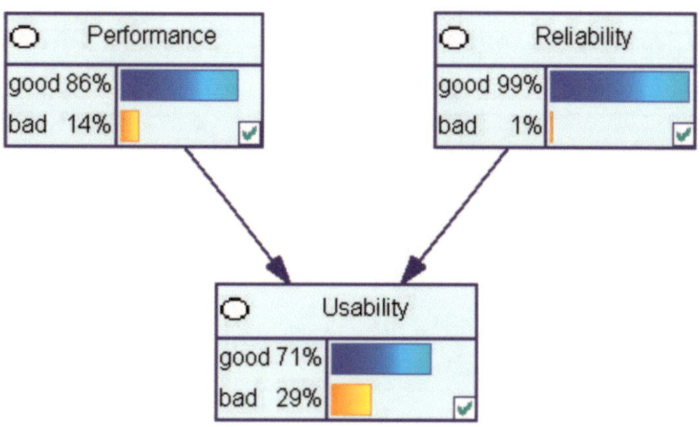

Fig. 6. The example BBN composed of *Performance, Reliability* and *Usability* nodes.

In this example only the Agent Based simulation paradigm was used to ensure simplicity and verifiability of simulation results by inspection. However, even at this level, the example shows that the knowledge of the specific system and context characteristics can be used to obtain, via simulation, numerical representation of the influences exerted by these characteristics onto the emergent system qualities. Specifically, the information about user behaviour, when combined with known system behaviour, reliability and performance, contributed towards the creation of the model for the emergent quality of usablility.

4.4 Guidance Data Interpretation and Analysis

The resultant Causal Quality Model can be used for analysis in three distinct ways:

1. The structure it exhibits can be analysed.
2. The conditional probabilities at each node can be used to explore the strength of inter-relationships between system qualities.
3. The underlying joint probability function can be used to propagate some *observational evidence* through the network.

The uses of these techniques individually, or in combination, can be classified into two groups: *diagnostic* and *predictive*. Although the belief network examined in the example above (Section 4.3) is simple in composition, this is merely a result of the limited amount of input information used to create the hybrid simulation that provided the learning data for the MMHC algorithm. It is therefore possible that, in the context of an optimisation problem applied to a complex system, the designer will have to draw upon a number of disparate information sources describing both the technical characteristics and the operational qualities of the system.
Given this potential breadth of information, a Causal Quality Model, like the one depicted in Figure 6, could be used as a *diagnostic* tool to uncover the presence and strength of relationships between qualities and their contributing factors. Knowledge regarding the existence and effects of these relationships can then be exploited to guide the choice of design change heuristics to either remove or strengthen some relationships or increase the likelihood of higher measurement for some qualities of interest.

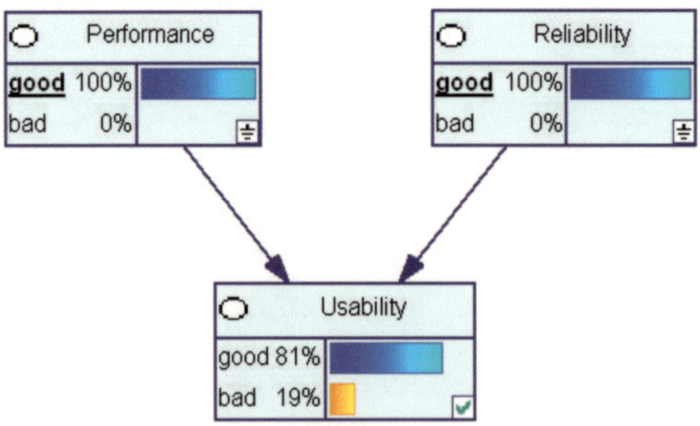

Fig. 7. The example BBN composed of *Performance*, *Reliability* and *Usability* nodes with evidence set for *Performance* and *Reliability* nodes.

As mentioned above, a Causal Quality Model can be used in a *predictive* capacity. In order to do this, the designer must provide *evidence* for one or more nodes of the BNN. In other

words, set the likelihood value for a specific state to 100%. This can be done in two ways: by manipulating either the parent or the child nodes in the causal interaction. A version of the former, as applied to the example BBN (Figure 6), is shown in Figure 7, while the latter is depicted in Figure 8.
In a case when the evidence regarding the states of parent nodes in a causal interaction is provided the Causal Quality Model can be used to determine the *likely* observed state of the child nodes. The Figure 7 shows the likelihood of 81% for the "Usability" variable to be in a state marked as 'good' when the variables for "Performance" and "Reliability" are observed to be within the bounds of what is considered 'good' for both of those qualities. This information can serve as a starting point for predicting changes that can advance the qualitative boundaries of the system.

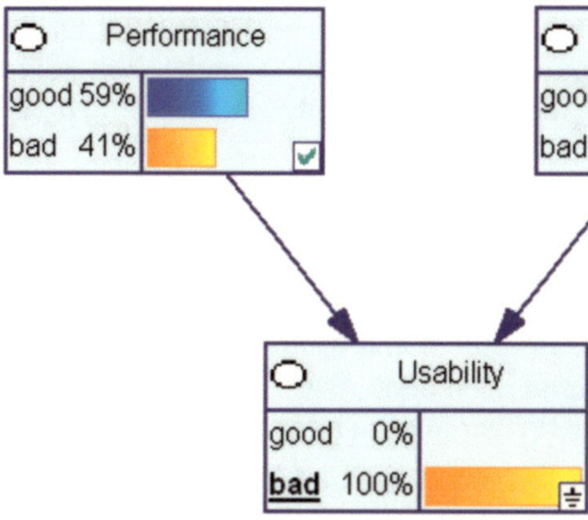

Fig. 8. The example BBN composed of *Performance*, *Reliability* and *Usability* nodes with evidence set for *Usability* node and calculated *posterior* probabilities.

However, when it comes to consideration of system qualities it may be more interesting to consider the negative case. Figure 8 depicts the posterior probabilities of observations for system's "Performance" and "Reliability" given evidence that "Usability" is observed to be 'bad'. It can be seen that, for low values of "Usability", while measure for "Performance" drops considerably, the measure for "Reliability" shows little change: only 2%. Based on these results the designer can conclude that under the current configuration, and based on known user behaviour patterns, the "Usability" of the system is highly sensitive to the "Reliability" of the system. The combined information of the original BBN and the results of setting evidence shown in Figures 7 and 8 give the designer a new perspective on the possible direction of potential design modifications.
Overall, the *diagnostic* and *predictive* uses for the Causal Quality Model can provide a valuable insight into the aspects of the problem and their relationship to the design change options made available to the designer. As a result, it is possible for the designer to understand clearly which system features contribute to the overall qualities of the system. Furthermore, the designer is able to identify which changes to those system features are most likely to render beneficial improvements in the system qualities, the goals of the optimisation process. This clarity also provides the designer with a new from of system representation, one which may prove useful both in decision making and as a communication tool capable of relating technical concepts to the non-technical stakeholders of the system.

5. Conclusion and Future work

The design optimisation guidance methodology aims to aid the designer in directing the overall system optimisation process. One of the major difficulties of providing such guidance is the nature by which this optimisation process is advanced. Specifically, the designer is essentially incapable of affecting the qualities directly. Instead, he or she is forced to consider a set of choices targeting the specific features of the design contributing towards achievement of desirable system qualities. As a result, since a single choice could affect multiple qualities, this introduces a requirement for guidance to provide the designer with understanding of the causal relationships existing in the system.

Achieving this involves the study of assumptions held by the designer and other stakeholders, the relevance of existing knowledge and the accuracy of possible predictions. The fusion of simulation modelling and the BBNs can serve as tool of such study as its aim is to provide a tangible link between the way in which the system is structured and its observed levels of quality. Additionally, by combining the hybrid simulation model with BBN discovery algorithm we managed to obtain a much more repeatable output that is validated against encoded assumptions and is less prone to human error.

However, the method's success relies greatly on validity of the model and clarity of the BBN representation. To this end we have found that the simulation model should be built in an incremental manner using a variety of information sources and explicit encoding of assumptions help by the participants. Consequently, the extracted BBN plays a dual role both as a guidance tool and a model verification tool as the conditional probabilities it displays can quickly highlight inconsistencies within the model.

The results presented herein warrant further investigation along four major axis:

- further research is needed to help the designer with choice of quality factors and criteria that contribute to the nodes of the CQM;
- a taxonomy of simulation primitives needs to be developed to aid the designer with construction of hybrid simulation models;
- additional research is needed to examine how various BBN discovery algorithms perform on the types of simulation output produced by models of systems from different domains;
- studies should be conducted into the various stochastic methods of optimisation such as Cross-Entropy (Caserata & Nodar, 2005) that could be implemented based on the outcomes of BBN use for qualitative applied over a succession of system development cycles.

Finally, the development of this approach to guidance should be used to construct a fully fledged decision support and optimisation framework described in Section 3.

6. References

Almeder, C. & Preusser, M. (2007). A hybrid simulation optimization approach for supply chains, *Proceedings EUROSIM 2007*, Ljubljana, Slovenia.

Borshchev, A. (2005). System dynamics and applied agent based modeling, *In Proceedings of International System Dynamics Conference*, Boston, MA.

Borshchev, A. & Filippov, A. (2004). From system dynamics and discrete event to practical agent based modeling: Reasons, techniques, tools, *In Proceedings of The 22nd International Conference of the System Dynamics Society*, Oxford, England.

Bosch, J. & Bengtsson, P.-O. (2001). Assessing optimal software architecture maintainability, *Proceedings of the Fifth European Conference on Software Maintenance and Reengineering*, IEEE Computer Society, p. 168.

Bucci, G. & Maio, D. (Sep 1982). Merging performance and cost-benefit analysis in computer system evaluation, *IEEE Computer* **15**(9): 23–31.

Bucci, G. & Streeter, D. N. (1979). A methodology for the design of distributed information systems, *Commun. ACM* **22**(4): 233–245.

Bunge, M. (1979). *Treatise on Basic Philosophy: Volume 4: Ontology II: A World of Systems*, 1st ed. edn, Springer.

Caserata, M. & Nodar, M. C. (2005). A Cross-Entropy Based Algorithm for Combinatorial Optimization Problems, *European Journal of Operational Research* .

Chen, H.-M., Kazman, R. & Garg, A. (2005). Bitam: An engineering-principled method for managing misalignments between business and it architectures, *Science of Computer Programming* **57**(1): 5–26.

Coit, D. & Konak, A. (2006). Multiple weighted objectives heuristic for the redundancy allocation problem, *Reliability, IEEE Transactions on* **55**(3): 551–558.

Coit, D. W. (2001). Cold-standby redundancy optimization for nonrepairable systems, *IIE Transactions* **33**(6): 471–478.
URL: *http://www.springerlink.com/content/q96wkpuqme2fe3ju*

Coit, D. W. & Smith, A. E. (1996). Reliability optimization of series-parallel systems using a genetic algorithm, *IEEE Transactions on Reliability* **45**: 254–260.

Denford, M., Leaney, J. & O'Neill, T. (2004). Non-functional refinement of computer based systems architecture, *Engineering of Computer-Based Systems, 2004. Proceedings. 11th IEEE International Conference and Workshop on the*.

Diaconescu, A. & Murphy, J. (2005). Automating the performance management of component-based enterprise systems through the use of redundancy, *ASE*, pp. 44–53.

Folmer, E. & Bosch, J. (2005). Case studies on analyzing software architectures for usability, *EUROMICRO '05: Proceedings of the 31st EUROMICRO Conference on Software Engineering and Advanced Applications*, IEEE Computer Society, Washington, DC, USA, pp. 206–213.

Forrester, J. W. (1991). *The Systemic Basis of Policy Making in the 1990s*, Sloan School of Management, MIT, Boston, MA, chapter System Dynamics and the Lessons of 35 Years.

Gokhale, S. S. (2004). Cost constrained reliabilty maximization of software systems, *Reliability and Maintainability, 2004 Annual Symposium - RAMS* pp. 195–200.

Grunske, L. (2003). Transformational patterns for the improvement of safety properties in architectural specification, *Proceedings of The 2nd Nordic Conference on Pattern Languages of Programs (VikingPLoP 03)*.

Grunske, L. (2006). Identifying "good" architectural design alternatives with multi-objective optimization strategies, *ICSE '06: Proceedings of the 28th international conference on Software engineering*, ACM, New York, NY, USA, pp. 849–852.

Grunske, L., Geiger, L., Zündorf, A., Van Eetvelde, N., Gorp, P. V. & Varro, D. (2005). *Model-driven Software Development - Volume II of Research and Practice in Software Engineering*, Springer, chapter Using Graph Transformation for Practical Model Driven Software Engineering.

Haddawy, P. (1999). An overview of some recent developments in bayesian problem solving techniques, *AI Magazine Special Issue on Uncertainty in AI*.

Kazman, R., Barbacci, M., Klein, M., Carrière, S. J. & Woods, S. G. (1999). Experience with performing architecture tradeoff analysis, *ICSE '99: Proceedings of the 21st international conference on Software engineering*, IEEE Computer Society Press, Los Alamitos, CA, USA, pp. 54–63.

Leaney, J., Denford, M. & O'Neill, T. (2004). Enabling optimisation in the design of complex computer based systems, *ECBS '04: Proceedings of the 11th IEEE International Conference and Workshop on Engineering of Computer-Based Systems*, IEEE Computer Society, Washington, DC, USA, p. 69.

Maxwell, C. (2007). *Representing Heuristics for Architectural Optimisation*, PhD thesis, University of Technology, Sydney, Faculty of Information Technology.

Maxwell, C., O'Neill, T. & Leaney, J. (2006). A framework for understanding heuristics in architectural optimisation, *13th Annual IEEE International Conference and Workshop on the Engineering of Computer Based Systems (ECBS 2006)*, Potsdam, Germany.

Maxwell, C., Parakhine, A., Denford, M., Leaney, J. & O'Neill, T. (2005). Heuristic-based architecture generation for complex computer systems, *12th Annual IEEE International Conference and Workshop on the Engineering of Computer Based Systems (ECBS 2005)*.

McCall, J. A. (1994). *Encyclopedia of Software Engineering*, Vol. 2 O-Z, John Wiley & Sons, New York, chapter Quality Factors, pp. 1085–1089.

MDA Guide (2003). omg/2003-06-01. Model-Driven Architecture Guide, Version 1.0.1.

Parakhine, A. (2009). *Architectural Optimisation Guidance of Complex Computer-Based Systems*, PhD thesis, University of Technology, Sydney, Faculty of Information Technology.

Pearl, J. (1986). A constraint-propagation approach to probabilistic reasoning, *Uncertainty in Artificial Intelligence*, Elsevier Science, North-Holland, Amsterdam, pp. 357–369.

Peterka, V. (1981). *Trends and Progress in System Identification*, Pergamon Press, chapter Bayesian Approach to System Identification, pp. 239–304.

Popkov, T., Karpov, Y. & Garifullin, M. (2006). Using Simulation Modeling for IT Cost Analysis, *Technical report*, Distributed Computing and Network Department, St.Petersburg State Technical University, St.Petersburg, Russia.

Rowe, D., Leaney, J. & Lowe, D. (1998). Defining systems evolvability - a taxonomy of change, *International Conference and Workshop: Engineering of Computer-Based Systems (ECBS '98)*.

Sharma, V. S. & Trivedi, K. S. (2005). Architecture based analysis of performance, reliability and security of software systems, *WOSP '05: Proceedings of the 5th international workshop on Software and performance*, ACM Press, New York, NY, USA, pp. 217–227.

Spirtes, P., Glymour, C. & Scheines, R. (2000). *Causation, Prediction, and Search*, The MIT Press.

Trendowicz, A. & Punter, T. (2003). Applying bayesian belief networks for early software quality modeling, *IESE-Report No. 117.03/E* . Fraunhofer IESE.

Tsamardinos, I., Aliferis, C. F. & Statnikov, A. (2003). Algorithms for large scale markov blanket discovery, *The 16th International FLAIRS Conference*, St. Augustine, Florida, USA.

Tsamardinos, I., Brown, L. E. & Aliferis, C. F. (2006). The max-min hill-climbing bayesian network structure learning algorithm, *Machine Learning* **65**(1): 31–78.

van Gurp, J. (2003). *On The Design & Preservation of Software Systems*, PhD thesis, Rijksuniversiteit Groningen.

van Gurp, J. & Bosch, J. (2000). Automating software architecture assessment, *In Proceedings of Nordic Workshop on Programming and Software Development Environment Research 2000 (NWPER 2000)*, Lillehammer, Norway.

8

Method of multi-source testing information fusion based on bayesian networks

Tang Xue-mei and Hu Zheng-dong
National Key Laboratory of Science&Technology on Complex Systems Simulation-Π,
China

1. Introduction

Large Equipment System Acceptance (such as reliability and accuracy, etc.) is a complicated system engineering. In order to check the performance of large equipment, multi-perspective and more-approach tests are adopted to get a variety of test information. These information are related to many aspects, such as the information in different phases of design, development, pilot, production and application phase, test information of products in different levels (systems, subsystems, components) and the history information of test-related products. These different but still interrelated information have brought a great more reference to analysis and assessment of large-scale equipment, and meanwhile those uncertain information would brought more risk to the assessment of decision-making. How to integrate these information of multiple sources effectively to make an objective evaluation on the performance of large-scale equipments has been a great challenge to the engineering researchers.

For example, in assessing the reliability of weapons systems, the cost of system-level testing is often too much which limits the number of test times. In that condition the test information of various equipment and subsystems are urged to be fully utilized. Similarly, in order to improve the practical accuracy of INS (inertial navigation system), a series of checking from the test phase to application phase such as ground calibration tests, vehicle-loaded tests, aircraft-loaded tests and missile-loaded tests, are requisite and the outcome should function to the error coefficients estimation. In the circumstance of sufficient test data, the classical approach of comprehensive assessment to reliability has been widely used; while in contrast when test data are insufficient and moreover they present multi-stage and multi-level properties, the classical approach is in effectiveness challenge. With the development of computer technology and improvement of Bayes methods, especially the emergence of MCMC (Markov chain Monte Carlo) methods and WinBUGS (Bayesian inference Using Gibbs Sampling) software, the Bayesian network is more and more popularized in the application of multi-source information fusion [1~3].

The Bayesian network is a causal network, which is used as an inference engine for the calculation of beliefs or probability of events given the observations of other events in the same network. It does not only make good use of model information and sample data, but also integrates the unknown parameters in the overall distribution of information. Besides, it

has overcome defects of traditional static model being incapable of handling emergencies. These flexible, easy to adapt to external changes features can make up for the shortage of insufficient poor quality samples brought to traditional statistical methods, so it is more suitable for prediction and reality reveal to models. The most attractive feature of the Bayesian network is given an observation for one node, the statistical information for all nodes would be updated. This feature is very valuable in the context of model validation, when experimental observations may not be available on the final model output but may be available on one or more intermediate quantities.
This paper presents a new approach of information fusion used Bayesian network and is organised as follows. The background of this research, especially for the application in reliability assessment and precision evaluation, is introduced in section 1. In section 2, the fundamental of Bayesian network is stated and how to establish networks for a typical case are then illustrated. In section 3, it is emphasized in utilizing Bayesian networks to integrate multi-source testing information obtained from different layers, states and environments, where the examples of reliability parameters estimation for weapon system and information conversion for inertial navigation system error model are simulated to show the effectiveness of the scheme presented. Finally, some conclusions are given in the end.

2. Bayesian networks

2.1 Bayesian inference

The basic idea of Bayesian inference is to express the uncertainty of all the unknown parameters of the model by probability distributions [4]. This means that an unknown parameter is modeled as a random parameter beforehand. are in the text the random parameters of our interest is denoted as $\mathbf{\Theta} = (\Theta_1, ..., \Theta_n)$, where the index of n is presumed finite and the set of variables are observable. Random variables are expressed as $\mathbf{X} = (X_1, ..., X_m)$ with finite number of m. The observable variables X_j, may consist of statistical observations or various experts judgments.
The observed variables, or the evidence $\mathbf{x} = (x_1, ..., x_m)$, are modeled by their joint distribution, i.e. the likelihood function $f(\mathbf{x} \mid \boldsymbol{\theta})$, which can be described as the probability to observe the evidence **x**. Before observations are obtained, the uncertainty about the value of the random parameter $\mathbf{\Theta}$ is modeled by a prior probability distribution of $f(\boldsymbol{\theta})$. Given the evidence that the posterior distribution is the conditional distribution of $\mathbf{\Theta}$, it would be denoted as $f(\boldsymbol{\theta} \mid \mathbf{x})$. The evidence **x** provides additional information about $\mathbf{\Theta}$, and the posterior distribution is updated by using the Bayes' rule

$$f(\boldsymbol{\theta} \mid \mathbf{x}) = \frac{f(\mathbf{x} \mid \boldsymbol{\theta}) f(\boldsymbol{\theta})}{\int f(\mathbf{x} \mid \boldsymbol{\theta}) f(\boldsymbol{\theta})\, d\boldsymbol{\theta}} \tag{1}$$

or

$$f(\theta_1,...,\theta_n \mid x_1,...,x_m) = \frac{f(x_1,...,x_m \mid \theta_1,...,\theta_n) f(\theta_1,...,\theta_n)}{\int_{\theta_1} \cdots \int_{\theta_n} f(x_1,...,x_m \mid \theta_1,...,\theta_n) f(\theta_1,...,\theta_n) d\theta_1 ... d\theta_n} \quad (2)$$

2.2 Bayesian networks

In practice, many models under interest are usually complex which are related to the multi-layer Bayesian problems. For example, suppose the observable variable Y is normally distributed with mean parameter θ and standard deviation parameter σ_1 as following

$$\mathrm{Y} \mid \theta \sim N(\theta, \sigma_1^2) \quad (3)$$

where θ is also normal distributed with parameters α and σ_2

$$\theta \mid \mu \sim N(\alpha, \sigma_2^2), \quad \alpha = H\mu \quad (4)$$

and the prior distribution of the random variable μ is known as

$$\mu \sim N(\beta, \sigma_3^2) \quad (5)$$

Note that only $\sigma_1, \sigma_1, \sigma_1, H, \mu$ are constants. Thus, with the observations $y_1,..., y_n$, how to get the posterior estimation of θ and μ is a typical multi-layer Bayesian problem. To do this, we have to model the overall uncertainty by postulating the joint distribution of the all random variables of the model

$$f(\theta, \mu, Y) = f(Y \mid \theta, \mu)\ f(\theta \mid \mu)\ f(\mu) \quad (6)$$

in which we have assumed that the appropriate conditional distributions are available.

Actually the joint distribution model described in equation (6) consists of network of conditional dependencies between random variables. Such networks are often called Bayesian networks. A Bayesian network can be represented as a directed acyclic graph, in which elliptic nodes correspond to random variables and rectangular nodes represent constants and directed arcs between the nodes describe the dependence between the parameters. Moreover, a solid arrow indicates a stochastic dependence while a hollow arrow indicates a logical function. As an example the graphical representation of the hierarchical model described by equations (3)~(6) is depicted as a Bayesian network in Figure 1.

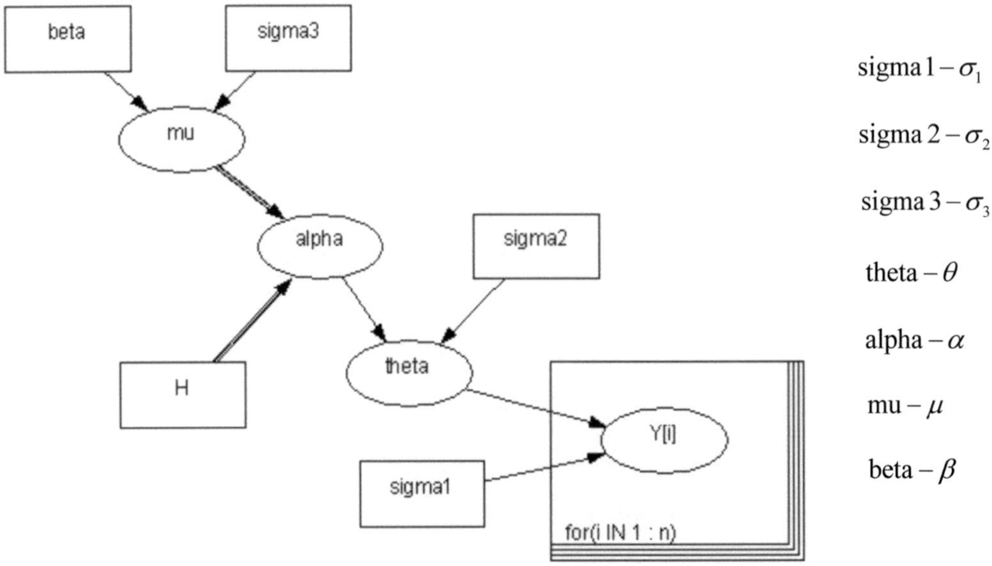

Fig. 1. Example of a Bayesian network

2.3 Implementation of the proposed method

In Bayesian models, where we are interested in the relationships of a large number of variables, Bayesian network becomes an appropriate representation. A Bayesian network is a graphical model that efficiently encodes the joint probability distribution for a large set of variables. However, determining the conditional posterior distributions for the parameters of interest is usually not a simple task in Bayesian networks. To obtain an analytic result for the conditional posterior distribution the denominator of the Bayes formula, which normalizes the conditional posterior distribution to unity, must be evaluated. A proportional result for the posterior distribution can be obtained without resolving the denominator, but the integral for the numerator is only one dimension less. For analytic result, or at least for a good approximation of the result, the integrals have to be determined in a way or another. For simple models the integrals can be evaluated using conventional numerical techniques, but in most applications the Bayesian network contain tens and hundreds of parameters and the analytic evaluation of the integrals by conventional numerical techniques is impossible.

Therefore, an MCMC [5] approach is used for obtaining the posterior distribution. In MCMC methods, Monte Carlo estimates of probability density functions and expected values of the desired quantities are obtained using samples generated by a Markov chain whose limiting distribution is the distribution of interest. Thus one can generate samples of multiple random variables from a complicated joint probability density function without explicitly evaluating or inverting the joint cumulative density function.

Several schemes such as Metropolis-Hastings algorithm, Gibbs sampling, etc. are available to carry out MCMC simulations [6]. Gibbs sampling is commonly utilized due to its simplicity in the implementation. Let $\mathbf{x}$ denote a vector of k random variables $(x_1,\ldots,x_k)$, with a joint density function $g(\mathbf{x})$. Then let $\mathbf{x}_{-i}$ denote a vector of k-1 variables, without the *i*th variable, and the full conditional density for the *i*th component is defined as

$g(x_i \mid \mathbf{x}_{-i})$. To sample quantities from the full conditional density of the ith variable, the following relationship is used:

$$g(x_i \mid \mathbf{x}_{-i}) = \frac{g(x_i, \mathbf{x}_{-i})}{\int \cdots \int g(x_i, \mathbf{x}_{-i})\, d\boldsymbol{\theta}} \tag{7}$$

Gibbs sampling can then be used to sequentially generate samples from the joint probability density function using the full conditional densities, as below:

Step 1: Initialize $\mathbf{x}^0 = \left\{x_1^0, x_2^0, \ldots, x_k^0\right\}$, $j = 1$;

Step2: Generate $x_1^j \sim g(x_1 \mid x_2^{j-1}, x_3^{j-1}, \ldots, x_k^{j-1})$,

$x_2^j \sim g(x_2 \mid x_1^j, x_3^{j-1}, \ldots, x_k^{j-1})$,

...

$x_k^j \sim g(x_k \mid x_1^j, x_2^j, \ldots, x_{k-1}^j)$;

Step3: $j = j + 1$;

Step4: End if j reaches the maximum number of runs, or else, return to step 2.

Gibbs sampling has been shown to have geometric convergence of order N (number of runs) [4]. Exact full conditional densities may not always be available. In such cases, the Gibbs sampling procedure is supplementary to the Metropolis-Hastings algorithm. During each run, the full conditional density function $g(x_i \mid \mathbf{x}_{-i})$ is constructed by taking the product of terms containing x_i in the joint probability density function. A rejection sampling technique is then used to obtain a sample x_i from $g(x_i \mid \mathbf{x}_{-i})$. A large number of samples of all the random variables can be repeatedly generated using these full conditional density functions. The marginal density function for any random variable x_i can be obtained by collecting the samples of that particular random variable.

3. Testing information fusion using Bayesian networks

Since Bayesian networks can easily establish the uncertainty relationships among parameters and update all the prior distributions of the random variables once new observations come out, it is a effective solution to multi-source information fusion. In this section, two representative applications of Bayesian networks to weapon system reliability evaluation and INS testing information conversion under different circumstance are discussed as illustration. Note that the modeling and simulations in this paper are carried out using the WinBUGS program, and so all the Bayesian networks presented below are depicted in the WinBUGS format. For closer review about the WinBUGS program, see Spiegelhalter et al. [7].

3.1 Reliability evaluation of weapon system

Since a great deal of manpower and material resources are requisite in system-level tests to reliability evaluation for such complex weapon system, whereas much more convenience would be obtained if in unit-level test case, the engineering practice usually adopts reliability information of composition units to analyze the reliability of the entire system. These unit-level test information make up for the lack of information on system-level test, and reduce the number of tests in the premise of sustaining its confidence effectively.

Obviously, to evaluate weapon system reliability in Bayes method is a kind of information fusion. More clearly, reliability test information about unit and system should be fused into the posterior distribution of system reliability first, and based on it the Bayesian statistical inference could then be carried out. To facilitate following discussion, suppose the weapon system is pass-fail series system.

3.1.1 Reliability analysis of pass-fail unit

The pass-fail unit likelihood function is

$$L(R) = R^{n-f}(1-R)^{f}, \quad 0 \le R \le 1 \tag{8}$$

In the discussion of binomial distribution, the prior distribution of reliability is often in Beta distribution, i.e.

$$\pi(R) = \frac{R^{a-1}(1-R)^{b-1}}{\beta(a,b)}, \quad 0 \le R \le 1 \tag{9}$$

where a and b are auxiliary parameters,

$$\beta(a,b) = \frac{\Gamma(a)\Gamma(b)}{\Gamma(a+b)} \tag{10}$$

Since these auxiliary parameters reflect the full utilization of prior information, selection of a and b are very critical for reliability analysis. Martz et al [8] displayed the empirical Bayesian parameters estimation for a and b.

Suppose there are m groups of tests information, where l_i $(i = 1, 2, \cdots m)$ denotes the test number and R_i represents the point estimation of reliability to each group, therefore

$$a + b = \frac{m^2\left(\sum_{i=1}^{m} R_i - \sum_{i=1}^{m} R_i^2\right)}{m\left(m\sum_{i=1}^{m} R_i^2 - K\sum_{i=1}^{m} R_i\right) - (m-K)\left(\sum_{i=1}^{m} R_i\right)^2} \tag{11}$$

$$a = (a+b)\overline{R} \tag{12}$$

where

$$K = \sum_{i=1}^{m} l_i^{-1}, \quad \overline{R} = \frac{\sum_{i=1}^{m} R_i}{m} \tag{13}$$

When m is small, the sampling error may yield negative value in (11) of $(a+b)$, so that it is amended as

$$a+b = \left(\frac{m-1}{m}\right)\left(\frac{m\sum_{i=1}^{m} R_i - (\sum_{i=1}^{m} R_i)^2}{m\sum_{i=1}^{m} R_i^2 - (\sum_{i=1}^{m} R_i)^2}\right) - 1 \tag{14}$$

Once the auxiliary parameters of prior distribution are determined, using Bayes' rule there would be

$$\pi(R \mid D) = \frac{R^{n-f+a-1}(1-R)^{b+f-1}}{\beta(n-f+a, b+f)} \tag{15}$$

where D is the experimental data, n is the number of tests, f indicates the number of failure. It is obviously that the posterior probability density function $\pi(R \mid D)$ for R is still in Beta distribution.

The Bayesian analysis method for unit reliability is deduced in above discussion. The following would proceed to system reliability calculation for pass-fail series system.

3.1.2 Series system synthesis

Assume all the reliability tests of system or units considered is in pass-fail type. The series system consists of p units, and denotes Θ_i as the reliability of the constituent units, thus the prior distribution of Θ_i is

$$\pi_i(\theta_i) = B^{-1}(a_i, b_i) \cdot \theta_i^{a_i - 1} \cdot (1-\theta_i)^{b_i - 1}, \quad 0 < \theta_i < 1 \tag{16}$$

Denote n_i as the number of unit tests, x_i as the number of success, so the system reliability is

$$\Theta = \prod_{i=1}^{p} \Theta_i \tag{17}$$

Assume Θ_i is independent of each other, and prior distribution of $\pi_i(\theta_i)$ is in Beta form. if $\Theta_i = \theta_i$, and x_i are subject to binomial distribution with parameters of n_i and θ_i, in response the posterior probability density function is Beta distributed, where $a_i + x_i$ and

$b_i + n_i - x_i$ are the distribution parameters. The posterior probability density function of the system's reliability Θ is induced as

$$g(\theta) = K_p \cdot \theta^{b_p - 1} (1-\theta)^{\nu - 1} \cdot \sum_{r=0}^{\infty} \sigma_r^{(p)} \cdot (1-\theta)^r \big/ \Gamma(\nu + r) \tag{18}$$

where

$$K_p = \prod_{i=1}^{p} B^{-1}(a_i + x_i, b_i + n_i - x_i)$$

$$b_p = \alpha_p + x_p$$

$$\nu_k = \sum_{i=1}^{k} (b_i + n_i - x_i), \quad k = 1, 2, \cdots, p$$

$$\nu = \nu_p$$

and $\sigma_r^{(p)}$ satisfies the following recursive relationship

$$\sigma_r^{(k)} = \left[\Gamma(\nu_{k-1} + r) \big/ \Gamma(\nu_k + r)\right] \sum_{s=0}^{r} \left[s \cdot B(n_k + a_k + b_k - a_{k-1} - x_{k-1}, s)\right]^{-1} \cdot \sigma_{r-s}^{(k-1)}$$

$$r = 0, 1, \cdots; \; k = 2, \cdots, p$$

$$\sigma_0^{(1)} = 1 \big/ \Gamma(b_1 + n_1 - x_1)$$

$$\sigma_r^{(1)} = 0, \; r = 1, 2, \cdots$$

If there are few units in series system, the above formula for the series' system reliability evaluation is feasible; while if the cell number is in great many, the calculations could not be sustainable any more. The system encountered in engineering practice used to be composed in many units, and the information obtained are comprised of unit reliability test information and system reliability information, in this condition the abovementioned method is incapable in handling complex tests information. The following would introduce the reliability analysis method for this kind of complex system using Bayesian network.
Assume the distribution of system reliability is also subject to Beta, thus the posterior joint probability density function of unit's reliability Θ_i and system reliability Θ can be rewritten as:

$$\pi(\Theta_1, \Theta_2, \ldots \Theta_p; \Theta \mid D_1, D_2, \ldots D_p; D) = \prod_{i=1}^{p} \frac{\theta_i^{n_i - f_i + a_i - 1} (1-\theta_i)^{b_i + f_i - 1}}{\beta(n_i - f_i + a_i, b_i + f_i)} \cdot \frac{\theta^{n - f + a - 1} (1-\theta)^{b + f - 1}}{\beta(n - f + a, b + f)} \tag{19}$$

Since the above form of joint distribution is too complex, Bayesian network of the system reliability is established in assistance. In this Bayesian network, MCMC sampling method is employed to update the network graph, hence the analysis to posterior distribution of reliability could be implemented as soon as Markov chain is stabilized. Take the three-numbered pass-fail series as an example, the Bayesian network is developed as below.

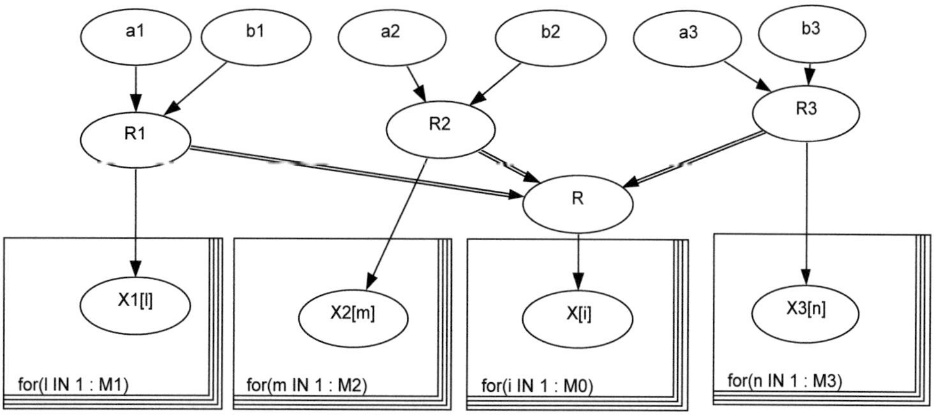

Fig. 2. Bayesian network of system reliability

In Figure 1, R, R_1, R_2, R_3 represent the system's reliability and units' reliability; $a_1, b_1, a_2, b_2, a_3, b_3$ indicate prior distribution parameters respectively, and X, X1, X2, X3 are the units and system test samples respectively. If the unit reliability parameters of prior distribution are set to be in normal distribution, experimental data (n and $n_i, i = 1,2,3$ are the experimental times; f and $f_i, i = 1,2,3$ are the failure times) of $n_1 = 12, f_1 = 0$, $n_2 = 12, f_2 = 1, n_3 = 12, f_3 = 2, n = 12, f = 3$ are derived.

reliability	prior distribution		posterior distribution				
	mean	standard deviation	mean	Standard deviation	2.5% percentile	Median percentile	97.5% percentile
R_1	0.99	0.1	0.9621	0.03783	0.8622	0.9729	1.0
R_2	0.99	0.1	0.8739	0.07813	0.689	0.8862	0.9855
R_3	0.90	0.1	0.8082	0.08715	0.6146	0.8163	0.9521
R	/	/	0.6781	0.08928	0.4928	0.6836	0.8391

Table 1. Prior and posterior statistical properties of units and system reliability

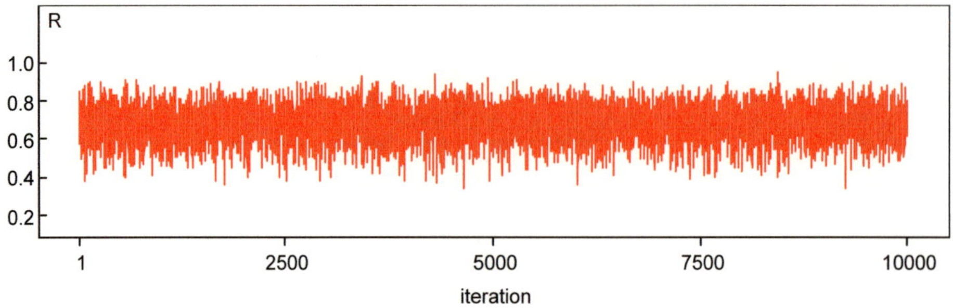

Fig. 3. Sample sequence of system reliability R

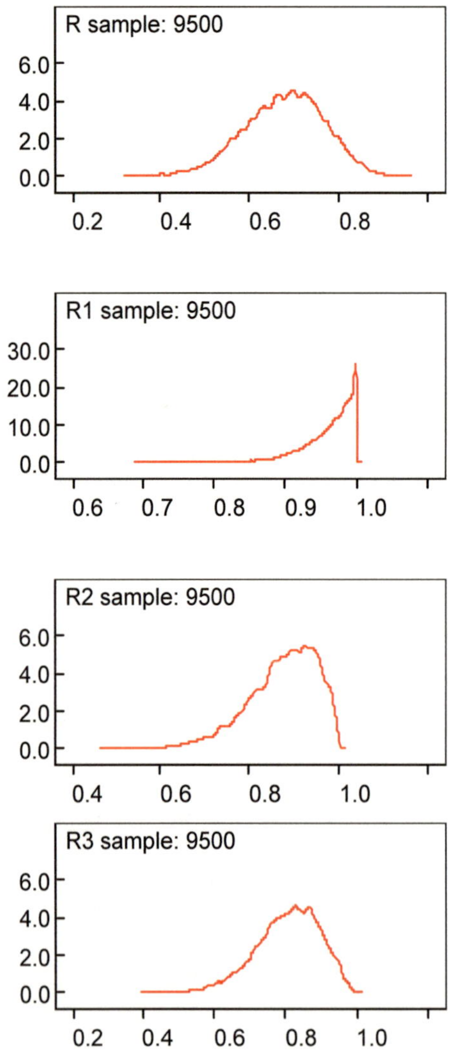

Fig. 4. Posterior density distribution of units and system reliability

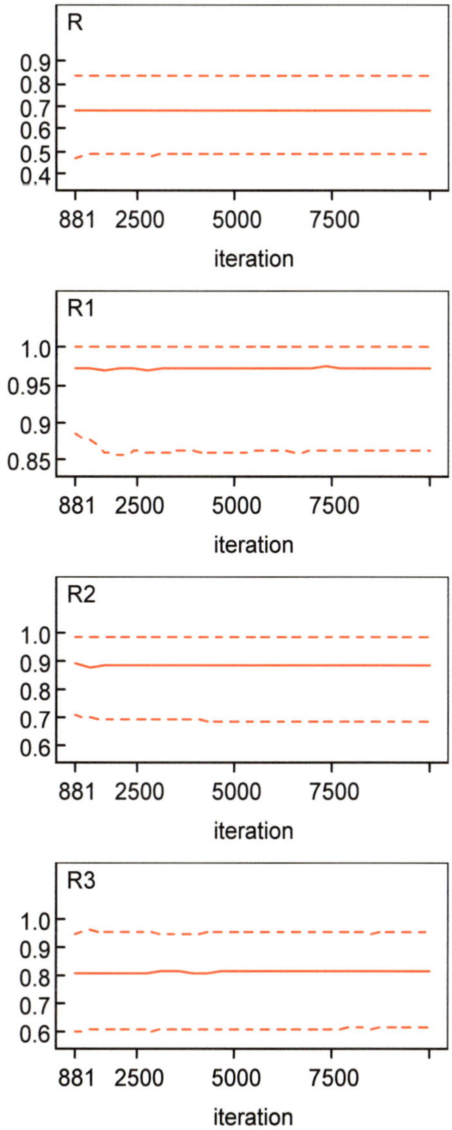

Fig. 5. Percentile statistics sequence of the unit and system reliability

Using MCMC method of sampling for 10,000 times in the Bayesian network, and implementing statistical analysis to the sample sequence in steady-state Markov chain, the prior and posterior statistical characteristics of units and system reliability are computed out at last, see Table 1. In the sample sequence of system reliability of Figure 3, the Markov chain has been shown fused completely and furthermore reached steady state in sampling 2,000 times. Figure 4 shows the profile of posterior density distribution estimation of reliability which is depicted in consistent with Beta distribution explicitly.

Bayesian network integrates test information and prior information about the units and system, and the information in each node is then disseminated to the entire network through the directed link, therefore the integrated inference about the test information is realized. The advantage of this approach is that reliability statistical analysis of the system in it would be more accurate. And furthermore, the percentile information such as the upper bound of reliability are also obtained through MCMC sampling, in result the system reliability analysis is more comprehensive and effective.

3.2 Testing information fusion of INS

When the tests are implemented under different technical conditions, the error coefficients of inertial navigation system may have different statistical characteristics. This paper presents a method of multi-source testing information fusion for inertial navigation system based on Bayesian network, which might provide a new idea to the precision evaluation work. Firstly, the testing information of all sorts is interrelated to each other by circumstance-conversion-factor, and then a graphic mapping model is constructed to represent the relationship of all variables by using Bayesian network. With testing information, the post statistical characteristics of variables such as circumstance-conversion-factor can be rapidly inferred by MCMC algorithm applied in Bayesian network, and consequently information conversion of inertial navigation system error model could be carried out between different testing conditions.

As the test information are related to the temperature, pressure, humidity and other circumstance factors, a standard state for each type of test should be selected beforehand. In this condition, all the information would be conversed to be the one in the corresponding standard state first, and then conversed in the reference of standard state information.

3.2.1 Inference of circumstance conversion factor

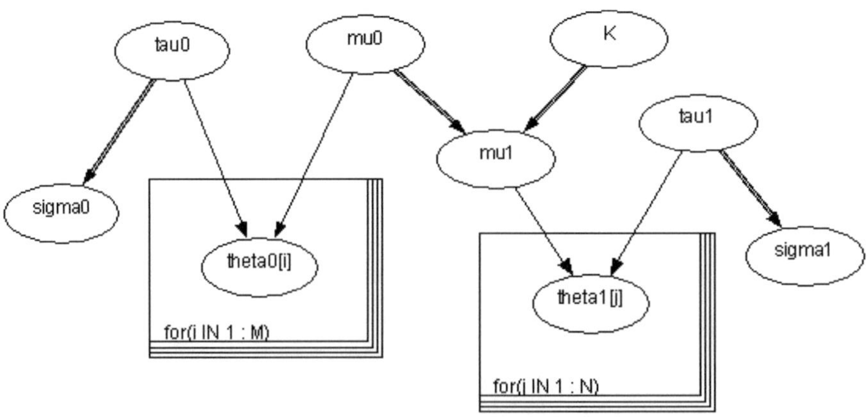

Fig. 6. Bayesian network for testing information fusion

Suppose an error coefficient to be a normally distributed random variable $\theta_0 \sim N(\mu_0, \sigma_0^2)$ in the ground calibration tests, and another normally distributed random variable $\theta_1 \sim N(\mu_1, \sigma_1^2)$ in vehicle tests. Treat the circumstance factor K, mean μ_i and standard deviation σ_i in calibration tests and vehicle tests to be unknown random variables, where

K and μ_0 are normal distributed, and $\tau_i = 1/\sigma_i^2$ is subject to Gamma distribution, in which way the established Bayesian network for information fusion is depicted as Figure 6.

For the sake that the tests of INS could not be a great many, there are only 10 groups of ground calibration and 5 groups of vehicle-loaded estimates to error factor generated through the simulation, see Table 2. Note that a new data generation would accompany a set of mean and standard deviation production.

Ground calibration						true distribution of variables in data production
θ_0	1.8609	1.8781	2.1273	1.9274	1.9414	$K \sim N(1.2, 0.1^2)$ $\mu_0 \sim N(2, 0.05^2)$ $\tau_0 \sim Gamma(100, 1)$ $\tau_1 \sim Gamma(10, 1)$
	1.9230	2.1637	1.9988	1.8513	1.9506	
vehicle-loaded estimates						
θ_1	2.8879	2.5901	2.2561	2.1137	2.4323	

Table 2. Simulation data

Before conducting statistical inference in Bayesian networks, the prior distribution of random variable nodes are required to set up. On the assumption that there are no prior information

$$K_{prior} \sim N(1, 1000^2)$$
$$\mu_{0\,prior} \sim N(0, 1000^2)$$
$$\tau_{0\,prior} \sim Gamma(10^{-6}, 10^{-6})$$
$$\tau_{1\,prior} \sim Gamma(10^{-6}, 10^{-6})$$

Given initial random nodes and set the number of iterations of 20,000 times, Bayesian network could get updated by MCMC based on the test data θ_0 and θ_1. Iteration process and posterior kernel density estimates of some variables are shown in Figure 7 to 14.

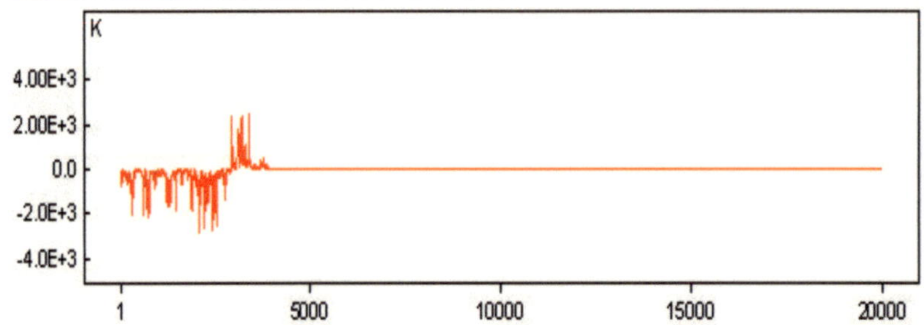

Fig. 7. Track of variable K in iteration

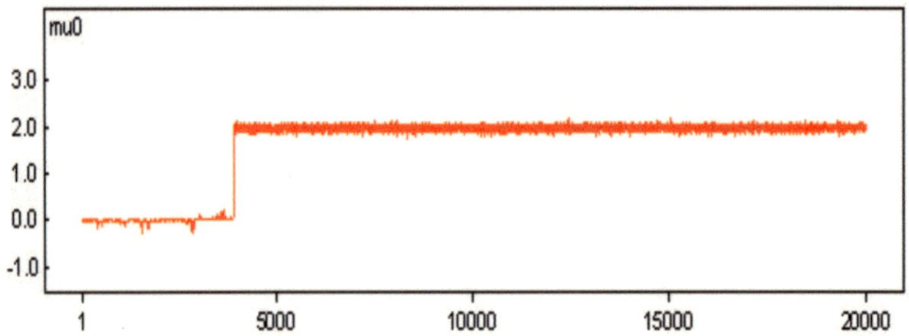

Fig. 8. Track of variable μ_0 in iteration

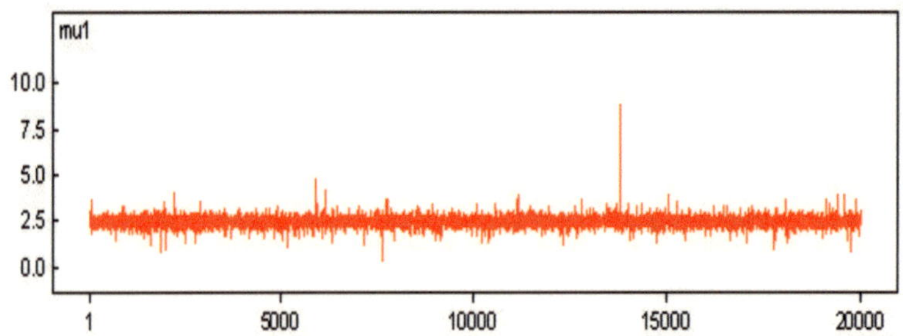

Fig. 9. Track of variable μ_1 in iteration

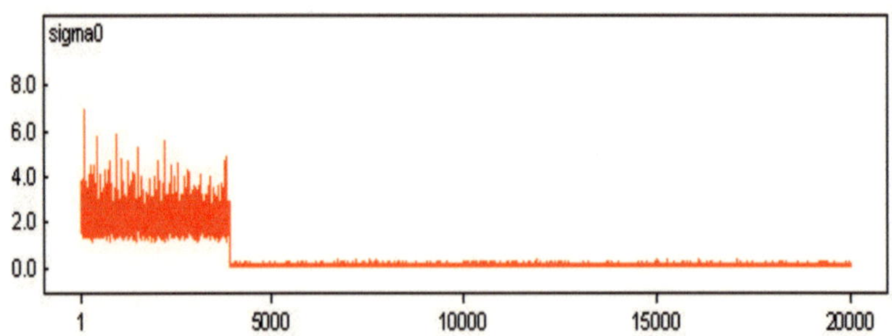

Fig. 10. Track of variable σ_0 in iteration

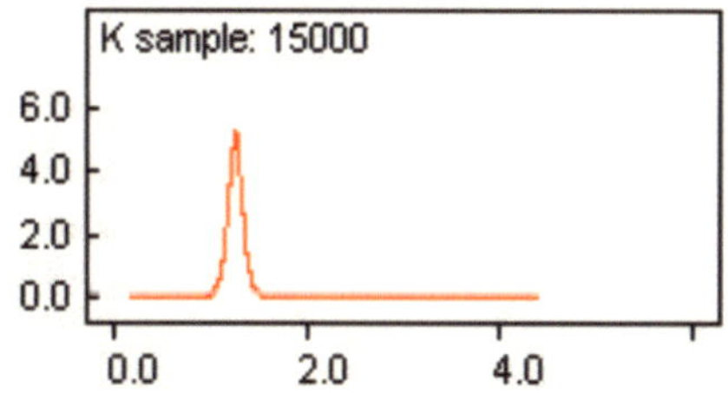

Fig. 11. Posterior distribution estimate of K

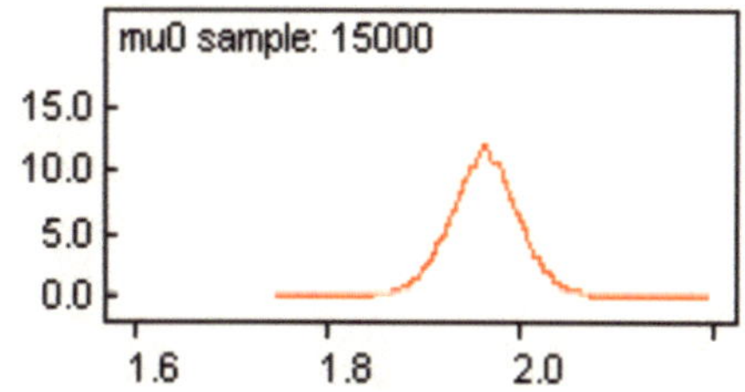

Fig. 12. Posterior distribution estimate of μ_0

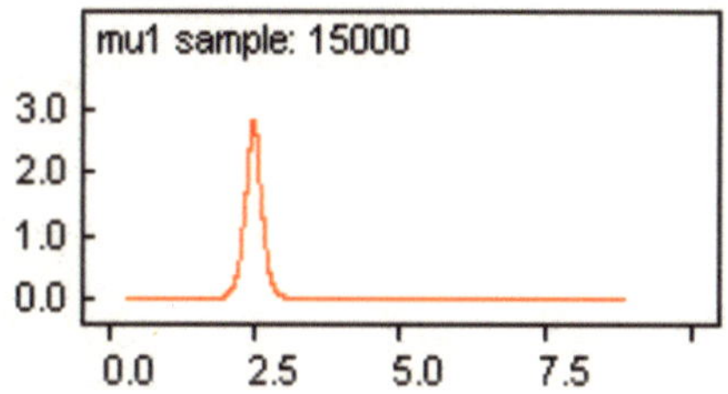

Fig. 13. Posterior distribution estimate of μ_1

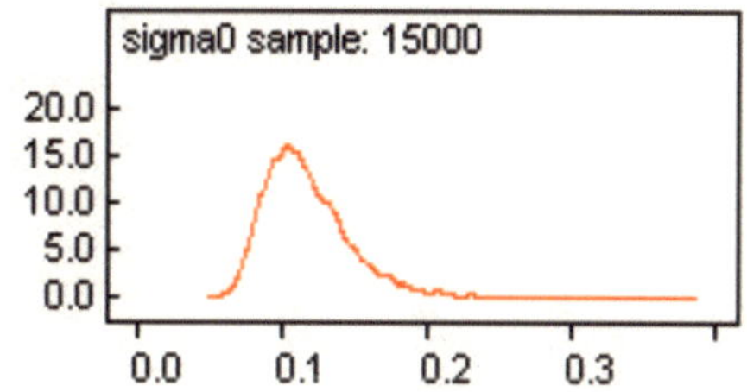

Fig. 14. Posterior distribution estimate of σ_0

From the iterative trajectories of variables, it is known that MCMC algorithm converges in about 4000 steps. Therefore, abandoning the former 5000 iterations, and utilizing the latter 15,000 values of samples to infer variables' posterior statistical characteristics. Compare the prior, posterior and the true statistical characteristics of parameters comprehensively, and get the results summarized in Table 3, where the true distribution characteristics (mean and standard deviation) of μ_1, σ_0 and σ_1 may be computed out from those of other variables.

Obviously, in the case of 15 groups of observational data, Bayesian network has effectively fused the information obtained from calibration tests and vehicle-loaded tests. In comparison with prior distribution, the characteristics of the posterior distribution of all variables whether mean or standard deviation is much closer to those in real situation. Summarized from posterior statistical properties, estimates of the mean for each variable is slightly better than that of standard deviation, while the posterior inference to τ_i (or σ_i) is shown inferior to that of K and μ_i. Given the limited sample size, accomplish system-level test data fusion utilizing Bayesian network is still quite effective despite of the errors exist between posterior inference and the true data.

variable	true distribution		prior distribution		posterior distribution				
	mean	standard deviation	mean	standard deviation	mean	standard deviation	2.5% percentile	Median percentile	97.5% percentile
K	1.2	0.1	1	1000	1.2520	0.1025	1.0580	1.2520	1.4470
μ_0	2	0.05	0	1000	1.9620	0.0387	1.8830	1.9620	2.0390
μ_1	2.4	0.2089	/	/	2.4550	0.1958	2.0840	2.4550	2.8210
σ_0	0.1004	0.0050	/	/	0.1179	0.0315	0.0739	0.1121	0.1953
σ_1	0.3287	0.0552	/	/	0.3772	0.1972	0.1790	0.3306	0.8537
τ_0	100	10	1	1000	86.020	40.800	26.220	79.540	183.40
τ_1	10	3.1623	1	1000	11.050	7.9590	1.3740	9.1510	31.270

Table 3. Prior, posterior and true statistical characteristics of random variables

3.2.2 Testing information conversion

System-level test information fusion does not only purpose to induce posterior statistical properties of variables, what's more important is through the acknowledgement of different types of circumstance factors, the test information about the INS error model are transmitted among those tests, which realizes the conversion of different types of testing error coefficients. For the sake Bayes method deals with the error coefficient as a random variable, so this kind of "conversion" is essentially that of variables' statistical properties.

In actual project, the true statistical characteristics of error coefficient may vary with the improvement of inertial navigation system manufacturing techniques. Therefore, an assumption should be made before converting the error coefficient in different types of tests: the variance of the true expectation about this coefficient remained proportional in different types of tests, that is to say the statistical characteristics of circumstance factor K remains almost unchanged.

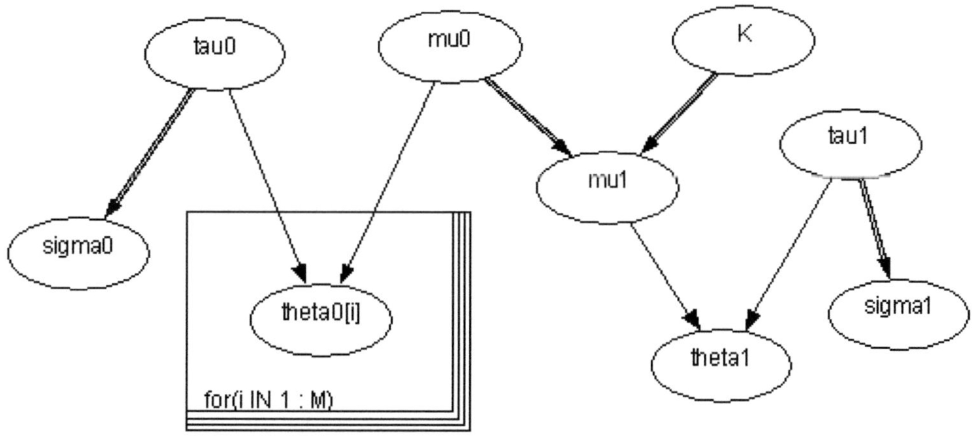

Fig. 15. Error factors' conversion from different types of tests in Bayesian network

Still take the case of data fusion between calibration test and vehicle-loaded test aforementioned as an example. Suppose the statistical properties of each variable have been inferred from the calibration test data and vehicle test data. After that the inertial measurement system accepted technique improvement and whereby another set of test data θ_0' about error coefficient were induced from ground calibration tests (see Table 4). In this condition, we wish to infer the statistical characteristics of error coefficient represents θ_1' in vehicle-loaded test in the light of circumstance factor K. The links among the variables are depicted in Bayesian network of Figure 15.

Ground calibration tests data						true distribution of variables in data generation
θ_0'	1.0524	1.2588	1.2021	1.3972	1.2134	$\mu_0 \sim N(1.2, 0.04^2)$ $\tau_0 \sim Gamma(150, 1)$

Table 4. Test data in improved technique

As long as our purpose is to do the error factor conversion, so we focus on the posterior inference to θ_1', K and μ_0 merely. According to the using means of prior information, two different conversion methods are adopted respectively.

A. Conversion method 1

At first, deal with the posterior statistical characteristics of the variables as the prior information of current coefficient in conversion, by making use of the data fusion results prior to technology improvement merely. Then update Bayesian network as shown in Figure 15 based on the improved calibration test data. Statistical inference results are obtained and displayed in Table 5.

variable	true distribution		prior distribution		posterior distribution				
	mean	standard deviation	mean	standard deviation	mean	standard deviation	2.5% percentile	Median percentile	97.5% percentile
θ_1'	/	/	/	/	2.4190	0.4962	1.4930	2.4230	3.3510
K	/	/	1.2520	0.1025	1.2520	0.1028	1.0490	1.2510	1.4510
μ_0	1.2	0.04	1.9620	0.0387	1.9340	0.0388	1.8590	1.9340	2.0120

Table 5. Statistical results in use of prior information merely

As indicated from Table 5, Circumstance factor of K is hardly changed, mean of error coefficient μ_0 differs slightly, so there is a big gap between the posterior statistical characteristics and the real states. Although the posterior distribution of the test data is slightly "pulled back" to the real state by the novel test data, the effect is still not very obvious. That's because compared to the prior distribution on one hand, the prior distribution is more certain (standard deviation is small), and on the other the sample information is too limited, so that prior information plays a leading role in the posterior statistical inference. The novel test information is greatly weakened by the prior, which yields inferior posterior inference of μ_0. In the presence of large deviations of posterior inference, the statistical results about the conversion value of error factor is not that credible in this occasion.

B. Conversion method 2

Allowing for the impact of prior information to posterior statistical inference, especially in the occasion of small samples, the error coefficient conversion is dealt with improved prior information. These improvements include two aspects. At first, in spite the prior information of error coefficient may vary from the state of current system due to technical progress, the variation is not too much so that the mean of prior distribution could be remained. Second, by increasing the standard deviation of the prior distribution, the prior information could be "fuzzed up" so that "over- conservative" posterior inference from "over-certain" prior characteristic would be avoided; but note that the standard deviation should not be set too large, otherwise the system would tend to non-informative prior and lose the useful information.

The statistical inference results from improved prior information are shown in Table 6.

In contrast to the results in Table 5, the posterior statistical inference in Table 5 is significantly closer to the true distribution, so the statistical characteristics of θ_1' can be used as the conversion of error factor from the ground calibration tests to the one in vehicle-loaded tests.

variable	true distribution		prior distribution		posterior distribution				
	mean	standard deviation	mean	standard deviation	mean	standard deviation	2.5% percentile	Median percentile	97.5% percentile
θ_1'	/	/	/	/	1.5320	0.4563	0.6356	1.5310	2.4140
K	/	/	1.2520	0.1025	1.2520	0.1028	1.0531	1.2530	1.4530
μ_0	1.2	0.04	1.9620	5	1.2260	0.0548	1.1180	1.2260	1.3350

Table 6. Statistical results in use of improved prior information

The error coefficient conversion has been done in two methods from preceding texts analysis. Since very few samples are available, the first method takes more advantage of prior information so that the impact of test information to posterior inference is very weak; while the second method of "fuzzes up" prior information to reduce its impact on posterior inference by increasing the standard deviation, in return the impact of test information is got increased. In the premise of small but capable of accurately reflecting the true statistical characteristics samples, the above examples prove the second method is better in error coefficient conversion than the first. However, when large deviations exist between sample information and the real distribution, the risk of the using the second method increases in accompany; therefore, it is not reasonable to say that the second method is certainly better than the first.

4. Conclusions

MCMC technology has brought a revolutionary breakthrough to the development and application of Bayes statistical theory. Especially the emergence and further promotion of WinBUGS software, which gets the Bayesian network inference of model parameters out of complicated high-dimensional integral calculations, has routinized the analysis and application of Bayesian network. This paper has discussed the reliability assessment of weapon systems and the conversion of inertial navigation test information, which provides model reference and possible solutions to the Bayesian network based multi-source information fusion methods.

5. References

[1] Helminen A. Reliability Estimation of Safety-critical Software-based Systems Using Bayesian Networks. STUK-YTO-TR 178. Radiation and Nuclear Safety Authority, Helsinki 2001:1-23.

[2] Littlewood B, Fenton N E, Neil M. Applying Bayesian Belief Networks to Systems Dependability Assessment. In: 4th Safety Critical Systems Symposium, 1996, Leeds. Proceedings of the conference, Springer Verlag, 1996:71-93.

[3] Fenton N E, Littlewood B, Neil M, Strigini L, Wright D. Bayesian Belief Network Model for the Safety Assessment of Nuclear Computer-based Systems. In: DeVa ESPRIT Long Term Research Project No. 20072-2nd Year Report, City University, London, 1997:1-28.

[4] Cai H, Zhang S F, Zhang J H. Bayesian Test Analysis and Evaluation. Changsha: National University of Defence Technology Publishing Press, 2004.

[5] Gilks G R, Richardson S, Spiegelhalter D J. Markov Chain Monte Carlo in Practice. Interdisciplinary Statistics. London: Chapman & Hall, 1996.

[6] Sankaran M, Ramesh R. Validation of Reliability Computational Models Using Bayes Networks. Reliability Engineering and System Safety, 2005,87:223-232.

[7] Spiegelhalter D J, Thomas A, Best N G. WinBUGS Version 1.4 User Manual. Cambridge, UK, MRC Biostatistics Unit, 2002.

[8] Martz H F, Waller R A. Bayesian Reliability Analysis. John wiley & Sons (New York), 1982.

Dynamic data feed to Bayesian network model and SMILE web application

Nipat Jongsawat, Anucha Tungkasthan and Wichian Premchaiswadi
Graduate School of Information Technology in Business, Siam University
Thailand

1. Introduction

Constructing Bayesian network models is a complex and time consuming task. It is difficult to obtain complete and consistent models but to get the correct and reliable probability data for the designed models is much more difficult. Normally, there are two methods to enter the probability values into the chance node of a Bayesian network model. The first method is to consult an expert for the probability values and enter them into the models. The second method is to obtain probability values from statistical or learned data (Druzdzel et al., 2001). Both methods use static data, not dynamic data. The second method acts like dynamic data but it is actually not. The statistical data from a database need to be loaded and processed each time to get the probability values. This works similar to batch processing. Finally, users still need to enter probability values into the model by manual feeding the data by hand. It is not possible to have real-time processing. The probability values are fed to every node of the model and the joint probability distribution is computed at the final stage when the model is performing Bayesian updates. The disadvantage of using manually fed data or static data is that it cannot be performed in using real-time processing, monitoring, and updating.

In this article, we propose a technique for feeding data into the Bayesian network model dynamically. A case study of several factors that have an impact on students for making a decision in enrollment is selected as the case for an application implementation of a Bayesian network model. The probability values for each node are calculated from student's data and then transferred into the model dynamically. A SMILE web-based application provides a user friendly web interface for Bayesian inference. It provides the feature set of Bayesian diagnosis for the user. The SMILE web-based application was developed based on SMILE (Structural Modeling, Inference, and Learning Engine) and SMILE.NET. SMILE is a reasoning engine that is used for graphical probabilistic models and provides functionality to perform diagnosis. SMILE.NET is used for accessing the SMILE library from the web-based interface. Using SMILE application, users can also perform Bayesian inference in the model and they can compute the impact of observing values of a subset of the model variables on the probability distribution over the remaining variables based on real-time data. Using the other BN software tools for constructing a Bayesian network model, there are some limitations such as dependent platform and is unusable on a global basis. Fig. 1

shows a generic implementation for dynamic data feed to Bayesian network model and SMILE web application.

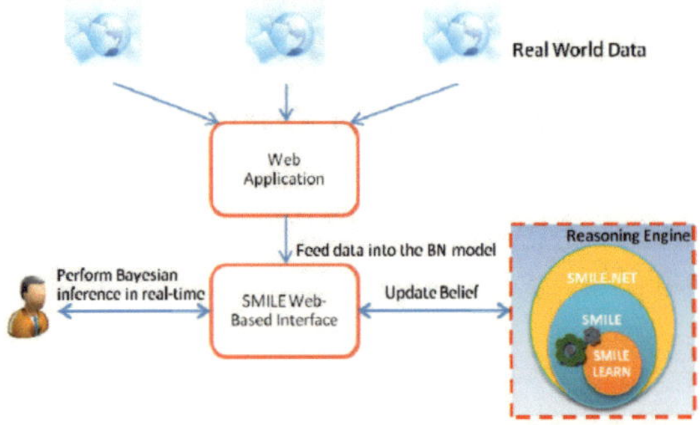

Fig. 1. A Generic implementation for dynamic data feed to BN model and SMILE web application

2. Fundamentals

This section is intended to describe the fundamentals and techniques for implementing a Bayesian network model in general. They are the followings:

2.1 Bayesian Network

Bayesian networks (also called belief networks, Bayesian belief networks, causal probabilistic networks, or causal networks) (Pearl, 1988) are acyclic directed graphs in which nodes represent random variables and arcs represent direct probabilistic dependencies among them. The structure of a Bayesian network is a graphical, qualitative illustration of the interactions among the set of variables that it models. The structure of the directed graph can mimic the causal structure of the modeled domain, although this is not necessary. When the structure is causal, it gives a useful, modular insight into the interactions among the variables and allows for prediction of the effects of external manipulation.
Nodes of a Bayesian network are usually drawn as circles or ovals. The following simple Bayesian network, shown in Fig. 2, represents two variables, Curriculum and Enrollment, and expresses the fact that they are directly dependent on each other.

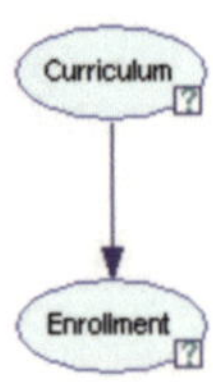

Fig. 2. An example of Bayesian network

A Bayesian network also represents the quantitative relationships among the modeled variables. Numerically, it represents the joint probability distribution among them. This distribution is described efficiently by exploring the probabilistic independence among the modeled variables. Each node is described by a probability distribution conditional on its direct predecessors. Nodes with no predecessors are described by prior probability distributions. For example, the node Curriculum shown in Fig. 2 will be described by a prior probability distribution over its two outcomes: Impact and NoImpact. See Fig. 3 below.

Impact	0.7
NoImpact	0.3

Fig. 3. Prior probability distribution for a curriculum node

The enrollment node will be described by a probability distribution over its outcomes (Enroll, NotEnroll) conditional on the outcomes of its predecessor (node Curriculum outcomes, Impact and NoImpact). See Fig. 4 below.

Curriculum	CurriculumImpact	CurriculumNoImpact
Enroll	0.7	0.4
NotEnroll	0.3	0.6

Fig. 4. Conditional probability values for an enrollment node

Both the structure and the numerical parameters of a Bayesian network can be elicited from an expert. They can also be derived from data, as the structure of a Bayesian network is simply a representation of independencies in the data and the numbers are a representation of the joint probability distributions that can be inferred from the data. Finally, both the structure and the numerical probabilities can be a mixture of expert knowledge, measurements and objective frequency data.

2.2 Bayesian Updating

Bayesian updating, also referred to as belief updating, or somewhat less precisely as probabilistic inference is based on the numerical parameters captured in the model (Cooper, 1990). The structure of the model which is an explicit statement of the independencies in the domain helps in making the algorithms for Bayesian updating more efficient (Dagum & Luby, 1997). All algorithms for Bayesian updating are based on a theorem proposed by Rev. Thomas Bayes (1702-1761) and is known as Bayes Theorem.

Belief updating in Bayesian networks is computationally complex. In the worst case, belief updating algorithms are NP-hard (Cooper, 1990). There exist several efficient algorithms, however, that make belief updating in graphs consisting of tens or hundreds of variables tractable. Pearl developed a message-passing scheme that updates the probability distributions for each node in a Bayesian network in response to observations of one or more variables (Pearl, 1986). Lauritzen and Spiegelhalter, Jensen et al, and Dawid proposed an efficient algorithm that first transforms a Bayesian network into a tree where each node in the tree corresponds to a subset of variables in the original graph (Lauritzen & Spiegelhalter, 1988; Jensen et al., 1990; Dawid, 1992). The algorithm then exploits several mathematical properties of this tree to perform probabilistic inference.

Several approximate algorithms based on stochastic sampling have been developed. Of these, best known are probabilistic logic sampling (Henrion, 1988), likelihood sampling (Shachter & Peot, 1989; Fung & Chang, 1989), and backward sampling (Fung & del Favero,

1994), Adaptive Importance Sampling (AISBN) (Cheng & Druzdzel, 2000), and Approximate Posterior Importance Sampling (APIS-BN) (Yuan & Druzdzel, 2003). Approximate belief updating in Bayesian networks has also been shown to be worst case NP-hard (Dagum & Luby, 1993).

2.3 SMILE and SMILE.NET

The core reasoning engines of the SMILE web-based application development capability consist of SMILE and SMILE.NET. SMILE is a reasoning engine that is used for graphical probabilistic models and provides functionality to perform diagnosis. SMILE.NET is used for accessing the SMILE library from the web-based interface. This section provides some more detailed information about SMILE and SMILE.NET wrapper.

SMILE (Structural Modeling, Inference, and Learning Engine) is a fully platform independent library of functions implementing graphical probabilistic and decision-theoretic models, such as Bayesian networks, influence diagrams (IDs), and structural equation models (Druzdzel, 1999). Its individual functions, defined in the SMILE Application Programmer Interface (API), allow creating, editing, saving, and loading graphical models, and using them for probabilistic reasoning and decision making under uncertainty. SMILE can be embedded in programs that use graphical probabilistic models as their reasoning engines. Models developed in SMILE can be equipped with a user interface that best suits the user of the resulting application. SMILE is written in C++ in a platform-independent manner and is fully portable. Model building and the reasoning process are under full control of the application program as the SMILE library serves merely as a set of tools and structures that facilitates them. The sample source code below is the main function of SMILE that contains the core functions of the implemented model SMILE.

```
int main()
{
   CreateNetwork();
   InfereceWithBayesNet();
   UpgradeToInfluenceDiagram();
   InferenceWithInfluenceDiagram();
   ComputeValueOfInformation();
   return(DSL_OKAY);
};
```

SMILE.NET is a library of .net classes for reasoning about graphical probabilistic models, such as Bayesian networks and influence diagrams. It can be embedded in programs that use graphical probabilistic models as a reasoning engine. It is a wrapper library that enables access to the SMILE and SMILEXML C++ libraries from .net applications. SMILE.NET is not limited to stand-alone applications. It can also be used on the back-end side of a multi-tiered application.

2.4 GeNIe

The GeNIe's name and its uncommon capitalization originate from the name Graphical Network Interface, given to the original simple interface to SMILE, the library of functions for graphical probabilistic and decision-theoretic models (Druzdzel, 1999). GeNIe is a development environment for building graphical decision-theoretic models. It is implemented in Visual C++ and draws heavily on MFC (Microsoft Foundation Classes). It allows for building models of any size and complexity, limited only by the capacity of the available memory of the computer. The original interface was designed for SMILE which is described in a previous section. It may be seen as an outer shell to SMILE. It provides numerous tools for users such as an interface to build Bayesian network models or influence diagrams, to learn the causal relationships of a model using various algorithms, and to perform model diagnosis. In order to use GeNIe efficiently, the GeNIe software must be installed and the user should have some background knowledge about probabilistic graphical models and become familiar with the tools provided in GeNIe. Fig. 5 shows the main interface of GeNIe program.

Fig. 5. The main GeNIe interface

3. Graphical Bayesian Network Model

3.1 Bayesian network model in GeNIe

In the first phase, we develop and test the graphical Bayesian network model in GeNIe as shown in Fig. 6. The students' attitude on several factors in an enrollment decision has been proposed as a case study for the model. This model contains ten variables or nodes. There are nine parent nodes thus there are no predecessor nodes and one child or predecessor node. The outcomes of each parent node are identical. It consists of impact and no impact values. There are also two outcomes for the child node (the enrollment node), enroll and not enroll values. The probability values for each parent node and the values for each state combination with an enrollment node are further defined by an expert.

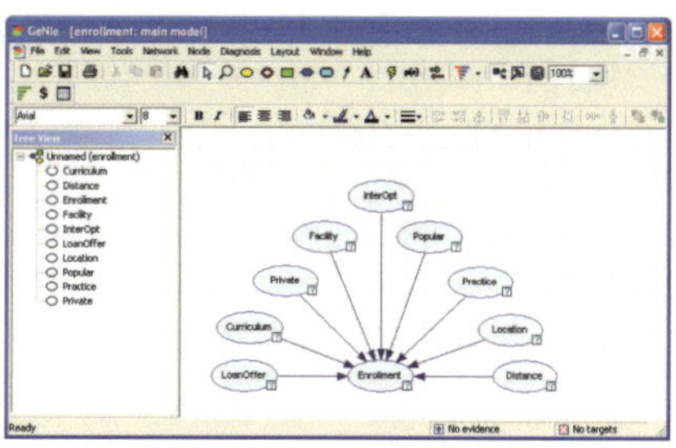

Fig. 6. Graphical Bayesian network model in GeNIe

When the specified outcome of each node and their probability values are defined, the belief updating is ready. The belief update allows for performing Bayesian inference. It is used to compute the impact of observing values of a subset of the model variables on the probability distribution over the remaining variables. Working with this model and performing Bayesian inference, we can answer simple questions. For example, the question: "What is the chance for the impact for every parent node if the expert judges the prospects for impact to be enroll?" The evidence for the enrollment variable is set at the value of “enroll” as shown in Fig. 7. We have observed a value of the enrollment variable and ask it to update its probability distribution over all parent variables. The result is shown in Fig. 8.

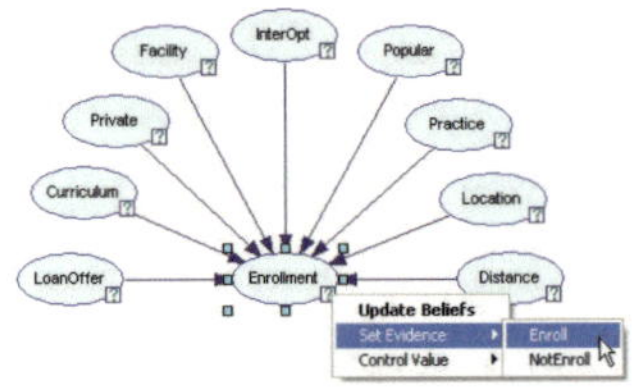

Fig. 7. Setting evidence at enroll outcome for an enrolment node

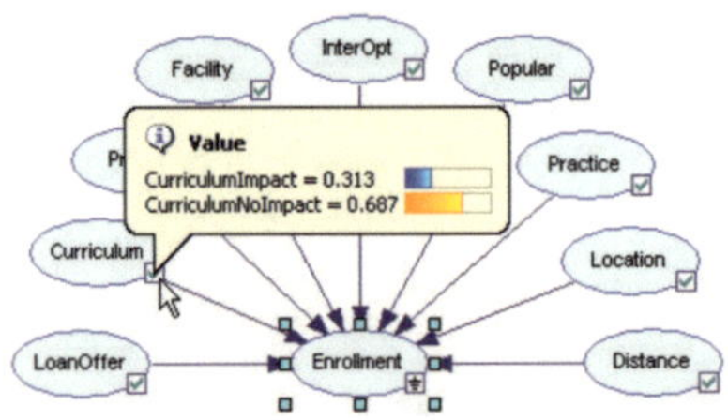

Fig. 8. The posterior probability distribution over a curriculum node

Constructing a Bayesian network model in GeNIe is simply done. There are a lot of tools provided in GeNIe for working and implementing a model but GeNIe has some limitations. Firstly, GeNIe only runs under the Windows operating systems. GeNIe is implemented in Visual C++ and draws heavily on the MFC (Microsoft Foundation Classes), which runs only on a Windows platform. It does not support cross-platform, web or an Internet-based application environment so that there are some limitations for its use on a worldwide basis. Secondly, the probability value of each variable node must be entered manually. This means that the probability determination method must be done before using GeNIe. The probability values can be obtained by asking the experts, statistical methods, or learned data from a database. However, the probability values are still put into the model by hand because GeNIe itself cannot support real-time or dynamic data. Thirdly, a graphical presentation such as pie chart or bar chart in GeNIe is intentionally designed for displaying an individual node. It does not present an overview or comparison for similar outcomes of all nodes. Lastly, the model in GeNIe is static, not dynamic. The model needs to be loaded, have some values changed, and observe the results after updating beliefs one at a time.

3.2 Client/server architecture for SMILE web application

To overcome these limitations of GeNIe mentioned in 3.1. We designed the SMILE web application that works similar to GeNIe. GeNIe is the interface to SMILE for a windows platform. The SMILE web application is the interface of SMILE on the web or an Internet-based platform. It means that the SMILE web application can support real-time data processing that GeNIe cannot. It also supports a dynamic data feed into the model. See Client/Server Architecture of the SMILE web application in Fig. 9.

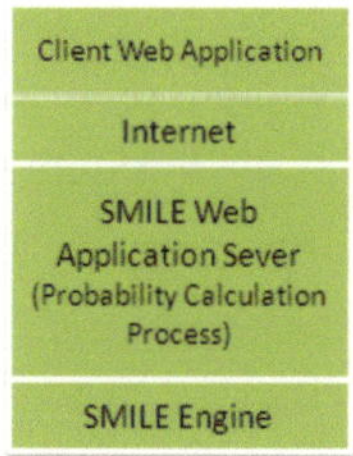

Fig. 9. Client/server architecture of SMILE web

In the client/ server architecture of the SMILE web application, the client web application is designed in order to collect data from students through an online questionnaire. The data from the client is sent over the Internet to the server. The server web application or SMILE web is designed to handle incoming data, calculate probability values and put them into each chance node, construct the Bayesian network model in .xdsl file format, feed the calculated probability values into the model, call the core functions of SMILE, read and update probability values for each node in database, send all parameters to SMILE, receive values from SMILE and visualize the results. Both the client and server web application are implemented in the ".NET" environment. Web pages are created by ASP.NET and the code behind is developed in visual C#.net. The code behind the web server application contains the core functions of SMILE such as CreateNetwork(), InfereceWithBayesNet(), and ComputeValueOf Information(). A CreateNetwork function is mainly used for creating the Bayesian network model. This function creates chance nodes, adds arcs from one node to other nodes, and fills in the conditional probability distribution for all nodes in the model. An InfereceWithBayesNet function is used to read the .xdsl file or model, specify the clustering algorithm, update the network or update beliefs, set an evidence for each node and obtain the returned result values. The clustering algorithm in the second function works in two phases: (1) compilation of a directed graph into a junction tree, and (2) probability updating in the junction tree. It has been a common practice to compile a network and then perform all operations in the compiled version. The clustering algorithm is the fastest known exact algorithm for belief updating in Bayesian networks. The clustering algorithm is the SMILE web default algorithm and should be sufficient for most applications. When networks become very large and complex, the clustering algorithm may not be fast enough. In that case, it is suggested that the user choose an approximate algorithm, such as one of the stochastic sampling algorithms. The "ComputeValueOf Information" function is used to compute an expected value of information for the model.

4. Implementation

According to the Client/Server Architecture of SMILE Web mentioned in section 3, SMILE web is designed to work in a more flexible manner for analyzing and diagnosing reasoning. It is designed for worldwide users, who can access the Internet for diagnosing the model. It overcomes platform dependent, limitations on graphical presentation, and the manual data entry for a Bayesian network model found in GeNIe. To implement SMILE web, there are four main components according to the client/server architecture as follows: 1) Client Web Application, 2) SMILE Server Web Application, 3) Probability Calculation Process, and 4) SMILE Engine.

The first part, client web application, is an online questionnaire designed for prospective students. They are asked to fill out the questionnaire before downloading an application form from the university website. See Fig. 10.

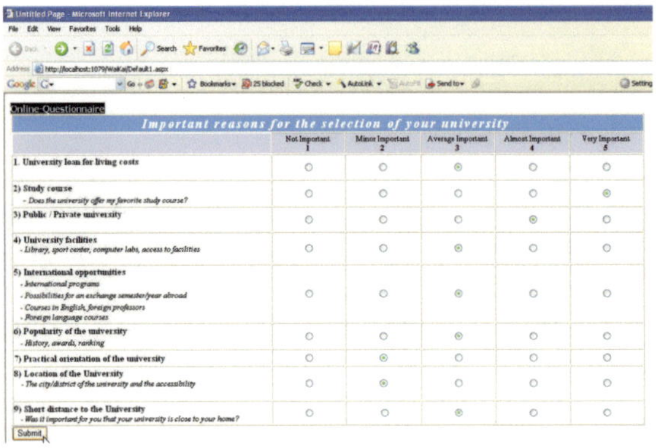

Fig. 10. Online questionnaire for prospected students

The second part, SMILE Server Web Application, is designed for the reasoning aspect of the web user interface for SMILE. Users can update beliefs and perform diagnosis through the SMILE web application as GeNIe did, See Fig. 11 and Fig. 12. The third part, Probability Calculation Process, is actually a probability calculation function in the SMILE web application. It receives the data from client web application (online questionnaire) and processes the probability values in real-time. Moreover, it is responsible for feeding the probability values into the model dynamically. The advantage of this function is that we can get real-time data and probability values for the model that GeNIe could not do. The last part, the SMILE Engine, receives data from the SMILE web application. SMILE's functions such as CreateNetwork(), InfereceWithBayesNet(), and ComputeValueOfInformation () are called to perform according to its operation. The resulting values are sent back to the SMILE web application. The SMILE engine is written in C++ in a platform-independent fashion and is fully portable. The web application's interface is defined in terms of a collection of C++ classes that form the "body" of the library and can be used within an application program. These classes allow building graphical models, editing, saving and loading them, and using them for probabilistic reasoning and decision making under uncertainty.

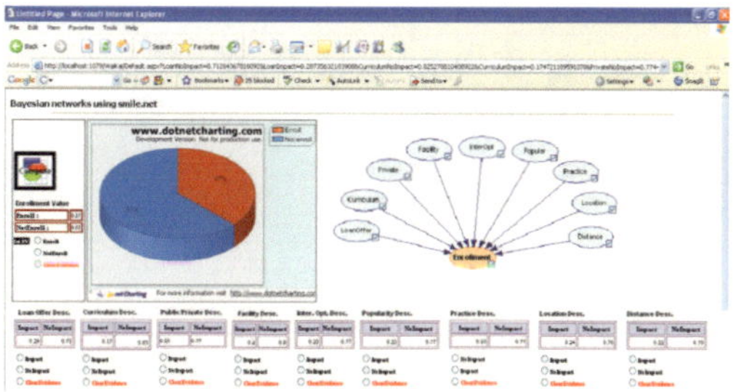

Fig. 11. SMILE web application

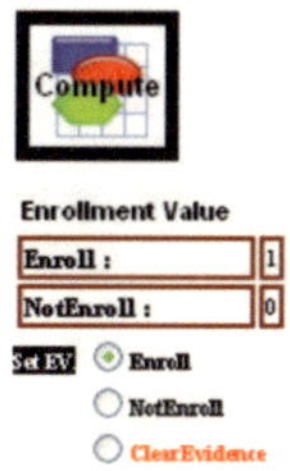

Fig. 12. Setting evidence at enroll outcome for an enrolment node

Users are allowed to perform diagnosis by setting evidence at one variable or node and exploring the probabilistic independencies among the modeled variables. See the sample variables, Public/Private University, Facilities, and International Opportunity, in Fig. 13.

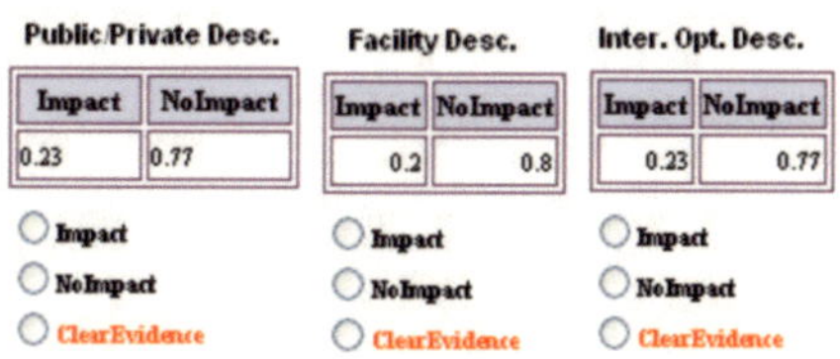

Fig. 13. Three sample nodes for observing values.

The Clear Evidence option is also provided for canceling the diagnosis and going back to use the original values in the calculation. Users can set and clear the evidence at every node in the model in order to perform diagnosis. The graphical representation of SMILE web is shown in Fig. 14, 15, and 16.

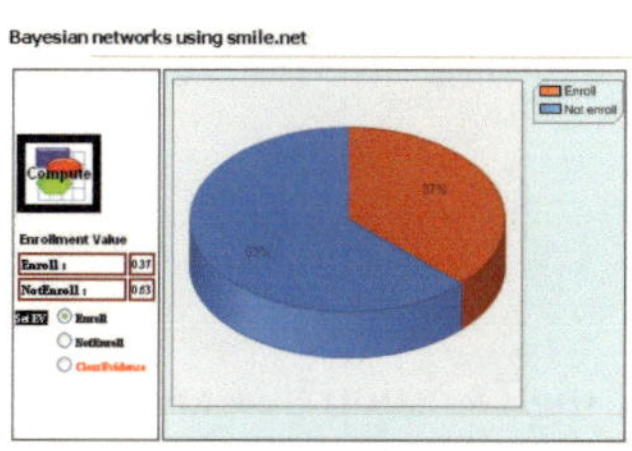

Fig. 14. Pie chart for enrollment node

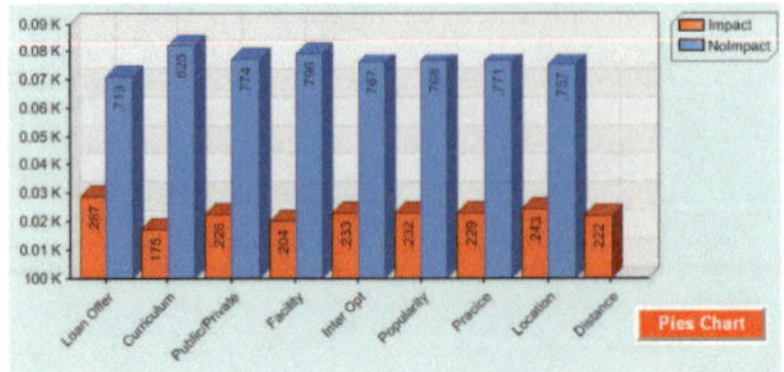

Fig. 15. Bar chart for parent nodes

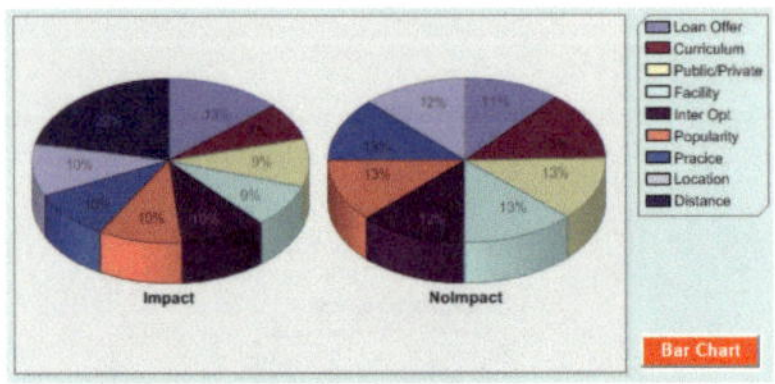

Fig. 16. Pie chart for parent nodes

5. Conclusion

GeNIe, Graphical Network Interface, is designed for a windows environment. It works well on a windows platform. It cannot be run on a web or Internet-based platform. That is why there is some limitation for its use on a worldwide basis. Another thing is that it does not support is real-time data processing. To overcome the limitations of GeNIe, the SMILE web application was designed and implemented on a client/server architecture mentioned in section 3. GeNIe is an outer shell of SMILE. SMILE web is also the outer shell of SMILE. The difference is that the SMILE web application is basically constructed in a web-based environment. SMILE web calls and submits parameters to the core functions of SMILE directly. After processing, SMILE returns all computed values back to SMILE web. SMILE web represents the Bayesian network model on a website. It is the model that users, who access the Internet, can utilize to perform diagnosis. They can update the probability distributions for each variable in a Bayesian networks in response to observations of one or more variables. SMILE web also provides a function to handle dynamic data, compute probability values in real-time, and enter them into the model. This article presents the first step for developing SMILE web application. The next step is to enhance the efficiency of SMILE web by improving the SMILE web interface, including more functions, and increasing the flexibility for model creation. The final phase for SMILE web development will be to enable it to handle influence diagrams and structural equation models. Users can use SMILE web for choosing a decision alternative that has the highest expected gain or utility.

6. Acknowledgement

The authors would like to thank the Decision Systems Laboratory, University of Pittsburgh for supporting documents, and source file of the engines: Structural Modeling, Inference, and Learning Engine (SMILE), SMILEarn, and SMILE.NET wrapper. All necessary files and documentations have been obtained from the Decision Systems Laboratory's web site. It is available at http://genie.sis.pitt.edu.

7. References

Agniezka O., Druzdzel, M. J., Hanna W., & Warsaw. (2001).Learning Bayesian Network parameters from Small Data Sets", *International Journal of approximate Reasoning,* 27(2), p. 165-182.

Cheng, J. & Druzdzel, M. J. (2000). AIS-BN: An Adaptive Importance Sampling Algorithm for Evidential Reasoning in Large Bayesian Networks. *Journal of Artificial Intelligence Research (JAIR),* Vol. 13, p. 155-188.

Cooper, G. F. (1990). The Computational Complexity of Probabilistic Inference using Bayesian Belief Networks, *Artificial Intelligent,* Vol. 42, No. 2-3, p. 393-405.

Dagum, P. & Luby, M. (1997). An Optimal Approximation Algorithm for Bayesian Inference, *Artificial Intelligence,* Vol.93, p.1-27.

Dagum, P., & Luby, M. (1993). Approximate probabilistic reasoning in Bayesian belief network is NP-Hard. *Artificial Intelligence,* Vol. 60, p. 141-153.

Druzdzel, M. J. (1999). SMILE: Structural Modeling, Inference, and Learning Engine and GeNIe: A Development Environment for Graphical Decision-Theoretic Models. *In Proceedings of the Sixteenth National Conference on Artificial Intelligence (AAAI–99),* p. 902-903, Orlando, FL.

Druzdzel M. J., & Roger R. F. (2002). Decision Support Systems. *Encyclopedia of Library and Information Science,* Second Edition.

Henrion, M. (1989). Some practical issues in constructing belief networks. *In L. N. Kanal, T. S. Levitt, and J. F. Lemmer, editors, Uncertainty in Artificial Intelligence,* 3, p. 161–173.

Gregory, F. C. (1990). The computational complexity of probabilistic inference using Bayesian belief networks. *Artificial Intelligence,* 42(2–3), p. 393–405.

Jensen, F. V.; Olesen, K. G. and Andersen, S. K. (1990). An Algebra of Bayesian Belief Universes for Knowledge-Based Systems. *Networks: Special Issue on Influence Diagrams,* Vol.20, No. 5, August 1990, p.637-659.

Lauritzen, S. L. & Spiegelhalter, D. J. (1988). Local Computations with Probabilities on Graphical Structures and their Application to Expert Systems (With Discussion). *Journal of the Royal Statistical Society Series B,* Vol. 50, No 2, p.157-224.

Pearl, J. (1986). Fusion, Propagation, and Structuring in Belief Networks. *Artificial Intelligence,* Vol. 29, No. 3, p. 241-288.

Pearl, J. (1988). Probabilistic Reasoning in Intelligent Systems. *Networks of Plausible Inference,* San Mateo, CA, Morgan Kaufmann Publishers.

Pieter, C., Kraaijeveld, & Druzdzel, M. J. (2005). GeNIeRate: An Interactive Generator of Diagnostic Bayesian Network Models. Air Force Office of Scientific Research under grant F49620-03-1-0187 and by Intel Research.

Shachter, R. D. & Peot, M. A. (1989). Simulation Approaches to General Probabilistic Inference on Belief Networks. *In Uncertainty in Artificial Intelligence,* p. 221-231, New York, N.Y., Elsevier Science Publishing Company, Inc.

Yuan, C. & Druzdzel, M. J. (2003). An Importance Sampling Algorithm Based on Evidence Pre-propagation. *Nineteenth International Conference on Uncertainty in Artificial Intelligence,* Acapulco, Mexico, p. 624-631.

http://genie.sis.pitt.edu
http://genie.sis.pitt.edu/wiki/Probabilistic_Decision_Support_System:_Bayesian_Networs
http://genie.sis.pitt.edu/wiki/SMILE:_Probabilistic_Inference_in_Bayesian_Networks
http://genie.sis.pitt.edu/wiki/Appendices:_XDSL_File_Format__XML_Schema_Definitios

10

Markovian approach to time transition inference on bayesian networks

Adamo L. Santana, Diego L. Cardoso,
João C. W. Costa and Carlos R. L. Francês
Federal University of Pará
Brazil

1. Introduction

Also known in literature as belief networks, causal networks or probabilistic networks, Bayesian networks (BN) can be seen as models that codify the probabilistic relationships between the variables that represent a given domain (Chen, 2001); being one of the most prominent when considering the easiness of knowledge interpretation achieved. These models possess as components a qualitative (representing the dependencies between the nodes) and a quantitative (conditional probability tables of these nodes) structure, evaluating, in probabilistic terms, these dependencies (Pearl, 1988). Together, these components provide an efficient representation of the joint probability distribution of the variables of a given domain (Russel and Norvig, 2003).

An abundance of papers in literature study BNs and the many aspects and characteristics of their inherent architecture. The development of these studies have led the BNs to be known in many areas out of their original scope, and their application and capabilities are still being passed on to many other areas and domains. BNs are more known and popularised by the name of Bayesian networks.

The wide study and evolution of BNs has led not only to a spread in their usability, but also, perhaps most importantly, to their development and improvement. Their features (e.g. graphical modelling, representation, inference, analysis, diagnosis, etc.) have been carefully studied, and provided us with both enhanced quality and performance. The studies heretofore are, however, still a fraction of what can still be accomplished; for, as it holds true similarly as in many other models, there is plenty of room for improvements, whether it is on particular aspects, discovering new applications, creating new hybrid systems or models with its theoretical principles, etc.

The fact remains that BNs are now widely used in the most varied areas of study, and their use has spread such that nowadays they are not only limited to researchers, but also used by regular users, perhaps even unaware of the theory and mathematics behind them. Free and commercial versions of programs implementing their algorithms are easily available and accessible.

This paper is mainly focused on the inference and representation of BNs. The main objectives are as follows: (i) present a time analysis approach for BNs based on Markovian models by using a graphical representation to model the networks' attributes and transitions; (ii) allow to directly model the effect of inferences in all the attributes of the network within their state space and instances of time; and (iii) to make possible for analyses of inferences considering the order that they are applied.

In section 2, some concepts of probabilistic networks are presented. The theoretical model proposed here is presented in section 3. The description of the model is target of section 4. Section 5 shows a case study application. In section 6, the final remarks are presented.

2. Probabilistic networks

A probabilistic network is composed of several nodes, with each node representing a variable (i.e. an attribute of the domain); arcs connecting them and whose direction implies in the relation of dependency between the variables; and probability tables for each node.

One of the major advantages of BNs is their semantics, which facilitates, given the inherent causal representation of these networks, the understanding and the decision making process for the users of these models [2]. This is basically because the relations between the variables of the domain can be visualised graphically, besides providing an inference mechanism that allows quantifying, in probabilistic terms, the effect of these relations.

We will consider here the notation for the probability of an event b given the evidence of a as $P(b \mid a)$, where $a \in A$ and $b \in B$, and A, B are variables of the BN. To calculate the posterior probability, the Bayes' Rule (1) is used.

$$P(b \mid a) = \frac{P(a \mid b)P(b)}{\sum_{b'} P(a \mid b')P(b')} \tag{1}$$

The analysis of BN presented here excludes the initial activity of creation of the BN graphical structure, assuming it has been previously made. This step is, however, of extreme importance, being when the independence relations are discovered (whether automatically or with the help of a domain expert).

The learning of the network's model is also complemented by the learning its parameters (i.e. the associated probabilities of the attributes), thus creating the structure representation (qualitative and quantitative). We will abstain to further detail this aspect here, but there are many papers in literature that study the learning of graphical representation of the PN and its details, among them (Cooper and Herskovitz, 1992), (Li et al., 2004), (Santana et al., 2007), (Spirtes et al., 1994) and (Zheng and Kwoh, 2004).

It can, then, be seen that BN represent a time variant model, representing the relations between the variables of a domain. Such relations are thus modelled in an architecture composed of nodes and directed arcs, and the direction of these arcs represent a relation of cause and effect; which, by definition implies on a relation of time, however brief it might be.

3. Background and theoretical model

In most works presented in literature, time analysis is made by using time series models. However, techniques such as dynamic Bayesian networks (DBN) (Murphy, 2002), hidden Markov models (HMM) (Rabiner and Juang ,1986) or Kalman filters (Kamlman, 1960) are more appropriate when there is a need to study the dependencies between variables, adding also a probabilistic reasoning. Hidden Markov models and Kalman filters can also be considered as particular cases of dynamic Bayesian networks (Nilsson, 1998).

The model presented here differs from the application of temporal or dynamic Bayesian networks, in which the time constraints are seen differently. While we observe each directed arc as the representation of a given instance of time *t*; in a DBN, the full network structure is considered, remaining unchanged for each *t*, which is held separately.

The data model for a time series can be represented as a structure formed by a time scale with a number of k cases, where $k = 1,2,\ldots,t$; a number of j attributes $j = 1,2,\ldots,p$, usually divided into i discrete objects (or time intervals) which repeat throughout the studied period of time. Figure 1 presents the time series model according to the data cube representation (Dillon and Goldstein ,1984).

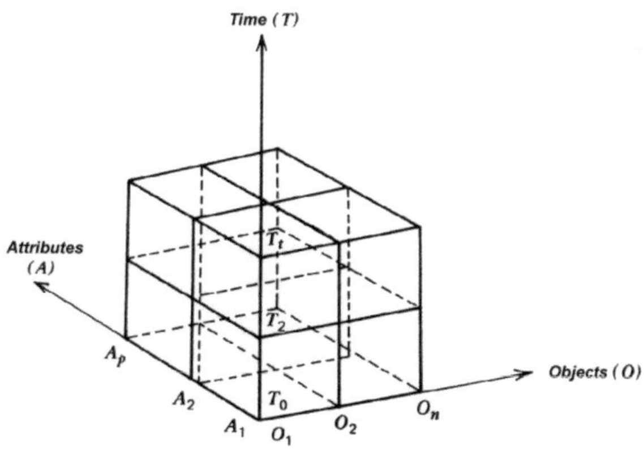

Fig. 1. Data cube structure.

A classical initial problem when working with BNs in the time would be the existing need to built conditional probability tables for each discrete unit of time analysed. Thus, a stationary random process is often assumed.

In the work described here, the time analysis and transition preceding from the BN are modeled into a discrete time Markov chain. Providing with means to compute, for example, the effect of a given inference after *n* units of time or how many units of time would take to achieve desirable probabilistic states for the attributes.

The approach presented uses the qualitative and quantitative data of the BN by modelling, for a given variable, a Markovian time transition matrix according to a first-order process; but also intrinsically considering the other variables of the domain, which might also influence in the behaviour of this attribute. This is because a BN can be seen as an array of attributes that might influence on one another over time.

To exemplify the model, a simple example of a BN can be considered, composed by only two variables: *Grade* and *Study;* where the grade obtained on a given test depends on the amount of study applied. It is also assumed that the tests are taken on a monthly time scale. It is considered as possible values for the attributes the following: Study (Hard, Medium, Little); and Grade (Excellent, Good, Regular).

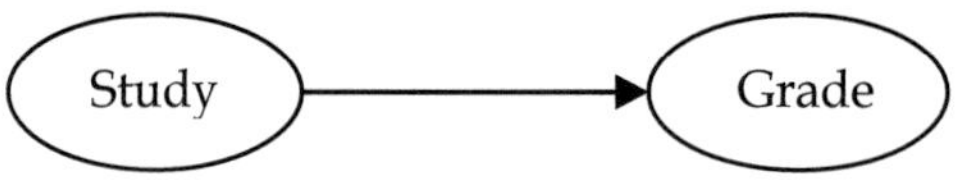

Fig. 2. Bayesian network for variables Grade and Study.

In this sense, the BN would also present the values of initial and conditional (for *Grade* only, given that it is the only attribute that possesses a parent attribute, that is, a dependence relation of the *Grade* given the *Study*) probabilities. The dependency model and the probability tables would represent all the data the BN could offer us.

Following the Markovian modeling, what we are seeking to obtain is the time instant that, given an inference, a determined probability configuration of an attribute would happen (e.g. considering our example, given that we study *Hard*, when we would obtain a grade *Excellent* with probability of 70%, *Good* with 25% and *Regular* with 5%).

Given that what we seek is in fact the new configuration of a determined attribute, what we end up needing is to set up the Markovian transition matrix of this attribute. This is done by mapping the transition probabilities for the states of the attribute onto the matrix, based on the conditional probabilities that it possesses given its dependencies with the other attributes (e.g. also considering the example, we must map the transition probabilities of *Grade* for: *Excellent* and pass to *Good, Excellent* to *Regular, Excellent* and achieving *Excellent* again etc). That is, we would have to compute the transition probabilities for the states of a given variable, which Markovianly speaking we can anagously see as the transition probability to achieve a state N_{t+1} based on N_t. Hence we seek to find the probability $P(N_{t+1} = s_y \mid N_t = s_x) = p_{xy}$; thus creating a Markov transition matrix, according to the model in Table 1.

Grade\Grade	*Excellent*	*Good*	*Regular*
Excellent	p_{EE}	p_{EG}	p_{ER}
Good	p_{GE}	p_{GG}	p_{GR}
Regular	p_{RE}	p_{RG}	p_{RR}

Table 1. Model of the Markov transition matrix to be mounted

However, considering only the factor of study in relation to the grade is not enough to verify the relation of the variable *Grade* with itself and to make the transition between its states, as the Markov transition matrix would immediately converge to the stationary state. So, we must also consider the value of the attribute *Grade* at a previous point of time, acting together with the variable *Study* and thus obtaining the transition relations for the variable *Grade.*

For such, the first record in the existing historical database is ignored so that we can insert in the analysis, analogously to a 1st order Markovian process, the Previous Grade obtained. Tables 2 and 3 present the marginal and conditional (Study, Grade and the Grade in the previous period) probabilities of the Current Grade considering the Study and the Previous Grade (Grade-1).

Study	
Hard (Ha)	0,133
Medium (Me)	0,534
Little (Li)	0,333

Grade	*Grade*	*Grade-1*
Excellent (E)	0,210	0,333
Good (G)	0,467	0,333
Regular (R)	0,323	0,333

Table 2. Initial probabilities of the Bayesian network.

Study ∩ G-1\Grade	E	G	R
Ha ∩ E	0,934	0,033	0,033
Ha ∩ G	0,333	0,333	0,333
Ha ∩ R	0,333	0,333	0,333
Me ∩ E	0,491	0,491	0,018
Me ∩ G	0,033	0,934	0,033
Me ∩ R	0,018	0,491	0,491
Li ∩ E	0,333	0,333	0,333
Li ∩ G	0,018	0,491	0,491
Li ∩ R	0,033	0,033	0,934

Table 3. Conditional probabilities of the Bayesian network - P(Grade | Study ∩Grade-1).

The calculations for the Markov transition matrix would follow:

$$p_{EG} = P(E) \times \left[P(G \mid Ha \cap E)P(Ha) + P(G \mid Me \cap E)P(Me) + P(G \mid Li \cap E)P(Li) \right] \quad (2)$$

Generalizing, the Markovian transition matrix (Table 4) will be computed by mapping the transition probabilities of the states of a given variable; that is, the transition probability to achieve N_{t+1} based on N_t, being $P(N_{t+1} = s_y \mid N_t = s_x) = p_{xy}$.

$$\begin{array}{c} \\ x_1^t \\ x_2^t \\ \vdots \\ x_n^t \end{array} \begin{array}{c} \begin{array}{cccc} x_1^{t+1} & x_2^{t+1} & \cdots & x_n^{t+1} \end{array} \\ \begin{bmatrix} p_{x_1} & p_{x_1x_2} & \cdots & p_{x_1x_n} \\ p_{x_2x_1} & p_{x_2} & \cdots & p_{x_2x_n} \\ \vdots & \vdots & \ddots & \vdots \\ p_{x_nx_1} & p_{x_nx_2} & \cdots & p_{x_n} \end{bmatrix} \end{array}$$

Table 4. General model of the Markov transition matrix.

The probabilities p_{xy} for the transition matrix are calculated according to:

$$p_{xy} = \frac{\sum_{i=1}^{n} P(s_y \mid s_x \cap Pa_i) \times P(Pa_i)}{\sum_{j=1}^{m} \sum_{k=1}^{n} P(s_j \mid s_x \cap Pa_k) \times P(Pa_k)} \quad (3)$$

where s represents the observed variable and its respective states; Pa is the variable that represents the parents of variable s; m is the number of states the attribute can assume; and n is the number of possible states and/or combinations that the parents of this attribute can assume.

Consisting the denominator of the equation only as a normalized function (α), we have:

$$p_{xy} = \alpha \sum_{i}^{n} P(s_y \mid s_x \cap Pa_i) \times P(Pa_i) \quad (4)$$

Calculating from (4), we obtained the Markov transition matrix (represented by the letter P), presenting the transition probabilities for the states of the variable studied. For the considered example, we would have (Table 5):

Grade\Grade	*Excellent*	*Good*	*Regular*
Excellent	0.497	0.378	0.125
Good	0.068	0.707	0.225
Regular	0.065	0.318	0.618

Table 5. Markov transition matrix obtained.

Furthermore, to find the probability vector at a given time *n*, we need only to calculate the *n*th power of the probability matrix $P^{(n)}$, as described by the Equations of Chapman - Kolmogorov (Bolch et al., 1988).

$$P^{(n)} = P^{(m)} \times P^{(m-n)} \tag{5}$$

where $P^{(n)}$ is the transition matrix in the step *n*; and thus $P^{(n)} = P^{n}$.

Thus, following on the example, if the unit of time is discretized in months and if we wanted to obtain the probabilities for the grades occurrence three months from now, we would have to find the power P^3 of the matrix (Table 6).

Grade\Grade	*Excellent*	*Good*	*Regular*
Excellent	0.1878	0.5274	0.2851
Good	0.1085	0.5561	0.3359
Regular	0.1071	0.4976	0.3974

Table 6. States transition matrix in the step $n = 3$.

The analysis presented (Tables 5 and 6), considered the behavior of the domain, given the available data, in time without any inference being made. Such analysis, however, can also be made, thus providing make the analysis in time given the evidence of a determined state of a variable, being able, as well, to consider its impact in a given time step. As example, considering as fact that the level of *Study* applied to make the test was *Medium*, we would have (Table 7):

Grade\Grade	*Excellent*	*Good*	*Regular*
Excellent	0.491	0.491	0.018
Good	0.033	0.934	0.033
Regular	0.018	0.491	0.491

Table 7. Transition matrix considering the inference made - Study: Medium.

Thus, considering the inference made, we would have in a step $n = 3$ the following matrix (Table 8).

Grade\Grade	*Excellent*	*Good*	*Regular*
Excellent	0.150	0.805	0.045
Good	0.054	0.892	0.054
Regular	0.045	0.805	0.150

Table 8. Transition matrix in the step $n = 3$ considering the inference made - Study: Medium.

To go back from the Markovian transition matrix to the marginal probabilities of the variable we apply (6).

$$P(s_x^t) = \sum_i^n p_{ix} \times P(s_i^{t-1}) \tag{6}$$

From (6), the probabilities for each state of the attribute *Grade* in a time period $n = 3$ given the inference of *Medium Study* can be found. The probabilities for the attribute *Grade* considering the example given here are as follow: *Excellent* 0.083, *Good* 0.834 and *Regular* 0.083.

4. Model description

In order to keep track of the whole network, and allow to directly model the effect of inferences in all the attributes of the network – and not just one at a time, as it was initially specified in [4] – we will first ascertain the diagram representation, to which we will map the BN.

We take a simple example of a BN (Fig. 3), for sake of simplicity, from which we will explain and build our model. The BN consists of six binary variables $X = [A, B, C, D, E, F]$.

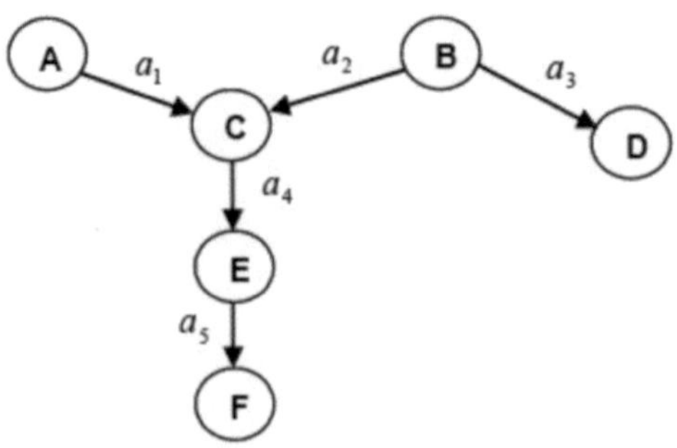

Fig. 3. Bayesian network example.

In Fig. 3 we see the existence of five arcs (a_1 to a_5) connecting the six variables of the BN, considering τ as the set of all r arcs in a BN, whereas $\tau = [a_1, a_2, \ldots, a_r]$ and each arcs connects two nodes of the network. Notably, each arc of τ can represent a different time instant in the domain's transition timeline, from which an event (cause) inferred in the network will take to present an impact (effects) in the node directly connected to it.

We insert here the definition of *eras*. While this concept might be familiar to some, and has been applied in the literature of quantum networks (Tucci, 1998), we use it here with some different considerations, pending toward the analysis of each node.

The set of eras E, where $\mathrm{E} = \varepsilon_1, \varepsilon_2, \mathrm{K}, \varepsilon_n$, could be specified by removing successive layers of nodes [16], either internally (staring from the root nodes) or externally (starting from the leaf nodes). Considering the network in Fig. 3, by removing each layer of root nodes one after the other, we would have the eras as depicted in Fig. 4a. Similarly, by removing the layers of external nodes, the schemata would be as shown in Fig. 4b.

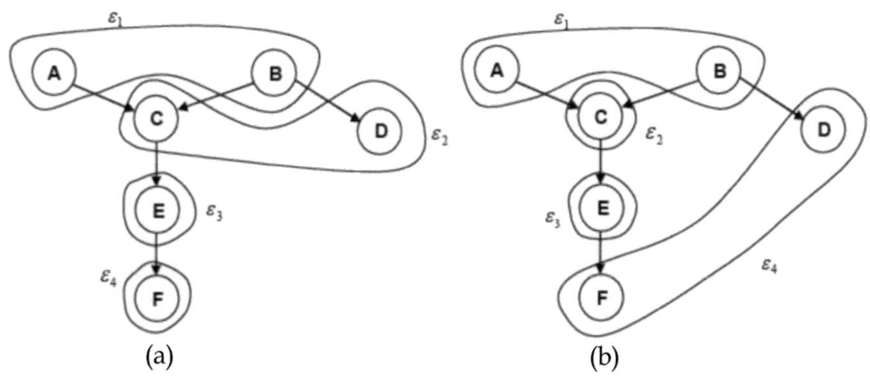

Fig. 4. (a) BN separated by eras considering root nodes removal; (b) BN separated by eras considering external nodes removal.

In the model proposed here, the eras can also be built starting from either the root or leaf nodes. For each era, separated space instants are drawn for each of the nodes held therein. For each of these spaces, a sub-network is placed, consisting of the node and its parents. From the BN graphical structure in Fig. 3, a temporal structure for the theoretical model presented here is built, as presented in Fig. 5.

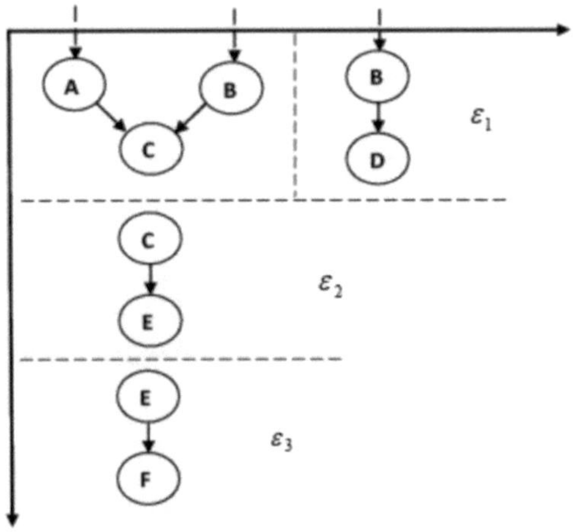

Fig. 5. Diagram structure for the BN.

In the diagram shown in Fig. 5 we separate the nodes of the network according to their dependencies and order of transitions. To visualize the network in this manner is useful when we consider that, for the Markovian time analysis that is induced, the transition matrix is calculated based on the individual node and the other nodes it is directly correlated.

The criteria for the subnets differs, however, for the one of a Markov blanket (Lauritzen , 1996), which, for any node in a BN, represents the set of nodes comprising the parents, the children, and the parents of the children of the node of interest (Chang et al., 2000); consisting here of a given node and its directly related parents as root nodes. To account for the diagnosis type of evidence analysis, however, which involves the backward flow in the active trail of the inference, the consideration of all correlated nodes involved would be necessary for the calculation of the transition matrix (thus making use of the entire Markov blanket space).

As described previously, the model presented here focuses on the analysis and inference method of the BN. The former is applied here by building a corresponding time specific model characterized by the transition of successive eras, from which the latter will take place by following classical search parameters (Pearl, 1988) for the inference calculation processing, defined by (4).

According to the probability rules of which BNs are based, and together with their graphical structure, it can be easily seen the reason for which an order of occurrence for multiple evidences in a BN cannot be defined. We can, however, make assumptions here to account the impact of such ordering, and to consider the evidences as being simultaneous or successive.

As it was stated, we can see the more than intrinsic relation between both, that is, that causal influence determines temporal relationship - as cause regularly precedes its contiguous effect (Hume, 1975). Thus the temporal order is determined by the causal order (Carrier, 2003). Again, considering every arc in τ as a time transition instant.

The idea that the order of occurrence for multiple evidences cannot be defined in BNs presents as a faculty toward simultaneity, in which mutual evidences in the system are tied together. Such need for ordering might be irrelevant though, as the single probabilistic effect of each evidence can be differentiated by the path of causal arcs and the probability values.

So, while their evidence is made simultaneously, their effect can be seen as successive. The exception might be for the inferences made on the parent nodes of the given variable. In fact, unless the events are bound to the same space, their simultaneity in a given frame do not imply a simultaneity in another. Hence their grouping in subnets, for states that are simultaneous in experience stand in interaction with one another and are mutually tied together (Kant, 1787).

But what if, disregarding their causality order of the BN structure, a temporal analysis: as to an evidence e_a is made, and only after *n* units of time an evidence e_b is considered?

Taking the example of Fig. 5, and the evidence $P(E=e \mid C=c \wedge B=b)$, being e_a and e_b as $(C=c)$ and $(B=b)$, respectively, and $n=2$. In this example, the influence of evidence e_a over C is continuous long before e_b is applied. A successive application of (4) would account on such matters of differently temporal instantiation of inferences (7).

$$P(a \mid e_1, \ldots, e_n) = \alpha \sum_t^n P(e_t \mid a) P(e_t) \tag{7}$$

It is important to notice that such assumption is only possible on some domain analysis. In the sense that, considering a model in which e_a would bring to an ultimate absorbing state (e.g. death, destruction, a deadlock in the system, etc.), no size of *n* or evidence e_b would cause any change of state.

The model presented here can also address such matter of evidence ordering; using (3) for building the transition matrix and thereon applying inferences according to the order of evidences e_t and their given time instant *t*. The transition model for our curretn example (Fig. 3 and 5) can be seen as described in Fig. 6.

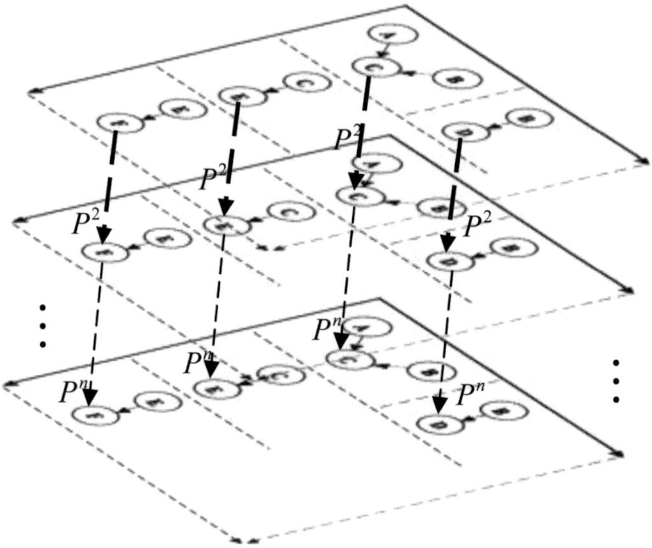

Fig. 6. Time transition model.

We are thus able to calculate the probabilities of the attributes in their time transitions, visualising the impact of the inferences as they progress. It is also elastic enough to allow the insertion of a new evidence at any time frame, visualising the influence of the evidences according to the order that they occur.

5. Case study application

An example of application of the proposed model in a case study in the area of power systems is presented next, ratifying the applicability of the method.

The analysis presented is part of a study made in (Rocha et al, 2006), to establish prospections for the consumption of energy in a given region. One of the most desired aspects for power suppliers is the acquisition/sale of energy for a future demand. However, power consumption forecast is characterized not only by the variables of the power system itself, but also related to socio-economic and climatic factors.

Since the methods for load forecast use only the consumption data, it was necessary to offer means to analyze the correlations. Hence the use of Bayesian networks to codify the probabilistic relations of the variables and to make inferences on the conditions of the power system from the historical consumption and its correlation with the climatic and socio-economic data.

We present an application of the model for the power suppliers to project and correlate these parameters, studying the progression of their behaviour through time.

The data used in the work referred to a study of correlations for the consumption of energy of the city of Oriximiná - Pará and the climatic factors, established in a monthly time scale.

The database is composed by ten variables, with eleven arcs connecting them in the BN (Figure 7). The attributes denote the observed types of power consumption (residential, commercial, industrial and public) and climatic factors (temperature, relative humidity and pluviometric rate).

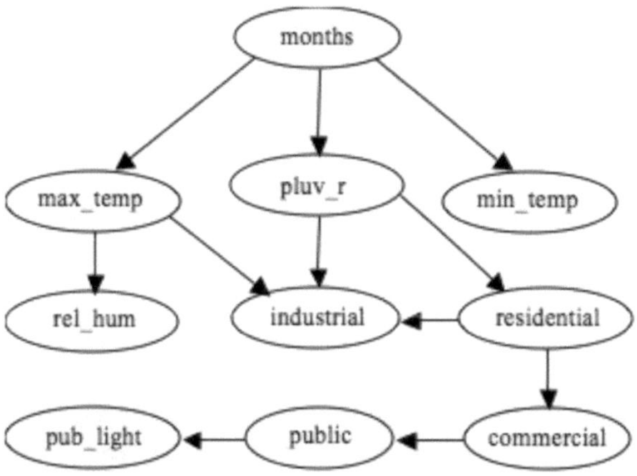

Fig. 7. Bayesian network correlating the Power consumption and the climatic factors

The analysis considered for example intends to study the changes occurred in the probabilities of the variable of commercial consumption (commercial), given an inference in the increase of the pluviometric rate, assuming this constant increment in a period of six months. The attribute of pluviometric rate (pluv_r), used to infer in the BN model, is a continuous variable by nature; its values, however, are presented as discretized in five states, according to the frequency of their values, which vary from a value of 1.479 to a maximum of 315.292 mm; the variable commercial, which represents the power consumption (in MW) in the commercial sector, had its values discretized in five states as well, varying from 126,918 to 219,649. The discretized states are displayed in Table 9.

Pluv_r
$[1.497 \rightarrow 32.408)$
$[32.408 \rightarrow 43.422)$
$[43.422 \rightarrow 88.154)$
$[88.154 \rightarrow 161.583)$
$[161.583 \rightarrow 315.292]$

Commercial
$[126{,}918 \rightarrow 148{,}047)$
$[148{,}047 \rightarrow 160{,}840)$
$[160{,}840 \rightarrow 174{,}684)$
$[174{,}684 \rightarrow 195{,}908)$
$[195{,}908 \rightarrow 219{,}649]$

Table 9. Discretized states of the variables pluv_r and commercial

The progression of the commercial consumption given the established hypothesis is computed according with the Equation (4), thus obtaining the Markovian transition matrix for the observed variable, as presented in Table 10. The discretized states (range of values), pointed in Table 9, are, for simplification, represented by labels from C_1 to C_5, according to the increasing values of its states.

$$P = \begin{array}{c} \\ C_1 \\ C_2 \\ C_3 \\ C_4 \\ C_5 \end{array} \begin{array}{c} \begin{array}{ccccc} C_1 & C_2 & C_3 & C_4 & C_5 \end{array} \\ \left[\begin{array}{ccccc} 0.371 & 0.371 & 0.086 & 0.086 & 0.086 \\ 0.319 & 0.191 & 0.391 & 0.049 & 0.049 \\ 0.049 & 0.238 & 0.427 & 0.143 & 0.143 \\ 0.078 & 0.078 & 0.205 & 0.360 & 0.278 \\ 0.116 & 0.116 & 0.116 & 0.301 & 0.351 \end{array} \right] \end{array}$$

Table 10. Markovian transition matrix for the variable of power consumption

Its equivalent obtained after the sixth iteration, that is, the Markovian matrix representing the transition probabilities after a six months period, is presented in the following table.

$$P^6 = \begin{array}{c} \\ C_1 \\ C_2 \\ C_3 \\ C_4 \\ C_5 \end{array} \begin{array}{c} \begin{array}{ccccc} C_1 & C_2 & C_3 & C_4 & C_5 \end{array} \\ \left[\begin{array}{ccccc} 0.181 & 0.205 & 0.266 & 0.176 & 0.171 \\ 0.179 & 0.203 & 0.265 & 0.177 & 0.172 \\ 0.177 & 0.201 & 0.264 & 0.181 & 0.175 \\ 0.175 & 0.199 & 0.262 & 0.184 & 0.177 \\ 0.176 & 0.199 & 0.262 & 0.183 & 0.177 \end{array} \right] \end{array}$$

Table 11. Markovian transition matrix after the transition of six time units

Applying Equation (6), the marginal probabilities for the given analysis can be obtained again, identifying the following distributions for the commercial variable: $C_1 = 0.1776$; $C_2 = 0.2014$; $C_3 = 0.2638$; $C_4 = 0.1802$; and $C_5 = 0.1744$. Resulting in an update in the probabilities of the events and a presents the evidence of a higher consumption in the intermediate state, which ranges the values from 160,840 to 174,684 MW.

6. Remarks on the presented work

This work described a Markovian approach to represent the variables in a probabilistic network and their behavior throughout time, providing with a method for visualising and capturing their correlations.

The use of a Markovian model introduces advantages from its mathematical basis: the assumption that the present state depends only of its previous state and, adding to it, the fact that the Markovian models possess relatively simple solutions compared to its computational effort and to the mathematical complexity involved; which stimulates and facilitates its use.

A Markovian approach for time transition is shown, highlighting the use of the network's structure, that alone expresses the relations of dependence and causality among the variables; and the probabilities associated to them, which serve as a basis for the creation of the Markovian transition matrix. Thus providing means for the study of the probabilistic transitions of the observed events, considering or not inferences, throughout the time.

The model also provides for the analysis of inferences considering the order in time that that they are applied in the network. This fact allows extending the interpretability of the probabilistic networks and adjusting them even further for applications of the real world.

7. References

Bolch, G. ; Greiner, S. ; Meer, H. and Trivedi ; K. S. (1998) *Queuing Networks and Markov Chains: Modeling and Performance Evaluation with Computer Science Applications.* John Wiley & Sons, Inc, New York, USA.

Carrier, M. (2003) How to Tell Causes from Effects: Kant's Causal Theory of Time and Modern Approaches, *Studies in History and Philosophy of Science*, 34, (2), 59-71.

Chang, K. C. ; Fung, R. ; Lucas, A. ; Oliver, R. and Shikaloff, N. (2000) Bayesian networks applied to credit scoring, *IMA Journal of Management Mathematics*, 11 (1), 1-18.

Chen, Z. (2001) *Data Mining and Uncertain Reasoning - an Integrated Approach.* John Wiley Professional.

Cooper, G. and Herskovitz, E. (1992) A Bayesian Method for the Induction of Probabilistic Networks from Data, *Machine Learning*, 9, 309-347.

Dillon, W. R. and Goldstein, M. (1984) *Multivariate Analysis - Methods and Applications.* John Wiley & Sons.

Hume, D. (1975)*An Enquiry Concerning Human Understanding.* Oxford University Press.

Kalman, R. E. (1960) A New Approach to Linear Filtering and Prediction Problems, *Transactions of the ASME - Journal of Basic Engineering*, vol. 82, 35-45.

Kant, I. (1787) *Critique of pure reason.* P. Guyer, & A. W. Wood - Ed. & Trans., Cambridge University Press.

Lauritzen, S. L. (1996) *Graphical Models.* Oxford University Press.

Li, X. ; Yuan, S. and He, X. (2004)Learning Bayesian networks structures based on extending evolutionary programming, *Machine Learning and Cybernetics, Proceedings of 2004 International Conference on*, 3, 1594-1598.

Murphy, K. (2002) *Dynamic Bayesian Networks: Representation, Inference and Learning*, PhD Thesis, Computer Science Division, UC Berkeley.

Nilsson, N. (1998) *Artificial Intelligence: A New Synthesis.* Morgan Kaufmann Publishers.

Pearl, J. (1988) *Probabilistic Reasoning in Intelligent System.* Morgan Kaufmann Publishers.

Rabiner, L. R. and Juang, B. H. (1986) An introduction to Hidden Markov Models, *IEEE ASSP Magazine* 3 (1), 4-16.

Rocha, C. ; Santana, A. L. ; Frances, C. R. ; Rego, L. ; Costa, J. ; Gato, V. and Tupiassu, A. Decision Support in Power Systems Based on Load Forecasting Models and Influence Analysis of Climatic and Socio-Economic Factors. SPIE, v. 6383, 2006.

Russel, S. and Norvig, P. (2003) *Artificial Intelligence.* Prentice Hall.

Santana, A. ; Frances C. and Costa, J. (2007) Algorithm for Graphical Bayesian Modeling Based on Multiple Regressions, *Lecture Notes in Computer Science*, 4827, 496-506.

Santana, A. ; Frances C.; Rocha, C. ; Rego, L. ; Vijaykumar, N. ; Carvalho, S. and Costa, J. (2007) Strategies for Improving the Modeling and Interpretability of Bayesian Networks, *Data & Knowledge Engineering*, 63, 91-107.

Spirtes, P. R. ; Shcheines, R. and Clark, G. (1994) *TETRAD II: Tools for Discovery.* Lawrence Erlbaum Associates, Hillsdale, NJ, USA.

Tucci, R. (1998) How to compile a quantum Bayesian net, arXiv, quant-ph/9805016.

Zheng, Y. and Kwoh, C. K. (2004) Improved mdl score for learning of Bayesian networks, *Proceedings of the 2nd International Conference on Artificial Intelligence in Science and Technology, AISAT*, 98-103.

11

miniTUBA: a web-based dynamic Bayesian network analysis system and an application for host-pathogen interaction analysis

Yongqun He and Zuoshuang Xiang
Unit for Laboratory Animal Medicine, Department of Microbiology and Immunology, Center for Computational Medicine and Bioinformatics, and Comprehensive Cancer Center, University of Michigan, Ann Arbor, MI 48109, USA

1. Introduction

A Bayesian network (BN) is a representation of a joint probability distribution over a set of random variables (Friedman, 2004). A BN includes two components: (i) a directed acyclic graph (DAG) with vertices representing variables and edges indicating conditional dependence relations, and (ii) a set of conditional distributions for each variable, given its parents in the graph. A dynamic Bayesian network (DBN), an extension of BN, describes how variables influence each other over time. Mathematically, a DBN is a discrete time approximation of a stochastic differential equation or as a Markov chain model with possibly many states. DBN analysis has been considered as a powerful method to analyze and interpret heterogeneous, fluctuating time course data for many systems including biomedical data (Dean and Kanazawa, 1988; Friedman, et al., 2000; Korb and Nicholson, 2004). Compared to static Bayesian networks, DBN captures time varying parameters and predicts a time course of biological progression (*e.g.*, disease). DBNs also permit temporal cycles between variables allowing the user to interpret connections as temporal causation—a more clinically relevant definition of causation for many clinicians. A key advantage of DBNs over static Bayesian network analysis is that the relationships described in DBNs always have an unambiguous direction of causality.

DBNs have been used to analyze and interprete various data in different systems including clinical data (Li, et al., 2007; Neapolitan, 2003; Peelen, et al., 2010; Watt and Bui, 2008). Most medical/clinical inference data are dynamic dataset. However, investigators in the field largely depended upon multiple static comparisons to determine which clinical or experimental variables may represent potential targets for prediction or prevention of deleterious outcomes in patients. However, a clinical event (*e.g.*, sepsis) is typically a complex, heterogeneous, and dynamic process. Multiple factors may influence its outcome. This static comparison approach has led to numerous failed clinical trials which targeted single inflammatory mediators in patients without careful consideration of their clinical states in specific clinical courses (Remick, 2003). DBNs provide a means by which these dynamic and often noisy datasets can be analyzed and interpreted to predict the impact of a complete network of clinical and experimental variables in an individual patient at any given time point. This then provides a clinician or researcher the ability to diagnose a disease state and intervene before overt clinical evidence is present.

DBNs have been applied to analyze gene expression data (Dojer, et al., 2006; Husmeier, 2003; Imoto, et al., 2006; Kim, et al., 2003; Ong, et al., 2002; Pe'er, et al., 2001; Rau, et al., 2010; Yu, et al., 2004; Zou and Conzen, 2005). DBNs have generated insights that could not be obtained from static Bayesian analysis. Unlike BNs which are acyclic, DBNs allow for cycles and more closely reflect the biological realities. In addition, DBNs can improve the ability to predict causal relationships based on the temporal nature of the data. For example, Ong *et al.* used DBNs to model regulatory pathways among 169 genes in *E. coli* to physiological changes that affect tryptophan metabolism (Ong, et al., 2002). Kim et al. applied DBNs to study a 45-gene subnetwork of the cell cycle system in *Saccharomyces cerevisiae*, a species of budding yeast (Kim, et al., 2003). More recently, Rau et al. introduced an iterative empirical Bayesian procedure with a Kalman filter that estimates the posterior distributions of DBN parameters (Rau, et al., 2010). This empirical method consumes considerably less computational time and was used to analyze human T-cell activation data with 58 genes over 10 time points (Rangel, et al., 2004). However, in general DBNs have not been widely used for gene expression data analysis. Its general usefulness in modeling reliable pathways and predicting testable hypotheses remain to be demonstrated (Xia, et al., 2004).

miniTUBA is a web-based dynamic Bayesian network analysis and Gibbs simpling prediction system (http://www.miniTUBA.org), with the goal of learning and simulating biomedical networks using temporal data from experimental and clinical investigations (Xiang, et al., 2007). The miniTUBA modelling system allows clinical and biomedical researchers to perform complex medical/clinical inference and prediction. This system is designed for easy data management, and the results are displayed in a way that is easily interpretable by a clinical or biomedical investigator. The miniTUBA implementation does not require a local installation or significant data manipulation. Using synthetic data and laboratory research data, our previous publication (Xiang, et al., 2007) demonstrates that miniTUBA accurately identifies regulatory network structures from temporal data.

This chapter will describe many updated features of the miniTUBA system, and provide general instructions and pitfalls involved in miniTUBA DBN modeling. The study of host-pathogen interactions is critical to understand microbial pathogenesis and host immune responses against pathogen infections. To our knowledge, dynamic Bayesian network analysis has not been applied to study host-pathogen interactions using microarray gene expression data. In this chapter, we will report how to apply miniTUBA DBN analysis to understand the immune networks in murine macropahges responded to different *Bruccella* infections.

2. miniTUBA design and implementation

2.1 Overall miniTUBA system design

The miniTUBA software allows users to continuously update their data, fill out missing data, choose different analysis settings (e.g., data discretization, Markov lags and prior topology), perform DBN, and visualize results (Fig. 1). miniTUBA can also make temporal predictions using Gibbs sampling to suggest interventions based on an automated learning process pipeline using all data provided. Different graphic supports are provided (Fig. 1).

miniTUBA currently runs on two Dell Poweredge 2580 servers running the Redhat Linux operating system (Redhat Enterprise Linux ES 4) and Apache HTTP Server. Data are stored in miniTUBA using a MySQL database and the interface is constructed using a variety of scripts including PHP and Perl.

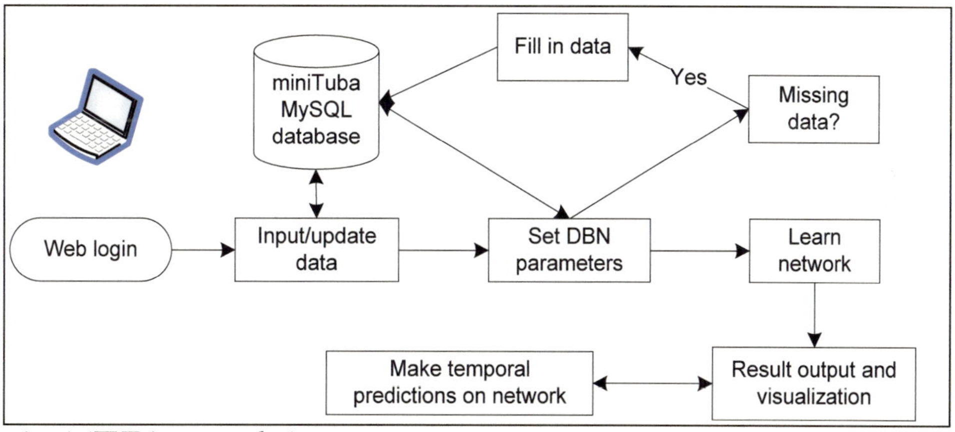

Fig. 1. miniTUBA system design.

The detailed tutorial for running miniTUBA is available at: http://www.minituba.org/docs/tutorial.php. We provide a miniTUBA Sandbox Demo (http://www.minituba.org/sandbox/index.php). This Sandbox Demo is developed for first time users to get familiar with the system by exploring the features using built-in user account and some simple data.

miniTUBA is a project-oriented web-based system. One or more projects can be created by a registered user. Currently, each project needs to go through an internal review process. This review process ensures that the computational resource is properly used since an approved project can run analyses that take up to 144 hrs – representing a significant computational investment. Once approved, a user can submit/update data, set up DBN settings and run each analysis. For each project, a user can run multiple analyses and these analyses will be stored in miniTUBA for later use.

By June 14 2010, miniTUBA has 66 registered users and hosted 82 projects including 32 testing projects.

2.2 miniTUBA DBN parameter selections

Optimal DBN parameter selections are critical to successful DBN analysis. Different DBN parameters can be set in miniTUBA to specify how the data are pre-processed and how constraints are set on the DBN learning algorithm. The DBN settings for each analysis will be stored in the miniTUBA database and can be reused for future analyses. Different DBN settings may result in different results. Default DBN settings follow the best practices described elsewhere (Yu, et al., 2004), but can be changed by the user if desired.

Depending on the purpose of a particular DBN simulation, in many cases only parts of experimental data and/or variables may be used (Fig. 2). A subset of the available experiment units has different meanings in different use cases. For microarray experiments, this may mean inclusion of selected microarray chips. In a clinical setting, it may mean analysis of only some of the patients. Certain variables can be excluded in a particular analysis. Those variables included in a study can be set as parents only (i.e., they do not act as children) or children only (i.e., they cannot be parents) (Fig. 2). For example, a drug treatment is usually considered as the start point of an experiment and hence can only have children, and a survival status is usually the final outcome of an event and hence cannot have any other child variable.

In many cases, not all required data points have experimental data. For example, most clinical data are not gathered at a consistent sampling interval. Therefore, a data fitting method is needed to fill in predicted data for these missing time points. Spline fitting is a general method of generating new data points within the range of a discrete set of known data points. Such a spline fitting approach has been shown by Yu et al to yield good behavior for reasonably smooth temporal data (Yu, et al., 2004). miniTUBA uses the R function splinefun (Forsythe GE, et al., 1977) to interpolate missing data across time.

Discretization is the process of transferring continuous variables into discrete counterparts. For efficient learning, the experimental data are discretized into a finite number of bins. Two basic discretization methods are equal interval method and equal number of variables (or quantile) method. The miniTUBA discretization allows 2 to 10 bins of interval discretization or 2 to 10 bins of quantile discretization. Alternatively, the users can make 2 to 10 customized bins. For example, an assignment of "2, 5" represents three bins with values <2, 2-5, and >5, respectively.

One advantage of DBN analysis is that structural priors can be easily defined and enforced. In some cases, a user may know that some edges between variables must or must not be present. These constraints can be included in the analysis in miniTUBA as structural priors (Fig. 2).

For DBN execution, a miniTUBA user can specify the time of DBN execution ranging from 1 minute to 144 hours. To speed up the searching process, a DBN execution job can be performed in parallel by using 1-16 analysis instances, each to be run on a separate node of the backend cluster.

Since the identification of the highest-scoring Bayesian network model for a given set of data is known to be NP-complete (Chickering, 1996), heuristic rather than exhaustive search strategies are used. Two optimization algorithms are available in miniTUBA for users to choose to learn the underlying DBN: simulated annealing and greedy learning. Simulated annealing is a learning process based on successive update steps (either random or deterministic). The update step length is proportional to an arbitrarily set parameter which can play the role of a temperature. In analogy with the annealing of metals, the temperature is made high in the early stages of the process for faster minimisation or learning, is then reduced for greater stability (Ispolatov and Maslov, 2008). The greedy random algorithm makes the locally optimal choice at each stage with the hope of finding the global optimum (Diniz-Filho, et al., 2005). Simulated annealing was found to consistently find the highest scoring Bayesian network models while greedy random algorithm does not (Hartemink, et al., 2002). Therefore, the simulated annealing approach is set as the default.

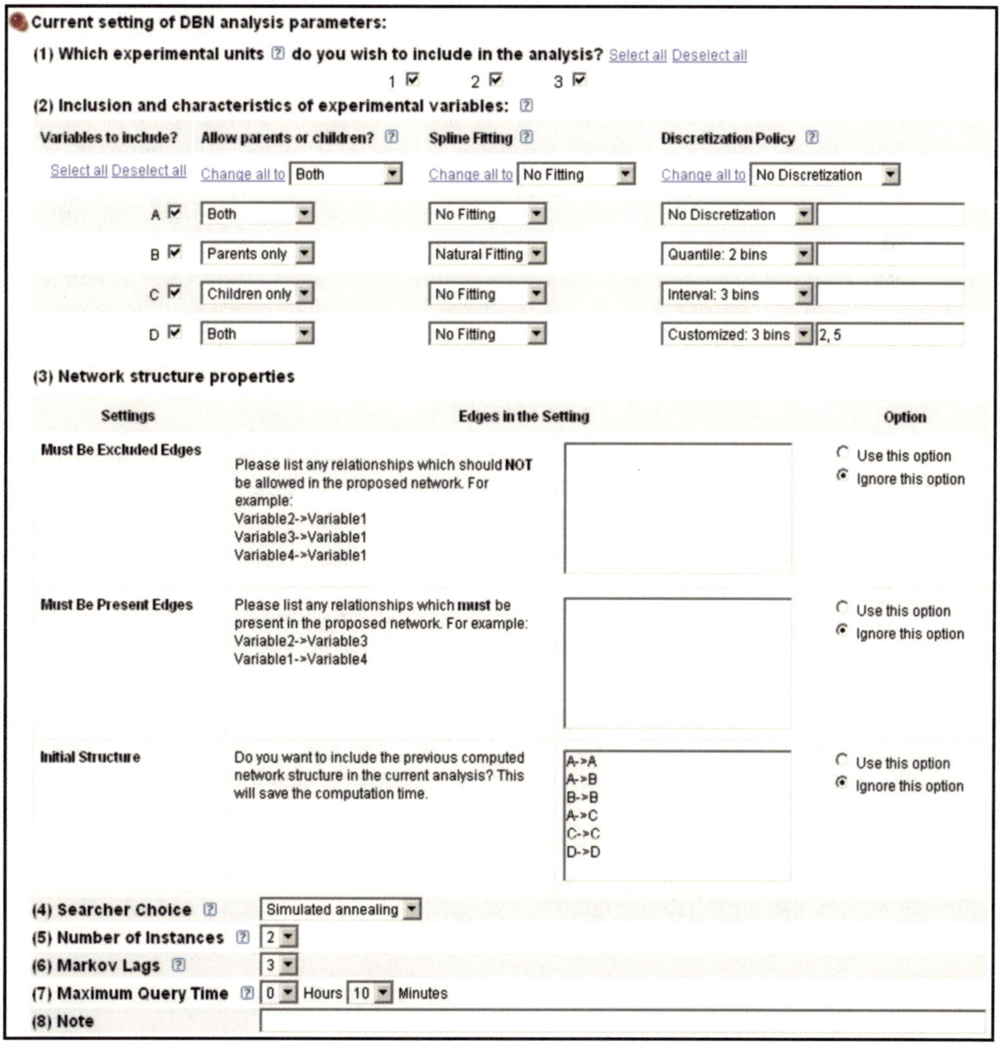

Fig. 2. An example setting for a miniTUBA DBN analysis. This example contains four nodes (A-D) and data of three experiments (i.e., three experimental units). Different nodes can have different settings.

Markov lag is the time interval (or lag) between the start of an event and its effect. For example, for a project with hourly data sets, Markov lag 1 implies that perturbations made now will have a measurable effect in one hour, while a Markov lag of 2 means that the effect will be observable after two hours. Depending on the nature of the experimental units and purpose of the experiment, a user may need to try different Markov lags to find out the optimal Markov lag. This time interval or lag can be easily changed in miniTUBA to examine influential relationships at different time intervals of interest. Although not used here, it is also possible to create models that cover a range of Markov lags. These more complete models are not included in miniTUBA as the results can be difficult to interpret mechanistically.

miniTUBA uses a modified version of the software package BANJO (http://www.cs.duke.edu/~amink/software/banjo/) developed under the direction of Dr. Alexander J. Hartemink at Duke University for dynamic Bayesian network learning for DBN learning (Smith, et al., 2006). DBN analysis does not require as much time as static BN requires due to its time restriction. However, generally BDN still requires much computationl power. In miniTUBA, parallel computation is implemented for DBN execution. The learning jobs are distributed to a 44 node cluster of Apple G5 computers. In parallel jobs, each processor begins network learning from either a random DBN topology or uses a different random seed when learning with a stochastic method such as simulated annealing. Due to the embarrassingly parallel nature of network structure learning, this approach results in a nearly linear decrease in computing time as additional nodes are added. Because initial network learning can take hours to days to complete, miniTUBA alerts registered users by email when a job completes.

2.3 Temporal predictions based on Gibbs sampling in miniTUBA

It is possible to predict the values of future time points given a DBN, conditional probabilities generated from experimental data, and initial values. A prediction module was written that combines a Gibbs sampler to sample future values and a bootstrapping step to de-discretize the predictions. To perform the Gibbs sampling prediction, the data are first discretized (e.g. low, medium, high) and a conditional probability table was generated for each variable. The associated observations for each condition are also recorded. Gibbs sampling is then used to predict future states for each variable by sampling from the conditional probability distribution (Korb and Nicholson, 2004). Bootstrapping is used to de-discretize the states to continuous numerical values by sampling from the associated observations of the predicted states. In prediction mode, miniTUBA repeats this process of sampling and bootstrapping 10,000 times. For numerical variables, the mean and the standard error calculated from the 10,000 predictions are plotted along with the initial values. In miniTUBA, a probability table is provided for variables with nominal values and a probability curve is shown for every such variable.

This feature can suggest interventions based on an automated learning process pipeline using all data provided. This is very useful in a clinical setting. Based on previous data, a doctor may suggest some interventions to stop a disease trend. Similar cases may occur in laboratory research.

2.4 miniTUBA output and visualization

The top scoring and consensus networks generated by the DBN learning process are visualized using Graphviz (http://www.graphviz.org/) (Gansner and North, 2000). The top 10 scoring network graphs are shown in the results page. A consensus network among the top 10 scoring networks can be generated to show edges that are present in all 10 networks, indicating relationships that are present with high confidence. A simple edge confidence is calculated based on the frequency of the edge present among the top 10 scoring networks. While other metrics for edge confidence are possible, such as p-values and probability of conservation, we have found from user studies that these more quantitative metrics tend to overwhelm most non-computational users and end up making the result less useful.

Once a node in a miniTUBA top network is clicked, a conditional probability table calculated based on the input dataset is displayed, together with proposed causal relationships associated with the node. To assess how much better or worse a network is than the others among the top 10 scoring networks, a plot of the Bayes score distribution for these networks can also be displayed in the results page. To simply and intuitively interpret the relationships predicted by the DBN engine, a module is developed to allow user to generate 2D/3D scatter plots by clicking on a variable node with 1 or 2 other variable nodes as parents. The R "plot" command and the LiveGraphics3D package (http://www.vis.uni-stuttgart.de/~kraus/LiveGraphics3D/) are used to draw 2D and 3D plots respectively. The 3D scatter plot can be rotated or zoomed in/out for users to find better angle or resolution.

3. Application of miniTUBA in analysis of macrophage responses to *Brucella* infections

Brucella is a facultative intracellular Gram-negative bacterium that causes a zoonotic disease called brucellosis in swine, cattle, wild life, other animals and "undulant fever" in humans. (Schurig, et al., 2002). Human brucellosis remains the most common zoonotic disease worldwide with more than 500,000 new cases annually (Pappas, et al., 2006). *B. suis* cause brucellosis mainly in swine and humans (Pappas, et al., 2006). The interaction between macrophages and *B. suis* is critical for the establishment of a chronic *Brucella* infection. Smooth virulent *B. Suis*, which contains intact lipopolysaccharide (LPS), prevents macrophage cell death. However, rough and attenuated strain *B. suis* strain VTRS1, which is deficient in the O antigen of LPS due to a *wboA* gene mutation (Winter, et al., 1996), was able to induce strong programmed cell death of infected macrophages. To further investigate the mechanism of VTRS1-induced cytotoxicity in infected macrophages, microarrays were used to analyze temporal transcriptional responses of murine macrophage-like J774.A1 cell line following infection with strain 1330 or VTRS1.

In this section, we demonstrate the application of miniTUBA dynamic Bayesian network analysis in analyzing the immune network in macrophages infected with *Brucella suis* live virulent strain 1330 or attenuated vaccine candidate strain VTRS1 (Winter, et al., 1996). The results indicate that miniTUBA can be used to predict novel targets in a programmed cell death pathway.

3.1 Microarray experimental design

This study contained a DNA microarray experiment using the Affymetrix 430 2.0 array technology. In total, 42 microarray chips were included with seven time points. The basic protocol is as follows: J774.A1 mouse macrophages were plated in T75 at 8 x 10^6 cells per flask one day prior to infection, and then infected with *B. suis* S1330 or VTRS1 at a MOI of 200:1. Total RNAs were isolated by TRIzol and further purified using Qiagen RNeasy Mini Kit (Qiagen, Valencia, CA) at 0 h, 1 h, 2 h, 4 h, 8 h, 24 h, and 48 h post infection. The RNA samples were stored at -80 °C until an Agilent 2100 BioAnalyzer (Agilent Technologies, Palo Alto, CA) was used to assess the concentrations and quality of RNA samples. Total RNA (20 µg) per sample was used for hybridization with Affymetrix mouse GeneChip 430 2.0 array. Preparation of cDNA, hybridization, quality controls and scanning of the GeneChip 430 2.0 arrays were performed according to the manufacturer's protocol (Affymetrix, Santa Clara, CA) (He, et al., 2006).

The data has been deposited in the GEO datbase with the accession number GSE21117.

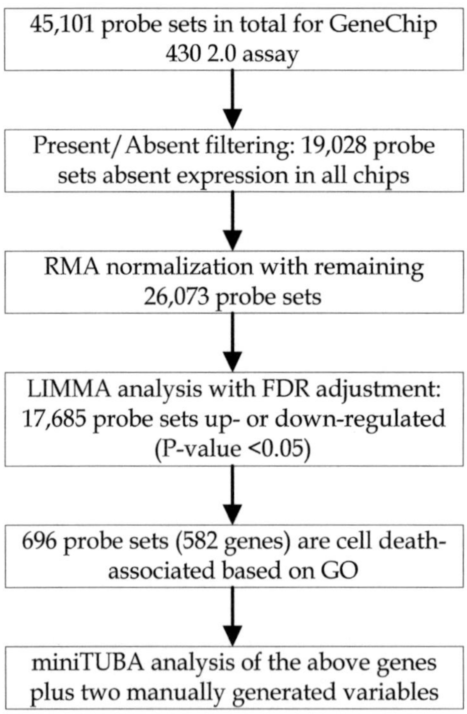

Fig. 3. Microarray data analysis workflow.

3.2 Data preprocessing prior to miniTUBA DNB analysis

For the probe sets that passed the Present/Absent filtering criterion, Robust Multi-array Average (RMA) normalization procedure was performed (Irizarry, et al., 2003). The log2 based gene expression values that have been background adjusted, normalized, and summarized were collected in the process. LIMMA (Linear Models for Microarray Data) with a false discovery rate (FDR) adjustment was further used to analyze up- or down-regulated genes (Smyth, 2005). The moderated F-statistic obtained from LIMMA was used to determine any effects of strain, time, or possible interaction between strain and time. Cell death-associated genes were determined using the Gene Ontology (GO) annotation (Ashburner, et al., 2000). Those genes that are related to cell death and significantly regulated were used for Bayesian network analysis with an internally developed software program miniTUBA (Xiang, et al., 2007). The microarray data analysis workflow and relevant results are shown in Fig. 3.

3.3 miniTUBA DBN analysis of macrophage responses to *Brucella* infections

A miniTUBA DBN analysis was performed to analyze the immune network with a key purpose to determine why strain 1330 prevents programmed cell death of infected macrophages, while strain VTRS1 induced a cell death. The simulation used those genes that significantly regulated and belong to different cell death pathways (Fig. 3).

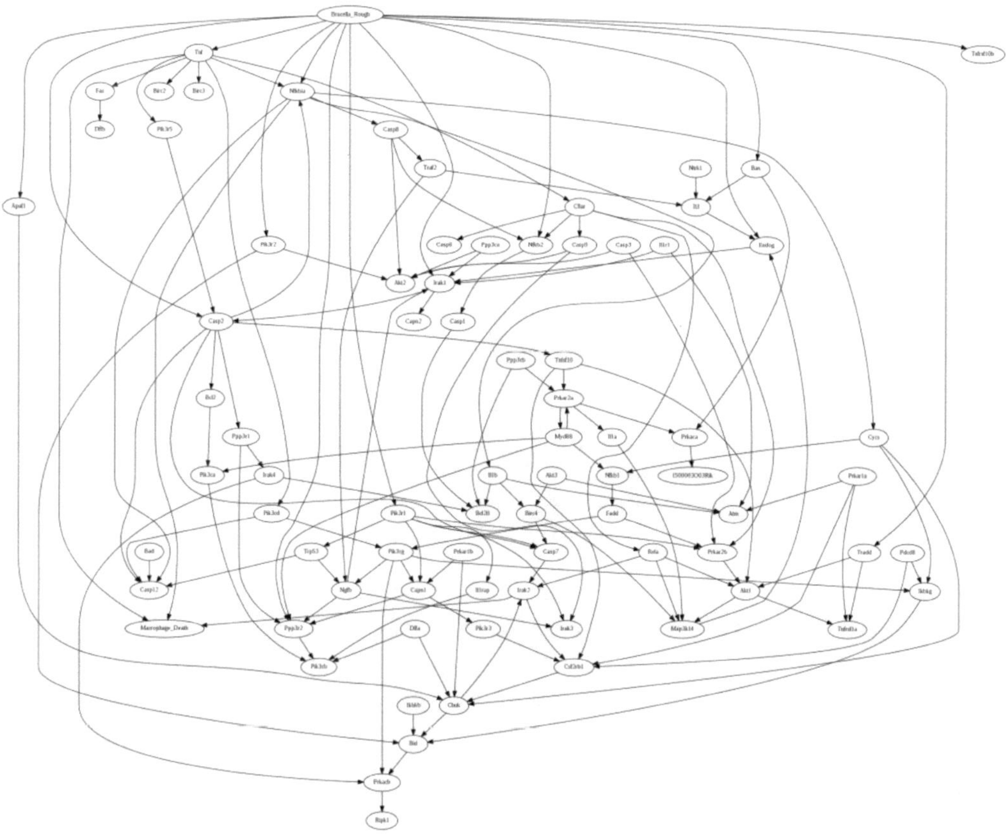

Fig. 4. Conserved network among top 10 DBN results. Four genes were screened out after the conservation. Ninety two edges were screened out for top 1 scoring network.

To perform miniTUBA analysis, the transcriptional data of those genes associated with cell death were extracted and formatted into miniTUBA input data format. The time points 0 h, 1 h, 2 h, 4 h, and 8 h post infection were used for the simulation. The Markov lag is set as 1 hour. Since the time points 3 h, 5 h, 6 h, and 7 h were missing, spline fitting was used to fill in the missing data. We did not use 24 h and 48 h since it would require filling in too many missing time points between 8 h and 24 h and between 24 h and 48 h. Using so many missing data would make the DBN analysis not reliable. The setting of quantile 3 bins was used for variable data discretization. It means that the gene expression values of each gene are seperated into three bins (low, medium, and high), and each bin contains one third of values. The two manually generated variables are "Brucella_Rough" (*i.e.*, rough or smooth *Brucella* strain) and "Macrophage_Death" (*i.e.*, live or dead macrophages). These two variables were used to represent the bacterium strains and the cell death phenotypes. The variable "Brucella_Rough" was set to have no parent, meaning no other variable pointing to this variable. The variable "Macrophage_Death" was set to have no child, meaning this variable cannot point to any other variable. In total, 16 instance runs were performed using 16 nodes. Each run took one hour. In total 16 hours of execution time was used.

After the miniTUBA DBN simulations, 10 top networks were saved. Fig. 4 demonstrates the conserved network among all top 10 networks. Those edges that are not conserved in all top 10 networks are removed. Those variables that do not have any connection to other variables are also removed from the conserved network.

Each node in a top network can be clicked to show more details about this node (Fig. 5-6). For each node, its conditional probability table is displayed in details with the edges between the node and the parent nodes. When the node has only one parent (Fig. 5) or two parents (Fig. 6), a 2D or 3D scatter plot is displayed, respectively. The data and result visualization provide users direct information for a specific variable and its relations with other associated variables.

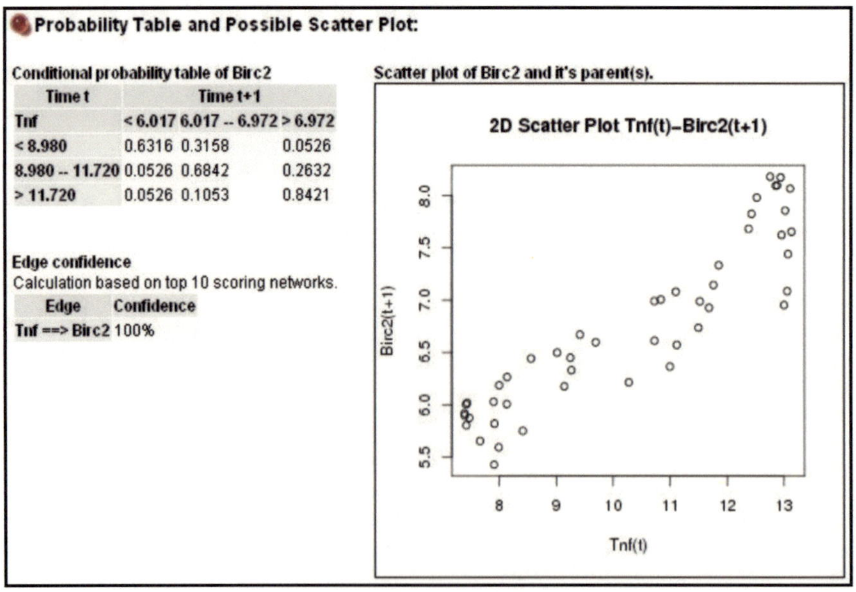

Fig. 5. The conditional probability table of Birc2 and the scatter plot for TNF and Birc2. The expression values of each gene was discretized into three bins with indicated ranges. The relation is based on the top 1 scoring network. A clear correlation between these two genes is observed. Based on miniTUBA DBN analysis, the edge Tnf -> Birc2 is identified.

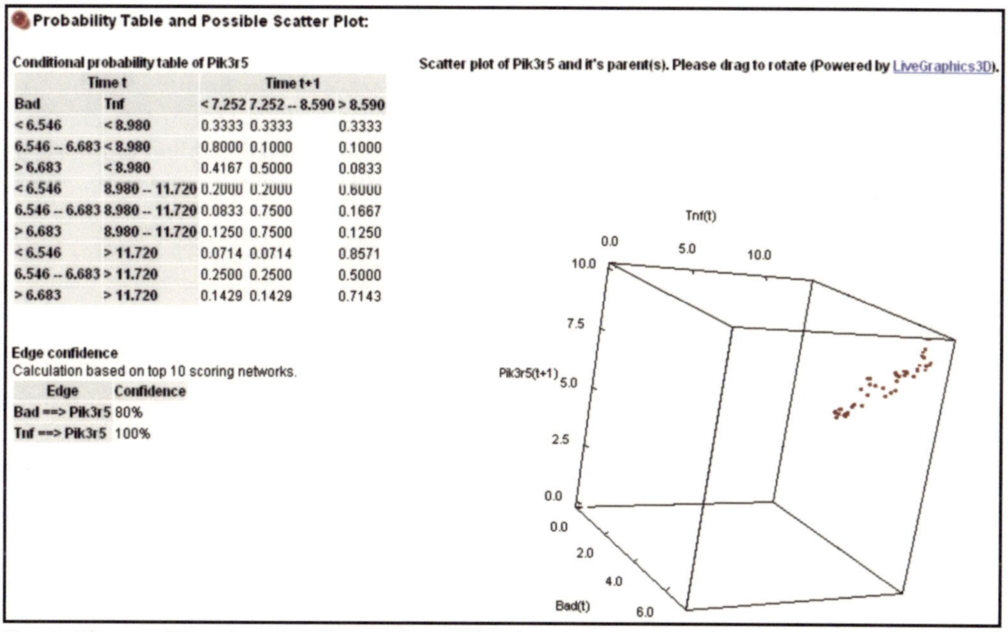

Time t		Time t+1		
Bad	Tnf	< 7.252	7.252 -- 8.590	> 8.590
< 6.546	< 8.980	0.3333	0.3333	0.3333
6.546 -- 6.683	< 8.980	0.8000	0.1000	0.1000
> 6.683	< 8.980	0.4167	0.5000	0.0833
< 6.546	8.980 -- 11.720	0.2000	0.2000	0.6000
6.546 -- 6.683	8.980 -- 11.720	0.0833	0.7500	0.1667
> 6.683	8.980 -- 11.720	0.1250	0.7500	0.1250
< 6.546	> 11.720	0.0714	0.0714	0.8571
6.546 -- 6.683	> 11.720	0.2500	0.2500	0.5000
> 6.683	> 11.720	0.1429	0.1429	0.7143

Edge	Confidence
Bad ==> Pik3r5	80%
Tnf ==> Pik3r5	100%

Fig. 6. The conditional probability table of Pik3r5 and the scatter plot for TNF, Bad, and Pik3r5. The expression values of each gene was discretized into three bins with indicated ranges. The relation is based on the top 1 scoring network. The plot was rotated to a position easily viewable.

3.4 Experimental verification of predicted miniTUBA results:

Our key question is the mechanism of why VTRS1 induced macrophage cell death while its virulent parent strain S1330 prevents the cell death of infected macrophages. To address this question, we have specifically analyzed the possible genes in the network that starts from "Brucea_Rough" and ends at "Macrophage_Death". A portion of the predicted results is shown in Fig. 7. This small network demonstrates that rough *Brucella* induced high level of expressions of proinflammatory gene TNF and NF-κB pathway gene Nfkbia, both contribute to the death of infected macrophages. The caspase-1 (Casp1) appears not to play a role in the macropahge death (Fig. 7).

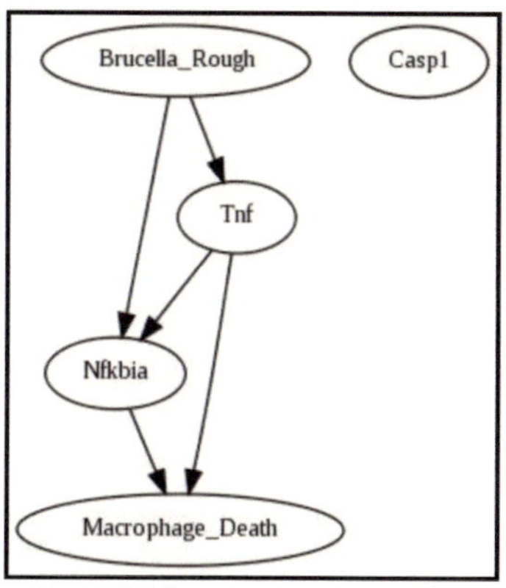

Fig. 7. Predicted network among 5 variables.

Our further experiments verified these predictions. A high level of proinflammatory response was induced by VTRS1 but not by its parent virulent strain S1330. The important roles of TNF-α and Nfkbia in VTRS1-induced macrophage death were further confirmed by individual inhibition studies (data not shown). While Casp1 plays an important role in Casp1-dependent proinflammtory pyroposis (Bergsbaken, et al., 2009), an inhibition study using a Casp1 inhbitor (Z-WEHD-FMK) indicated that Casp1 did not play an obvious role in the VTRS1-induced macrophage cell death (data not shown). We previously found that rough attenuated vaccine strain RB51 and a *wboA* mutant RA1 induced Casp2-mediated cell death (Chen and He, 2009). In this study, miniTUBA also predicted a critical role of Casp2 in the rough attenuated *B. suis* strain VTRS1-induced macrophage cell death, which was later experimentally verified (data to be published).

These studies further demonstrate that the miniTUBA DBN analysis was able to predict important factors in a biological pathway using high throughput microarray gene expression data, which successfully guide the experimental evaluations.

3.5 Demonstration of Gibbs sampling prediction:

While the prediction of future events is not designed or important for this microarrray study, we can use the same data sets to demonstrate how miniTUBA performs prediction and display predictive results (Fig. 8). In this demostration, we showed that using the data from early time points, the results in time points 5 h and 6 h post infection could be predicted with an apparent success. More detailed verification of this method is described in the original miniTUBA publication (Xiang, et al., 2007).

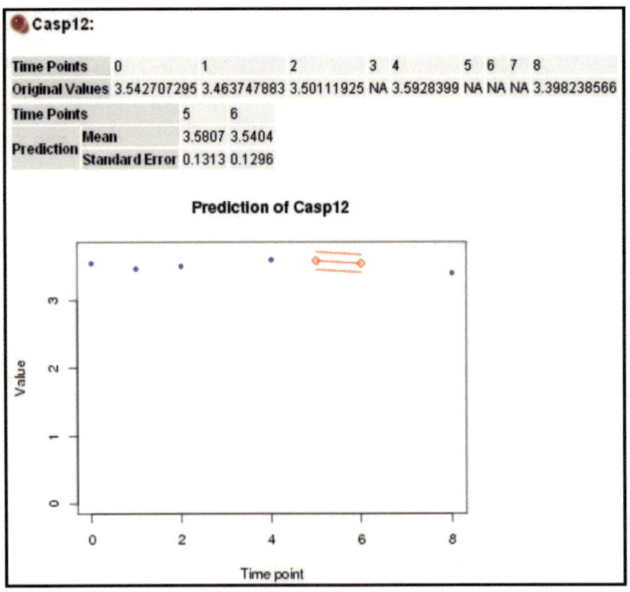

Fig. 8. Predicted results for Casp12 for the 5th and 6th time points. The round points without error ranges are the values from previously known time points and the diamond shaped points with error ranges are the predicted values and their associated standard errors. The 8th point was not used for prediction.

4. Conclusion and discussion

In this chapter, we have introduced in details the miniTUBA system, and how to apply the miniTUBA dynamic Bayesian network (DBN) approach to analyze a typical use case in the areas of host-pathogen interactions using high throughput microarray data.

The DBNs are powerful to model the stochastic evoluation of a set of random variables over time. Since the biological processes and various measurement errors are stochastic in nature, DBN has been considered as a suitable technique to study biological networks and pathways. Bayesian networks (BNs) and DBNs are based on a multinomial distribution. This distribution is very flexible, and each node has a different parameterization. Therefore, it is very feasible to use DBNs to model the dynamics of biological systems and responses to parameter perturbations. However, although a few applications for both Bayesian network and DBNs to modeling gene expression data have been discussed and reported, their usefullness remains to be shown with more well-understood pathways (Xia, et al., 2004). Programmed cell death (*i.e.*, apoptosis) pathways are well studied and important for all plant and animal organisms. We first demonstrated in this report how the DBN analysis can be used to predict crucial genes for a cell death pathway, which led to correct experimental verification.

Two major challenges in DBN analysis for biological network modeling exist. First, continuous gene expression data has to be descretized, leading to the loss of information. The descretization simplifies the computation and stablizes the predicated results. However, current equal quantile and interval descretization methods do not often reflect the biological realities. The customized descretization method is too time consuming and may not correlate with the unknown truth either. Therefore, alternative approaches will need to be explored to improve the descretization and minimize loss of information. How to find reliable ways to model continuous data remains to be a major challenge in the DBN and other modeling studies. Second, it is a big challenge to identify the correct time steps (*i.e.*, Markov lags) for a DBN modeling. By default, we require all variables have the same time step size. However, it might be possible to allow a mixture of different time step sizes. The time scale likely differ between variables. To identify the relevant time scale, we may allow different discretization schemes. While more finely discretized variables offer slower changes, it might be difficult to determine how many are appropriate. The generation of very large sizes of discretizations is also time consuming. One solution is to allow mixtures of time steps in the learning step. However, it is in practice very difficult because the current step depends on a range of past experiences. If the previous time steps are not multiplies of each other, a complex splining function is usually needed to dynamically interpolate the missing data. Alternatively, we can explicitly search for an optimum informative time step. A DBN search will favor small time steps because it means more data to be used. However, if the data represents only more interpolated data, it would not help. While DBN analysis can be improved in different directions, the two areas of DBN research with the largest impact are probably the discretization and correct time step setting.

Besides addressing the above challenges, dynamic Bayesian networks can further be improved through different directions: (i) those strong links (or edges) are conserved among top networks and can be detected by consensus analysis (Fig. 4). (ii) cross-species comparison may further help to reveal the conserved core network across different species (Gholami and Fellenberg, 2010). (iii) it is possible to learn dynamic regulatory networks by incorporating multiple data types (e.g., functional classification, shared motif motifs, protein-DNA binding, protein-protein interaction) (Bernard and Hartemink, 2005). (iv) incorporation with additional quantitative measurements (Xia, et al., 2004). (v) integration of DBN gene expression data analysis with literature-based network discovery (Ozgur, et al, 2010). (vi) Dynamic Bayesian network expansion for identification of new pathway elements as shown in a similar approach with static Bayesian network (Hodges, et al, 2010). Finally, all these new directions will need to be integrated in a proper way for accurate reconstruction and prediction of biological and medical networks. Such a network analysis approach is likely appliable for study of other networks (*e.g.*, social networks).

5. Acknowledgements

This work was sponsored by the NIH/NIAID Regional Center of Excellence for Bio-defense and Emerging Infectious Diseases Research (RCE) Program. The authors wish to acknowledge membership within and support from the Region V 'Great Lakes' RCE (NIH award 1-U54-AI-057153). This project was also supported by the NIAID R-21 grant (1R21AI057875) to Y. He.

6. References

Ashburner, M., *et al.* (2000) Gene ontology: tool for the unification of biology. The Gene Ontology Consortium, *Nat Genet*, **25**, 25-29.

Bergsbaken, T., Fink, S.L. and Cookson, B.T. (2009) Pyroptosis: host cell death and inflammation, *Nature reviews*, **7**, 99-109.

Bernard, A. and Hartemink, A.J. (2005) Informative structure priors: joint learning of dynamic regulatory networks from multiple types of data, *Pac Symp Biocomput*, 459-470.

Chen, F. and He, Y. (2009) Caspase-2 mediated apoptotic and necrotic murine macrophage cell death induced by rough Brucella abortus, *PLoS One*, **4**, e6830.

Chickering, D.M. (1996) Learning Bayesian Networks is NP-Complete. In Fisher, D. and Lenz, H.J. (eds), *Learning from Data: Artificial Intelligence and Statistics V (Lecture Notes in Statistics)*. Springer-Verlag, pp. 121-130.

Dean, T. and Kanazawa, K. (1988) *Probabilistic temporal reasoning*. Proceedings AAAI. AAAI Press / The MIT Press, St. Paul, MN, USA.

Diniz-Filho, J.A., *et al.* (2005) Priority areas for anuran conservation using biogeographical data: a comparison of greedy, rarity, and simulated annealing algorithms to define reserve networks in cerrado, *Braz J Biol*, **65**, 251-261.

Dojer, N., *et al.* (2006) Applying dynamic Bayesian networks to perturbed gene expression data, *BMC Bioinformatics*, **7**, 249.

Forsythe GE, Malcolm MA and CB., M. (1977) *Computer Methods for Mathematical Computations.* Prentice Hall Upper Saddle River, New Jersey.

Friedman, N. (2004) Inferring cellular networks using probabilistic graphical models, *Science*, **303**, 799-805.

Friedman, N., *et al.* (2000) Using Bayesian networks to analyze expression data, *J Comput Biol*, **7**, 601-620.

Gansner, E. and North, N.C. (2000) An open graph visualization system and its applications to software engineering, *Software-practice and experience*, **30**, 1203-1233.

Gholami, A.M. and Fellenberg, K. (2010) Cross-species common regulatory network inference without requirement for prior gene affiliation, *Bioinformatics*, **26**, 1082-1090.

Hartemink, A.J., *et al.* (2002) Combining location and expression data for principled discovery of genetic regulatory network models, *Pac Symp Biocomput*, 437-449.

He, Y., *et al.* (2006) Brucella melitensis triggers time-dependent modulation of apoptosis and down-regulation of mitochondrion-associated gene expression in mouse macrophages, *Infect Immun*, **74**, 5035-5046.

Hodges, A.P., et al. (2010) Bayesian network expansion identifies new ROS and biofilm regulators. *PLoS ONE*. **5(3)**:e9513.

Husmeier, D. (2003) Sensitivity and specificity of inferring genetic regulatory interactions from microarray experiments with dynamic Bayesian networks, *Bioinformatics*, **19**, 2271-2282.

Imoto, S., *et al.* (2006) Computational strategy for discovering druggable gene networks from genome-wide RNA expression profiles, *Pac Symp Biocomput*, 559-571.

Irizarry, R.A., *et al.* (2003) Exploration, normalization, and summaries of high density oligonucleotide array probe level data, *Biostatistics*, **4**, 249-264.

Ispolatov, I. and Maslov, S. (2008) Detection of the dominant direction of information flow and feedback links in densely interconnected regulatory networks, *BMC Bioinformatics*, **9**, 424.

Kim, S.Y., Imoto, S. and Miyano, S. (2003) Inferring gene networks from time series microarray data using dynamic Bayesian networks, *Brief Bioinform*, **4**, 228-235.

Korb, K.B. and Nicholson, A.E. (2004) *Bayesian artificial intelligence*. Chapman & Hall/CRC Press, London, UK.

Li, J., Wang, Z.J. and McKeown, M.J. (2007) A framework for group analysis of fMRI data using dynamic Bayesian networks, *Conf Proc IEEE Eng Med Biol Soc*, **2007**, 5992-5995.

Neapolitan, R.E. (2003) *Learning Bayesian Networks*. Prentice Hall.

Ong, I.M., Glasner, J.D. and Page, D. (2002) Modelling regulatory pathways in E. coli from time series expression profiles, *Bioinformatics*, **18 Suppl 1**, S241-248.

Ozgur, A., et al. (2010) Literature-based discovery of IFN-γ and vaccine-mediated gene interaction networks. *Journal of Biomedicine and Biotechnology*. **Volume 2010**, Article ID 426479, 13 pages.

Pappas, G., *et al.* (2006) The new global map of human brucellosis, *Lancet Infect Dis*, **6**, 91-99.

Pe'er, D., *et al.* (2001) Inferring subnetworks from perturbed expression profiles, *Bioinformatics*, **17 Suppl 1**, S215-224.

Peelen, L., *et al.* (2010) Using hierarchical dynamic Bayesian networks to investigate dynamics of organ failure in patients in the Intensive Care Unit, *J Biomed Inform*, **43**, 273-286.

Rangel, C., *et al.* (2004) Modeling T-cell activation using gene expression profiling and state-space models, *Bioinformatics*, **20**, 1361-1372.

Rau, A., *et al.* (2010) An empirical Bayesian method for estimating biological networks from temporal microarray data, *Stat Appl Genet Mol Biol*, **9**, Article 9.

Remick, D.G. (2003) Cytokine therapeutics for the treatment of sepsis: why has nothing worked?, *Curr Pharm Des*, **9**, 75-82.

Schurig, G.G., Sriranganathan, N. and Corbel, M.J. (2002) Brucellosis vaccines: past, present and future, *Vet Microbiol*, **90**, 479-496.

Smith, V.A., *et al.* (2006) Computational inference of neural information flow networks, *PLoS computational biology*, **2**, e161.

Smyth, G.K. (2005) limma: Linear Models for Microarray Data In, *Statistics for Biology and Health*. Springer, New York, pp. 397-420.

Watt, E.W. and Bui, A.A. (2008) Evaluation of a dynamic bayesian belief network to predict osteoarthritic knee pain using data from the osteoarthritis initiative, *AMIA Annu Symp Proc*, 788-792.

Winter, A.J., *et al.* (1996) Protection of BALB/c mice against homologous and heterologous species of Brucella by rough strain vaccines derived from Brucella melitensis and Brucella suis biovar 4, *Am J Vet Res*, **57**, 677-683.

Xia, Y., *et al.* (2004) Analyzing cellular biochemistry in terms of molecular networks, *Annu Rev Biochem*, **73**, 1051-1087.

Xiang, Z., *et al.* (2007) miniTUBA: medical inference by network integration of temporal data using Bayesian analysis, *Bioinformatics*, **23**, 2423-2432.

Yu, J., *et al.* (2004) Advances to Bayesian network inference for generating causal networks from observational biological data, *Bioinformatics*, **20**, 3594-3603.

Zou, M. and Conzen, S.D. (2005) A new dynamic Bayesian network (DBN) approach for identifying gene regulatory networks from time course microarray data, *Bioinformatics*, **21**, 71-79.

12

Joining Analytic Network Process and Bayesian Network model for fault spreading problem

Gábor Szűcs and Gyula Sallai
Budapest University of Technology and Economics
Hungary

1. Introduction

This chapter deals with fault spreading (fault tree) in infocommunication networks (e.g. computer network, wired or wireless telecommunication network). The probabilistic approach of fault trees is in the focus, where faults can occur in the inner part of the network, spread step by step and can appear at the front end (observable directly by end users) of the network. The probabilities of the different inner faults and conditional probabilities of the steps of spreading are given. At the front end of the network many different errors caused by inner faults can occur; some of them may be serious, others may be not. Serious errors cause large damages in the operation of the network, e.g. material damage (an equipment breaks down), economical cost, human resource loss (expert should prepare it); less significant errors cause only little damages, inconveniences. There is not easy to decide which error is serious and which is not, and how much is significant. The task is to analyse the inner causes of errors at front end, and investigate the relative effects of these causes.

For this problem a method has been developed by joining the Analytic Network Process and the extended Bayesian Network. The generalized Bayesian Network with vector extension has been outlined in a previous work (Szűcs & Sallai, 2008). In this chapter the usefulness, correctness of the new elaborated method will be demonstrated by a numerical example.

The chapter is organized as follows: in section 2 an introduction about Analytic Network Process is given, section 3 summarizes the Bayesian Network and shows its generalization (Vector Bayesian Network, VBN). In section 4 a new concept for solving complex multi-criteria engineering decision problems is presented by combining the Analytic Network Process and VBN. Section 5 describes an example in the area of fault analysis in infocommunication networks. Section 6 summarizes the results and draws the conclusions.

2. Analytic Network Process

2.1. Multi Criteria Decision Analysis Methods

Analytic Hierarchy Process (AHP) is a well-known and one of the most comprehensive procedures in Multi Criteria Decision Analysis (MCDA) area. The AHP - has been introduced by Thomas L. Saaty (Saaty, 1980) - for decision-making is a theory of relative measurement based on paired comparisons used to derive normalized absolute scales of numbers whose elements are then used as priorities.

The Analytic Network Process (ANP) (Saaty, 2001) is the generalization of AHP for decision making with dependence and feedback allowing inclusion of all the factors and criteria. ANP has been proposed as a suitable MCDA tool to evaluate the alternatives during the conceptual planning and design in many areas, e.g. in economical problems, in engineering. The ANP (Saaty, 2005) provides a way to input judgments and measurements to derive ratio scale priorities for the distribution of influence among the factors and groups of factors in the decision. Both the AHP and the ANP derive ratio scale priorities by making paired comparisons of elements on a common property or criterion. Even though the ANP is a new method, there are many validation examples of the Analytic Network Process (Whitaker, 2007).

2.2. Steps of ANP Procedure

An Analytic Network Model of a problem may consist of a single network (or a number of networks), where a network is structured of clusters (i.e. groups of nodes), nodes (any aspect of the problem, e.g. alternative, attribute) and links (connection between nodes). The stages of creating of an ANP model are the following:

a) Selection of logical groupings of nodes and clusters, which would best describe the problem.
b) Building a cluster first, and then creating the nodes within it.
c) Examination of influences.
d) Creating connections between nodes.
e) Clusters are linked automatically when nodes are linked.
f) Pair-wise comparison judgments on nodes and clusters.

After pair-wise comparison judgments the algorithms in ANP take calculations and at the end give the decision: which is the best alternative for the problem. The algorithms solve the problem by supermatrices. There are three supermatrices associated with each network: the *Unweighted Supermatrix*, the *Weighted Supermatrix* and the *Limit Supermatrix*.

The *unweighted supermatrix* contains the local priorities derived from the pair-wise comparisons throughout the network. A component is defined as a block determined by a cluster name/identity at the rows and a cluster name/identity at the columns in a supermatrix. The *weighted supermatrix* is obtained by multiplying all the elements in a component of the unweighted supermatrix by the corresponding cluster weight. Cluster weights come from cluster comparisons. If there are only two clusters, then cluster comparisons cannot be executed, in this case the weighted and unweighted supermatrices are the same. The *limit supermatrix* is obtained by raising the weighted supermatrix to powers by multiplying it many times itself. When the column of numbers is the same for every column, the limit matrix has been reached and the matrix multiplication process is halted. The priorities, as outputs of ANP for all the nodes can be read from any column, because the columns of the limit supermatrix are all the same.

AHP – as special case of ANP – can be used for many decision situations; its application area is wide: economy, business, engineering management and other areas, where the problems lead to multi criteria decision making. This can be applied in solving the technological decision problems as well, e.g. in network selection procedure for an integrated cellular/wireless local area network (WLAN) system to guarantee mobile users being always best connected. AHP helps to decide the relative weights of evaluative criteria set according to user preferences, network condition and service applications (Wei et al., 2007). Not only AHP, but ANP can help to take important decisions in network/telecommunication technology (Lee et al., 2009; Büyüközkan, 2007) 0or in media informatics (Chang, 2007). ANP can be applied in managerial practices as well (Chen, 2007; Wu & Lee, 2007). ANP in a little while becomes classical method, many publications deal with improvement, refinement (Saaty, 2007), further development (Yu & Cheng, 2007; Yu & Tzeng, 2006; Levy & Taji, 2007), supplement - e.g. with fuzzy (Dağdeviren et al., 2008; Promentilla et al., 2008) or with integer linear programming (Demirtas & Üstün, 2008) - of this.

3. Vector Bayesian Network

3.1. Bayesian Network Model

A Bayesian Network (BN) is a probabilistic graphical model for representing causal relationship among variables (Judea, 1982; Speigelhalter et al., 1993). This is a very important research topic in artificial intelligence and decision support area (Liu et al., 2009; Cheon et al., 2009; Correa et al., 2009). It consists of a set of nodes and directed arcs. The nodes represent variables and the arcs represent the directed causal influences between linked nodes. The arc starts from the parent node (A) to the child node (B). The child node is dependent on its parent node, but it is conditionally independent of others. The condition probability P(A|B) - showing how a given parent node A can influence the probability distribution over its child node B - is calculated using Bayes' Theorem:

$$p(A \mid B) = \frac{p(A \,\&\, B)}{p(B)} \tag{1}$$

$$p(A \mid B) = \frac{p(B \mid A) * p(A)}{p(B \mid A) * p(A) + p(B \mid \overline{A}) * p(\overline{A})} \tag{2}$$

BNs can be used for investigation of system in reliability analysis of engineering, there are some works (Wilson & Huzurbazar, 2007; Huang et al., 2006; Kohda & Cui, 2007) deal with it, but these do not handle with other aspects (e.g. financial costs) of the system.

3.2. Extension of Bayesian Network

In a Bayesian Network dependencies are generally complicated, so some preliminary formulas are required to handle the probabilities and variables. In Fig. 1. can be seen two typical types of BN pattern (part of graph), which can be used for building large networks. A such situation can be seen in Fig. 1/a, where more than one parent nodes have the same child node. In this case the conditional probability is:

$$p({}_1X \mid Y) = \left\{p(Y|{}_1X,\ {}_2X)\cdot p({}_1X)\cdot p({}_2X) + p(Y|{}_1X,\ \overline{{}_2X})\cdot p({}_1X)\cdot p(\overline{{}_2X})\right\} / \left\{p(Y|{}_1X,\ {}_2X)\cdot p({}_1X)\cdot p({}_2X) + p(Y|{}_1X,\ \overline{{}_2X})\cdot p({}_1X)\cdot p(\overline{{}_2X}) + p(Y \mid \overline{{}_1X},\ {}_2X)\cdot p(\overline{{}_1X})\cdot p({}_2X) + p(Y \mid \overline{{}_1X},\ \overline{{}_2X})\cdot p(\overline{{}_1X})\cdot p(\overline{{}_2X})\right\} \tag{3}$$

If aggregated conditional probabilities (e.g. $p(Y|{}_1X)$ instead of $p(Y|{}_1X,\ {}_2X)$) are given, this can be written more generally: Let us denote $\{S_j\}$ the partition of the event space (S), (i.e. $\cup S_j=S$, $\cap S_j=0$), the conditional probabilities can be formalized as:

$$p(S_i \mid Y) = \frac{p(Y \mid S_i)\cdot p(S_i)}{\sum_{j=1}^{n} p(Y \mid S_j)\cdot p(S_j)} \tag{4}$$

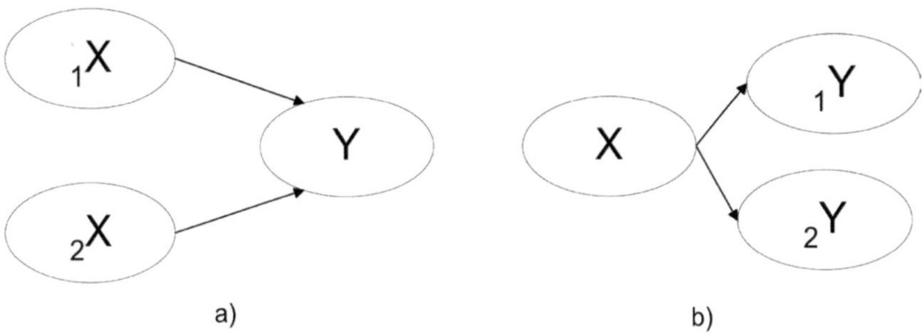

Fig. 1. Structure Cases in Bayesian Network

In Fig. 1/b. can be seen such situation, where one parent node has more than one child. In this case the conditional probability is:

$$p(X|{}_1Y) = \frac{p({}_1Y \mid X)\cdot p(X)}{p({}_1Y \mid X)\cdot p(X) + p({}_1Y \mid \overline{X})\cdot p(\overline{X})} \tag{5}$$

or generally:

$$p(X|_iY)=\frac{p(_iY\,|\,X)\cdot p(X)}{p(_iY\,|\,X)\cdot p(X)+p(_iY\,|\,\overline{X})\cdot p(\overline{X})} \tag{6}$$

Vectors are introduced for each $_iN$ node, $_iW=[_iW_1, {}_iW_2,\ldots,{}_iW_m]$ as weight of different effects. By this vector an extension of Bayesian Network is introduced (so called Vector Bayesian Network: VBN), where nodes contain not only probabilities, but these weights (representing any information, e.g. importance, relative effect, cost). The weight vectors at leafs (having no children) in Vector Bayesian Network are given, and the weights at parent nodes should be determined.

In the VBN graph different structures of relationships can occur as well, e.g. parents can have more than one child: *case a* (see Fig. 1/a.), or more parent nodes may be the parents of the same node: *case b* (see Fig. 1/b.). The unknown weights can be calculated in every case.

3.3. Weight Calculations in VBN

The weight vector of $_1X$ node in *case a* is $_1w=[_1w_1, {}_1w_2,\ldots, {}_1w_m]$, which depends on the weight vector of Y node $w=[w_1, w_2,\ldots, w_m]$. (The numbers of the dimensions of all vectors are equal.) The ratio of the weight vector of the parent node and the child node is the conditional probability. So each element of the weight vector $_1X$ node can be calculated as follows:

$$\begin{aligned}{}_1w_k &= p(_1X\,|\,Y)\cdot w_k = \\ &= w_k\cdot\left\{p(Y|_1X,\ {}_2X)\cdot p(_1X)\cdot p(_2X)+p(Y|_1X,\ \overline{_2X})\cdot p(_1X)\cdot p(\overline{_2X})\right\}/ \\ &\left\{p(Y|_1X,\ {}_2X)\cdot p(_1X)\cdot p(_2X)+p(Y|_1X,\ \overline{_2X})\cdot p(_1X)\cdot p(\overline{_2X})+\right. \\ &\left.p(Y\,|\,\overline{_1X},\ {}_2X)\cdot p(\overline{_1X})\cdot p(_2X)+p(Y\,|\,\overline{_1X},\ \overline{_2X})\cdot p(\overline{_1X})\cdot p(\overline{_2X})\right\}\end{aligned} \tag{7}$$

for every $1 \le k \le m$. The $_2X$ nodes and other $_iX$ nodes - in case of more than two parent nodes - can be calculated by similar way.

There is a different situation, when the number of parent node is one and this node has more child nodes, like in *case b*. The weight vector of X node is denoted $w=[w_1, w_2,\ldots,w_m]$. The weight vectors of the child nodes ($_1Y$, $_2Y$, etc.) are denoted $_1w=[_1w_1, {}_1w_2,\ldots, {}_1w_m]$, $_2w=[_2w_1, {}_2w_2,\ldots, {}_2w_m],\ldots, {}_nw=[_nw_1, {}_nw_2,\ldots, {}_nw_m]$, where *n* is the number of the child nodes. The elements of the weight vector - as representing the effects - at the parent node are cumulated from child node ones. So the weight vector X node can be calculated as follows:

$$w_k=\sum_{i=1}^{n} p(X|_iY)\cdot{}_iw_k=\sum_{i=1}^{n}\frac{p(_iY\,|\,X)\cdot p(X)}{p(_iY\,|\,X)\cdot p(X)+p(_iY\,|\,\overline{X})\cdot p(\overline{X})}\cdot{}_iw_k \tag{8}$$

for every $1 \le k \le m$.

4. Combination of ANP and VBN Methods

The problem described in the introduction can be solved by joining Analytic Network Process and extended Bayesian Network such way, that the weights, the results of ANP will give the input of VBN (as can be seen in Fig. 2.). Bayesian Network (and VBN also) can get inputs in many points, but in our combination VBN adopts the weights, as inputs in leafs of the graph; and these weights are organized in vector format.

Our joint method is able to analyse the reasons and spreading of faults in infocommunication network by the following way. At first step the network (human) expert defines the different types of the front end faults, and the criteria (features) which influence the importance (at the given goal) of the fault type. E.g. if the aim is to minimize the total cost, the criteria should involve the cost of fault repairing, the scope of the fault, the length of the repairing time. The network expert may declare other criteria for another aim. Then the expert compares the fault types with each other (pair-wise comparison with all), and the criteria (pair-wise comparison as well in order to get the criterion relative ratios). After the expert judgements the ANP method calculates the weights of the fault types, at case of economic aim this gives the total costs for each fault type.

At the next step the network expert draws the inner structure of the infocommunication network by nodes and directed edges such way, that the edges should be in the directions of the front end of the network. The expert describes the possible paths of the fault spreading with these directed edges. The arising of faults is modelled in the nodes (included inner nodes and front end nodes), the expert should give the probability of fault arising in every node. The nodes, where the directed edges "only come from", are *fault sources*, the other nodes can be considered as *fault spreading nodes*. The probabilities of fault arising at fault sources are unconditional ones, and probabilities of fault arising at nodes of fault spreading are conditional probabilities. The graph worked out by the above mentioned way will be the structure of the VBN model, furthermore the unconditional and conditional probabilities will be the parameters of VBN.

The serial of relative weights of fault types calculated by ANP method can be written as a vector, this vector will be the input of VBN at the front end nodes. This vector is a consequence, and the reasons of the consequence are the questions, so the expert users would like to know the origins. The VBN method gives the contribution of the given node to the different types of front end faults for every node in the graph (representing e.g. infocommunication network) as most important result of the joint procedure.

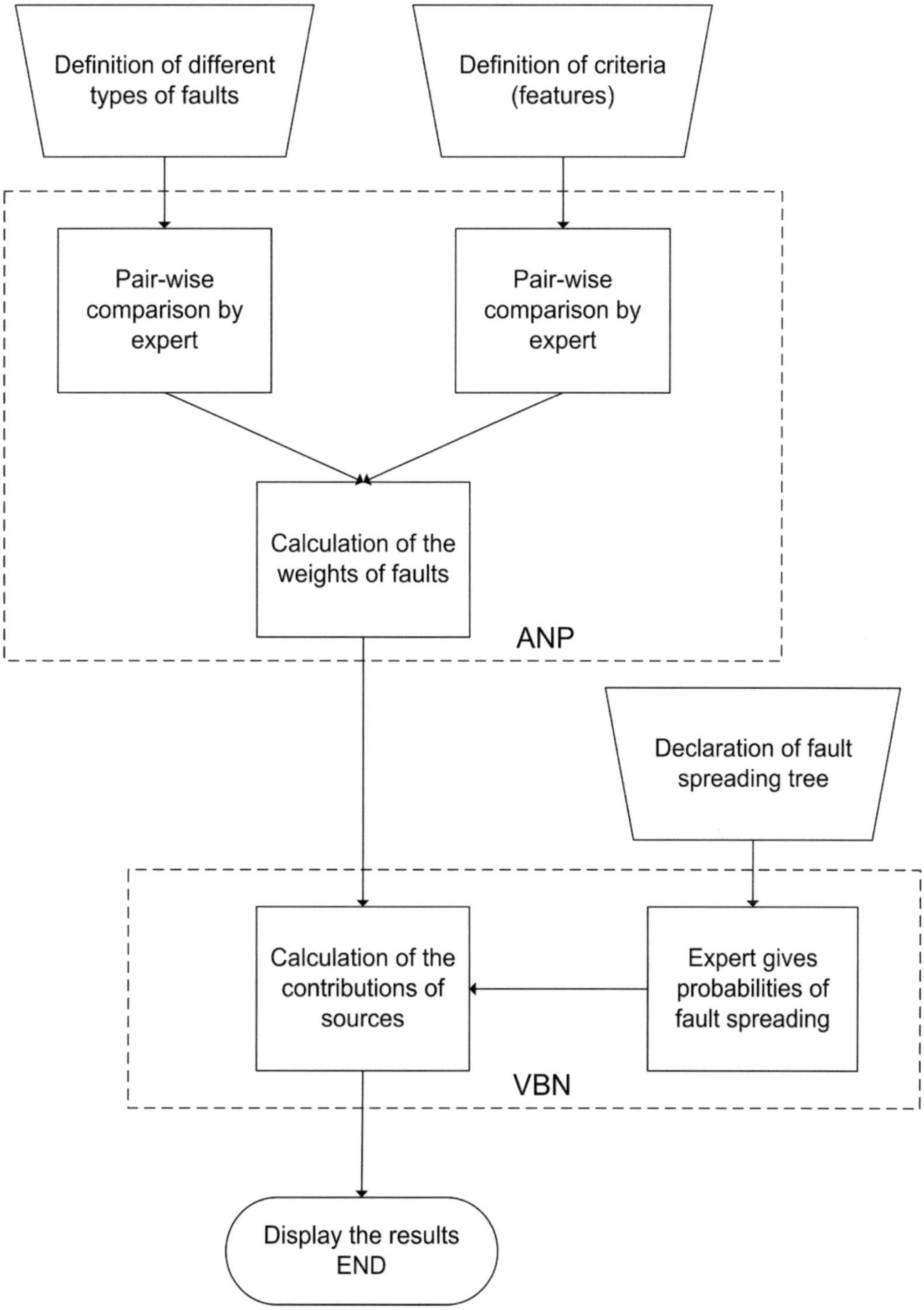

Fig. 2. Block of the Solution Procedure by ANP and VBN Methods

5. Case Study

5.1. Example for Fault Spreading

There are three errors occur at the front end of the network: i) breakdown event at an average user, ii) malfunction at an average user, iii) error at administrator site. These errors are considered as alternatives (A1, A2, A3) in ANP. The relative significance of them depends on many points of views, these criteria are in our example: priority of user, scope of the error, cost of reparation, time of restore (F1,…, F4).

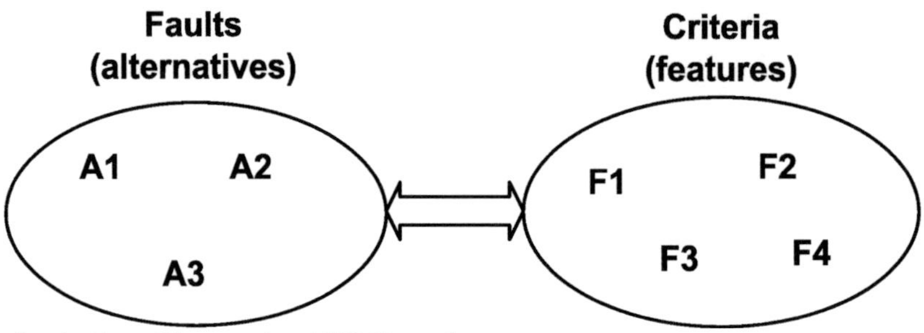

Fig. 3. Nodes in Two Clusters for ANP Procedure

The Fig. 3. shows the structure of the ANP model with two clusters (Faults and Criteria) and a connection between the clusters. This connection means that there are relationships between every node in cluster *Faults* and every node in cluster *Criteria*, but there is no inner-relationship in a cluster. The clusters contain the following nodes:
A1: breakdown event at an average user,
A2: malfunction at an average user,
A3: error at administrator site,
F1: priority of user
F2: scope of the error,
F3: cost of reparation,
F4: time of restore.

5.2. Numerical Example for ANP

A network expert can compare the features of errors based on these criteria. E.g. the 'scope of the error' feature of A_1 alternative is twice larger important than feature of A_2 alternative (see Table 2). The expert should execute all pair-wise comparisons in each criterion. The following matrices (Table 1-4) contain a possible judgment of expert's opinions.

	A1	A2	A3
A1	1,00000	1,00000	0,25000
A2	1,00000	1,00000	0,25000
A3	4,00000	4,00000	1,00000

Table 1. Comparisons based on Priority of User

	A1	A2	A3
A1	1,00000	2,00000	0,16667
A2	0,50000	1,00000	0,12500
A3	6,00000	8,00000	1,00000

Table 2. Comparisons based on Scope of the Error

	A1	A2	A3
A1	1,00000	6,00000	0,33333
A2	0,16667	1,00000	0,11111
A3	3,00000	9,00000	1,00000

Table 3. Comparisons based on Cost of Reparation

	A1	A2	A3
A1	1,00000	0,33333	0,16667
A2	3,00000	1,00000	0,25000
A3	6,00000	4,00000	1,00000

Table 4. Comparisons based on Time of Restore

	F1	F2	F3	F4
F1	1,00000	2,00000	0,25000	1,00000
F2	0,50000	1,00000	0,16667	0,50000
F3	4,00000	6,00000	1,00000	3,00000
F4	1,00000	2,00000	0,33333	1,00000

Table 5. Comparisons of Faults from the Viewpoint of A1

	F1	F2	F3	F4
F1	1,00000	2,00000	0,33333	1,00000
F2	0,50000	1,00000	0,20000	0,50000
F3	3,00000	5,00000	1,00000	3,00000
F4	1,00000	2,00000	0,33333	1,00000

Table 6. Comparisons of Faults from the Viewpoint of A2

	F1	F2	F3	F4
F1	1,00000	2,00000	0,40000	1,00000
F2	0,50000	1,00000	0,20000	0,50000
F3	2,50000	5,00000	1,00000	2,50000
F4	1,00000	2,00000	0,40000	1,00000

Table 7. Comparisons of Faults from the Viewpoint of A3

Having a comparison matrix the priority vector can be computed, which is the normalized eigenvector of the matrix, e.g. eigenvector of F1 (priority of user) matrix are: 0.166667, 0.166667, 0.666667. The other eigenvectors are also calculated by *SuperDecisions* software (realization of ANP theory helping by Thomas Saaty) and written to the corresponding cells of the supermatrix. This unweighted supermatrix (containing 4 components: Crit.-Crit., Crit.-Faults, Faults-Crit., Faults-Faults) can be seen in Fig. 4., and because of only two clusters this matrix is equivalent to weighted supermatrix.

Super Decisions Main Window: NetworkFaultsAllCon...

Cluster Node Labels		Criteria				Faults		
		F1	F2	F3	F4	A1	A2	A3
Criteria	F1	0.000000	0.000000	0.000000	0.000000	0.164653	0.185424	0.200000
	F2	0.000000	0.000000	0.000000	0.000000	0.088255	0.097141	0.100000
	F3	0.000000	0.000000	0.000000	0.000000	0.570584	0.532012	0.500000
	F4	0.000000	0.000000	0.000000	0.000000	0.176509	0.185424	0.200000
Faults	A1	0.166667	0.146760	0.278461	0.091399	0.000000	0.000000	0.000000
	A2	0.166667	0.083999	0.058471	0.217645	0.000000	0.000000	0.000000
	A3	0.666667	0.769241	0.663067	0.690956	0.000000	0.000000	0.000000

Done

Fig. 4. Unweighted and Weighted Supermatrix

The Limit supermatrix is calculated based on ANP theory and the matrix can be seen in Fig.5. In the last 3 rows can be seen the importance values of alternatives in the supermatrix, these can be normalized in its cluster, thus the final results are: A1: 20.81%, A2: 11.24%, A3: 67.95%.

Super Decisions Main Window: NetworkFaultsAllCon...

Cluster Node Labels		Criteria				Faults		
		F1	F2	F3	F4	A1	A2	A3
Criteria	F1	0.095503	0.095503	0.095503	0.095503	0.095503	0.095503	0.095503
	F2	0.048617	0.048617	0.048617	0.048617	0.048617	0.048617	0.048617
	F3	0.259144	0.259144	0.259144	0.259144	0.259144	0.259144	0.259144
	F4	0.096736	0.096736	0.096736	0.096736	0.096736	0.096736	0.096736
Faults	A1	0.104055	0.104055	0.104055	0.104055	0.104055	0.104055	0.104055
	A2	0.056208	0.056208	0.056208	0.056208	0.056208	0.056208	0.056208
	A3	0.339737	0.339737	0.339737	0.339737	0.339737	0.339737	0.339737

Done

Fig. 5. Limit Supermatrix

5.3. Calculations in VBN

The example is continued with the relative weights, which are equal the importance multiplied by the probabilities. The fault spreading and probabilities can be seen in the Fig.6., where A is the link defect, B is the node breakdown, C is the fault in central part, and D_1-D_3 are the front end errors (D_1: breakdown event at an average user, D_2: malfunction at an average user, D_3: error at administrator site). D_i probabilities can be calculated: 0.0639, 0.0213, 0.0229, so the weight vector in D node is $_dw$=[0.013298, 0.002394, 0.015561].

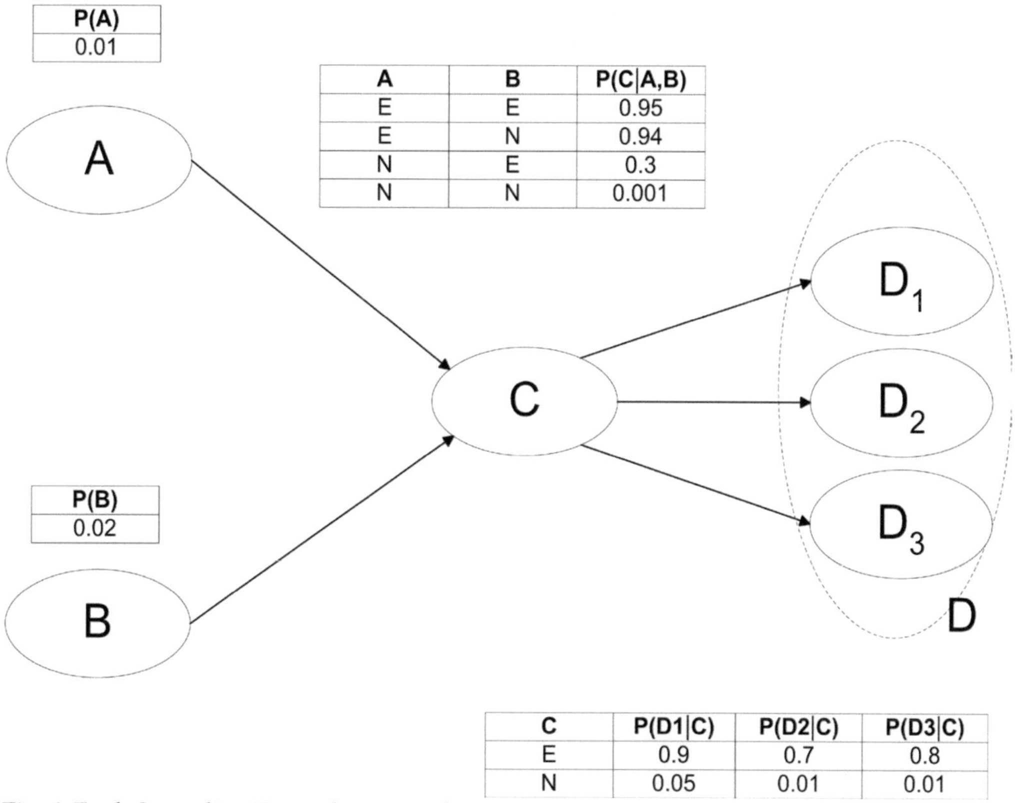

P(A)
0.01

P(B)
0.02

A	B	P(C|A,B)
E	E	0.95
E	N	0.94
N	E	0.3
N	N	0.001

C	P(D1|C)	P(D2|C)	P(D3|C)
E	0.9	0.7	0.8
N	0.05	0.01	0.01

Fig. 6. Fault Spreading Example in an Infocommunication Network

5.4. Numerical Results

Fig. 6. shows the unconditional (at node A and B) and conditional (at node C and D) probabilities of faults. In VBN the parent vector weights can be determined by the formulas shown above. The weight vectors, as final results at A, B, C node are $_aw$=[0.001769, 0.000742, 0.005118], $_bw$=[0.001293, 0.000499, 0.003428], $_cw$=[0.003057, 0.001213, 0.008320] respectively. These values represent the aggregated information about the average effect of faults at the front end. For example $_aw_1$ shows that inner link defect causes average 0.1769% damage in 'breakdown event at an average user' provided the damage of once occurrence of this front end error is 20.81% of total damage of the system.

6. Conclusion

The ANP has been applied to a large variety of decisions: marketing, medical, political, military, social, prediction and many others. ANP is able to take analysis of benefits, opportunities, costs, and risks (BOCR) (Wijnmalen, 2007).

Bayesian Networks are applied in query languages in scientific area of information retrieval (Cheng & Yang, 1999)0, in environmental modelling (Uusitalo, 2007). There are some improvements or combined versions of BN, e.g. combination with fuzzy (Li & Kao, 2005), and many authors deal with further development.

There are some complex (decisional and engineering) problems, where neither ANP nor BN could help to solve alone. Some of these problems are usually such sophisticated, which involve human opinions with uncertainty, causal relationships, and uncertainties in the occurrence of events. These tasks can be solved by the proposed method constructed by combination of ANP and extended version of Bayesian Networks, i.e. by joining these two methods in cascade. Vector Bayesian Networks (VBN) is a generalized BN, which able to handle not only the probabilities, but any numerical value attached to nodes. This extension is able to calculate spreading of effects in any network. The combined method is particularly useful to investigate fault spreading problem in infocommunication networks.

7. References

Büyüközkan, G. (2007). Determining the mobile commerce user requirements using an analytic approach, *Computer Standards & Interfaces*, Volume 31, Issue 1, (January 2009) Pages 144-152.

Chang, C. W.; Wu, C. R.; Lin, C. T.; & Lin, H. L. (2007). Evaluating digital video recorder systems using analytic hierarchy and analytic network processes, *Information Sciences*, Volume 177, Issue 16, (August 2007) Pages 3383-3396.

Chen, S. H. & Lee, H. T. (2007). Performance evaluation model for project managers using managerial practices, *International Journal of Project Management*, Volume 25, Issue 6, (August 2007) Pages 543-551.

Cheng, P. J. & Yang, W. P. (1999). A new content-based access method for video databases, *Information Sciences*, Volume 118, Issues 1-4, (September 1999) Pages 37-73.

Cheon, S. P.; Kim, S.; Lee, S. Y. & Lee, C. B. (2009). Bayesian networks based rare event prediction with sensor data, *Knowledge-Based Systems*, Volume 22, Issue 5, (July 2009) Pages 336-343.

Correa, M.; Bielza, C. & Pamies-Teixeira, J. (2009). Comparison of Bayesian networks and artificial neural networks for quality detection in a machining process, *Expert Systems with Applications*, Volume 36, Issue 3, Part 2, (April 2009) Pages 7270-7279.

Dağdeviren, M.; Yüksel, İ. & Kurt, M. (2008). A fuzzy analytic network process (ANP) model to identify faulty behavior risk (FBR) in work system, *Safety Science*, Volume 46, Issue 5, (June 2008) Pages 771-783.

Demirtas, E. A. & Üstün, Ö. (2008). An integrated multiobjective decision making process for supplier selection and order allocation, *Omega*, Volume 36, Issue 1, (February 2008) Pages 76-90.

Huang, H. Z.; Zuo M. J. & Sun, Z. Q. (2006). Bayesian reliability analysis or fuzzy lifetime data, *Fuzzy Sets and Systems*, Volume 157, Issue 12, (June 2006) Pages 1674-1686.

Judea, P. (1982) Reverend Bayes on inference engines: A distributed hierarchical approach, in *Proceedings of the National Conference on Artificial Intelligence (AAAI-82)*, Pittsburg, Pennsylvania, Morgan Kaufmann, , pp. 133-136.

Kohda, T. & Cui, W. (2007). Risk-based reconfiguration of safety monitoring system using dynamic Bayesian Network, *Reliability Engineering & System Safety*, Volume 92, Issue 12, (December 2007) Pages 1716-1723.

Lee, H.; Kim, C.; Cho, H. & Park, Y. (2009). An ANP-based technology network for identification of core technologies: A case of telecommunication technologies, *Expert Systems with Applications*, Volume 36, Issue 1, (January 2009) Pages 894-908.

Levy, J. K. & Taji, K. (2007). Group decision support for hazards planning and emergency management: A Group Analytic Network Process (GANP) approach, *Mathematical and Computer Modelling*, Volume 46, Issues 7-8, (October 2007) Pages 906-917.

Li, H. L. & Kao, H. Y. (2005). Constrained abductive reasoning with fuzzy parameters in Bayesian Networks, *Computers & Operations Research*, Volume 32, Issue 1, (January 2005) Pages 87-105.

Liu, G.; Feng, W.; Wang, H.; Liu, L. & Zhou, C. (2009). Reconstruction of Gene Regulatory Networks Based on Two-Stage Bayesian Network Structure Learning Algorithm, *Journal of Bionic Engineering*, Volume 6, Issue 1, (March 2009) Pages 86-92.

Promentilla, M. A. B.; Furuichi, T.; Ishii, K. & Tanikawa, N. (2008). A fuzzy analytic network process for multi-criteria evaluation of contaminated site remedial countermeasures, *Journal of Environmental Management*, Volume 88, Issue 3, (August 2008) Pages 479-495.

Saaty, T. L. (1980). *The Analytic Hierarchy Process*, McGraw-Hill International, New York

Saaty, T. L. (2001). *The Analytic Network Process: Decision Making with Dependence and Feedback*, RWS Publication, Pittsburgh, PA.

Saaty, T. L. (2005). *Theory and applications of the analytic network process*, RWS publications, Pittsburgh

Saaty, T. L. (2007). Time dependent decision-making; dynamic priorities in the AHP/ANP: Generalizing from points to functions and from real to complex variables, *Mathematical and Computer Modelling*, Volume 46, Issues 7-8, (October 2007) Pages 860-891.

Speigelhalter, D. J.; Dawid, A. P.; Lauritzen, S. L. & Cowell, R. G. (1993). Bayesian analysis in expert systems, *Statistical Science* 8 (3)

Szűcs, G. & Sallai, Gy. (2008). Combination of Analytic Network Process and Bayesian Network Model for Multi-Criteria Engineering Decision Problems, *Proceedings of IEMC2008 (International Engineering Management Conference)*, IEEE, Estoril, Portugal, June 28-30, 2008. pp. 141-145.

Uusitalo, L. (2007). Advantages and challenges of Bayesian Networks in environmental modeling," *Ecological Modelling*, Volume 203, Issues 3-4, (May 2007) Pages 312-318.

Wei, Y. F.; Hu, Y. H. & Song, J. D. (2007). Network selection strategy in heterogeneous multi-access environment, *The Journal of China Universities of Posts and Telecommunications*, Volume 14, Supplement 1, (October 2007) Pages 16-20.

Whitaker, R. (2007). Validation examples of the Analytic Hierarchy Process and Analytic Network Process, *Mathematical and Computer Modelling*, Volume 46, Issues 7-8, (October 2007) Pages 840-859.

Wijnmalen, D. J. D. (2007). Analysis of benefits, opportunities, costs, and risks (BOCR) with the AHP–ANP: A critical validation, *Mathematical and Computer Modelling*, Volume 46, Issues 7-8, (October 2007) Pages 892-905.

Wilson, A. G. & Huzurbazar, A. V. (2007). Bayesian Networks for multilevel system reliability, *Reliability Engineering & System Safety*, Volume 92, Issue 10, (October 2007) Pages 1413-1420.

Wu, W. W. & Lee, Y. T. (2007). Selecting knowledge management strategies by using the analytic network process, *Expert Systems with Applications*, Volume 32, Issue 3, (April 2007) Pages 841-847.

Yu, J. R. & Cheng, S. J. (2007). An integrated approach for deriving priorities in analytic network process, *European Journal of Operational Research*, Volume 180, Issue 3, (August 2007) Pages 1427-1432.

Yu, R. & Tzeng, G. H. (2006). A soft computing method for multi-criteria decision making with dependence and feedback, *Applied Mathematics and Computation*, Volume 180, Issue 1, (September 2006) Pages 63-75.

13

Monitoring of complex processes with Bayesian networks

Sylvain Verron, Teodor Tiplica and Abdessamad Kobi
LASQUO/ISTIA - University of Angers
France

1. Introduction

Industrial processes are more and more complex and include a lot of sensors giving measurements of some attributes of the system. A study of these measurements can allow to decide on the correct working conditions of the process. If the process is not in normal working conditions, it signifies that a fault has occurred in the process. If no fault has occurred, thus the process is in the fault-free case. An important research field is on the Fault Detection and Diagnosis (FDD) (Isermann (2006)). The goal of a FDD scheme is to detect, the earliest possible, when a fault occurs in the process. Once the fault has been detected, the other important step is the diagnosis. The diagnosis can be seen as the decision of which fault has appeared in the process, what are the characteristics of this fault, what are the root causes of the fault.
One can distinguish three principal categories of methods for the FDD (Chiang et al. (2001)): the knowledge-based approach, the model-based approach and the data-driven approach. The knowledge-based category represents methods based on qualitative models (FMECA - Failures Modes Effects and Critically Analysis; Fault Trees; Decision Trees; Risk Analysis) (Dhillon (2005); Stamatis (2003)). For the model-based methods, an analytical model of the process is constructed based on the physical relations governing the process (Patton et al. (2000)). The model gives the normal (fault free) value of each sensor or variable of the system for each sample instant, then residuals are generated (residuals are the differences between measurements and the corresponding reference values estimated with the model of the fault-free system). If the system is fault free, residuals are almost nil, and so their evaluations allow to detect and diagnose a fault. Theoretically, the best methods are the analytical ones, but the major drawback of this family of techniques is that a detailed model of the process is required in order to monitor it efficiently. Obtaining an effective detailed model can be very difficult, time consuming and expensive, particularly for large-scale systems with many variables. The last category of methods are the process history (or data-driven) methods (Venkatasubramanian et al. (2003)). These techniques are based on rigorous statistical developments of process data. In literature, we can find many different data-driven techniques for FDD. For the fault detection of industrial processes many methods have been submitted: univariate statistical process control (Shewhart charts) (Montgomery (1997)), multivariate statistical process control (T^2 and Q charts) (Westerhuis et al. (2000)), and some Principal Component Analysis (PCA) based techniques (Jackson (1985)). Kano et al. (2002) make comparisons between these different techniques. For the fault diagnosis techniques we can cite the book of Chiang et al.

(2001) which presents a lot of them (PCA based techniques, Fisher Discriminant Analysis, PLS based techniques, etc).
The purpose of this article is to present application of a promising tool for the Fault Detection and Diagnosis: the Bayesian network. The aim of the paper is to demonstrate that some FDD techniques can be modeled very simply in a Bayesian network, with very good performances. The article is structured in the following manner. In section 2, we introduce different notions (theoretical and practical) about Bayesian network. The section 3 presents how to model multivariate control charts in a Bayesian network, in order to make an effective way for the fault detection by the Bayesian network. In the same way, section 4 presents the modeling of discriminant analysis by Bayesian network for fault diagnosis of systems. The section 5 presents an evaluation of the proposed method for detection and diagnosis of faults on the benchmark Tennessee Eastman Problem. Finally, we conclude on this method and present some perspectives.

2. Bayesian network

A Bayesian Network (BN) (Pearl (1988)) is a probabilistic graphical model where each variable is a node. Edges of the graph represent dependences between linked nodes. A formal definition of Bayesian network (Jensen (1996)) is a couple $\{\mathbf{G}, \mathbf{P}\}$ where:

$\{\mathbf{G}\}$ is a directed acyclic graph, whose nodes are random variables $\boldsymbol{X} = \{X_1, X_2, \ldots, X_n\}$ and whose missing edges represent conditional independences between the variables,

$\{\mathbf{P}\}$ is a set of conditional probability distributions (one for each variable): $P = \{p(X_1|pa(X_1)), \ldots, p(X_n|pa(X_n))\}$ where $p(X_i|pa(X_i))$ is a table defined by $p(X_i = x_i^j|pa(X_i))$ with $x_i^j \in Dom(X_i) = x_i^1, x_i^2, ..., x_i^{n_i}$ where $Dom(X_i)$ is the set of modalities of variable X_i and n_i is the number of these modalities. The joint probability should read like the following equation:

$$p(x) = \prod_{i=1}^{n}(X_i|pa(X_i)) \tag{1}$$

with $x = (x_1^{j_1}, x_2^{j_2}, ..., x_n^{j_n})$.

Theoretically, variables $X_1, X_2, \ldots, X_n$ can be discrete or continuous. However, in practice, for exact computation, only the discrete and the Gaussian case can be treated. Such a network is often called Conditional Gaussian Network (CGN). In this context, to ensure availability of exact computation methods, discrete variables are not allowed to have continuous parents (see Lauritzen & Jensen (2001); Madsen (2008)).
In concrete terms, the conditional probability distribution is described for each node by his Conditional Probability Table (CPT). In a CGN, three cases of CPT can be found. The first one is for a discrete variable with discrete parents. For example, we take the case of two discrete variables A and B of respective dimensions a and b (with $a_1, a_2, \ldots, a_a$ the different modalities of A, and $b_1, b_2, \ldots, b_b$ the different modalities of B). If A is parent of B, then the CPT of B is represented in table 1.
As we can see, the goal of the CPT is to condense the information about the relations of B with his parents. We can denote that the dimension of this CPT (number of conditional probabilities) is $a \times b$. In general the dimension of the CPT of a discrete node (dimension a) with p parents (discrete) $Y_1, Y_2, \ldots, Y_p$ (dimension $y_1, y_2, \ldots, y_p$) is $a \times \prod_{i=1}^{p} y_i$.

	B			
A	b_1	b_2	$\dots$	b_b
a_1	$P(b_1\|a_1)$	$P(b_2\|a_1)$	$\dots$	$P(b_b\|a_1)$
a_2	$P(b_1\|a_2)$	$P(b_2\|a_2)$	$\dots$	$P(b_b\|a_2)$
$\vdots$	$\vdots$	$\vdots$	$\ddots$	$\vdots$
a_a	$P(b_1\|a_a)$	$P(b_2\|a_a)$	$\dots$	$P(b_b\|a_a)$

Table 1. CPT of a discrete node with discrete parents

The second case of CPT is for a continuous variable with discrete parents. Assuming that B is a Gaussian variable, and that A is a discrete parent of B with a modalities, the CPT of B can be represented as in the table 2 where $P(B|a_1) \sim \mathcal{N}(\mu_{a_1}, \Sigma_{a_1})$ indicates that B conditioned to $A = a_i$ follows a multivariate normal density function with parameters μ_{a_i} and Σ_{a_i}.

A	B
a_1	$P(B\|a_1) \sim \mathcal{N}(\mu_{a_1}, \Sigma_{a_1})$
a_2	$P(B\|a_2) \sim \mathcal{N}(\mu_{a_2}, \Sigma_{a_2})$
$\vdots$	$\vdots$
a_a	$P(B\|a_a) \sim \mathcal{N}(\mu_{a_a}, \Sigma_{a_a})$

Table 2. CPT of a Gaussian node with discrete parents

The third case occurs when a continuous node B has a continuous parent A. In this case, we obtain a linear regression and we can write, for a fixed value a of A, that B follows a Gaussian distribution $P(B|A = a) \sim \mathcal{N}(\mu_B + \beta \times a; \Sigma_B)$ where β is the regression coefficient. The three different cases of CPT enumerated can evidently be combined for different cases where a continuous variable has several discrete parents and several continuous (Gaussian) parents. The classical use of a Bayesian network (or Conditional Gaussian Network) is to enter evidence in the network (an evidence is the observation of the values of a set of variables). Therefore, the information given by the evidence is propagated in the network in order to update the knowledge and obtain a posteriori probabilities on the non-observed variables. This propagation mechanism is called inference. As its name suggests, in a Bayesian network, the inference is based on the Bayes rule. A lot of inference algorithms (exact or approximate) have been developed, but one of the more exploited is the junction tree algorithm (Jensen et al. (1990)).

Bayesian network classifiers are particular BN (Friedman et al. (1997)). They always have a discrete node C coding the k different classes of the system. Thus, other variables $X_1, \dots, X_p$ represent the p descriptors (variables) of the system.

A famous Bayesian classifier is the Naïve Bayesian Network (NBN), also named Bayes classifier (Langley et al. (1992)). This Bayesian classifier makes the strong assumption that the descriptors of the system are class conditionally independent. Assuming the hypothesis of normality of each descriptor, the NBN is equivalent to the classification rule of the diagonal quadratic discriminant analysis. But, in practice, this assumption of independence and non-correlated variables is not realistic. In order to deal with correlated variables, several approaches have been developed. We can cite the Tree Augmented Naïve Bayesian networks (TAN) (Friedman et al. (1997)). These BNs are based on a NBN but a tree is added between the descriptors. An other interesting approach is the Kononenko one (Kononenko (1991)), which represents some variables in one node. As in (Perez et al. (2006)) the assumption we will make is that this variable follows a normal multivariate distribution (conditionally to the class) and we will refer to this kind of BN as Condensed Semi Naïve Bayesian Network (CSNBN).

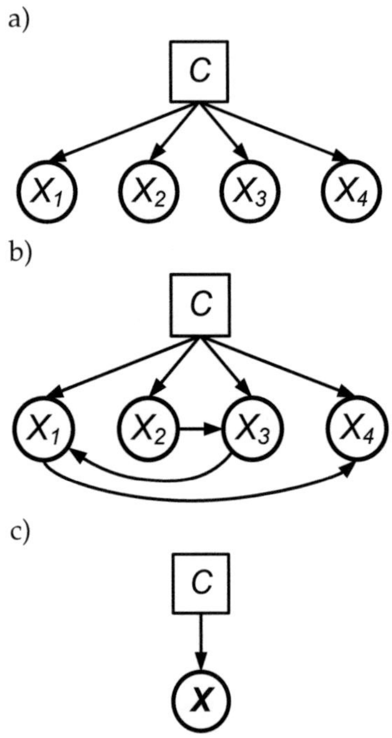

Fig. 1. Different bayesian network classifiers: NBN (a), TAN (b) and CSNBN (c).

3. Fault detection with Bayesian network

In previous work (Verron et al. (2007b)), we have demonstrated that a T^2 control chart Hotelling (1947) could be modeled with a Bayesian network. For that, we use two nodes: a Gaussian multivariate node $\boldsymbol{X}$ representing the data and a bimodal node E representing the state of the process. The bimodal node E has the following modalities: IC for "in control" and OC for "out-of-control". Assuming that $\boldsymbol{\mu}$ and $\boldsymbol{\Sigma}$ are respectively the mean vector and the variance-covariance matrix of the process, we can monitor the process with the following rule: if $P(IC|\boldsymbol{x}) < P(IC)$ then the process is out-of-control. This Bayesian network is represented on the Figure 2, where the conditional probabilities tables of each node are given.
In Figure 2, we can observe that a coefficient c is implicated in the modeling of the control chart by Bayesian network. This coefficient is the root (different of 1) of the following equation:

$$1 - c + \frac{pc}{CL}\ln(c) = 0 \tag{2}$$

where p is the dimension (number of variables) of the system to monitor, and CL is the control limit of the equivalent T^2 control chart. The demonstration of the computation of c is given in A. In numerous cases, CL is equal to $\chi^2_{\alpha,p}$, the quantile at the value α of the distribution of the

E	
IC	OC
$P(IC)$	$P(OC)$

E	$\boldsymbol{X}$
IC	$\boldsymbol{X} \sim N(\boldsymbol{\mu}, \boldsymbol{\Sigma})$
OC	$\boldsymbol{X} \sim N(\boldsymbol{\mu}, c \times \boldsymbol{\Sigma})$

Fig. 2. T^2 control chart in a Bayesian network

χ^2 with p degrees of freedom (Montgomery (1997)). α allows us to tune the false alarm rate of the control chart.
The application of this network to a two variables process is given in figure 3.

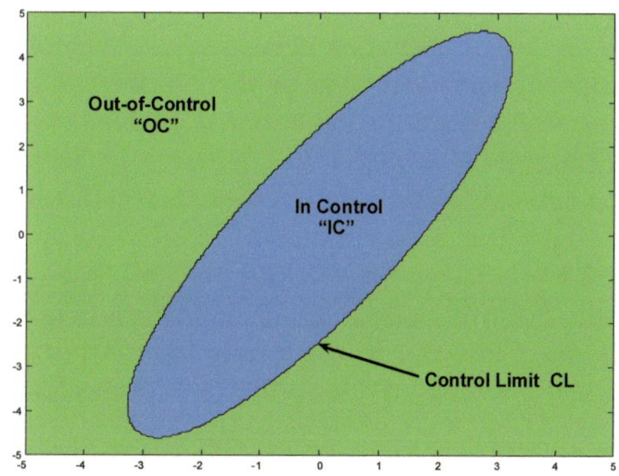

Fig. 3. Detection area of the Bayesian network

A particular interest of the modeling of control chart in a Bayesian network is that a MEWMA control chart (Lowry et al. (1992)) can also be modeled in the same way. The principle of the MEWMA control chart is to take into account the process evolution in weighting past observations extracted from the process. The MEWMA variable $\boldsymbol{y}_t$ is computed recursively, for each sample, by the equation 3 where the initialization is given by $\boldsymbol{y}_0 = \boldsymbol{\mu}$.

$$\boldsymbol{y}_t = \lambda \boldsymbol{x}_t + (1 - \lambda)\boldsymbol{y}_{t-1} \tag{3}$$

In the same way that the T^2 control chart, we can also monitor the process with a MEWMA control chart modeled by the Bayesian network of the figure 2.
We can precise that performances of the MEWMA control chart are function of λ. Indeed, a small λ allows a performing detection of small magnitude shifts, but a higher λ will be more adapted for large magnitude shifts. So, the choice of λ will be function of the magnitude shift that one wants to detect. A particular case of the MEWMA control chart is the case where $\lambda = 1$. In this case, the MEWMA chart is equivalent to the T^2 control chart.

E	
IC	OC
$P(IC)$	$P(OC)$

E	$\boldsymbol{Y}$
IC	$\boldsymbol{Y} \sim \mathcal{N}(\boldsymbol{\mu}, \left(\frac{\lambda}{2-\lambda}\right) \boldsymbol{\Sigma})$
OC	$\boldsymbol{Y} \sim \mathcal{N}(\boldsymbol{\mu}, c \times \left(\frac{\lambda}{2-\lambda}\right) \boldsymbol{\Sigma})$

Fig. 4. MEWMA control chart in a Bayesian network

4. Bayesian network for fault diagnosis

Once a problem (fault) has been detected in the evolution of the process by the mean of a detection method, we need to identify (diagnosis) the belonging class of this fault. Thereby, the diagnosis problem can be viewed as the task to correctly classify this fault in one of the predefined fault classes. The classification task needs the construction of a classifier (a function allocating a class to the observations described by the variables of the system). Two types of classification exist: unsupervised classification which objective is to identify the number and the composition of each class present in the data structure; supervised classification where the number of classes and the belonging class of each observation is known in a learning sample and whose objective is to class new observations to one of the existing classes. For example, given a learning sample of a bivariate system with three different known faults as illustrated in the figure 5, we can easily use supervised classification to classify a new faulty observation. A feature selection can be used in order to select only the most informative variables of the problem (Verron et al. (2008)). In this study, we will use the Bayesian network as a supervised classification tool.

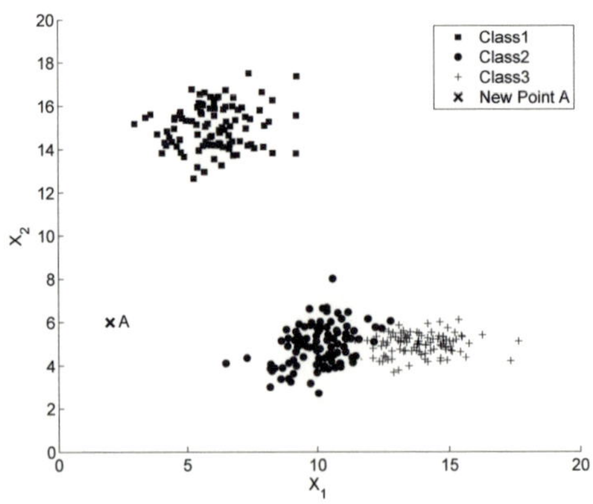

Fig. 5. Bivariate system with three different known faults

In the context of the diagnosis of industrial systems, Bayesian networks and Conditional Gaussian Networks have been already used and they give convenient results compared to other classification tools like support vector machines, neural networks or k-nearest neighborhoods (Pernkopf (2005); Perzyk et al. (2005); Tiplica et al. (2006); Verron et al. (2007a;c)). As the performances of the CGN have been previously demonstrated (Verron et al. (2007a;c)), we choose this classifier in this article which is equivalent to a Discriminant Analysis (DA). Therefore, we name the class node DA, and the observation node $\boldsymbol{X}$ (a normal multivariate node). The figure 6 presents the CGN equivalent to a discriminant analysis, with the probability tables associated to each node. To simplify, the a priori probability of each class F_i is fixed to $p(F_i) = \frac{1}{k}$, where k is the number of known faults. The node $\boldsymbol{X}$ follows the different normal probability densities ($\mathcal{N}$) conditionally to the class of DA, where $\boldsymbol{\mu}_i$ is the mean vector of the fault F_i, $\boldsymbol{\Sigma}_i$ is the covariance matrix of the fault F_i. $\boldsymbol{\mu}_i$ and $\boldsymbol{\Sigma}_i$ are estimated on the fault database by Maximum Likelihood Estimation (MLE) (Duda et al. (2001)). In the mere example of the figure 5, the CGN gives the different areas of classification of the figure 7.

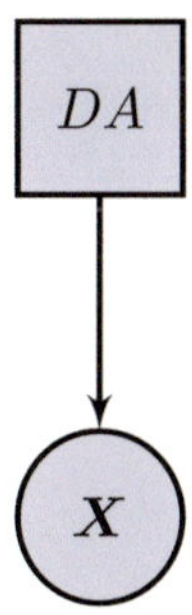

DA		
F_1	...	F_k
$\frac{1}{k}$	...	$\frac{1}{k}$

DA	$\boldsymbol{X}$
F_1	$\boldsymbol{X} \sim \mathcal{N}(\boldsymbol{\mu}_1, \boldsymbol{\Sigma}_1)$
...	...
F_k	$\boldsymbol{X} \sim \mathcal{N}(\boldsymbol{\mu}_k, \boldsymbol{\Sigma}_k)$

Fig. 6. Conditional Gaussian Network equivalent to a discriminant analysis

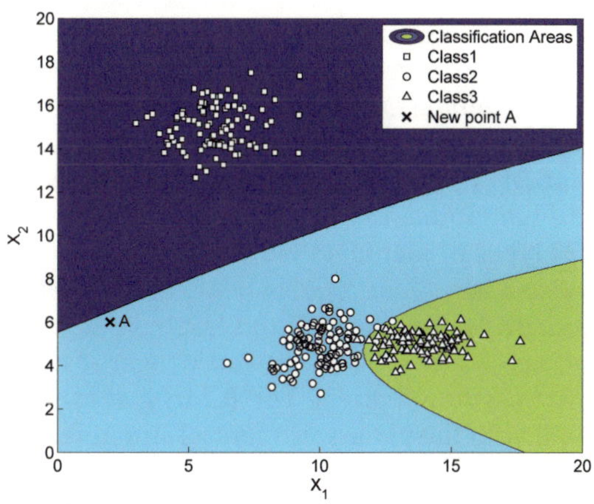

Fig. 7. Classification areas of the bivariate system

5. Application to the TEP

Now, we are going to study an application of the Bayesian network approach on a benchmark problem: the Tennessee Eastman Process (figure 8).

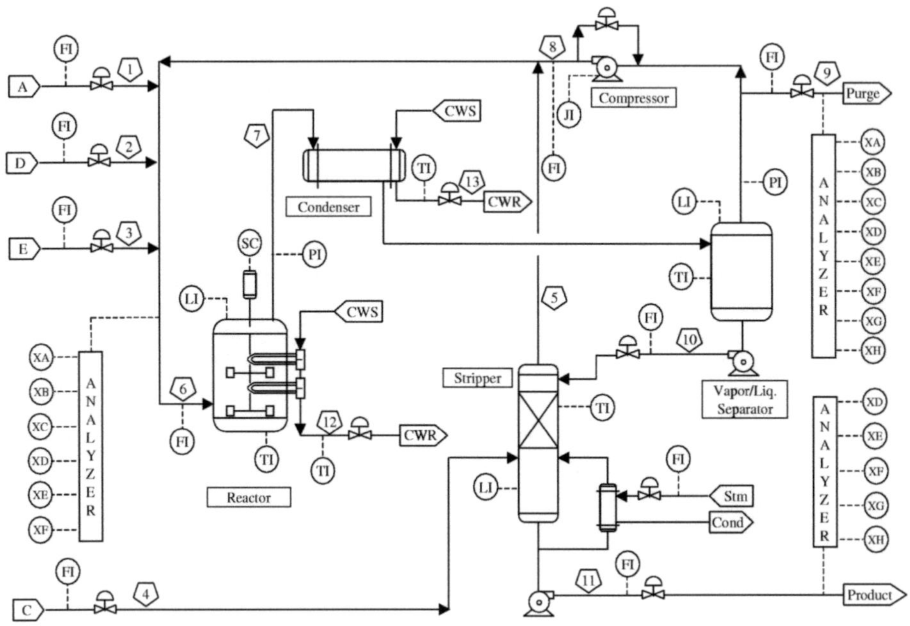

Fig. 8. Process flowsheet of the TEP

5.1 Presentation of the TEP

We have tested our approach on the Tennessee Eastman Process. The Tennessee Eastman Process (TEP) is a chemical process. It is not a real process but a simulation of a process that was created by the Eastman Chemical Company to provide a realistic industrial process in order to evaluate process control and monitoring methods. The article of Downs & Vogel (1993) entirely describes this process. The authors also give the Fortran code of the simulation of the process. Ricker (1996) has implemented the simulation on Matlab. The TEP is composed of five major operation units: a reactor, a condenser, a compressor, a stripper and a separator. Four gaseous reactants A, C, D, E and an inert one B are fed to the reactor where the liquid products F, G and H are formed. This process has 12 input variables and 41 output variables. The TEP has 20 types of identified faults. This process is ideal to test monitoring methods. However, it is also a benchmark problem for control techniques because it is open-loop unstable. A lot of articles present the TEP and test their approaches on it. For example, in fault detection, we can cite Kano et al. (2002) and Kruger et al. (2004). Some fault diagnosis techniques have also been tested on the TEP (Chiang et al. (2001; 2004); Kulkarni et al. (2005); Maurya et al. (2007)) with the plant-wide control structure recommended in Lyman & Georgakis (1995).
As indicated in the table 3, each type of fault is composed of 2 datasets: a training sample and a testing sample, containing respectively 480 and 800 observations. We precise that in the next part of this paper all computations have been made on Matlab with the BNT (BayesNet Toolbox) developed by Murphy (2001).

5.2 Detection

In order to test the performances of the Bayesian network approach for the detection, we set an acceptable false alarm for the detection of 0.01 (1%). As the detection is modeled with

Class	Train data	Test data
Fault free	480	800
Fault 1	480	800
Fault 2	480	800
...	...	...
Fault k	480	800
...	...	...
Fault 20	480	800

Table 3. Data of the TEP

two control chart, the local false alarm rate is set to 0.005. The table 4 presents the results of the Bayesian network dedicated to the detection, composed of the modeling of the T^2 and MEWMA control charts.

Fault	First detection instant	Detection rate
1	3	99.75
2	13	98.5
3	34	35
4	1	100
5	1	100
6	1	100
7	1	100
8	18	97.75
9	7	15.88
10	18	97
11	7	90.88
12	2	99.88
13	37	95.5
14	1	100
15	146	30.5
16	9	99
17	20	97.5
18	57	92.38
19	2	96.5
20	65	91.88
Mean	22.15	91.38

Table 4. Detection results

In table 4, we can affirm that faults F3, F9 and F15 are very difficult to detect. Chiang et al. (Chiang et al. (2004)), using PCA (Principal Component Analysis) based method, on the same data, have made the same conclusions on these 3 faults. However, without these 3 faults, the mean detection rate of the other faults is more than 97.44% and proves the efficiency of the Bayesian network for the fault detection task.

5.3 Diagnosis

Always on the same data, we have applied the method proposed in section 4. After the learning of the parameters of the Bayesian network, we have presented 16 000 observations to the network (800 observations of each 20 faults). The network has given probabilities of each observation to come from each known faults. The decision of the fault has been taken for the fault with the maximum a posteriori probability. Results of the 16 000 observations are given in the table 6 of appendix B. A more readable table of results is given in table 5.

Fault	Diagnosis rate
1	97,5
2	98,12
3	22
4	82,37
5	98
6	100
7	100
8	97
9	22,62
10	86,87
11	75,5
12	98,25
13	76,12
14	98,75
15	23,5
16	80,62
17	85
18	68,5
19	96,12
20	87,37
Mean	79.71

Table 5. Diagnosis results

In the table 5, we can observe that, like for the fault detection, the faults F3, F9 and F15 are difficult to diagnose. Indeed, these three faults are very similar to the fault free conditions, and so they are difficult to detect and difficult to diagnose. However, for the other faults, we can notice that a lot of observations are correctly classified, and without the 3 difficult faults (F3, F9 and F15), the mean diagnosis rate increase to 90%.

6. Conclusion

In this chapter, we have studied the application of Bayesian networks (and more particularly of Conditional Gaussian networks) for the fault detection and diagnosis. The fault detection is made by a modeling of multivariate control charts (T^2 and MEWMA) with Bayesian network. On the same way, the fault diagnosis is similar to a supervised classification task. A Bayesian network is able to discriminate between different faults of a system. For that, we have modeled a discriminant analysis directly in the Bayesian network. The performances of the proposed approach are evaluated on the benchmark problem of the Tennessee Eastman

Process, demonstrating that fault detection and fault diagnosis can be made with Bayesian network. Outlooks of this work are on the use a Bayesian network as a causal model of a process, in order to realize fault isolation of the different variables implicated in a fault.

A. Coefficient c demonstration

This appendix presents the demonstration of the equation 2.
As in the case of the T^2 control chart (Montgomery (1997)), we will fix a threshold (Control Limit CL for the control chart) on the a posteriori probabilities allowing to take decisions on the process: if, for a given observation $\boldsymbol{x}$, the a posteriori probability to be allocated to F_i ($P(F_i|\boldsymbol{x})$) is greater than the a priori probability to be allocated to F_i ($P(F_i)$), then this observation is allocated to F_i. This rule can be rewritten as: $\boldsymbol{x} \in F_i$ if $P(F_i|\boldsymbol{x}) > P(F_i)$, or equivalently $\boldsymbol{x} \in \overline{F_i}$ if $P(\overline{F_i}|\boldsymbol{x}) < P(\overline{F_i})$. The objective of the following developments is to define c in order to obtain the equivalence between the CGN and the multivariate T^2 control chart.
We want to keep the following decision rule:

$$\boldsymbol{x} \in F_i \quad if \quad T^2 < CL \tag{4}$$

with this decision rule:

$$\boldsymbol{x} \in F_i \quad if \quad P(F_i|\boldsymbol{x}) > P(F_i) \tag{5}$$

We develop the second decision rule:

$$\begin{aligned}
P(F_i|\boldsymbol{x}) &> P(F_i) \\
P(F_i|\boldsymbol{x}) &> (P(F_i))(P(F_i|\boldsymbol{x}) + P(\overline{F_i}|\boldsymbol{x})) \\
P(F_i|\boldsymbol{x}) &> P(F_i)P(F_i|\boldsymbol{x}) + P(F_i)P(\overline{F_i}|\boldsymbol{x}) \\
P(F_i|\boldsymbol{x}) - P(F_i)P(F_i|\boldsymbol{x}) &> P(F_i)P(\overline{F_i}|\boldsymbol{x}) \\
P(F_i|\boldsymbol{x})(1 - P(F_i) &> P(F_i)P(\overline{F_i}|\boldsymbol{x}) \\
P(F_i|\boldsymbol{x})P(\overline{F_i}) &> P(F_i)P(\overline{F_i}|\boldsymbol{x}) \\
P(F_i|\boldsymbol{x}) &> \frac{P(F_i)}{P(\overline{F_i})}P(\overline{F_i}|\boldsymbol{x})
\end{aligned}$$

However, the Bayes law gives:

$$P(F_i|\boldsymbol{x}) = \frac{P(F_i)P(\boldsymbol{x}|F_i)}{P(\boldsymbol{x})} \tag{6}$$

and

$$P(\overline{F_i}|\boldsymbol{x}) = \frac{P(\overline{F_i})P(\boldsymbol{x}|\overline{F_i})}{P(\boldsymbol{x})} \tag{7}$$

As a consequence, we obtain:

$$\begin{aligned}
\frac{P(F_i)P(\boldsymbol{x}|F_i)}{P(\boldsymbol{x})} &> (\frac{P(F_i)}{P(\overline{F_i})})\frac{P(\overline{F_i})P(\boldsymbol{x}|\overline{F_i})}{P(\boldsymbol{x})} \\
(\frac{P(F_i)}{P(\overline{F_i})})P(\boldsymbol{x}|F_i) &> (\frac{P(F_i)}{P(\overline{F_i})})P(\boldsymbol{x}|\overline{F_i}) \\
P(\boldsymbol{x}|F_i) &> P(\boldsymbol{x}|\overline{F_i})
\end{aligned} \tag{8}$$

In the case of a discriminant analysis with k classes C_i, the conditional probabilities are computed with the following equation 9, where ϕ represents the probability density function of the multivariate Gaussian distribution of the class.

$$P(\boldsymbol{x}|C_i) = \frac{\phi(\boldsymbol{x}|C_i)}{\sum_{j=1}^{k} P(C_j)\phi(\boldsymbol{x}|C_j)} \tag{9}$$

Equation 8 can be written as:

$$\phi(\boldsymbol{x}|F_i) \quad > \quad \phi(\boldsymbol{x}|\overline{F_i}) \tag{10}$$

We recall that the probability density function of a multivariate Gaussian distribution of dimension p, of parameters $\boldsymbol{\mu}$ and $\boldsymbol{\Sigma}$, of an observation $\boldsymbol{x}$ is given by:

$$\phi(\boldsymbol{x}) = \frac{e^{-\frac{1}{2}(\boldsymbol{x}-\boldsymbol{\mu})^T\boldsymbol{\Sigma}^{-1}(\boldsymbol{x}-\boldsymbol{\mu})}}{(2\pi)^{p/2}|\boldsymbol{\Sigma}|^{1/2}} \tag{11}$$

If the law parameters are $\boldsymbol{\mu}$ and $c \times \boldsymbol{\Sigma}$, then the density function becomes:

$$\phi(\boldsymbol{x}) = \frac{e^{-\frac{1}{2c}(\boldsymbol{x}-\boldsymbol{\mu})^T\boldsymbol{\Sigma}^{-1}(\boldsymbol{x}-\boldsymbol{\mu})}}{(2\pi)^{p/2}|\boldsymbol{\Sigma}|^{1/2}c^{p/2}} \tag{12}$$

In identifying the expression $(\boldsymbol{x}-\boldsymbol{\mu})^T\boldsymbol{\Sigma}^{-1}(\boldsymbol{x}-\boldsymbol{\mu})$ as the T^2 of the observation $\boldsymbol{x}$, we can write:

$$\begin{aligned}
\phi(\boldsymbol{x}|F_i) &> \phi(\boldsymbol{x}|\overline{F_i}) \\
\frac{e^{-\frac{T^2}{2}}}{(2\pi)^{p/2}|\boldsymbol{\Sigma}|^{1/2}} &> \frac{e^{-\frac{T^2}{2c}}}{(2\pi)^{p/2}|\boldsymbol{\Sigma}|^{1/2}c^{p/2}} \\
e^{-\frac{T^2}{2}} &> \frac{e^{-\frac{T^2}{2c}}}{c^{p/2}} \\
-\frac{T^2}{2} &> -\frac{T^2}{2c} - \frac{p\ln(c)}{2} \\
T^2 &< \frac{T^2}{c} + p\ln(c) \\
T^2 &< \frac{p\ln(c)}{1-\frac{1}{c}} \qquad (13)
\end{aligned}$$

However, we search the value(s) of c allowing the equivalence with the control chart decision rule: $\boldsymbol{x} \in F_i \;\; if \;\; T^2 < CL$. So, we obtain the following equation for c:

$$\frac{p\ln(c)}{1-\frac{1}{c}} = LC \tag{14}$$

Or, equivalently:

$$1 - c + \frac{pc}{LC}\ln(c) = 0 \tag{15}$$

B. Fault diagnosis detailed results

	F1	F2	F3	F4	F5	F6	F7	F8	F9	F10	F11	F12	F13	F14	F15	F16	F17	F18	F19	F20
F1	780	0	0	0	0	0	0	2	0	0	0	0	0	0	0	0	0	0	0	0
F2	0	785	0	0	0	0	0	0	0	0	0	0	0	0	0	0	0	0	0	0
F3	0	0	176	0	0	0	0	2	201	8	18	0	16	0	118	15	1	38	6	17
F4	0	0	0	659	0	0	0	0	0	0	27	0	0	0	0	0	0	0	0	0
F5	0	0	0	0	784	0	0	0	0	0	0	3	0	0	0	0	0	0	0	0
F6	0	0	0	0	0	800	0	0	0	0	0	0	0	0	0	0	0	0	0	0
F7	0	0	0	0	0	0	800	0	0	0	0	0	0	0	0	0	0	0	0	0
F8	18	5	1	0	0	0	0	776	0	4	0	1	109	0	1	26	0	0	0	0
F9	0	8	171	0	0	0	0	11	181	25	24	1	4	0	233	15	7	13	9	34
F10	0	0	48	0	0	0	0	0	40	695	9	0	0	0	64	48	0	3	5	6
F11	0	0	43	141	0	0	0	3	42	5	604	0	2	1	43	2	30	3	2	3
F12	0	0	0	0	16	0	0	4	0	6	0	786	41	0	4	10	0	168	0	23
F13	0	0	0	0	0	0	0	2	0	3	0	1	609	0	3	4	0	0	0	3
F14	0	0	17	0	0	0	0	0	10	3	28	0	0	790	20	4	71	0	0	1
F15	0	1	215	0	0	0	0	0	221	12	34	0	11	0	188	6	9	9	2	7
F16	0	1	85	0	0	0	0	0	39	35	5	0	2	0	82	645	1	10	4	3
F17	1	0	7	0	0	0	0	0	6	3	42	0	0	9	1	3	680	0	1	2
F18	0	0	0	0	0	0	0	0	0	0	0	8	5	0	0	0	0	548	1	0
F19	1	0	32	0	0	0	0	0	54	1	7	0	1	0	38	16	1	1	769	2
F20	0	0	5	0	0	0	0	0	6	0	2	0	0	0	5	6	0	7	1	699

Table 6. Confusion matrix

C. References

Chiang, L. H., Russell, E. L. & Braatz, R. D. (2001). *Fault detection and diagnosis in industrial systems*, New York: Springer-Verlag.

Chiang, L., Kotanchek, M. & Kordon, A. (2004). Fault diagnosis based on fisher discriminant analysis and support vector machines, *Computers and Chemical Engineering* **28**(8): 1389–1401.

Dhillon, B. (2005). *Reliability,Quality,and Safety for Engineers*, CRC Press.

Downs, J. & Vogel, E. (1993). Plant-wide industrial process control problem, *Computers and Chemical Engineering* **17**(3): 245–255.

Duda, R. O., Hart, P. E. & Stork, D. G. (2001). *Pattern Classification 2nd edition*, Wiley.

Friedman, N., Geiger, D. & Goldszmidt, M. (1997). Bayesian network classifiers, *Machine Learning* **29**(2-3): 131–163.

Hotelling, H. (1947). Multivariate quality control, *Techniques of Statistical Analysis* : 111–184.

Isermann, R. (2006). *Fault Diagnosis Systems An Introduction from Fault Detection to Fault Tolerance*, Springer.

Jackson, E. J. (1985). Multivariate quality control, *Communication Statistics - Theory and Methods* **14**: 2657 – 2688.

Jensen, F. V. (1996). *An introduction to Bayesian Networks*, Taylor and Francis, London, United Kingdom.

Jensen, F. V., Lauritzen, S. L. & Olesen, K. G. (1990). Bayesian updating in causal probabilistic networks by local computations, *Computational Statistics Quaterly* **4**: 269–282.

Kano, M., Nagao, K., Hasebe, S., Hashimoto, I., Ohno, H., Strauss, R. & Bakshi, B. (2002). Comparison of multivariate statistical process monitoring methods with applications to the eastman challenge problem, *Computers and Chemical Engineering* **26**(2): 161–174.

Kononenko, I. (1991). Semi-naive bayesian classifier, *EWSL-91: Proceedings of the European working session on learning on Machine learning*, pp. 206–219.

Kruger, U., Zhou, Y. & Irwin, G. (2004). Improved principal component monitoring of large-scale processes, *Journal of Process Control* **14**(8): 879–888.

Kulkarni, A., Jayaraman, V. & Kulkarni, B. (2005). Knowledge incorporated support vector machines to detect faults in tennessee eastman process, *Computers and Chemical Engineering* **29**(10): 2128–2133.

Langley, P., Iba, W. & Thompson, K. (1992). An analysis of bayesian classifiers, *National Conference on Artificial Intelligence*.

Lauritzen, S. & Jensen, F. (2001). Stable local computation with conditional gaussian distributions, *Statistics and Computing* **11**(2): 191–203.

Lowry, C. A., Woodall, W. H., Champ, C. W. & Rigdon, S. E. (1992). A multivariate exponentially weighted moving average control chart, *Technometrics* **34**(1): 46–53.

Lyman, P. & Georgakis, C. (1995). Plant-wide control of the tennessee eastman problem, *Computers and Chemical Engineering* **19**(3): 321–331.

Madsen, A. (2008). Belief update in clg bayesian networks with lazy propagation, *International Journal of Approximate Reasoning* **49**(2): 503–521.

Maurya, M. R., Rengaswamy, R. & Venkatasubramanian, V. (2007). Fault diagnosis using dynamic trend analysis: A review and recent developments, *Engineering Applications of Artificial Intelligence* **20**(2): 133–146.

Montgomery, D. C. (1997). *Introduction to Statistical Quality Control, Third Edition*, John Wiley and Sons.

Murphy, K. P. (2001). The bayes net toolbox for matlab, *Computing Science and Statistics : Proceedings of Interface*, Vol. 33, pp. 33–40.

Patton, R. J., Frank, P. M. & Clark, R. N. (2000). *Issues of Fault Diagnosis for Dynamic Systems*, Springer.

Pearl, J. (1988). *Probabilistic Reasoning in Intelligent Systems: Networks of Plausible Inference*, Morgan Kaufmann Publishers.

Perez, A., Larranaga, P. & Inza, I. (2006). Supervised classification with conditional gaussian networks: Increasing the structure complexity from naive bayes, *International Journal of Approximate Reasoning* **43**: 1–25.

Pernkopf, F. (2005). Bayesian network classifiers versus selective k-nn classifier, *Pattern Recognition* **38**(1): 1–10.

Perzyk, M., Biernacki, R. & Kochanski, A. (2005). Modeling of manufacturing processes by learning systems: The naive bayesian classifier versus artificial neural networks, *Journal of Materials Processing Technology* **164-165**: 1430–1435.

Ricker, N. (1996). Decentralized control of the tennessee eastman challenge process, *Journal of Process Control* **6**(4): 205–221.

Stamatis, D. H. (2003). *Failure Mode and Effect Analysis: FMEA from Theory to Execution*, ASQ Quality Press.

Tiplica, T., Verron, S., Kobi, A. & Nastac, I. (2006). Fdi in multivariate process with naïve bayesian network in the space of discriminant factors, *International Conference on Computational Intelligence for Modelling, Control and Automation*, Sydney, Australia, p. 216.

Venkatasubramanian, V., Rengaswamy, R., Kavuri, S. & Yin, K. (2003). A review of process fault detection and diagnosis part iii: Process history based methods, *Computers and Chemical Engineering* **27**(3): 327–346.

Verron, S., Tiplica, T. & Kobi, A. (2007a). Fault diagnosis of industrial systems with bayesian networks and mutual information, *European Control Conference*, pp. 2304–2311.

Verron, S., Tiplica, T. & Kobi, A. (2007b). Multivariate control charts with a bayesian network, *4th International Conference on Informatics in Control, Automation and Robotics (ICINCO)*, pp. 228–233.

Verron, S., Tiplica, T. & Kobi, A. (2007c). Procedure based on mutual information and bayesian networks for fault diagnosis of industrial systems, *American Control Conference*, pp. 420–425.

Verron, S., Tiplica, T. & Kobi, A. (2008). Fault detection and identification with a new feature selection based on mutual information, *Journal of Process Control* **18**(5): 479–490.

Westerhuis, J., Gurden, S. & Smilde, A. (2000). Standardized q-statistic for improved sensitivity in the monitoring of residuals in mspc, *Journal of Chemometrics* **14**(4): 335–349.

14

Bayesian Networks for Network Intrusion Detection

Pablo G. Bringas and Igor Santos
University of Deusto
Spain

1. Introduction

The increasing use of Internet has dramatically contributed to the growing number of threats that inhabit within it. Seeking for a better protection, Computer Security and, specifically, Network Intrusion Detection Systems (NIDS) have risen to become a topic of research and concern in order to fight these threats.

More accurately, a NIDS is a type of computer software that is able to distinguish legitimate network users from malicious ones. Moreover, due to the rising complexity and volume of the attacks, NIDS are performed in an automated manner, so the NIDS software monitors system usage to identify behaviour breaking the security policy. Generally, NIDS are categorised based in their scope: misuse network detectors and anomaly detectors. On the one hand, misuse detection systems deal with menaces already known in beforehand. Basically, these systems manage a comprehensive attack base and their work consists of invigilating at all incoming traffic to detect any sequence that appears in that knowledge base.

On the other hand, anomaly detection systems are more ambitious and try to discover new unknown threats (the so-called zero-day attacks). To this extent, these systems model benign or legitimate system usage in order to thereafter obtain a certainty measure of potential deviations from that normal profile. Each deviation that is found significant enough will be considered anomalous and notified to a human operator. Research in network anomaly detection has applied several well-known Artificial Intelligence paradigms such as finite automata (Vigna et al., 2000), neural networks (Mukkamala et al., 2005), genetic algorithms (Kim et al., 2005), fuzzy logic (Chavan et al., 2004), support-vector machines (Mukkamala et al., 2005) or diverse data-mining-based approaches (Lazarevic et al., 2003).

Actually, these solutions, both misuse and anomaly, perform better or worse against a network attack. Misuse detection systems are overwhelmed since they cannot face menaces that have not been previously described in their rule base but they overcome very fast the ones that have. Unfortunately, anomaly detection itself may not be considered as the the perfect solution, as well. In this way, it is much less exact than misuse detectors with well-known attacks and, despite they do find zero-day threads, sometimes they also produce false positives (i.e. select as a menace what is perfectly right). Summarizing, each approach is clearly surpassed when it comes to the other's area of expertise and the goal is, thus, to find the way to integrate both system's benefits while reducing their weaknesses.

In this way, Bayesian networks (Pearl & Russell, 2000) represent the sort of tool that can help us to achieve this integration. Specifically, they are probabilistic models very helpful when facing problems that require predicting the outcome of a system consisting of a high number of interrelated variables. After a training period, the Bayesian network *learns* the behaviour of the model and, thereafter it is able to foresee its outcome. In this way, successful applications of Bayesian networks include for instance email classification for spam detection (Yang et al., 2006), failure detection in industrial production lines (Masruroh & Poh, 2007) (Liu & Li, 2007), weather forecasting (Abramson et al., 1996) (Cofiño et al., 2002), intrusion detection over IP networks (Krügel et al., 2003) (Faour et al., 2006) or reconstruction of traffic accidents (Davis & Pei, 2003) (Davis, 2006). In all cases, the respective target problem is modelled as a constellation of interconnected variables whose output is always the result of the prediction (e.g. spam found, failure detected, intrusion noticed and so on). Therefore, we can model a NIDS as a constellation of variables controlling the type of the traffic, information on packet headers, packet payload or their temporal relationships (i.e. to check whether they form a coordinated attack). If we connect this representation to an attack variable, we will be able, after a proper training, to predict when do incoming packets represent a menace to the system.

Given this background, we present ESIDE-Depian (Intelligent Security Environment for Detection and Prevention of Network Intrusions), the first inherently unified misuse and anomaly detector. Besides, we focus on the integration of anomaly and misuse and show how this goal can be achieved by using a Bayesian network. In addition, we test this integration with real network attacks and show ESIDE-Depian's efficiency both as misuse and as anomaly detection.

The remainder of the chapter is organised as follows. follows. Section 2 illustrates the differences between misuse and anomaly detections systems. Section 3 details the concept of a Bayesian network and describes the used in ESIDE-Depian. Section 4 describes how ESIDE-Depian integrates misuse and anomaly prevention. Section 5 presents the experiments to evaluate this integration and discuses their results. Section 6 concentrates on the problems appeared and the solution designed to solve them. Section 7 discusses related work and, finally, section 8 concludes and outlines the avenues of future work.

2. Misuse versus Anomaly Detection

Currently, misuse detection is the most extended approach for intrusion prevention, mainly due to its efficiency and easy administration (Bringas et al., 2009). Its philosophy is quite simple: based on a rule base that models a high number of network attacks, the system compares incoming traffic with the registered patterns to identify any of these attacks. Hence, it does not produce any false positive (since it always finds exactly what is registered) but it cannot detect any new threat. Further, any slightly-modified attack will pass unnoticed. Finally, the knowledge base itself poses one of the biggest problems to misuse detection: as it grows, the time to search on it increases as well and, finally, it may require too long to be used on real-time.

Anomaly detection systems, on the contrary, start not from malicious but from legitimate behaviour in order to model what it is allowed to do. Any deviation from this conduct will be seen as a potential menace. Unfortunately, this methodology is a two-sided sword since, though it allows to discover new unknown risks, it also produces false positives (i.e. packets or situations marked as attack when they are not). In fact, minimising false positives is one of the pending challenges of this approach (Kruegel, 2002). Moreover, misuse detection presents a constant throughput since its knowledge base does not grow uncontrollably but gets adapted to new situations or behaviours. Again, an advantage is also source of problems because it is theoretically possible to make use of this continuous learning to little by little modify the knowledge so it ends seeing attacks as proper traffic (in NIDS jargon, this phenomenon is known as session creeping). In other words, its knowledge tends to be unstable. Finally, anomaly detection, unlike misuse, demands high maintenance efforts (and costs).

In summary, both alternatives present notable disadvantages that demand a new approach for network intrusion prevention.

3. Bayesian-network-based intrusion detection

3.1 Background

Reverend Thomas Bayes pioneered with his work the research on cause-consequence relationships. The most important fruit of that investigation, known as the "Bayes' theorem" (Bayes, 1763) in his honour, is the basis of the so-called Bayesian inference, a statistical inference method that allows, upon a number of observations, to obtain or update (if the system is already working) the probability that a hypothesis may be true. In this way, Bayes' theorem adjusts the probabilities as new informations on evidences appear.
According to its classical formulation, given two events A and B, the conditional probability $P(A|B)$ that A occurs if B occurs can be obtained if we know the probability that A occurs, P(A), the probability that B occurs, $P(B)$, and the conditional probability of B given A, $P(B|A)$ (as shown in equation 1):

$$P(A|B) = \frac{P(B|A) \cdot P(A)}{P(B)} \tag{1}$$

More accurately, Bayesian Networks (Pearl & Russell, 2000) are defined as graphical probabilistic models for multivariate analysis. Specifically, they are directed acyclic graphs that have an associated probability distribution function (Castillo et al., 1996). Nodes within the directed graph represent problem variables (they can be either a premise or a conclusion) and the edges represent conditional dependencies between such variables. Moreover, the probability function illustrates the strength of these relationships in the graph (Castillo et al., 1996) (Figure 1).

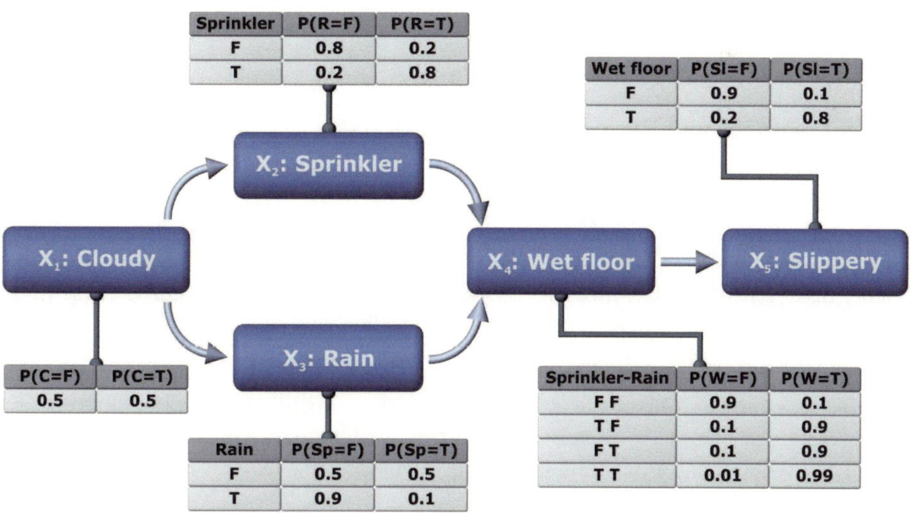

Fig. 1. Example of a Bayesian Network.

Formally, let a Bayesian Network B be defined as a pair, $B = (D, P)$, where D is a directed acyclic graph; $P = \{p(x_1|\Psi_2), ..., p(x_n|\Psi_n)\}$ is the set composed of n conditional probability functions (one for each variable); and Ψ_i is the set of parent nodes of the node X_i in D. The set P is defined as the *joint probability density function* (Castillo et al., 1996) (equation 2)

$$P(x) = \prod_{i=1}^{n} p(x_i|\Psi_i) \tag{2}$$

The most important capability of Bayesian Networks is their ability to determine the probability that a certain hypothesis is true (e.g., the probability of an e-mail to be spam or legitimate) given a historical dataset.

3.2 Bayesian Network Obtaining Process

The obtaining of the knowledge model in an automated manner can be achieved in an unsupervised or supervised way.

Typically, unsupervised learning approaches don not take into consideration expert knowledge about well-known attacks. They achieve their own decisions based on several mathematical representations of distance between observations from the target system, revealing themselves as ideal for performing Anomaly Detection.

On the other hand, supervised learning models do use expert knowledge in their making of decisions, in the line of Misuse Detection paradigm, but usually present high-cost administrative requirements. Therefore, both approaches present important advantages and several shortcomings. Being both ESIDE-Depian, it is necessary to set a balanced solution that enables to manage in an uniform way both kinds of knowledge.

Therefore, ESIDE-Depian uses not only Snort information gathering capabilities, but also Snort's decision-based labelling of network traffic. Thereby, the learning processes inside ESIDE-Depian can be considered as automatically-supervised Bayesian learning, divided into the following phases. Please note that this sequence only applies for the standard generation process followed by the Packet Header Parameter Analysis experts (see Figure 2).

We have divided the network traffic according to its type (TCP-IP, UDP-IP and ICMP-IP) and created three Bayesian networks (experts) to analyse their respective packet headers (which is an strategy already proven successful in this area (Alípio et al., 2003)). Moreover, in order to cover all possible kind of menaces, we also have to take into account the payload (i.e. body) of the packet and the potential temporal dependencies between packets. Therefore, we have added 2 further experts, the protocol payload and the connection tracking one, respectively. In each case, the Bayesian network is composed of several variables depending on the protocol and the expert; the value to induce is always the probability that the analysed packet is part of an attack.

Moreover, the creation and setting-up of each Bayesian network comprises the following phases:

- **Traffic sample obtaining.** First we need to stablish the information source in order to gather the sample. This set usually includes normal traffic (typically gathered from the network by sniffing, arp poisoning or so), as well as malicious traffic generated by the well-known arsenal of hacking tools (e.g. MetaSploit [1]).
- **Structural Learning.**
  The next step is devoted to define the operational model ESIDE-Depian should work within. With this goal in mind, we have to provide logical support for knowledge extracted from network traffic information. Packet parameters need to be related into a

[1] Metasploit: Exploit research. http://www.metasploit.org

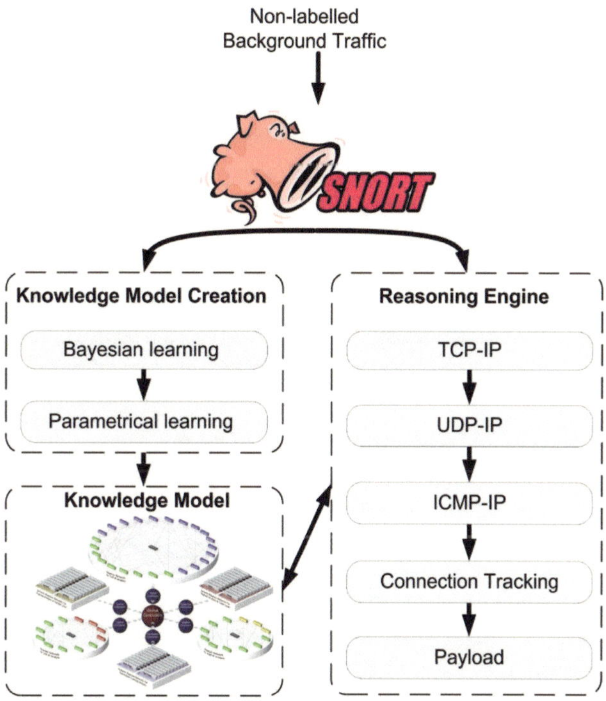

Fig. 2. ESIDE-Depian general architecture integrating misuse and anomaly detection.

Bayesian structure of nodes and edges, in order to ease the later conclusion inference over this mentioned structure.

In particular, the PC-Algorithm (Spirtes et al., 2001) is used here to achieve the structure of causal and/or correlative relationships among given variables from data. In other words, the PC-Algorithm uses the traffic sample data to define the Bayesian model, representing the whole set of dependence and independence relationships among detection parameters.

- **Parametric Learning.** The knowledge model fixed so far is a qualitative one. Therefore, the following step is to apply parametric learning in order to obtain the quantitative model representing the strength of the collection of previously learned relationships, before the exploitation phase began.

  Specifically, ESIDE-Depian implements maximum likelihood estimate (Kjærulff & Madsen, 2008) to achieve this goal. This method completes the Bayesian model obtained in the previous step by defining the quantitative description of the set of edges between parameters. This is, structural learning finds the structure of probability distribution functions among detection parameters, and parametric learning fills this structure with proper conditional probability values.

- **Bayesian Inference.** Next, every packet capture from the target communication infrastructure needs one value for the posterior probability of a badness variable, (i.e. the Snort [2] label), given the set of observable packet detection parameters.

  Hence, we need an inference engine based on Bayesian evidence propagation. More accurately, we use the Lauritzen and Spiegelhalter method for conclusion inference over junction trees, provided it is slightly more efficient than any other in terms of response time (Castillo et al., 1996). Thereby, already working in real time, incoming packets are analysed by this method (with the basis of observable detection parameters obtained from each network packet) to define the later probability of the attack variable.

  The continuous probability value produced here represents the certainty that an evidence is good or bad. Generally, a threshold-based alarm mechanism can be added in order to get a balance between false positive and negative rates, depending on the context.

- **Adaptation.** Usually, the system operation does not keep a static on-going way, but usually presents more or less important deviations as a result of service installation or reconfiguration, deployment of new equipment, and so on.

  In order to keep the knowledge representation model updated with potential variations in the normal behaviour of the target system, ESIDE-Depian uses the general sequential/ incremental maximum likelihood estimates (Castillo et al., 1996) (in a continuous or periodical way) in order to achieve continuous adaptation of the model to potential changes in the normal behaviour of traffic.

3.3 Connection Tracking and Payload Analysis Bayesian Experts Knowledge Model Generation

The Connection Tracking expert attends to potential temporal influence among network events within TCP-based protocols (Estevez-Tapiador et al., 2003), and, therefore, it requires an structure that allows to include the concept of time (predecessor, successor) in its model. Similarly, the Payload Analysis expert, devoted to packet payload analysis, needs to model state transitions among symbols and tokens in the payload (following the strategy proposed in (Kruegel & Vigna, 2003).

Generally, Markov models are used in such contexts due to their capability to represent problems based on stochastic state transitions. Nevertheless, the Bayesian concept is even more suited since it not only includes representation of time (in an inherent manner), but also provides generalization of the classical Markov models adding features for complex characterization of states.

Specifically, the Dynamic Bayesian Network (DBN) concept is commonly recognized as a superset of Hidden Markov Models (Ghahramani, 1998), and, among other capabilities, it can represent dependence and independence relationships between parameters within one common state (i.e. in the traditional static Bayesian style), and also within different chronological states.

Therefore, ESIDE-Depian implements a fixed twonode DBN structure to emulate the Markov-Chain Model (with at least the same representational power and also the possibility to be extended in the future with further features) because full-exploded use of Bayesian concepts can remove several restrictions of Markov-based designs. For instance, it is not necessary

[2] A well-known misuse detector. Available at: http://www.snort.org

to establish the first-instance structural learning process used by the packet header analysis experts since the structure is clear in beforehand.
Moreover, according to (Estevez-Tapiador et al., 2003; Kruegel & Vigna, 2003), the introduction of an artificial parameter may ease this kind of analysis. Respectively, the Connection Tracking expert defines an artificial detection parameter, named TCP-h-flags (which is based on an arithmetical combination of TCP flags) and the Payload Analysis expert uses the symbol and token (thus, in fact, there are two Payload Analysis experts: one for token analysis and another for symbol analysis).
Finally, traffic behaviour (and so TCP flags temporal transition patterns) as well as payload protocol lexical and syntactical patterns may differ substantially depending on the sort of service provided from each specific equipment (i.e. from each different IP address and from each specific TCP destination port). To this end, ESIDE-Depian uses a multi-instance schema, with several Dynamic Bayesian Networks, one for each combination of TCP destination address and port. Afterwards, in the exploitation phase, Bayesian inference can be performed from real-time incoming network packets.
In this case, the a-priori fixed structure suggests the application of the expectation and maximization algorithm (Murphy, 2001), in order to calculate not the posterior probability of attack, but the probability which a single packet fits the learned model with.

3.4 Naive Bayesian Network of the Expert Modules

Having different Bayesian modules is a twofold strategy. On the one hand, the more specific expertise of each module allows them to deliver more accurate verdicts but, on the other hand, there must be a way to solve possible conflicting decisions. In other words, an unique measure must emerge from the diverse judgements.
To this end, ESIDE-Depian presents a two-tiered schema where the first layer comprises the expert modules (TCP-IP, UDP-IP, ICMP-IP, Connection Tracking and Protocol Payload) and the second layer includes only one class parameter: the most conservative response of the experts (in order to prioritize the absence of false negatives in front of false positives). Both layers form, in fact, a naive Bayesian network.
Such a Naive classifier (Castillo et al., 1996) has already been proposed in network intrusion detection, mostly for anomaly detection (Amor et al., 2004). This approach provides a good balance between representative power and performance, and also affords interesting flexibility capabilities which allow, for instance, ESIDE-Depian's dynamical enabling and disabling of expert modules. Figure 3 details the individual knowledge models and how do they fit to conform the general one.

4. Integration of Misuse and Anomaly Detection

The internal design of ESIDE-Depian is principally determined by its dual nature. Being both a misuse and anomaly detection system requires answering to sometimes clashing needs and demands. In other words, it must be able to simultaneously offer efficient response against both well-known and zero-day attacks. The Bayesian network, according to the ability to extrapolate its knowledge and apply it to not-previously seen cases, is the ideal tool for these zero-day attacks. Still, we have to integrate detection of already registered threads and provide an efficient methodology to update and to continuously adapt to changes. ESIDE-Depian achieves this objectives in two ways. First, it incorporates Snort to the training of the Bayesian network. Second, already in working-time, Snort's opinion is passed to the experts so they can take this additional information into account.

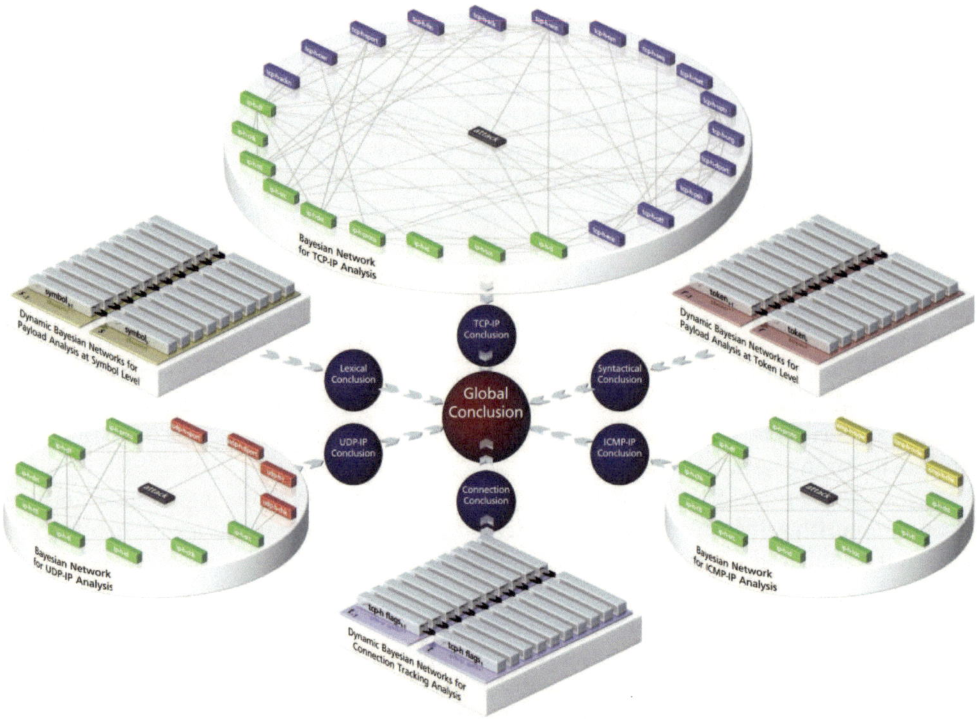

Fig. 3. ESIDE-Depian Final Knowledge Representation Model.

4.1 Snort-driven Automated Learning

The obtaining of the knowledge model in an automated manner can be achieved in an unsupervised or supervised way. In the training phase, Snort provides information regarding the legitimacy or malice of the network packets. Specifically, Snort's main decision about a packet is added to the set of detection parameters, receiving the name of attack variable. In this way, it is possible to obtain a complete sample of evidences, including, in the formal aspect of the sample, both protocol fields as well as Snort labelling information.

Therefore, it combines knowledge about normal behaviour and also knowledge about well-known attacks, or, in other words, information necessary for misuse detection and for anomaly detection.

4.2 Snort-labelled Network Traffic

Initial designs of ESIDE-Depian considered including Snort's opinion at the same level as experts' verdict in the naive Bayesian network but experiments showed that it biased the result too much. Therefore, we chose an strategy similar to the one used in the Bayesian network training (described in the previous section). Hence, already in real time, every packet gets Snort's opinion added as the badness variable mentioned before. In this way, experts know again the decision of Snort in beforehand and can act in consequence according to their knowledge model. Figure 2 illustrates how Snort is integrated within the different modules that conform ESIDE-Depian.

5. Evaluation and results

In order to asses the performance of ESIDE-Depian both as misuse and as anomaly detector, we have performed different kinds of experiments. Since Snort analyses only superficially the body of each packet, we have been forced to divide these tests into header-based and packet-body-based attacks in order to evaluate all of them more efficiently.

5.1 Header Parameter Analysis

Three are the Bayesian experts involved in this series of tests (though this does not mean that only one expert deals with the analysis; the naive Bayesian network considers all of them before obtaining the final verdict): TCP-IP, UDP-IP and ICMP-IP experts. The methodology applied intends to, first, demonstrate that the initial reference knowledge has been acquired, and second, that this reference knowledge has been superseded and exceed. In other words, we initially test the misuse detection capability and then, the anomaly detection ability.
The acquisition of the initial reference knowledge is performed already in the training phase. The BN is fed with a traffic sample basically based on the attack-detection rules battery provided by Snort. Therefore, the training acquaints the BN with either kind of traffic simultaneously, good and bad. Still, due to the disparity in the amount of packets belonging to one or another (see Table 1), traces containing attacks have to be fed several times (in the so-called presentation cycles) in order to let the BN learn to evaluate them properly. Table 1) summarises the results of testing the initial (Snort) reference knowledge acquisition. To this end, the BN was fed with a new sample traffic merging normal one extracted from a one hour capture at the University of Deusto and also malicious packets (crafted with the tool PackIt).

Traffic type	**TCP**	**UDP**	**ICPM**
Reference knowledge good/bad traffic ratio	699,560/42	5,130/11	1,432/95
Presentation cycles required	2943	2	2
Snort's hits	38	0	450
Analysed packets	100,000	10,000	5,000
Attacks detected by Snort	5	1	600
Attacks detected by ESIDE-Depian	5 (100%)	1 (100%)	600 (100%)

Table 1. Misuse Detection Tests Analysing Packets Headers.

ESIDE-Depian shows the same performance as Snort in these tree different traffic sorts. The high number of presentation cycles required by the TCPIP expert to grasp the initial reference knowledge is due to the very high good/bad traffic ratio, much lower in the cases of UDP and ICMP. Therefore, we can conclude that gaining the reference knowledge was completed successfully. Regarding going beyond this reference knowledge (i.e. the ability of ESIDE-Depian to find zero-day attacks) we have created artificial anomalies along to the proposal of Lee et al. (2001). In this way, table 2 shows some of the TCP-IP packets that we inserted in the traffic (crafted to this end again with PackIt).
Snort was not able to detect any of them, whereas ESIDE-Depian achieved a 100% of success. Table 2 shows 15 packets labelled as potential negatives, this is, packets marked as positive (i.e., attack) by ESIDE-Depian but not by Snort. All of them correspond to the artificial anomalies we inserted and ESIDE-Depian was able to find the 100% of them. Table 3 shows some of the modified packets for the UDP-IP traffic tests

Examples of Anomalies	
`packit -nnn -s 10.12.206.2 -d 10.10.10.100 -F SFP -D 1023`	
`packit -nnn -s 10.12.206.2 -d 10.10.10.100 -F A -q 1958810375`	
`packit -nnn -s 10.12.206.2 -d 10.10.10.100 -F SAF`	
Anomaly detection results	
Potential false positives (anomalous packets)	15
Anomaly detection rate	100%

Table 2. Anomaly Detection Tests for TCP-IP Traffic.

Examples of Anomalies	
`packit -t udp -s 127.0.0.1 -d 10.10.10.2 -o 0x10 -n 1 -T ttl -S 13352 -D 21763`	
`packit -t udp -s 127.0.0.1 -d 10.10.10.2 -o 0x10 -n 0 -T ttl -S 13353 -D 21763`	
`packit -t udp -s 127.0.0.1 -d 10.10.10.2 -o 0x50 -n 0 -T ttl -S 13352 -D 21763`	
Anomaly detection results	
Potential false positives (anomalous packets)	2
Anomaly detection rate	100%

Table 3. Anomaly Detection Tests for UDP-IP Traffic.

Again, in UDP-IP traffic Snort did not discover any anomaly, as expected. The 2 false positives reflected in table 3 belong again to the artificial anomalies fed by us (and crafted with PackIt). Table tbl:table4 summarises the results obtained with ICMP-IP traffic. Similarly to the previous cases, Snort failed to detect any of the attacks, whereas the 45 false positives that appear in table 4 are exactly the anomalies introduced by us in the traffic sample.

5.2 Connection Tracking and Payload Analysis

With the goal of evaluating these analysis capabilities of ESIDE-Depian in mind, we have followed a different strategy than in the case of header parameters: Snort is mainly focused on the analysis of the latter and covers the payload inspection by applying a set of regular expressions that do not provide any useful information to the Bayesian network (basically because it presents a different morpho-syntactical structure).

Moreover, the dynamic nature of the data these experts focus on, forces this change. Therefore, we have generated a brand new traffic sample to be used in the training phase. Then, only for test purposes, we have created yet another different one with some of its packet sequences modified by means of the tool NetDude (since PackIt only allows to change packets, not sequences).

Table 5 summarises the results achieved by ESIDE-Depian for the tests focused on the connection tracking and payload analysis.

Examples of Anomalies	
`packit -i eth0 -t icmp -n 666 -s 3.3.3.3 -d 10.10.10.2`	
`packit -i eth0 -t icmp -K 0 -s 3.3.3.3 -d 10.10.10.2`	
`packit  i cth0  t icmp  K 17 -C 0 -d 10.10.10.2`	
Anomaly detection results	
Potential false positives (anomalous packets)	45
Anomaly detection rate	100%

Table 4. Anomaly Detection Tests for ICMP-IP Traffic.

Analysis Type	Connection Tracking	Payload Analysis
Analysed network packets	226,428	2,676
Attacks contained in sample	29	158
ESIDE-Depian hits	29	158

Table 5. Connection Tracking and Payload Analysis Results.

6. Problems and solutions

This section gives account of the main problems that emerged during the design and test phase. More accurately, they were:

- **Integration of Snort:** The first difficulty we faced was to find an effective way of integrating Snort in the system.

  Our first attempt placed the verdict of Snort at the same level as those of the Bayesian experts in the Naive classifier. This strategy failed to capture the real possibilities of Bayesian networks since it simply added the information generated by Snort at the end of the process, more as a graft than a real integrated part of the model.

  The key aspect in this situation was letting the Bayesian network absorb Snort's knowledge to be able to actually replace it. Therefore, in the next prototype we recast the role of Snort as a kind of advisor, both in training and in working time.

  In this way, the Bayesian experts use Snort's opinion on the badness of incoming packets in the learning procedure and afterwards (as described in section 4) and manage to exceed Snort's knowledge (Penya & Bringas, 2008).

- **Different parameter nature:** The next challenge consisted on the different nature of the parameters that ESIDE-Depian has to control. Whereas TCP, UDP and ICMP are static and refer exclusively to one packet (more accurately to its header), the connection tracking and payload analysis experts are dynamic and require the introduction of the time notion.

  In this way, the connection tracking expert checks if packets belong to an organised sequence of an attack (Estevez-Tapiador et al., 2003), so time is needed to represent predecessor and successor events. In a similar vein, the payload analysis expert must model state transitions between symbols and tokens that appear on it.

Therefore, in the same way that different tests had to be performed, we had to prepare an special traffic sample tailored to the kind of traffic those expert should focus to inspect

- **Disparity between good and bad traffic amount:** Another problem to tackle was the composition of the traffic sample used to train the first group of experts (TCP, UDP, ICMP).

  In order to help the acquisition of the initial reference knowledge in the training phase, the BN is fed with a traffic sample basically based on the attack-detection rules battery provided by Snort. Therefore, the training acquaints the BN with either kind of traffic simultaneously, good and bad.

  Nevertheless, due to the disparity in the amount of packets belonging to one or another, traces containing attacks have to be fed several times (in the so-called presentation cycles) in order to let the BN learn to evaluate them properly.

- **Task parallelisation:** Bayesian networks require many computational resources. Hence, several of the tasks to be performed were designed in a parallel way to accelerate it. For instance, the structural learning was devoted concurrently in 60 computers. In this way, the traffic sample (about 900.000 packets) was divided in blocks of 10.000 of packets that were processed with the PC-Algorithm. In addition, already on real-time, each expert was placed in a different machine not only to divide the amount of resources consumed but also to prevent from having a single point of failure.

- **False positives and false negatives:** Finally, we coped with a usual problem related to anomaly detection systems: false positives (i.e. packets marked as potentially dangerous when they are harmless). In fact, minimising false positives is one of the pending challenges of this approach (Lundin, 2004).

  Nevertheless, the double nature of ESIDE-Depian as anomaly and misuse detector reduces the presence of false positives to a minimum. False negatives, on the contrary, did threaten the system and, in this way, in the experiments accomplished in ESIDE-Depian, security was prioritized above comfort, so quantitative alarm-thresholds were set upon the production of the minimum false negatives, in spite of the false positive rates.

  It is possible to find application domains, e.g., anti-virus software, in which false positive numbers are the target to be optimized, in order not to saturate the final user or the system administrator. Also in these cases ESIDE-Depian is able to manage the detection problem, simply by the specific setting up of the mentioned thresholds.

7. Related Work

Different approaches to develop network misuse detectors include expert systems (Alípio et al., 2003), intent-specification languages (Doyle et al., 2001), intelligent agent systems (Helmer et al., 2003) or rule-induction systems (Kantzavelou & Katsikas, 1997) (in (Kabiri & Ghorbani, 2005) the reader can obtain a detailed analysis of related work in this area).
Research in network anomaly detection has applied several well-known Artificial Intelligence paradigms such as support-vector machines (Mukkamala et al., 2005) or diverse data-mining-based approaches Lazarevic et al. (2003). Still, there is only one attempt to bring these two strands of work together.

More specifically, in Valdes & Skinner (2000), they achieve to combine anomaly and misuse but its analysis of network packets is too superficial to yield any good results in real life. In particular, despite the brilliant main contribution about integrating misuse-based and anomaly-based detection in one inherently unified and compact knowledge representation model, this work presents several shortcomings that prevent it from being applied in real scenarios: on the one hand, this approach only considers 7 detection parameters
Popular protocols as UDP connection-less protocol or the very-very problematic ICMP protocol are not taken into consideration. On the other hand, Bayesian Networks' full capabilities are not really used. Thus, one of the most important topics provided by the Bayesian approach, the structural learning concept, is not definitively applied. Instead, they propose the Naive approach, which assumes the (unrealistic) hypothesis that there is no statistical dependence among the collection of detection parameters.
Finally, time notion does not play any role in the analysis model, even under the focus achieved over the TCP target protocol, which is, of course, connection-oriented and, so, chronological dependence among events is sure to appear.

8. Conclusions

As the use of Internet grows beyond all boundaries, the number of menaces rises to become subject of concern and increasing research. Against this, Network Intrusion Detection Systems (NIDS) monitor local networks to separate legitimate from dangerous behaviours. According to their capabilities and goals, NIDS are divided into misuse detection systems (which aim to detect well-known attacks) and anomaly detection systems (which aim to detect zero-day attacks). So far, no system to our knowledge combines advantages of both without any of their disadvantages. Moreover, the use of historical data for analysis or sequential adaptation is usually ignored, missing in this way the possibility of anticipating the behaviour of the target system.
ESIDE-Depian, a Bayesian-networks-based misuse and anomaly detection system. In another work, we detailed the composition of the Bayesian network, its training methodology and showed general performance results. Here we have focused on evaluating the integration of misuse and anomaly detection. To this end, we have adopted Snort (a well-known misuse detector) as misuse detector trainer so the Bayesian Network of five experts is able to react against both misuse and anomalies. The Bayesian experts are devoted to the analysis of different network protocol aspects and obtain the common knowledge model by means of separated Snort-driven automated learning process
Since ESIDE-Depian has passed the experiments brilliantly, it is possible to conclude that ESIDE-Depian using of Bayesian Networking concepts allows to confirm an excellent basis for paradigm unifying Network Intrusion Detection, providing not only stable Misuse Detection but also effective Anomaly Detection capabilities, with one only flexible knowledge representation model and a well-proofed inference and adaptation bunch of methods.
On the other hand, the Bayesian approach also enables to implement powerful features over it, such as Dynamic-Bayesian-Network-based full representation of time, in order to accomplish totally-characterised connection tracking and low level chronological event correlation, or explanation tracking of the inferred cause-effect reasoning processes. Furthermore, contrary to other approaches such as Neural Networks, Bayesian networks allow administrative managing of inner information structures, so specific relationships among packet detection parameters and final conclusion can be explained, in a white-box manner. Moreover, it is not only possible to recover reasoning information, but also to act on both Bayesian network

structures and conditional probability parameters, in order to adjust the whole behaviour of the Network Intrusion Detection System to special needs or configurations.
Besides, dynamic regulation of knowledge representation model can be accomplished by using the sensibility analysis proposed by Castillo et al. (1996), so as to avoid denial of service attacks, automatically enabling or disabling expert modules by means of one combined heuristic measure which considers specific throughputs and representative features. In addition, it is also possible to perform model optimization, to obtain the minimal set of representative parameters, and also the minimal set of edges among them, with the subsequent increase of the general performance.
Moreover, approximate evidence propagation methods can also be applied, in order to improve inference and adaptation time of response. Current expert models only consider exact inference, but it is possible to find methods which provide fast responses, with only a small and affordable loss of accuracy.
In addition, Bayesian knowledge representation models present one further interesting capability in current Intrusion Detection state of art, the possibility to provide an ad-hoc method for IDS evaluation. Bayesian concept provides simulation of learned knowledge corresponding samples, so it is an ideal environment for artificial anomaly generation.
At last, also unifying of Host and Network Intrusion Detection paradigms can be accomplished at low level through the Dynamic Bayesian Network concept. Specifically, both sorts of event (i.e., basically, operating system syscalls and network packets) can be characterized in one single representation model, with a dynamic approach that can obtain, for example, the posterior probability of an exploitation of one specific host service due to one specific network packet (e.g. an Unix exec syscall from a shellcode inside a packet payload). Besides, not only inference can be afforded, but even prediction of next event, due
Future work will focus on further research on exploiting the aforementioned omni-directional inference capability of Bayesian networks to the prediction of the next event, as well as on comparing ESIDE-Depian to other cutting-edge intrusion detection systems.

9. References

Abramson, B., Brown, J., Edwards, W., Murphy, A. & Winkler, R. L. (1996). Hailfinder: A Bayesian system for forecasting severe weather, *International Journal of Forecasting* **12**(1): 57–71.

Alípio, P., Carvalho, P. & Neves, J. (2003). Using CLIPS to detect network intrusions, *Lecture Notes in Computer Science* pp. 341–354.

Amor, N., Benferhat, S. & Elouedi, Z. (2004). Naive bayes vs decision trees in intrusion detection systems, *Proceedings of the 2004 ACM symposium on Applied computing*, ACM New York, NY, USA, pp. 420–424.

Bayes, T. (1763). An essay towards solving a problem in the doctrine of chances, *Philosophical Transactions of the Royal Society* **53**: 370–418.

Bringas, P., Penya, Y., Paraboschi, S. & Salvaneschi, P. (2009). Bayesian-Networks-Based Misuse and Anomaly Prevention System, *Proceedings of the Tenth International Conference on Enterprise Information Systems (ICEIS)*.

Castillo, E., Gutiérrez, J. M. & Hadi, A. S. (1996). *Expert Systems and Probabilistic Network Models*, erste edn, Springer, New York, NY, USA.

Chavan, S., Shah, K., Dave, N., Mukherjee, S., Abraham, A. & Sanyal, S. (2004). Adaptive neuro-fuzzy intrusion detection systems, *Proceedings of the International Conference on Information Technology: Coding and Computing (ITCC'04) Volume*, Vol. 2.

Cofiño, A. S., Cano, R., Sordo, C. & Gutiérrez, J. M. (2002). Bayesian networks for probabilistic weather prediction., *Proceedings of the European Conference on Artificial Intelligence (ECAI)*, pp. 695–699.

Davis, G. (2006). Bayesian networks, falsification, and belief revision in accident reconstruction, *Proceedings of the Transportation Research Board 85th Annual Meeting*.

Davis, G. & Pei, J. (2003). Bayesian networks and traffic accident reconstruction, *Proceedings of the 9th international conference on Artificial intelligence and law (ICAIL)*, pp. 171–176.

Doyle, J., Kohane, I., Long, W., Shrobe, H. & Szolovits, P. (2001). Event recognition beyond signature and anomaly, *Proceedings of the 2001 IEEE Workshop on Information Assurance and Security*, pp. 170–174.

Estevez-Tapiador, J., Garcia-Teodoro, P. & Diaz-Verdejo, J. (2003). Stochastic protocol modeling for anomaly based network intrusion detection, *Proceedings of the 1st IEEE International Workshop on Information Assurance (IWIAS)*, pp. 3–12.

Faour, A., Leray, P. & Eter, B. (2006). A SOM and Bayesian network architecture for alert filtering in network intrusion detection systems, *Proceedings of the International Conference on Information and Communication Technologies*, pp. 3175–3180.

Ghahramani, Z. (1998). Learning dynamic Bayesian networks, *Adaptive Processing of Sequences and Data Structures* p. 168.

Helmer, G., Wong, J., Honavar, V., Miller, L. & Wang, Y. (2003). Lightweight agents for intrusion detection, *The Journal of Systems & Software* **67**(2): 109–122.

Kabiri, P. & Ghorbani, A. (2005). Research on intrusion detection and response: A survey, *International Journal of Network Security* **1**(2): 84–102.

Kantzavelou, I. & Katsikas, S. (1997). An Attack Detection System for Secure Computer Systems-Outline of the Solution, *Computers and Security* **16**(3): 207–207.

Kim, D., Nguyen, H. & Park, J. (2005). Genetic algorithm to improve SVM based network intrusion detection system, *Advanced Information Networking and Applications, 2005. AINA 2005. 19th International Conference on*, Vol. 2.

Kjærulff, U. B. & Madsen, A. L. (2008). *Bayesian Networks and Influence Diagrams: A Guide to Construction and Analysis*, Information Science and Statistics, Springer.

Kruegel, C. (2002). *Network Alertness-Towards an adaptive, collaborating Intrusion Detection System*, PhD thesis, Technical University of Vienna.

Kruegel, C. & Vigna, G. (2003). Anomaly detection of web-based attacks, *Proceedings of the 10th ACM conference on Computer and communications security*, ACM, pp. 251–261.

Krügel, C., Mutz, D., Robertson, W. K. & Valeur, F. (2003). Bayesian event classification for intrusion detection, *Proceedings of the 19th Annual Computer Security Applications Conference*, pp. 14–23.

Lazarevic, A., Ertoz, L., Kumar, V., Ozgur, A. & Srivastava, J. (2003). A comparative study of anomaly detection schemes in network intrusion detection, *Proceedings of the 3rd SIAM International Conference on Data Mining*, pp. 25–36.

Lee, W., Stolfo, S., Chan, P., Eskin, E., Fan, W., Miller, M., Hershkop, S. & Zhang, J. (2001). Real time data mining-based intrusion detection, *Proceedings of DARPA Information Survivability Conference and Exposition II*.

Liu, Y. & Li, S.-Q. (2007). Decision support for maintenance management using Bayesian networks, *Proceedings of the International Conference on Wireless Communications, Networking and Mobile Computing (WiCom 2007)*, pp. 5713–5716.

Lundin, E. (2004). *Logging for intrusion and fraud detection*, PhD thesis, University of Goteborg, Department of Computer Engineering, Sweden.

Masruroh, N. A. & Poh, K. L. (2007). A Bayesian network approach to job-shop rescheduling, *Proceedings of the IEEE International Conference on Industrial Engineering and Engineering Management*, pp. 1098–1102.

Mukkamala, S., Sung, A. & Abraham, A. (2005). Intrusion detection using an ensemble of intelligent paradigms, *Journal of Network and Computer Applications* **28**(2): 167–182.

Murphy, K. (2001). An introduction to graphical models, *Technical report*.

Pearl, J. & Russell, S. (2000). Bayesian networks, *Technical Report Tech. Rep. R-216*, Computer Science Department, University of California, Los Angeles.

Penya, Y. & Bringas, P. (2008). Integrating network misuse and anomaly prevention, *Proceedings of the 6th IEEE International Conference on Industrial Informatics (INDIN)*, pp. 586–591.

Spirtes, P., Glymour, C. & Scheines, R. (2001). *Causation, prediction, and search*, The MIT Press.

Valdes, A. & Skinner, K. (2000). Adaptive, model-based monitoring for cyber attack detection, *Recent Advances in Intrusion Detection*, Springer, pp. 80–93.

Vigna, G., Eckmann, S. & Kemmerer, R. (2000). The STAT tool suite, *Proceedings of the DARPA Information Survivability Conference and Exposition (DISCEX)*.

Yang, Z., Nie, X., Xu, W. & Guo, J. (2006). An approach to spam detection by naive Bayes ensemble based on decision induction, *Proceedings of the Sixth International Conference on Intelligent Systems Design and Applications (ISDA'06)*, IEEE Computer Society, pp. 861–866.

A novel probabilistic approach for analysis and planning of large capillarity broadband networks based on ADSL2+ technology

Diego L. Cardoso, Adamo L. Santana,
Claudio A. Rocha and Carlos R. L. Francês
Federal University of Pará
Brazil

1. Introduction

The increasingly spread of information through digital media raises new realities in the world's present scenario and, thus, new technologies have been emerging in order to streamline the process of disseminating information and providing quality access to such information by the population. The Next Generation Network (NGN) holds tremendous potential, with a promise to merge the transmission of data, voice, video and other media into a single network; unfortunately, several developing countries do not have the necessary infra-structure to implement NGN technology. The main concern in these networks is not the backbone or the transport layer, but in the last mile itself. Last mile has become a popular keyword to indicate the technology which connects the End User to the Network backbone.

In most of North America and Western Europe, Internet penetration is very high and nearly every citizen has access to the Internet. However, this is not true in many parts of the developing world, where only a small percent of the population has access, even if the bandwidth is significantly low and the cost is a substantial fraction of the user's income (Ambrosi et al., 2005). According to (IWS, 2009), more than 70% of the population of the developing countries does not have access to Internet due to lack of infrastructure ; furthermore, in countries like China, India and Brazil, with continental dimensions, the construction of a new telecommunications network, like optic fiber, becomes costly and impractical. In this context, new alternative technologies that can offer trade-off between performance and costs must be sought.

There are several approaches to deliver service to the end user (Xiao et al., 2007). An alternative with less time and cost would be to use a combination of existing infrastructures such as electrical grids or telephone networks, based on copper loops, which are widely available to end users in most developing countries (Papagianni et al., 2009). For areas where the network has low penetration, wireless network can be a better solution; it, however, requires a basic infrastructure (base stations, antennas, etc.).

Telephone access networks were originally built for analog voice communication, carrying voice-band signals up to 4 KHz in the frequency bandwidth, and not for digital data communication. We considered here a large capillarity broadband network because it uses a combination of the existing copper infrastructure and digital subscriber line transmission technologies, thus enabling a universal broadband access at a fraction of the cost and in a fraction of the time required for others access networks.

DSL remains the dominant access technology with 65% of the worldwide subscribers for broadband, compared with 33% of fiber optics connection. It is in the developing countries that the number of DSL connections for last mile really stands out, such as in India, representing 83% of broadband connections; and China, which continues to grow, reaching 93,549,000 subscribers (Point Topic, 2009).

The DSL (Digital Subscriber Line) is considered as the dominant broadband access technology, not only in Europe but also in Latin America and developing countries like India (Olsen et al., 2006) (Arena et al., 2006) (Faudon et al., 2006). In Latin America, DSL technology accounts for 77% of all broadband access. At the end of 2005 there were nearly 5,300,000 subscribers of ADSL (Asymmetric Digital Subscriber Line) in Latin America (Arena et al., 2006).

Particularly, Brazil was marked by a major growth in broadband access. In 2005, 52.1% of home users had dial-up, 41.2% broadband and 6.7% both forms of access. By 2008, the statistics had changed considerably; Broadband access rose to 58% against 31% of dial-up access; whereas the DSL access accounts for 23% of the total broadband access (CETIC, 2008).

It is now a fact that the broadband access has been changing the user's needs, which was initially only for accessing websites. Now, users are keen to use services such as video, voice and data separately, one at a time. Customers enjoy the convenience of receiving all three services they need today from one service provider, increasing demand for triple play services. Thus, telephone companies (Telcos) offer triple play by providing television service using IP (i.e. Internet Protocol Television - IPTV) in order to compete more effectively with cable television companies that have entered the voice and Internet access markets. Therefore, it is imperative to study the computer applications in such infrastructures that were not designed with this goal.

The last mile network maintenance is another important factor as it is currently performed with a mixture of help systems, manual testing by technicians, and automated tests that are developed for plain-old telephone service (POTS) lines, which ignore DSL frequencies above 4 kHz. Provisioning is based on rough estimates of the loop length and does not account for individual loop characteristics. There will be more complications in the maintenance when the DSLs start supporting triple-play services: the Internet, Voice-over-Internet Protocols (VoIPs), and Internet Protocol TVs (IPTVs).

It is important to use a test system that can accurately identify and inform the source of a problem in the network; whether this problem comes from the Internet service provider (ISP), the telephone central office (CO), the outside plant, the modem, or the user's PC. DSL testing is not only limited to measuring electrical parameters on the copper pair but also to include the comparative analysis of extracted data with previously known limits as well as comparing assigned configuration with discovered configurations (Kerpez & Kinney, 2008).

The convergence between existing and emerging broadband technologies has been regarded as a major challenge, especially with respect to supporting multimedia content in these technologies. This is because new applications, such as IPTV, require high bandwidths, which are usually not available due to long distances of DSL links, noise, data congestion, lack of protocols implemented for this new need, among others.

The IP Protocol is considered as the standard protocol for communication among different network types. Unfortunately, the IP network presents issues on the provision of end-to-end QoS. For this matter, some services use the transport protocol TCP (Transmission Control Protocol), which has been considered as the main communication protocol. However, for networks with high packet loss, this protocol is flawed, for it enables the congestion control unnecessarily, since some applications can tolerate losses in communication.

It is then observed that the current data communication, which primarily uses the TCP/IP, is appropriate for applications such as HTTP traffic; for it maintains compatibility with existing networks (routers, gateways, etc.), and controls the flow and congestion. However, for present day applications, referred to as triple play with flows that are sensitive to QoS, TCP/IP is inefficient.

The division of the TCP/IP into layers facilitates the implementation of new applications. This feature, however, has become a negative aspect, since TCP/IP does not implement means for interaction between its layers. New applications require minimum levels of QoS for their operation, and this requires a greater interaction between its layers, in order to maintain these minimum levels for specific applications over others.

The Crosslayer model aims to implement an interaction between the layers of the protocol, providing more quality of service to the user, who is less interested on the technology details, but demands more in terms of quality. Using this technique, it is possible to, for example, adjust in real time the performance parameters of the application layer, such as throughput and jitter, given that the transport layer provides information about lost packets, allowing an adaptation of the application that is being used and the characteristics of the environment in which the transmission is carried out.

It is therefore essential to investigate the transmission technologies (last mile, and protocols used) in order to achieve a better strategy to expand telecommunications services in regions with little infrastructure available to the typical end user among the various possible scenarios. These inferences of possible scenarios, as a rule, are performed using a combination of prototyping and modeling for performance evaluation.

The main motivation of this work is the need to provide a high quality service to users, by ensuring that objectives of the network service level will be maintained; the need to correlate events from the lower layers of management, in order to determine strategies in the upper layers (crosslayer); the need to increase the sophistication required for diagnosis, given the greater complexity of the system; and the need to quickly detect network failures and, if possible, provide automatic recovery. In order to achieve these objectives, we apply a hybrid model, combining the qualities of genetic algorithms (GA) for space search with a Bayesian probability model for inference.

The correlation of events and attributes is important to analyse the behavior and functionality of applications, and reduce costs with respect to the network maintenance, improve availability and performance of its services. Raw data are interpreted and analyzed, taking into account a set of predetermined criteria, or defined dynamically according to the management process.

The tests were implemented over the backbone of the Brazilian Telecommunications System (Telebrás) and based on specific standards of DSL communication.

2. Test Bed Architecture

In order to evaluate the triple play communication in a DSL network, a standard model must be used, such as the ones suggested by (Papagianni et al., 2009) (Kerpez & Kinney, 2008) and (Sadri et al., 2007) (Figure 1), which define the sequence of connected elements:

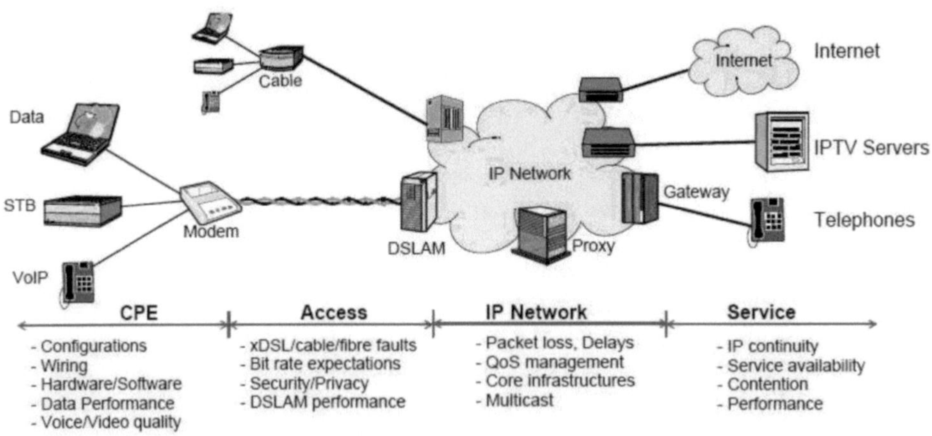

Fig. 1. Example of a Standard Test Bed

The architecture to be used will include the following items:

1. xDSL Modems, including ADSL, ADSL2, ADSL2+;
2. DSLAM (Digital Subscriber Line Access Multiplexer);
3. Simulator of cables for the European standard;
4. Protocols analyzer.

3. DSL Loop Length

With the recent rapid growth of high-speed DSL access subscriptions, there is a high demand in the telecommunication industry for equipment to accurately predict DSL access performance over a telephone subscriber line (also referred to as a local loop).

The subscriber line is a metallic twisted-pair network link between the customer and the telephone Central Office (CO). While some of the existing DSL analysis equipment is already capable of assessing the performance rate, it requires two-point operation (sending test signals from one end of the loop and measuring the signals at the other end) involving the dispatch of a service vehicle. This leads to expensive testing processes and it is therefore an undesirable solution for DSL access providers.

The use of DSL technology to transmit data at high speed enables quick service delivery, mainly due to the fact that there is an external network and cabling with twisted copper pair, with wide coverage in almost all areas and niche markets. This network is not homogeneous, co-existing with new systems and old networks of 30, 40, 50 years ago.

Particular items that can prevent the use of the service are: loop losses, bridge taps, specific noises of twisted pair links, long loops (distances from the central office to the user), among others, which can slow down the service. Usually the specifications are given for wiring 24 or 26 AWG (American Wire Gauge) over distances of 2090m (Telebras, 1997); it is, however, unknown exactly which parts are with each type of cable and their distances, making necessary for an extra effort to measure it. Such items are shown more specifically in the following section.

4. DSL Copper Impairments

As DSL uses relatively high spectrum frequencies, its signal is susceptible to external noise sources. Thus, the understanding about the behavior of different kinds of noise and their effects on network performance are extremely useful on the design of well established DSL systems (ADSL, ADSL2+) as well as those of upcoming generations (VDSL, VDSL2). During the past years, crosstalk has been considered the major impairment to DSL services. However, other types of noise have gained importance, such as radio frequency interference (RFI), impulsive noise (IN), repetitive electrical impulse noise (REIN), and isolated burst of electrical noise (IBEN) among others (Wallace et al., 2005) (Stolle, 2002).

Fundamental loop transmission impairments may not cause the highest number of DSL trouble calls; however, they can be very difficult to fix, and so, they result in a high DSL maintenance cost. Figure 2 illustrates the DSL copper impairments (Kerpez et al., 2003)(Starr et al., 2003), which are mainly loop and bridged tap loss, crosstalk, electromagnetic interference (EMI) radio ingress, impulse noise, harmonic distortion, and background noise.

Noise on phone lines normally occurs because of imperfect balance of the twisted pair. There are many types of noises that couple through imperfect balance into phone lines, the most common of which are crosstalk noise, radio noise and impulse noise.

Crosstalk is caused by electromagnetic radiation of other phone lines in close proximity, in practice, within the same cable. Such coupling increases with frequency and can be caused by signals traveling in the opposite direction, called near-end crosstalk (NEXT), and by signals traveling in the same direction, called far-end crosstalk (FEXT).

Radio noise is the remnant of wireless transmission signals coupling into phone lines, particularly AM radio broadcasts and amateur (HAM) operator transmissions.

Impulse noise is a nonstationary crosstalk from temporary electromagnetic events (such as the ringing of phones on lines sharing the same binder, and atmospheric electrical surges) that can be narrowband or wideband and that occurs randomly. Impulse noises can be tens of millivolts in amplitude and can last as long as hundreds of microseconds (Cioffi, 1999)(Starr, 1999).

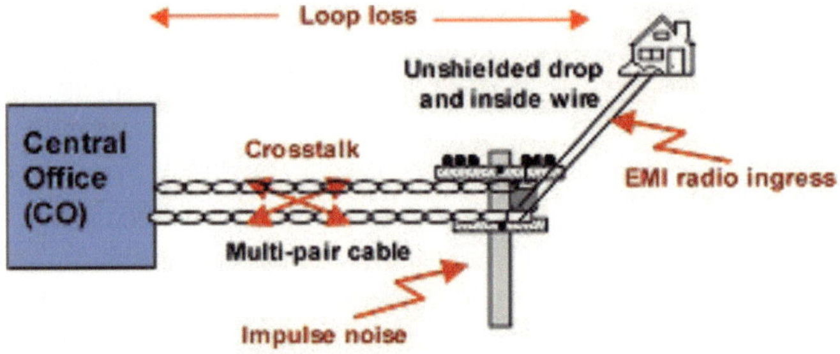

Fig. 2. DSL Copper Impairments (Kerpez & Kinney, 2008)

DSLs are generally provisioned to withstand a worst-case level of crosstalk; however, provisioning systems are approximate, and some older cables have poor crosstalk isolation. Moreover, after several DSLs are activated, some small percentage will actually exceed fundamental worst-case crosstalk engineering rules—however, this small percentage can translate into a high number of troubles.

In spite of conducting several investigations about the impact of non-stationary noise in DSL systems, just few studies have been conducted addressing their impact in terms of experimental analysis. This may be credited to the inaccessibility to a proper infrastructure to handle practical experiments.

5. Planning Methodology and Performance Results

This work implements, through crosslayer techniques, strategies for planning and evaluating the performance of ADSL2+ networks, which implement minimum levels of QoS for Triple Play applications. This approach will be achieved through a set of techniques such as: data measurement, modeling, optimization, simulation, etc. So this will enable creating an information framework that will guide the implementation of triple play applications and/or infrastructure for broadband networks.

The strategy and methodology to be used in the tests are divided into the following topics:

- Definition of architecture and equipment ;
- Definition of variables to be analyzed;
- Implementation of the testbed;
- Set up of equipments and preparation for the tests;
- Empirical tests;
- Analysis of the results;
- Correlation study using Bayesian networks.

With the performance measures and the help of a domain specialist, conjectures can be taken about the behavior and functionality of applications. This study, however, is not complete without considering factors such as the influence and correlation of all the attributes involved.

The correlation of events is important to reduce costs with respect to the network maintenance, improve availability and performance of network services. Raw data are interpreted and analyzed, taking into account a set of predetermined criteria, or defined dynamically according to the management process.

Among the computational intelligence techniques available for correlation analysis and uncertainty, we implement for this analysis the algorithm of Bayesian networks. Known for their models as components with a qualitative (representing the dependencies between the nodes) and quantitative (conditional probability tables - CPTs of the nodes) structures, evaluating, in probabilistic terms, these dependencies (Korb & Nicholson, 2003)(Chen, 2001). Together, these components provide an efficient representation of the joint probability distribution of the variables in a given field.

Bayesian networks are probabilistic graphical models for knowledge representation and reasoning in domains with uncertainty. Its unified nature makes it possible to compare different scenarios about the data, and the intuitive nature of its graphical formalism makes it one of the best analytical methods available for decision making (Rocha et al., 2007).

With Bayesian networks, the behavior of the attributes can be studied; propagating and evaluating hypothesis given certain evidences. Thus, from the Bayesian networks, one could predict how the triple play flow will behave in last mile networks or what are the physical characteristics that the network should have to meet this new need. This should provide a quantified indication in order to enable telecommunications companies invest safely, given the user's need for quality and efficiency in the service provision.

5.1 Definition of architecture and equipment

The test was implemented in the Laboratory of Technological Innovation in Telecommunications (LABIT), with a scenario consisting of modems, DSLAM, telecommunication cables, noise generator, and computers.

The generation of noise is made by the DSL 5500, a noise generator from Spirent Communications, in the operating range of ADSL2+ (4.3125 kHz to 2.208 MHz). A protocol analyzer from RADCOM (Radcom, 2009) was also used to filter the packets that will travel in the network, isolating specific flows to generate performance metrics.

DSLAM/EDA (Ethernet DSL Access) is the equipment available in the telephone central office, allowing the data communication via a DSL link. The computer connected to the DSLAM is responsible for generating video flows to be distributed to the clients via multicast.

A Wireline Simulator of ADSL2+ ETSI DLS 410E3 from Spirent Communications was used. It reproduces the AC and DC characteristics of twisted pair copper telephony cable using passive circuitry (R, L & C).

The methodology applied is conventionally used for benchmarking of high protocol layers, considering all types of data that can be transmitted; where the data to be changed are specific of the DSL technology, they are the loop length and the applications that are used.

5.2 Definition of variables to be analyzed

The performance measures obtained for this case study are divided by application:

- Voice flow: Jitter (Jitter_VoIP), loss of IP packets (Loss_VoIP), MOS - Mean Opinion Score (MOS_VoIP), number of successful attempts (Attempts_VoIP).
- Video flow: Jitter (Jitter_Video), vídeo throughput (Throughput_Video) and loss of IP packets (Loss_Video).
- Data flow (FTP): Delay (Delay_FTP), jitter (Jitter_FTP), loss of IP packets (FTP_Loss) and throughput (Throughput_FTP).

Where:

Def.1: We call it "throughput" the maximum bit rate, that allows end-to-end IP packet transmission without occurring any packet loss during the test (retransmission is not provided).

Def.2: The one-way IP packet delay is the time an IP packet (of a certain size) needs to travel from source to destination.

Def.3: The IP packet loss is the ratio between the number of lost packets and transmitted packets between source and destination over a long period of time.

Def.4: We have repeated the measurements for different loop distances (2500m, 3000m, 3500m, 4000m and 4500m) and cable type Ø=0.4mm PE. Simulating scenarios without any noise (named Case0), level of White Noise W= -140 dBm and 24 DSL (ISDN) Impairment (named Case1), level of White Noise W= -130 dBm, and 24 DSL (ISDN) Impairment (called Case2) and level of White Noise W= -120 dBm, and 24 DSL (ISDN) Impairment (called Case3). All noises recommended for (TR-048, 2002) and (ITU-T, 2005).

All tests, for each loop length, were repeated 10 times, with duration of 120 seconds.

5.3 Empirical Tests

For the analysis of this last mile technology, a typical scenario for IPTV transmission will be used, where services of voice, video and data will be available (Papagianni et al., 2009).

One of the challenges that VoIP carriers deal with early in their network planning is to choose the most appropriate voice coding standard in order to provide good voice quality and adequate network efficiency. From uncompressed G.711 at 64 kbps to G.726 at 16kbps, G.729 at 8kbps and the highly compressed G.723.1 at 5.3 kbps, the VoIP service providers can choose the level of voice compression that will be applied to their customers. In this particular study the G.711 codec is employed. G.711 is the international standard for encoding telephone audio on a 64 kbps channel.

For the voice transmission, Callgen (VoIP tool developed by the OpenH232 project) (OpenH323, 2007) was used. Besides being widely used for testing, Callgen supports the G.711 codec (Papagianni et al., 2009).

The video codec H.264 standard, which is being adopted by all major video service operators, is utilized in the performance evaluation of triple play service over xDSL access network (ITURec.H.264 & ISO/IEC14496-10, 2007). It was jointly developed by ITU Video Coding Experts Group (VCEG) and ISO Moving Picture Experts Group (MPEG). H.264 is used in fixed and wireless network environment.

H.264 has proven to be more resilient to error prone networks through the use of flexible macroblock ordering, slice interleaving and data partitioning. In addition, it attains enhanced compression performance; therefore it is a "network-friendly" standard. It is capable of providing good video quality at substantially lower bit rates than other standards. Compared to MPEG-2 video, it cuts down transmission bit rate by half, while the coding gain over H.263 and H. 263+ is in the range of 24% up to 47% (Kamaci & Altunbasak, 2003).

VLC (VideoLAN Client) (VLC, 2009) was used to generate the video traffic. VLC is a multimedia player that supports various video formats and streaming protocols, the RTP (Real Time Protocol) was used for the video transmission. The codec H264 was used for the video flow, with rate of 1.2 Mbps; for the audio, the AAC was used with rate of 192 Kbps.

IPERF tool (IPERF, 2009) was used to simulate the FTP traffic continuously, aiming to occupy the total available bandwidth. This tool is dedicated to the performance analysis of networks and widely used in testing (Rao et al., 2009) (Primet et al., 2002).

The tests performed are divided into: tests of the network capacity (Basic Connection); and tests of the applications behavior (Network Capacity Services), where the behavior of the protocols involved are studied.

5.4 Results Obtained

The Mean Opinion Score (MOS) is a numerical pattern (proposed by the ITU-T P.800) used to measure the quality of voice after the compression and/or transmission. Figure 3 shows the behaviour for the quality of the VoIP communication by enlarging loop lengths and inserting noise; we considered MOS equals to zero when the connection cannot be made or maintained. We can see that the noise of -120 dBm and 24D (case3) has a negative influence in the communication; and that the noise impact in the voice communication is higher in distances above 4000m, with quality loss up to 69% over a distance of 2500m.

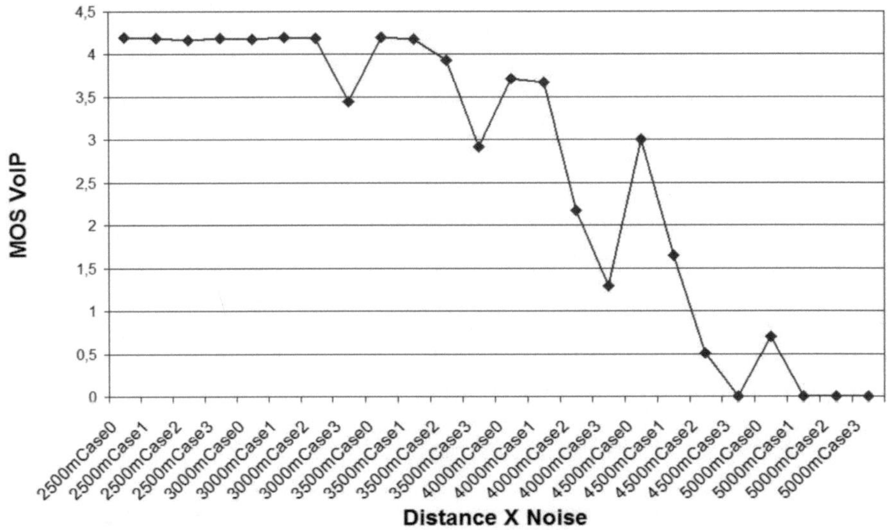

Fig. 3. VoIP MOS Behavior.

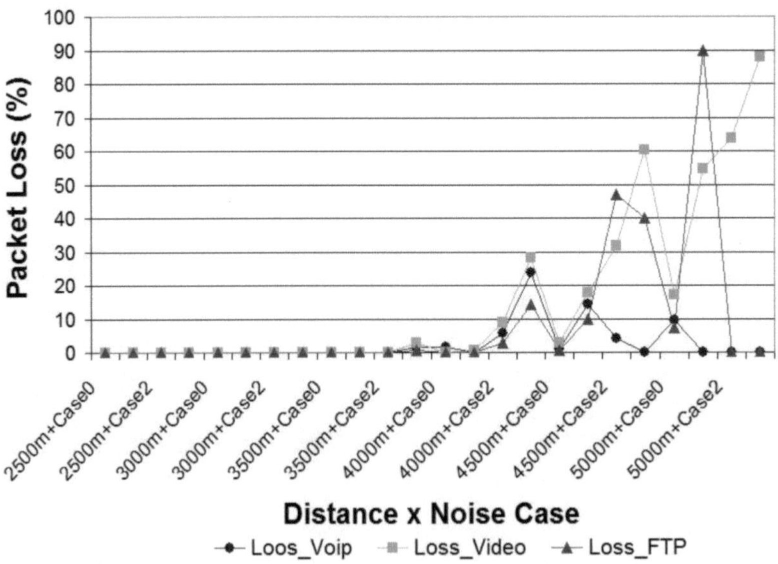

Fig. 4. Triple Play Packet Loss per flow.

Packet loss is one of the main aspects that affect the quality of triple play flows, particularly for applications not using reliable communication protocols (primarily voice and video). Data based applications, which uses reliable protocols that implement retransmissions, guarantee the arrival of information with integrity, even at low transmission rates. Figure 4 shows the behavior of applications considering the packet loss. The results illustrate the direct relationship between distance, noise and degradation of flows, especially at distances from 3500m to 4500m, which are more susceptible to noise. These distances measures are widely used in countries with large geographical area (such as Brazil, India and China) and an already established telephony infrastructure, which should now be adapted for digital transmission of data.

This fact can be better identified in Tables 1, 2 and Figure 5, which represent the behavior of applications (in percentage levels) when compared with a communication without noise. The voice application did not suffer packets loss in the entire range of noise, however, for a white noise with -130 dBm and 24D (Case2), a growth of 44% was seen in the jitter, which directly impacted on the MOS, causing a degradation of 6% in the quality of the communication. This impact was even greater when combining the white noise -120 dBm and 24D (Case3), which led to a degradation of 40% in the quality of communication, sending the MOS from, initially, 4.2 to an average of 2.9.

The video and data applications were barely impacted, with small variations (up to 3%), as shown in Table 2; with exception of the data flow, which suffered a drop of 2 Mbps (without noise in communication) to about 317 Kbps (white noise of -120 dBm and 24D - Case3), that is, a decrease of 600%.

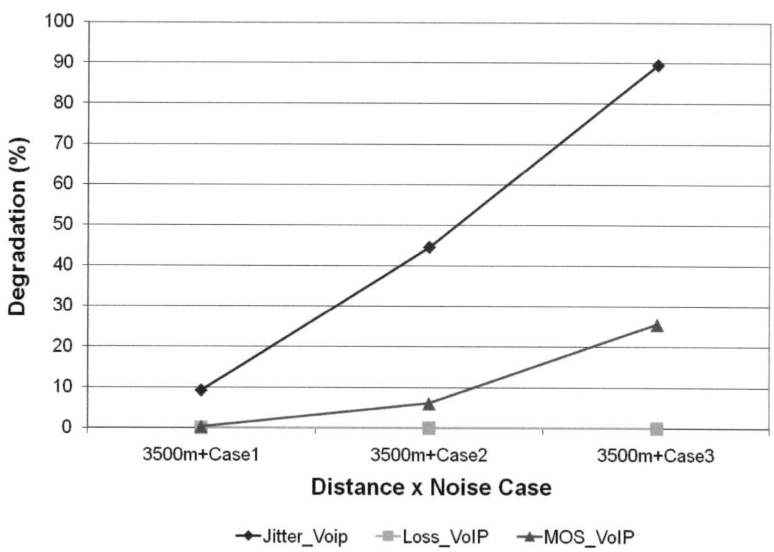

Fig. 5. VoIP Behavior for 3500m.

Metric / Impairment&Distance	Jitter Video (ms)	Loss Video (%)	Throughput Video (Mbps)	Jitter FTP (ms)	Loss FTP (%)	Delay FTP (ms)	Throughput FTP (Mbps)
3500m+(Case01)	7.9	0.02	1.41	5.07	0.11	0.073	2.06
3500m+(Case02)	7.39	0.03	1.42	5.37	0.18	0.079	1.98
3500m+(Case03)	4.55	2.7	1.39	12.01	0.56	0.28	0.317

Table 1. Video and Data metrics for 3500m.

For the distance of 4000m, it is observed that VoIP and FTP applications do not vary much with the insertion of a noise -130 dBm and -120 dBm; this is due to the inability to maintain the FTP transfer with these noises, enabling other applications to use all the available bandwidth. As the video application has a greater need for bandwidth, it is expected to show a greater variation than other applications, as shown in Table 2.

Metric / Impairment& Distance	Jitter Voip (ms)	Loss Voip (%)	MOS voip	Throughput Voip (Kbps)	voip Calls	Loss Video (%)	Jitter Video (ms)	Throughput Video (Mbps)	Jitter FTP (ms)	Loss FTP (%)	Delay FTP (ms)
4000m_sr	8.58	1.7	3.71	170	4	2.73	4.43	1.38	4.952	0.12	79.56
4000m rb140dBm&24D	19.56	0.14	3.67	170	4	17.81	5.18	1.15	8.093	0.13	163.58
4000m rb130dBm&24D	43.63	5.55	2.17	160	4	31.73	6.9	0.94	17.475	2.88	376.92
4000m rb120dBm&24D	39.26	23.87	1.29	141.66	4	60.33	12.7	0.54	30.729	14.3	364.19

Table 2. Video and Data metrics for 4000m.

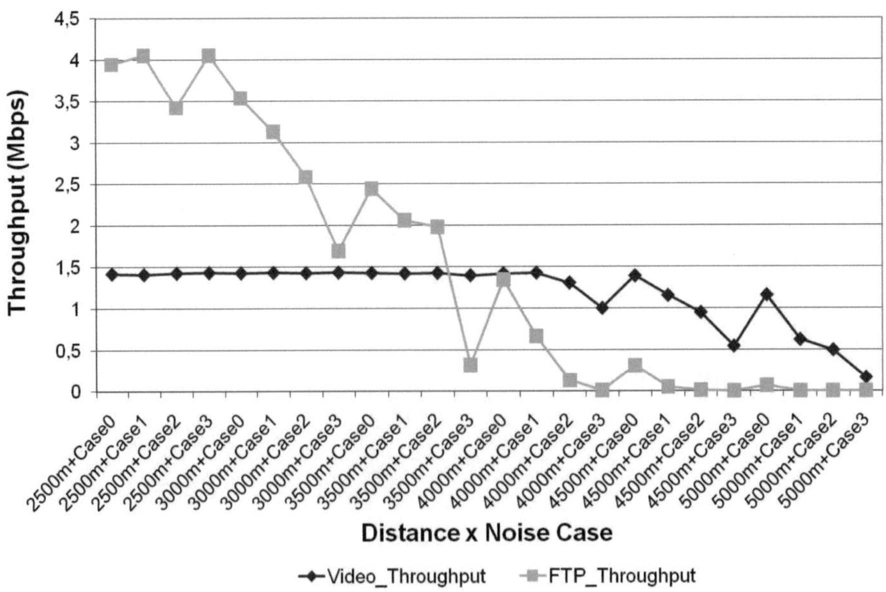

Fig. 6. Video and Data throughput behavior.

The impact of the loop distance and noise in the applications are shown in Figure 6; it shows the throughput behavior of the data and video applications. In case of video application, throughput remains constant until 4500m with noise -140 dBm and 24D (Case1), where a sharp drop is observed. The data application has an unstable behaviour due to the use of the TCP protocol, which is adaptive and uses the available bandwidth for transmission.

5.5 Bayesian Correlation study

Figure 7 shows the Bayesian network (BN) with all the attributes obtained from the empirical testing (see section 5.2). Each node has a conditional probability table associated with it (e.g. Delay_FTP); with the nodes, the dependencies are also represented, given the direction of their connecting arrows (e.g. the existence of noise in the communication influences the likelihood of a jitter variation in the VoIP application, and, in turn, in the VoIP MOS).

In the BN, all the attributes were discretized in twenty states, according to the frequency of their values, allowing us to verify the probability associated to each one of them, as well as the conditional probabilities existing among the variables.

When inferences are made in the network (e.g. it is evidenced from the occurrence of a white noise of -140 dBm and 24D in the communication), the impacts of these events are propagated, as a chain reaction, throughout the network, updating the probability values of the remaining nodes, in order to reflect their behavior; thus predicting how the network would perform given the occurrence of the instantiated event.

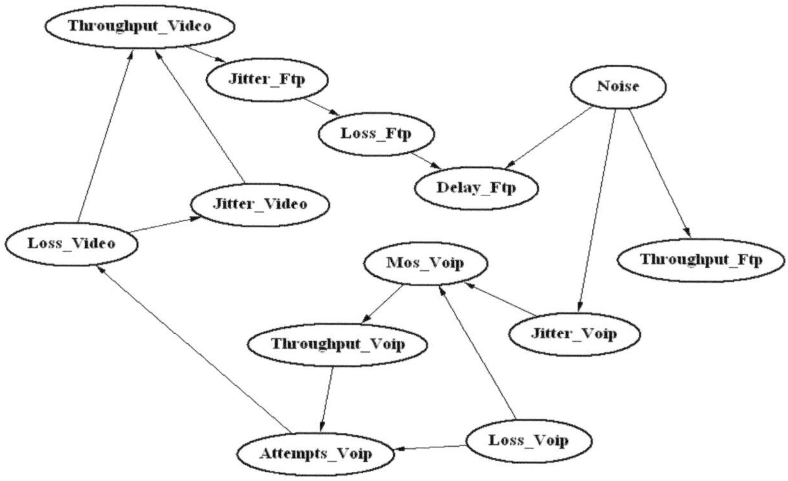

Fig. 7. Bayesian network for Triple Play applications over DSL last mile.

5.6 Scenario Analysis

Case studies were implemented to demonstrate the usability of this approach for network planning.

In the first case study minimum QoS parameters were used; these parameters are those of international standards, which define the quality of these applications and the expected performance measures. The objective is to find the maximum loop length and the set of noise that will enable us to effectively accomplish the quality in transfer of flows. So, Telcos could assess whether their links can support these applications. The results are compared with loop samples obtained from the Brazilian telecommunications networks (Telebras, 1997).

In the second case study, the inverse process is made; a certain loop distance is given as input (inferred), and the impact of this evidence is observed in the performance measures of the applications, by comparing the estimated results with their standards.

Test Case 1: VoIP Application

According to (Papagianni et al., 2009), a VoIP communication should have a minimum quality parameters, without which the VoIP communication is not feasible. Using the G.711 VoIP protocol, a VoIP communication is seen as feasible if it shows measures such as a 60ms jitter, 10% packet loss and up to 150 ms of delay. G.711 is the standard protocol for 64kbps communication, hence its use in the testing analyses.

VoIP communication requires a low bandwidth, but it is very susceptible to communication problems like bottlenecks, delays, losses and noise in communication, making it a major concern in a triple play communication. Therefore, from the inferences made in the BN, which considers and propagates the correlations between all the attributes of the domain, it was verified which behavior other applications (video and data) would have to present in order to maintain the quality required for voice communication.

Initially, with regard to the physical layer, as shown in Figure 8, the distances that enabled these parameters to be maintained were 3000m with white noise of -130 dBm and -24D, 3500 without any noise, or white noise of -140 dBm and 24D (Case1).

Since the average distance of telephone links, according to standard (Telebras, 1997) is 2090 m, it is observed that there is a possibility of extending that distance in 66.9%, i.e. 1404 m. Allowing smaller investments from Telcos in repeaters (bridge taps) which create new interference with communication or the exact definition of the distances necessary to meet these needs.

The video application has 88.7% probability of the bandwidth to be between 1.2 Mbps and 1.4 Mbps, with 99% jitter to be in the range below 20 ms and 87.9% of loss up to 10%. The implementation of FTP, which uses the adaptive TCP, has 91.2% chance of presenting average delays up to 100ms, 90.7% loss under 10% and flow rate between 1.6 Mbps and 2.4 Mbps. All levels are acceptable according to international standards (TR-126, 2006).

Test Case 2: Loop length

Here, the inverse analysis will be used, by setting a specific distance and analyzing the behavior of the triple play flow for this situation. The distance of 4500m was used, with white noise -120 dBm and 24D; that is, the influence of a high-intensity noise at a distance representing 11% (4 to 4.5 Km) of the existing loops of the telecommunication network (Telebras, 1997).

For this distance and noise scenario, it was noticed, from the correlations, that the VoIP communication was impossible due to a high competition for the channel and the lack of dedicated channels for each application. The TCP/IP protocol defines the dispute over the channel was modeled to be fair, in which all applications have the same chance to secure the channel; in this scenario, the video application manage to occupy, almost entirely, the available bandwidth, even with precariously.

The video application can obtain bandwidths from 400 to 550 Kbps, but will suffer a packet loss from 50% to 60% of the total, which is above the maximum stipulated by the TR-126 standard (TR-126, 2006), which is set to 10%. The use of FTP presents an unstable performance, with a throughput rate up to 400kbps and 10% of packet loss.

A solution to this situation is to implement QoS (Quality of Service) on the last mile, which would allow to establish routing priority for the packets; so the VoIP flow can be transmitted, even if other applications suffer from performance and/or from quality losses. The video application would have to be adapted to a new quality of both picture and sound, thus achieving the available bandwidth; codecs such as H.264 (Xiao et al., 2007) enable video compression with high quality and transmission with bandwidths from 256Kbps to 10Mbps.

Optimal State Configuration Search

Here we present the model used to search for the best strategy for implementing or expanding telecommunications services in regions with little infrastructure available to the typical end user, providing a high quality service to users by ensuring that objectives of the network service level will be maintained and correlating events from lower layers of management for determining strategies in the upper layers.

By applying a hybrid model developed using GAs and BNs (Rocha, 2009) we introduce an approach for implementing strategies for capacity planning of networks that were not originally designed for triple play applications. We show that the use of real measures and probabilistic analysis will enable the planning of communication networks, considering logical and physical parameters, such as noise, protocols and triple play applications.

The model characterizes the process of discovering scenarios that can lead to achieving a specific goal. It is aimed at identifying the best configuration, among the possible values (states of nodes in a BN) of variables in the domain, corroborating the achievement of a target value for one (or more) variable(s) in the domain in question.

The interaction between these two computational intelligence techniques (GA and BN) occurs as follows. As can be seen in Figure 8, the process of scenario discovery starts with supplying the BN, generated from the data, and its parameters; then, a GA is applied using as fitness function for the individuals (characterizing the possible scenarios available) the actual inference engine of the BN; at the end of its iterations, the optimal scenario to achieve a particular goal is obtained.

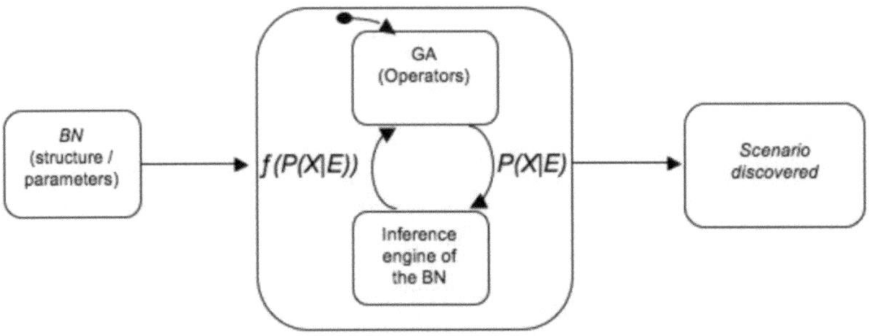

Fig. 8. Representation of the method for discovery of scenarios (Rocha, 2009).

The GA starts with the random generation of an initial population I (where each gene corresponds to a node in the BN), consisting of a set of candidate scenarios, which are then evaluated by the method of inference of the BN; in order to obtain the fitness of the scenarios, the probability of obtaining the target value for the queried variable X, given a particular configuration of states (scenario) of the variables of evidence E is calculated. The process continues with the selection of individuals, through the method of roulette. Next, we apply the operators of crossover, with crossover rate Tc; and mutation, with a mutation rate Tm. The process is repeated for n generations.

With this model we can obtain the best scenario that can derive in a specific target (or set of targets), as well as pointing the main variables and quantifying their contribution for achieving the given goal.

Using the variables obtained from the empirical tests and the BN structure defined, all the attributes were discretized in twenty states, according to the frequency of their values. We follow to apply the method described earlier to search for the best scenario, based on the network attributes, to achieve a desired behaviour for a given attribute.

Here, each of the individuals of the GA represents an inference configuration of the BN, generated randomly. Each individual is then, for its classification, submitted to the Bayesian inference module in order to verify the probability for the chosen behaviour to manifest; obtaining, at the end of the iterations, the best possible scenario of inferences on the BN to achieve desired behaviour for the chosen attribute(s).

Instead of a cost function to validate the individuals of the population, a Bayesian inference algorithm is implemented; that is, the BN is used as a cost function. This way, each of the individuals of the genetic algorithm represents an inference configuration of the BN, generated randomly (e.g. evidencing the variables noise with state 18, Jitter_VoIP with state 1, Loss_VoIP with 7 and Throughput with 4 generates the individual 2-1-7-4). Each individual is then, for its classification, submitted to the Bayesian inference module in order to verify the probability in which the chosen attribute(s) would be maximized, obtaining, at the end of the iterations, the best possible configuration of inferences on the BN for the maximization of the chosen attribute(s).

At the end of this step (after the genetic algorithm analysis) are obtained only the respective states for this maximization (for each attribute).

We point only one simulated scenario, where the loop length that connects the user to the CO has 4500m with level of noise W= -140 dBm and 24 DSL (ISDN) impairment; which needed a VoIP communication with an acceptable quality (MOS values from 3 to 4). Based on these needs, the attributes Noise and MOS_VoIP were defined with states 18 (4500m plus White Noise of -140 dBm and 24 DSL (ISDN) impairment) and 7 (MOS from 3 to 4).

Using the technique presented was obtained the results above:

Attribute	States
Jitter_VoIP	378 to 425 ms
Loss_VoIP	2.5 to 5 %
Throughput_VoIP	153 to 170 kbps
Attempts_VoIP	2
Loss_Video	0 to 8.7 %
Jitter_Video	28.5 to 30.78 ms
Throughput_Video	0.547 to 0.675 Mbps
Jitter_FTP	0 to 0.1 ms
Loss_FTP	54.6 to 63.7 %
Delay_FTP	0 to 0.1 ms
Throughput_FTP	1.6 to 2.02 Mbps

Table 3. Values of the attributes for the maximization of noise and MOS_VoIP.

The results obtained showed that the inference was possible, but defined some restrictions. For the VoIP communication, only 2 of the 4 VoIP calls made can be successfully maintained. For this, the video application will have a top available bandwidth of 600Kbps. With this, only videos with standard resolution can be transmitted. The FTP application will have 1.6 to 2.02 Mbps of available bandwidth, however with a packet loss from 54 to 63%, considered very high. The results showed the difficult or, in the worst scenario, impossibility of maintaining these applications, unless some kind of QoS (hardware or software) or adjustment in the loop is implemented. With this, the diagnosis needed for complex systems and quickly detection of network failures can be improved and an automatic recovery provided.

6. Final Remarks

DSL (Digital Subscriber Line) technology enables a universal broadband access at a reduced cost and time for implementation required for others access networks since it is considered a large capillarity broadband network, using a combination of the existing telephony infrastructure and digital subscriber line transmission technologies, which are widely available to end users in most developed countries. The environment and the flow to be transmitted must be analyzed and evaluated, given that the data obtained in this stage can prove applications to be infeasible or, at the very least, to require for an increased investment in infrastructure.

For this reason, the implementation of planning methods to aid in this process, and that take into account the current needs of applications (voice, video and data) are of major importance.

This paper implemented, with the use of crosslayer techniques, strategies for the planning and evaluation of ADSL2+ networks, which implement minimum levels of QoS for Triple Play applications.

The main contribution of this work was to apply computational intelligence methods to extract patterns in last mile DSL networks, studying the behaviour of Triple Play applications on future or already existing networks; especially those with long distances, common in countries with wide geographic area. It then becomes possible to establish more suitable contracts and/or investments with greater security; and provide government managers, in partnership with Telecommunications suppliers with subsidies to better formulate government programs for digital/social inclusion; since the expansion in the provision of Internet access, particularly when it comes to the Amazon region, which still has many areas with no basic communication infrastructure, is an essential factor of development.

7. References

Ambrosi A. ; Peugeot V. & Pimienta D. (2005). Word Matters: multicultural perspectives on information societies, C & F Editions.

IWS - Internet World Status (2009). Usage and Population Statistics, 2009. Available at http://www.internetworldstats.com/stats.htm.

Xiao Y. ; Du X. ; Zhang J. ; Hu F. & Guizani S. (2007). Internet protocol television (IPTV): the killer application for the next-generation internet. *IEEE Communication Magazine* 2007;45(11):126–34.

Papagianni C. A. ; Tselikas N. D. ; Kosmatos E. A. ; Papapanagiotou S. & Venieris I. S. (2009). Performance Evaluation Study For QoS-aware Triple Play Services Over Entry-level xDSL Connections, *Journal of Network and Computer Applications*, 32, 215-225.

Point Topic (2009). Broadband Forum announces broadband and IPTV statistics for Q2-2009. Available at: http://point-topic.com/content/dslanalysis/bbwfq209.html.

Olsen B. ; Katsianis D. ; Varoutas D.; Stordahl K. ; Harno J. ; Elnegaard N. ; Welling I. ; Loizillon F. ; Monath T. & Cadro P. (2006). Technoeconomic Evaluation of the Major Telecommunication Investment Options for European Players, *IEEE Network*, vol. 20, issue 4, pp.6-15, July/August.

Arenas D. ; Caldas C. ; Ramundo C. ; Vargas S. & Hostos L. (2006). Challenges to expanding Fixed Broadband Services in Latin America, White Paper, *Alcatel Telecommunications*, September.

Faudon V. ; Vleeschauwer D. ; Festraets E. & Ross P. (2006). End-User Services for Broadband uptake in High-Growth Economies, White Paper, *Alcatel Telecommunications*, September.

CETIC (2009). Research on the use of Information Technologies and Communication in Brazil - in Portuguese, 2008. Available at: http://www.cetic.br/usuarios/tic/2008-total-brasil/index.htm.

Kerpez K. J. & Kinney R. (2008). Integrated DSL Test, Analysis, and Operations, *IEEE Transactions On Instrumentation And Measurement*, Vol. 57, No. 4, April.

Sadri S. M. R. ; Harandi Y. N. ; Pirhadi M. ; Waskasi M. Y. ; Tabrizipoor A. I. & Mirzabaghi M. (2007). Test strategy for DSL Broadband IP Access Services, *In High Capacity Optical Networks and Enabling Technologies*, HONET, 2007.

Telebrás (1997). Telebrás 225-540-788 System Documentation, Abril, 1997.

Wallace W. ; Humphrey L. ; Kirkby R. & Pitt C. (2005). Enhanced DSL Algorithms - Deliverable number DB2.2, MUSE, (Multi-Service Access Everywhere) Project, December, 2005.

Stolle R. (2002). Electromagnetic Coupling of Twisted Pair Cables, *IEEE Journal on Selected Areas in Communications*, vol. 20, pp. 883-892, June, 2002.

Kerpez K. ; Waring D. L. ; Galli S. ; Dixon J. & Madon P. (2003). Advanced DSL management, *IEEE Communications Magazine*, vol. 41, no. 9, pp. 116–123, September, 2003.

Starr T. ; Sorbara M. ; Cioffi J. M. & Silverman P. J. (2003). *DSL Advances*. Upper Saddle River, NJ: Prentice-Hall, 2003.

Cioffi J. M. (1999). Very-high-speed digital subscriber lines, *IEEE Communications Magazine*, pp. 72–79, April, 1999.

Starr T. ; Cioffi J. M. & Siverman P. J. (1999). *Understanding Digital Subscriber Line Technology*, Englewood Cliffs, NJ: Prentice-Hall, 1999.

Korb K. B. & Nicholson A. E. (2003). *Bayesian Artificial Intelligence*, CRC PRESS, 2003.

Chen Z. (2001). *Data Mining and Uncertain Reasoning - an Integrated Approach*, John Wiley Professional, 2001.

Rocha C. A. ; Santana A. L. ; Frances C. R. L. & Rego L. (2007). Sistema de Suporte à Decisão para Predição de Cargas e Modelagem de Dependência em Sistemas Elétricos de Potência, *Anais do XXVI Congresso da SBC*, Campo grande, julho, 2007.

Radcom (2009). The State of Art, Available at http://www.radcom.com/.

TR-048 - DSL Forum Technical Report (2002). ADSL Interoperability Test Plan, April, 2002.

ITU-T Recommendation G.992.5 (2005). Asymmetric Digital Subscriber Line (ADSL) transceivers, Extended bandwidth ADSL2 (ADSL2+), January, 2005.

OpenH323 (2007). Open Phone Application, Available at http://www.openh323.org/.

ITU-TRec.H.264 & ISO/IEC14496-10 (2007). ITU-Tand ISO/IEC JTC1, Version1:May2003, Version2:May2004, Version3:March2005, Version4:September2005, Versions5 and 6:June2006, Version7:April2007, Version8(includingSVCextension): consented in July, 2007.

Kamaci N. & Altunbasak Y. (2003). Performance comparison of the emerging H.264 video coding standard with the existing standards. *In: Proceedings of IEEE International Conference on Multimedia and Expo,Baltimore*, p.345–8, 2003.

VLC (2009). VLC Media Player, Available at: www.videolan.org/vlc.

IPERF (2009). Available at: http://dast.nlanr.net/Projects/Iperf/ (2009).

Rao N. S. V. ; Poole S. W. ; Wing W. R. & Carter S. M. (2009). Experimental Analysis of Flow Optimization and Data Compression for TCP Enhancement, *In IEEE INFOCOM Workshops*, IEEE 19-25, April, 2009.

Primet P. ; Harakaly R. & Bonnassieux F. (2002). Experiments of Network Throughput Measurement and Forecasting Using the Network Weather, *In 2nd IEEE/ACM International Symposium on Cluster Computing and the Grid*, 21-24 May, 2002.

TR-126 - DSL Forum Technical Report (2006). Triple-play Services Quality of Experience (QoE) Requirements, December, 2006.

Rocha C. A. J. (2009). Doctorate Thesis, Optimization Strategy for Improving the Interpretability of Bayesian Networks: Applications in Power Systems - in Portuguese, UFPA, 2009.

16

Optimization strategies for improving the interpretability of bayesian networks: an application in power systems

Claudio A. Rocha, Diego L. Cardoso,
Adamo L. Santana and Carlos R. L. Francês
Federal University of Pará
Brazil

1. Introduction

The search for new methods, techniques and tools to support the decision-making processes is a subject that has aroused major interest in international research; with intelligent systems emerging as one of the most robust solutions.

Such studies characterize an area called Data Mining (DM), also known as Knowledge Discovery in Database (KDD), which represents a source of mature technologies, largely embedded in organizational processes of modern corporations. DM can be understood as an interactive and iterative process to identify understandable, valid, new and potentially useful patterns from large data sets.

This work presents an analysis to improve the comprehensibility of the patterns discovered in the DM process, which is related to the easiness of interpretation by the human being (Rezende, 2003). Thus, the use of DM techniques that provide mechanisms of presentation and visualization that simplify the analysis of the knowledge obtained can strongly contribute to the users to measure the quality of this knowledge.

Among the many DM techniques found in literature, Bayesian networks (BN) comes as one of the most prominent, when considering easiness for interpreting knowledge obtained from a domain with uncertainty. The reason is that it provides a mechanism for representing the causal model of a given dataset (Pearl, 1988), allowing qualitative and quantitative analyses from the variables of the domain; thus, providing support to the decision making process (Korb & Nicholson, 2003), (Russel & Norvig, 2003).

However, BNs present a restriction to establish the optimal combination of states for given variables (discrete or continuous) that would achieve a certain requirement (state of one or more variables of the domain). In many real applications, the search for situations which would lead to the attainment of certain goals is extremely important.

For example, to achieve a certain level of sales, it is necessary to find which set of factors that can influence in this progression and, thus, determine which are the conditions (states) of these factors that have greater impact on the sales rate obtained. In this work, we present a method to solve such problem, by combining the techniques of genetic algorithms (GAs) with the BNs, built from the domain's data. In light of these indicatives, we point, as contribution of this work, the development of new strategies to extend the power of interpretability of BN, implementing a strategy for the discovery of scenarios.

In summary, the model presented here characterizes the process of discovering scenarios that can lead to achieving a specific goal; for such, we use a novel hybrid model developed using GAs and BNs, that combines the qualities of evolutional algorithms for space search with a Bayesian probability model for inference. It is aimed at identifying the best configuration, among the possible values (states of nodes in a BN) of variables in the domain, corroborating the achievement of a target value for one (or more) variable(s) in the domain in question.

The main objectives are twofold: analysis and use of Bayesian methods for knowledge extraction, basically with respect to the creation of a method capable of extending the power of interpretability of the BN; proposal of a model to measure the causal relationship among the variables of a domain, by discovering the values that compose an optimal combination (configuration) of states for given variables of this domain.

This work is organized as follows: section 2 presents the main motivations for using BN in the data mining process and some related work to this study. In section 3, the method proposed for the search of the optimal configuration is presented, aiming at the improvement of the BN interpretability. As a case study, the method is applied in the power systems domain, as will be presented in section 4. Finally, section 5 presents the final remarks of the paper.

2. Bayesian Networks and related work

A BN represents a probabilistic model of the variables of a given domain, being able to represent the qualitative (dependencies), as well as the quantitative (conditional probabilities distribution) information. Together, these components propitiate an efficient representation of the joint probability distribution of the set of variables $X_i = \{X_1, X_2, \ldots, X_n\}$ of a given domain (Pearl, 1988).

Moreover, three factors have motivated the use of BN in DM processes (Heckerman, 1997): first, the effective manipulation of incomplete datasets; second, the learning of causal relationships among the variables of the domain, which facilitates the analysis of the domain; third, the BN allow the combination of prior knowledge of the domain with the data.

In order to corroborate with the importance and the applicability of BN in the electric sector, used as case study here, some related studies presented in literature are shown next.

BN is known to offer, given its knowledge representation formalism, a natural mechanism for modeling diagnosis. In the power systems domain, there is a massive application on fault diagnosis of equipment and operations.

In (Yongli et al., 2006), an application of BN is presented for the diagnosis of possible transmission faults in power systems. The main motivation presented for the use of this approach is the easiness with which relationships of cause-effect, particularly in domains with a high degree of uncertainty, can be modeled.

As a way to decrease the size of the probability tables used in the mentioned problem, a BN model is proposed with nodes Noisy-Or and Noisy-And. These nodes can be seen as a generalization for the conventional logical connector *or* and *and*, respectively. The idea is to use them in the networks as elements that can simplify the correlations among the variables of the system and their implication with respect to the appearing of transmission faults. Instead of directly establishing the relation of cause and effect between two variables, they imply to a node Noisy-Or or Noisy-And, whose connections are parameterized with the use of probabilities; this way quantifying the impact that each variable has for causing transmission faults.

In (Yonggiang et al., 2005), another application of BN in the context of fault diagnosis is presented, with emphasis in the possible defects that may occur in the functioning of an important class of electric equipment - the transformers. Given the uncertainty of this diagnosis, usually due to the complexity for configuring these equipments, it is necessary to use a method in order to assist the specialist in the analysis of possible defects.

Several other applications of BN in fault diagnosis are investigated in the literature, as presented in (Flores-Loredo et al., 2005).

In (Zhou et al., 2006) BN are used to predict the possibility of faults in the energy distribution, considering some climatic aspects. In this case, a BN is modeled to carry out the fault predictions (in 7 possible states) from the conditions of wind (in 4 states) and the possibility of occurrence of atmospheric discharges (2 states - yes or no).

With respect to the mechanisms for improving the comprehensibility of the patterns discovered by the BN, in most of the available literature, the technique of genetic algorithms is usually employed only for the process of learning the BN structure (Li et al., 2005), (Gamez et al., 2002), (Morales et al., 2004).

Some proposals, however, lean to a hybrid approach of computational intelligence methods to optimize and improve the process of knowledge extraction, in its post-processing stage. For example, in (Yang, 1997) a Bayesian-Fuzzy method is used to manipulate continuous values of evidence in the inference processes.

Although without employing a technique of optimization combined to the inference of BN, an interesting method to accomplish these inferences was proposed in (Andersen et al.,

1989) and is implemented in the Hugin software; allowing to identify the most likely configuration of values for the variables of a BN, given one or more evidences.

This method has two basic differences compared to the method proposed here. First, Hugin seeks to find the composition of states (configuration) of the variables studied, based on the evidence of a given variable. Here, the idea is to attain the states of the studied variables (our particular goal) that would allow to achieve a given state on other variable(s) of the BN. Another difference is related to the capacity of obtaining the continuous values, and not discretized range of values, of the studied variables, to achieve the desired value for the goal variable. In the particular case of power systems, this is primordial, given that a variation of 0.1% in the consumption can represent a considerable financial economy.

3. Optimal State Configuration Search

The objective of this model is to identify the best configuration, among the possible values of the existing variables in the domain, which maximizes a given attribute, identifying initially the other variables that present a dependency from it.

In contrast to the way genetic algorithms are used in the majority of the hybrid systems proposed in the literature, where they are adopted to optimize the process of learning the structure of BN, here, the technique is used for the discovery of the most probable values of the variables of a BN, given the value of a key attribute.

The discovery of scenarios that are conducive to achieving a particular goal is of utmost importance to support the process of decision making. For example, determine which socio-economic scenario corroborate with obtaining a target value of total energy consumption, defined by the user.

The method developed is aimed at subsidizing decision making users with methods to analyze, in advance, the scenarios that can lead to achieving a certain goal; identifying the best configuration, among the possible values of variables in the domain, corroborating the achievement of a target value for one(or more) variable(s) in the domain in question. For this, we used a hybrid method that combines the probabilistic and correlation power of BNs, with the ease of GAs for the incorporation of specific knowledge of the problem, in order carry out optimization tasks.

The interaction between these two computational intelligence techniques (GA and BN) occurs as follows. As can be seen in Figure 1, the process of scenario discovery starts with supplying the BN, generated from the data, and its parameters; then, a GA is applied using as fitness function for the individuals (scenarios) the actual inference engine of the BN; at the end of its iterations, the optimal scenario to achieve a particular goal is obtained.

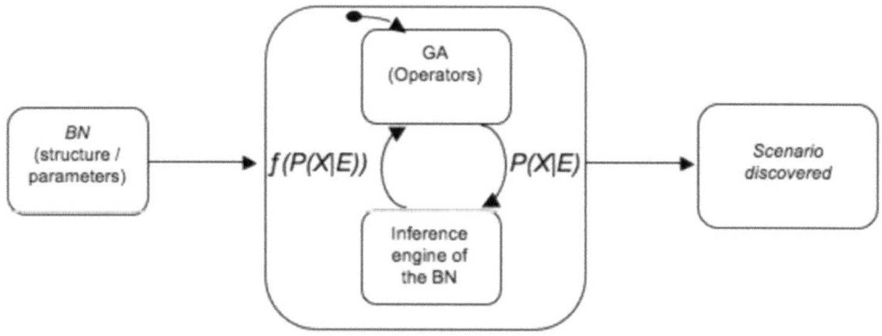

Fig. 1. Representation of the method for discovery of scenarios.

In Figure 1, *P(X | E)* represents the probability of obtaining a particular state of *X* (target variable), given the set of remaining variables in the domain E. Thus, the scenarios (configuration of states for variables *E*) represent the individuals of the GA, which are evaluated (fitness function) by the probability of obtaining the goal *X*. That is, the probability *P(X | E)* of occurrence of each scenario is provided as input to the BN method of inference, returning as output the value for this query. As mentioned previously, this value is used as fitness function for the individuals (scenarios) of the genetic algorithm (GA).

```
SCENARIO DISCOVERY (bn)
/* returns the scenario that best contribute to achieving the target
    value for a target variable of the domain */
// bn – Bayesian network that codifies the joint distribution P(X1, X2, ...Xn)
population ← GENERATE_RANDOM_POPULATION;
repeat
  initialize new_population
  for i ← 1 to SIZE(population) do
      a ← SELECT(population, APTITUDE_FUNCTION(INFERENCE_MODEL_I(bn))
      b ← SELECT(population, APTITUDE_FUNCTION(INFERENCE_MODEL_I(bn))
    if (CROSSOVER_RATE is met)
      child_ab ← CROSSOVER(a,b)
    if (MUTATION_RATE is met)
      child_ab ← MUTATION(child_ab)
    include child_ab in new_population
  end for
  population ← new_population;
until termination criteria is met
return the best scenario in the population, according to the
APTITUDE_FUNCTION
```

Fig. 2. Algorithm for the process of scenario discovery.

BNs and GAs are used in different subsystems that collaborate to reach a solution, i.e., the intelligent paradigms are independent, exchange information and perform separate functions to generate solutions as shown in Figure 1. Therefore, the method presented here can be considered in the category of intercommunicative hybrid methods. Figure 2 shows the algorithm for method presented.

The GA starts with the random generation of an initial population *I* (where each gene corresponds to a node in the BN), consisting of a set of candidate scenarios, which are then evaluated by the method of inference of the BN; in order to obtain the fitness of the scenarios, the probability of obtaining the target value for the queried variable *X* is calculated, given a particular configuration of states (scenario) of the variables of evidence *E*. The process continues with the selection of individuals, through the method of roulette. Next, we apply the operators of crossover, with crossover rate T_c; and mutation, with a mutation rate T_m. The process is completed following one of the following criteria:

- establishment of a predetermined number of generations, i.e. define, a priori, a number *n* of iterations;
- until the algorithm can find an acceptable scenario. The acceptance of the scenario is made based on a subjective quality model for evaluation, considering opinions and definitions of the domains experts.

One can notice that it is possible to employ any inference method (INFERENCE_MODEL_I) for the BN, exact or approximate; the probability is used to evaluate the quality of the individuals in the GA. We point that the parameters used to execute GA are defined by the user and vary according to the application domain.

In order to show the general interaction process of the GA and the BN inference, consider a BN *B*, generated from a dataset *D*. Consider also the general inference process over B, expressed by a set of query variables *X*, a set of E variables for inference, and a set of e observed states from *E*, and a set *Y* representing the remaining variables (not contained in *X* and *E*). *A* query *P(X | e)* can be expressed by:

$$P(X \mid e) = \alpha P(X,e) = \alpha \sum_{y} P(X,e,y) \tag{1}$$

Where α is a normalization constant, that ensures that the sum for the probability distribution of *P(X | e)* equals 1; and *y* are possible values for variables in the set *Y*.

Equation 1 can infer specific queries over *X* from any set of evidence variables *E*, considering for the calculations the state space of variables *Y*, (1). The method for discovery of scenarios can be viewed as a specialization of this equation, which aims to find which values (states) e from the set of variables E maximizes the probability of a given $x \in X$. In this case, *E* is formed by all variables of the domain, i.e. Y = ∅. Thus, we can write (1), as follows, considering the suitability of a particular individual in the GA that enables achieving the target value x_i,

$$P(x_i | e_1, e_2, ..., e_n) = P(x_i) \prod_{k=1}^{n} P(e_k | x_i) \quad (2)$$

Where:

$e_1, e_2, ..., e_n$ are the possible evidences;

and x_i is the event we want to observe.

The chromosomes in the GA are represented by decimal values, characterized by the state space for the variables used for inference, as shown in Figure 3, where e_1 represents any state of E_1, e_2 represents a certain state of E_2 and so forth.

e_1	e_2	e_3	e_4	e_5	...	e_m

Fig. 3. Representation of chromosomes of the genetic algorithm.

To calculate of the fitness, the chromosomes are submitted to the inference module of the BN, in order to calculate the probability of the queried variable to attain a given target certain value. The higher the probability, the fittest the considered individual will be. It is worth to mention that, more than a single query variable can be used for the discovery of scenarios.

To illustrate the operation of the method, consider the BN, showed in Figure 4 and its respective variables (nodes) and states (Table 1).

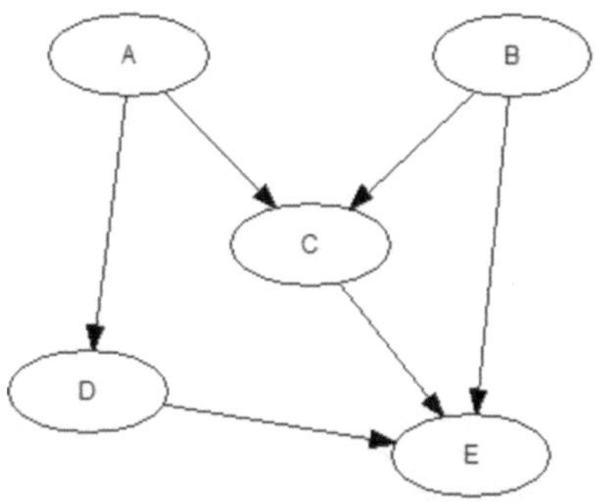

Fig. 4. Bayesian network to illustrate the discovery of scenarios.

Variable	States
A	a_1, a_2
B	b_1, b_2, b_3, b_4
C	c_1, c_2
D	d_1, d_2
E	e_1, e_2, e_3, e_4

Table 1. Nodes and states of the BN.

In the example, d1 is considered the target value, highlighting that it would be possible to choose any variable (or set of variables) of the BN. The GA acts on the inference method of the BN (e.g. the exact method Junction Tree) to find the scenario that maximizes the probability for d_1, to occur.

A possible candidate solution to this simple example could be the set {2,3,1,2}, in which the first position (gene) infers state a_2 of variable A, b_3 of variable B, c_1 of C and e_2 for E. The fitness evaluation, will be given by $P(d_1 \mid a_2, b_3, c_1, e_2)$. Thus, after application of GA operators (selection, crossover and mutation) and at the end of iterations (generations), the best configuration (scenario) for variables A, B, C and E, which maximize the probability of d_1, would be obtained.

4. Case study application

4.1. Motivation and Context of the Proposed Model

The analysis described here was originated from the demands of the research project "PREDICT - Support Decision Tool for Load Prediction of Electrical Systems". This project, a joint venture between the Government of the State of Pará and the Power Supplier of the State of Pará, aims at designing and implementing a decision support system, using mathematical and computational intelligence methods, to foresee the demand for energy purchase in the future market.

With that in mind, studies are usually made to measure the impact that many other variables (temperature, humidity, socio-economic factors etc.) influence over the consumption, so that it is possible to foresee scenarios where the operation of the power systems are economic, safe and reliable.

So, the consumption forecast and the correlation of some exogenous variables to the power system, specifically associated to climatic and socio-economic factors, served as basis for the project. In its first phase the project used methods of regression and artificial neural networks, to apply the forecasts, and BN to model the mentioned correlations.

However, throughout the development of the project, a series of demands for new inferences, necessary for a reliable and safe planning and operation of the power systems, were raised by the specialists (managers and engineers). Amongst these demands we point out the creation of indicators that influence the future performance of the power system, such as mechanisms that would optimize the consumption, given its relation with socio-economic and climatic variables.

To assist in these new demands, the BN were elected as models for representing these correlations. This proposal was elaborated in order to not only cover this domain of application, but also to enable its application in many other areas.

4.2. Description of the Optimization Model

The case study, proposed by the domain specialists of the power system market, and used for the optimization model was to discover under which circumstances the power consumption would be maximized. For this case, the optimization model was based on a few steps that are described as follows.

Firstly, identify which attributes, among those from the database, influence directly the power consumption by building the BN structure.

The Government of the State of Pará, from its State Executive Bureau of Budget and Finances Planning supplied a database with 15 years of monthly records of the State's socio-economic aspects, consisting 35 attributes.

Only the attributes selected by the specialists were used for the generation of the BN, according to their impact in the variation of power consumption; they are: number of employments in the sectors of the transformation industries and agriculture and cattle breeding, and the values of the total turnover and of the dollar. We point out that their influence reflects directly not only to the total power consumption in the State, but also to the many classes of consumption (residential, industrial, commercial etc).

Given the knowledge that the variables of number of employments in the transformation industries (*emp_ind*), employments in the agriculture and cattle breeding (*emp_agro*), value of the total turnover (*val_turn*) and the value of the dollar (*val_dol*) are the main influences in the variation of the power consumption, they were used in the next step, which consisted in the creation of a BN (Figure 5), using the search and score algorithm K2 (Cooper & Herskovitz, 1992).

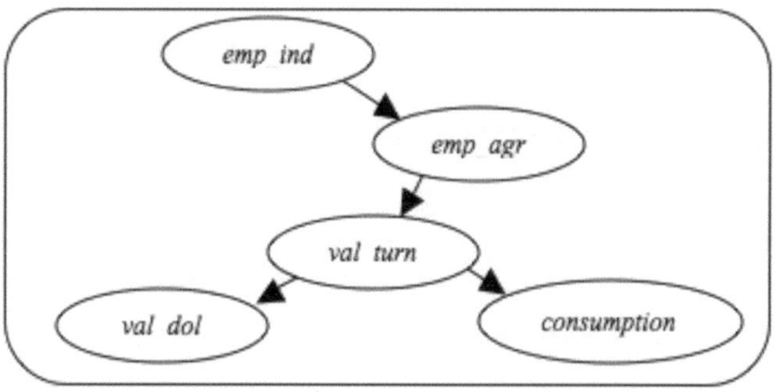

Fig. 5. BN created with the K2.

In the BN, all the attributes were discretized in ten states, according to the frequency of their values, allowing us to verify the probability associated to each one of them, as well as the conditional probabilities existing among the variables.

Once the network is set, the next step is, by making use of the data given by the BN, to search the network attributes for the states that would maximize the power consumption. In this stage we use a modified genetic algorithm.

Here, instead of a cost function to validate the individuals of the population, a Bayesian inference algorithm is implemented (Equation 2); that is, the BN is used as a cost function.

This way, each of the individuals of the genetic algorithm represents an inference configuration of the BN, generated randomly (e.g. evidencing the variables *emp_ind* with state 2, *emp_agro* with state 1, *val_turn* with 7 and *val_dol* with 4 generates the individual 2-1-7-4). Each individual is then, for its classification, submitted to the Bayesian inference module in order to verify the probability in which the power consumption attribute would be maximized, obtaining, at the end of the iterations, the best possible configuration of inferences on the BN for the maximization of the power consumption.

However, we would have at the end of this step (after the genetic algorithm analysis) only the respective states (i.e. band of values) for this maximization, instead of a single value (for each attribute), which is what we seek. Following this phase, we make use, again, of a genetic algorithm; but this time a traditional genetic algorithm, whose aptitude function we obtain from the data.

The function used for the genetic algorithm is obtained from a regression of multiple variables made over the attributes of the BN (Dillon & Goldstein, 1984), (Hair et al., 1998). The multivariate analysis is however made over the consumption data, but considering only the data instances located within the ranges found in the previous step. Thus, we obtain an equation (presented below) with a good representativity (approximately 0.9039) over the domain.

$$Y = 258{,}598{,}510.5 + 3{,}675.6834\,X_1 + 4{,}430.9036\,X_2 + \\ + 0.4701\,X_3 - 12{,}182{,}208.61\,X_4 \quad (3)$$

where Y represents the power consumption and X_1, X_2, X_3 and X_4 represent the values of the attributes *emp_ind, emp_agro, val_turn* and *val_dol*, respectively.

Based on Equation (3), the genetic algorithm is then used, thus obtaining the values, for each of the attributes that would maximize the power consumption. It is worth mentioning again that the individuals evaluated by the aptitude function (2) are only those within the range of values that maximize the value of consumption. Thus, in order to achieve the occurrence of the maximum consumption, it is necessary that the values in Table 2 are achieved, for the attributes *emp_ind, emp_agro, val_turn* and *val_dol*.

Attribute	**Value**
emp_ind	5.380
emp_agro	3.357
val_turn	R$ 100.752.576,00
val_dol	R$ 2,861

Table 2. Values of the attributes for the maximization of the consumption.

The genetic algorithms used were, basically, parameterized according to the values in Table 3. The representation used for the individuals, however, was different. The first genetic algorithm used a representation with size based on the number of possible states that the variables of the BN could assume; and the second one used a binary representation. Other tests specifying different values for the parameters in Table 3 were also made; the results obtained, however, did not present any significant alteration.

Parameters	**Values**
Initial population	50 individuals
Number of generations	1,000
Selection	Roulette
Crossover	One point
Crossover rate	98%
Mutation rate	0.1%
Elitism	Yes

Table 3. Parameters used in the algorithms.

It is worth mentioning that the optimization model used is restricted not only to the discovery of the maximum values of consumption, but can also be used to identify scenarios that cause a minimum, average or any other value to be achieved by the power supplier, given the variation of the considered economic aspects.

Moreover, it is important to emphasize that although this case study presented a reduced search space, the method can be applied for cases with a sparse number of variables, given the evolutionary heuristics presented.

5. Final Remarks

This paper presented a strategy to extend the potentialities of BN, with respect to their inference process. It also showed, as a motivation for this strategy, assistance to the demands of the electric sector. Among the main contributions of the proposed strategy, we can point out the following.

The extension of the power of interpretability of the BN through the discovery of the optimal combination of values, represents the possibility of quantifying the causal relationships among the socio-economic and electricity consumption variables, and allows to achieve a given goal or a key aspect;

The interest of those involved in this Project, in applying the functionalities of the model in many other scenarios, not only relative to the power consumption, but also for government actions (e.g. discovery of the variables, and their values, that would maximize the generation of employment and income), has encouraged the use of the proposed model. This interest can further seen by the current use of the model in other Brazilian states, whose energy is also provided by the same group of companies of which the power supplier of Pará belongs to.

We concludes pointing that by applying the hybrid model presented, the following analyses can be implemented:

1. Identify the variables that have the greatest impact in achieving a target value. This feature is particularly useful in situations where a given goal is established, but not all states of the variables are known or manageable. Besides being interesting in intractable high degree networks;
2. Find the singular values (when dealing with continuous variable) within the discretized ranges (states) of each evidence variables, that most contribute to achieving a target value for the queried variable;
3. Extending the method developed to obtain target values for more than one queries variable. This functionality allows to establish a query based on more than one goal, considering the isolated importance and impact of each;
4. Embedding expert knowledge, so that subjective criteria are used to evaluate the scenarios. This provides a stopping criterion guided by the degree of interest, measured from the belief of key aspects related to the target variable, set a priori by the specialist.

Moreover, it is important to point out that the solutions of problems involving the combination of techniques that can establish relations of cause and effect (BN) and of optimization (e.g. genetic algorithms) are not very well defined in the literature, particularly aiming at finding the states of given variables that can establish a desired condition, also influenced by these variables.

6. References

Andersen, S. L. ; Olsen, K. G. ; Jensen, F. V. & Jensen, F. Hugin - A shell for building Bayesian belief universes for expert system. *In Proceeding of the 11th International Joint. Conference on Artificial Intelligence.* Spring-Verlag, (1), (1989) 1080-1085.

Cooper, G. & Herskovitz, E. A Bayesian method for the induction of probabilistic networks from data. *Machine Learning*, (9), (1992) 309–347.

Dillon, W. R. & Goldstein, M. *Multivariate analysis - methods and applications.* John Wiley & Sons, 1984.

Flores-Loredo, Z. ; Ibarguengoytia, P. H. & Morales, E. F. On line diagnosis of gas turbines using probabilistic and qualitative reasoning. *Intelligent Systems Application to Power Systems.* Proceedings of the 13th International Conference on, (1), (2005) 297-301.

Gamez, J. A. ; Campos, L. M. & Moral, S. Partial abductive inference in Bayesian belief networks - an evolutionary computation approach by using problem-specific genetic operators. *Evolutionary Computation. IEEE Transactions*, 6 (2), (2002) 105–131.

Hair, J. F. ; Anderson, R. E. ; Tatham, R. L. & Black, W. C. *Multivariate data analysis.* Prentice-Hall. 1998.

Heckerman, D. Bayesian networks for Data Mining. *Data Mining and Knowledge Discovery*, Kluwer Academic Publishers, (1), (1997) 79-119.

Korb, K. B. & Nicholson, A. E. *Bayesian Artificial Intelligence*. Chapman & Hall/CRC, 2003.

Li, X. ; He, X. & Yuan, S. Learning Bayesian networks structures from incomplete data based on extending evolutionary programming. *Machine Learning and Cybernetics*, Proceedings of 2005 International Conference on, (4), (2005) 2039–2043.

Morales, M. M. ; Dominguez, R. G. ; Ramirez, N. C. ; Hernandez, A. G. & Andrade, J. L. A method based on genetic algorithms and fuzzy logic to induce Bayesian networks. *IEEE Proceedings of the Fifth Mexican International Conference in Computer Science*, (1), (2004) 176-180.

Pearl, J. *Probabilistic reasoning in Intelligent System*, Morgan Kaufmann Publishers, 1988.

Rezende, S. O. *Intelligent systems: concepts and application*, Manole, 2003.

Russel, S. & Norvig, P. *Artificial Intelligence – a modern approach*. Prentice Hall, 2003.

Yang, C. C. Fuzzy Bayesian inference. *Proceedings of IEEE International Conference on Systems, Man, and Cybernetics*, (1997) 2707-2712.

Yonggiang, W. ; Fangcheng, L. & Heming, L. The fault diagnosis method for electrical equipment based on Bayesian network. *Electrical Machines and Systems*, (ICEMS). Proceedings of the Eighth International Conference on Publication, (3), (2005) 2259-2261.

Yongli, Z. ; Limin, H. & Jinling, L. Bayesian networks-based approach for power systems fault diagnosis. *IEEE Transactions on Power Delivery*, (21) 2, (2006) 634-639.

Zhou, Y. ; Pahwa, A. & Yang, S. S. Modeling weather-related failures of overhead distribution lines. *Power Systems, IEEE Transactions*, 21 (4), (2006) 1683–1690.

17

Strategy for Wireless Local Area Networks Planning and Performance Evaluation through Bayesian Networks as Computational Intelligence Approach

Jasmine Araújo, Josiane Rodrigues, Simone Fraiha,
Hermínio Gomes, Gervásio Cavalcante and Carlos Francês
Federal University of Pará, Institute of Technology
Brasil

1. Introduction

The use of wireless local area networks (WLANs), as well as the proliferation of the use of multimedia applications has grown fast in recent years, mainly due to their mobility, easy configuration and low cost deployment, so they have became an interesting alternative for industries, enterprises, among others. This technology, usually, supports data traffic generated by applications such as web browsing. In recent years, however, it has been used for voice communication, especially in offices (Medepalli et al., year 2004). The VoIP technology provides the transmission packages of voice over IP protocol, used inside the Internet, reducing significantly the cost of calls when compared with those carried out by public switched telephone network (PSTN). However, the VoIP application requires that WLAN must be able to support rigid QoS specifications for the voice transmission, it has been established in ITU-T(International Telecommunication Union) G.114 recommendation and (Zhai et al., year 2006). Some factors affect the quality of service (QoS) received by the user. The interference is an example.

Thus, some works focuses on the problem of achieving a high coverage level in terms of received signal quality (Rodrigues et al., 2000),(Mateus et al., year 2001),(Kamenetsky & Unbehaun, 2002),(Unbehaun & Kamenetsky, year 2003),(Lee et al., 2002). QoS oriented criterion was considered and performance was studied in (Molina & Alonso, 2004),(Amaldi et al., 2005),(Prommak et al., 2002),(Bianchi, year 2000),(Heusse et al., 2003),(Lu & Valois, 2006). Moreover, in (Jaffres-Runser et al., year 2007) is proposed mono-objective and multi-objective formulations for the wireless local area network planning problem including the coverage, the interference level and the quality of service (in terms of data throughput per user). Metaheuristic methods are explored and the results show the assets of both approaches but mainly emphasize the benefit of the multi-objective search strategy that offers several alternative solutions. Finally, in (Bosio et al., year 2007) a framework for AP placement was developed with the maximization of network efficiency.

This chapter presents a strategy to evaluate performance based on hybrid approach that considers measuring and Bayesian inference applied to wireless networks, considering QoS parameters as power, jitter, packet loss, delay and PMOS. It differs from the other previous works developed by the fact that the model take into account a crosslayer vision of networks and the Bayesian network correlates aspects of the physical environment, on the signal propagation (power or distance) with aspects of VoIP applications (e.g., jitter and packet loss). Moreover, in order case studies were carried out considering two indoor environments and two outdoor environments one of them with important characteristics of the Amazon region (e.g., densely arboreous environments).

2. Measurement Campaigns

The indoor measurement campaigns were performed in two buildings at Federal University of Para (UFPA). The first one is a classrooms building, while the second is a building especially built for research laboratories and teachers' rooms. The outdoor measurements campaigns were performed in a square and at a parking lot of a university campus. The main differences between the aforementioned environments, are the equipments used because in the first ones, it was used an access point 802.11g Linksys@ WRT54G Router Speed Booster for indoor attendance and at the last, as they were an outdoor attendance, a HotZone Motorola@ equipment was used. In the next subsections these indoor environments will be presented.

2.1 Classromm Buildings

In this measurement campaign the metrics were collected in the second floor of a building of the Federal University of Pará. That building is made of bricks and concrete, with lateral glass windows while the other side there is a corridor along all the building (Fig. 1). In this building there are only classrooms, that are divided by walls built on bricks.

Fig. 1. Photograph of the building. In clockwise: external side with glass windows, classroom, corridor and external side with corridor along

2.2 Research Laboratories

The metrics are collected in a two-storey building made of bricks with rooms for Lectures, Computer and Telecommunication Labs and an anechoic camera, whose height occupies both the floors. The building has side glass windows with aluminum frames except the anechoic camera. The rooms are divided by walls built on bricks. The building is still empty and with no furniture (Fig. 2).

Fig. 2. Photograph of the research laboratories building: corridors of the second and first floor, and laboratory classroom

In the next subsections the outdoor environments will be presented.

2.3 Marituba's Square

The metrics were collected in a square. The presence of densely arboreous environment typically of the Amazon Region was found there. The testbed was done in a real system, because the Government of the State of Pará has a digital inclusion program. The Marituba city is the first digital network city of this program (NAVEGAPARA, 2010). Fig. 3 shows a picture of the square. The scenario is highlighted by the black rectangle and the black circle at the left side of the rectangle indicates the access point under study.

2.4 Parking Lot

The metrics were collected in a parking lot of a Federal University of Pará. Fig. 4 shows a picture of the parking lot. The scenario is highlighted by the red line and the blue circle at the right side of the picture indicates the access point under study.

3. Measurement Methodology

The methodology of measurement was done as described below:

- Measurement points and the access point positioning: some points were marked to perform the measurements. Their distances from the walls were also measured to the

Fig. 3. Photograph of the Marituba's square

Fig. 4. Photograph of the parking lot

indoor environments and to the outdoor environment the GPS coordinates were collected. Firstly, the network under study was installed.

- Connection of the Network under Study - the architecture of the network under study (channel 7, central frequency of 2.442GHz) it is shown in Fig. 5, where APS in Fig. 7 is connected, through a cable to the protocol analyzer ethernet port. The second ethernet port is connected to a computer. This computer was used as a VoIP receiver, using CallGen323 (Callgen, 2010) software;
- Traffic Generation at Network under Study - A notebook computer, located in the first plan in Fig. 5, was used to generate traffic in the WLAN network. Files were transferred to a server located at the cable network through APS.
- VoIP Transmitter - to transmit the VoIP calls another notebook was used. It was located on a cart Fig. 6, and it was positioned in the selected measurement points;
- Power Measurement - The cart carries also another notebook. The power measurement was done in each point, through the Network Stumbler@ software (Netstumbler, 2010). This notebook was necessary because the Network Stumbler, while in use doesn't allow the connection of the computer to any WLAN.

With the methodology and equipments described in the stages, the first phase of the measurement campaign was performed. In this case there was only a transmitter in the environment in study. During the measurements, the following parameters were stored: received power (through the Netstumbler software), distance transmitter-receiver, jitter, delay, packet loss and PMOS (measured by the protocol analyzer). After that first measurement phase, a second one was performed using the same procedure of the first, but now, with the presence of another network using the same channel of the network under study, called interference network. The access point of the interference network was positioned in the second floor in the same direction of the network under study, APT in Fig. 7. The Iperf program (Iperf, 2010) was used to generate traffic in the inteference network, it is allowed specify the time during which this traffic is generated. After this second measurement phase, the data were treated and compared to find a parameters variation in the presence of a interference network. The following section presents the results of those comparisons. The only difference among the procedures used at the two buildings is the application used to compete traffic with VoIP in the network under study. In Fig. 8 is showed the layout, the location of the measured points and the location of the access points to the classrooms building. After this measurement campaign, data were treated and the measurements were compared.

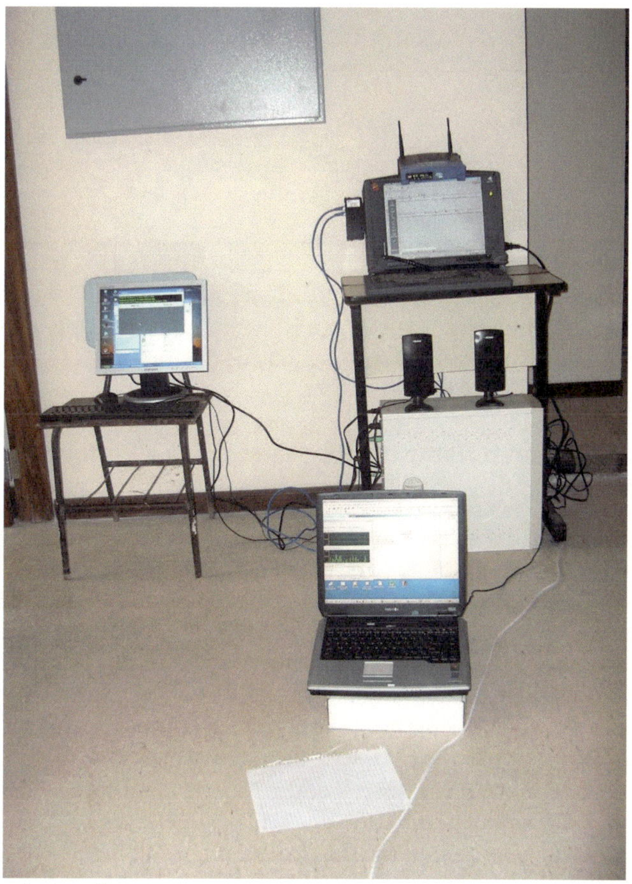

Fig. 5. Network under study

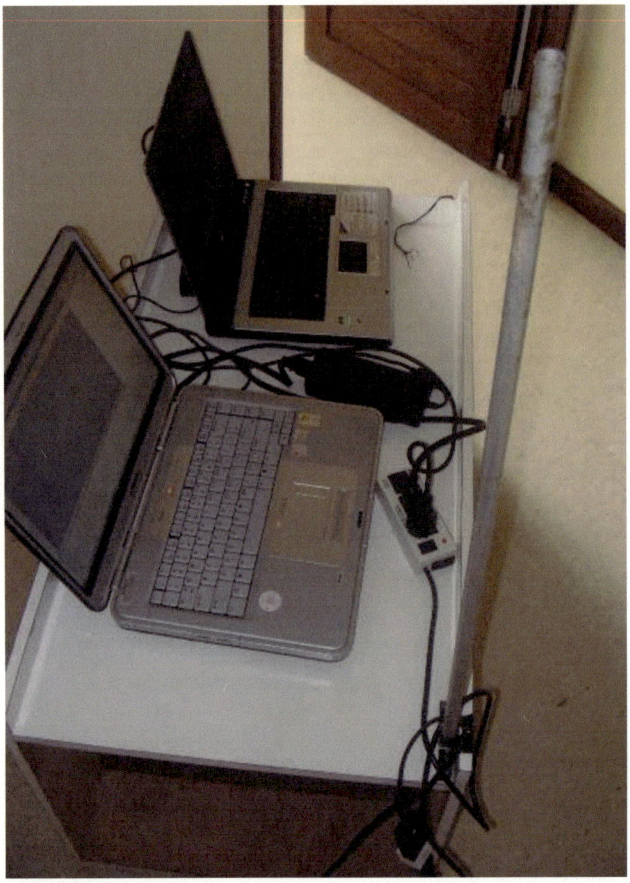

Fig. 6. The cart with notebook running Network NetStumbler (the lower side of the picture) and the notebook running VoIP calls (the upper side of the picture)

4. Strategy using Bayesian Networks for Planning and Performance Evaluation of Wireless Networks

The process of knowledge discovery in database (KDD) stands as a technology capable of widely cooperating in the search of existing knowledge in the data. Therefore, its main objective is to find valid and potentially useful patterns from the data. The extraction of knowledge from data can be seen as a process with, at least, the following steps: understanding of the application domain, selection and preparation of the data, data mining, evaluation of the extracted knowledge and consolidation and the use of the extracted knowledge. Once in the data mining stage, considering the core of the KDD process, methods and algorithms are applied for the knowledge extraction from the database. This stage involves the creation of appropriate models representing patterns and relations identified in the data. The results of these models, after the evaluation by the analyst, specialist and/or final user are used to predict the values of attributes defined by the final user based on new data. In this work, the computational intelligence algorithm used for data mining was based on Bayesian networks.

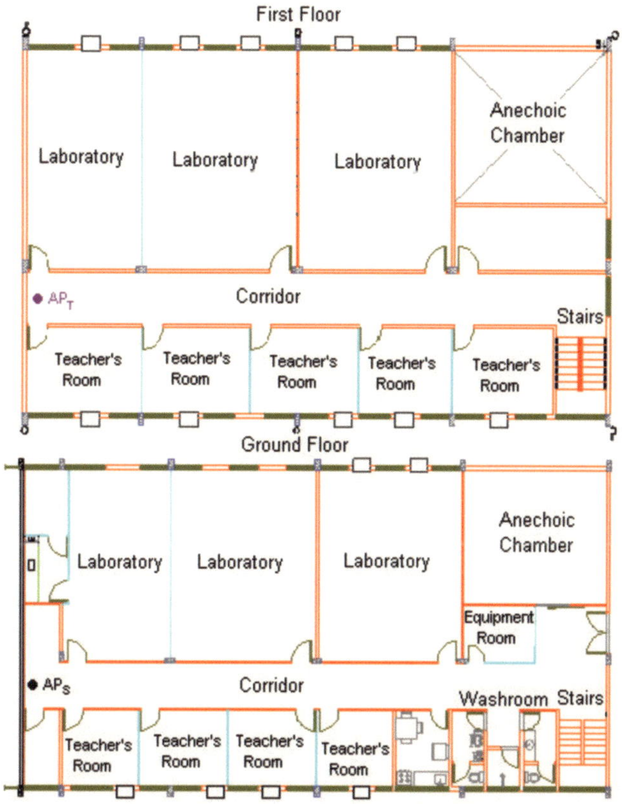

Fig. 7. Layout of the research laboratories(first and second floor) with the location of the measurements points and APs

A Bayesian network is composed of several nodes, where each node of the network represents a variable, that is, an attribute of the database; directed arcs connecting them implies in the relation of dependency that the variable can possess over the others; and finally probability tables for each node.The Bayesian networks can be seen as coding models of the probabilistic relationships between the variables that represent a given domain. These models possess as components a qualitative representation of the dependencies between the nodes and a quantitative (conditional probability tables of these nodes) structure, that can evaluate, in probabilistic terms, these dependencies. These components together provide an efficient representation of the joint probability distribution of the variables of a given domain.
One of the major advantages of the Bayesian networks is their semantics, which facilitates, given the inherent causal representation of these networks, the understanding and the decision making process for the users of these models. Basically, due to the fact that the relations between the variables of the domain can be visualized graphically, besides providing an inference mechanism that allows quantifying, in probabilistic terms, the effect of these relations (Santana et al., year 2007).

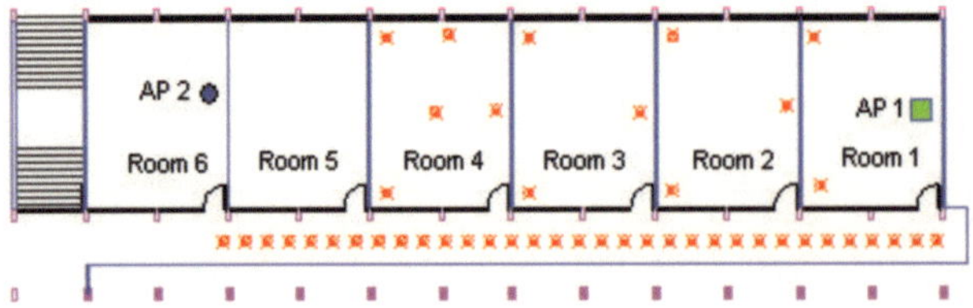

Fig. 8. Layout of the classrooms building with the location of the measurements points and APs

4.1 Bayesian Inference Results

This section discusses the measurements of the application and physical layers as well as the results obtained by using Bayesian networks. The study involved treating the measured data acquired with the novel strategy using any intelligence computational technique, i.e. the Bayesian network technique (Araújo et al., 2007). In any process of knowledge discovery, there is a pre-analysis phase of treatment (soft mining) of the data where information that is not going to contribute to the final result are removed. Hence, the input fields for the Bayesian network were obtained from the protocol analyzer after the pre-analysis. They worked as input to the free version Bayesware Discoverer(BDD) commercial software (Discoverer, 2010).

4.1.1 Indoor Environment - Research Lab

According to Fig. 9, the inference results related to distance with the best value are presented. The probability of throughput lying within 142760.0 to 149180.0 bps is 67.7%. The results for other metrics are described as follows: in the case of packet loss, the probability of loss lying within 0 to 0.14% is 60.0%. This value added to the second interval of larger probability (31.6%) results in the probability of 91.6% of packet loss for lying within 0 to 0.55% (recommended less than 1%). Considering now the jitter, its probability for lying within 0.86 to 2.72 ms is 75.5% (maximum recommended 30 ms). Finally, the PMOS probability values for lying within 4.0 and 4.9(Good) is 94.1% (the values of PMOS were codified in agreement with ITU-T Recommendation P.800 (ITU-TP800, 1996)).

Another inference performed is the selection of the lowest throughput in the Fig. 10. The packet loss for the network with inference of lowest throughput is 32.8% lies within 2.15% to 7.67%. The jitter probability to be greater than 8.4 ms is 35.7% and smaller is 64.3%. The PMOS has the probability value of 62.7% for lying within 3 and 3.9 (Fair). Finally, the distance metric presents relevance values for this second inference scenario. Its probability is 48.9% to be located beyond 19 meters of the access point (distances less than 19 meters can be guaranteed acceptable QoS parameters for half of times).

4.1.2 Indoor Environment - Classrooms Building

According to Fig. 11, the inference results related to best power. In the case of throughput, the probability value of 63.3% for lying within 152110 to 152520 bps. Considering now the packet loss, its probability of loss being equal to zero was 59.4%. The jitter probability value was 75.1% for lying within 2.2107 ms and 4.3643 ms. PMOS had the probability value of 52.8% for lying within 3.9386 and 4.1095. Finally, the delay has a 80.9% probability of lying within 74.032 ms and 150.87 ms.

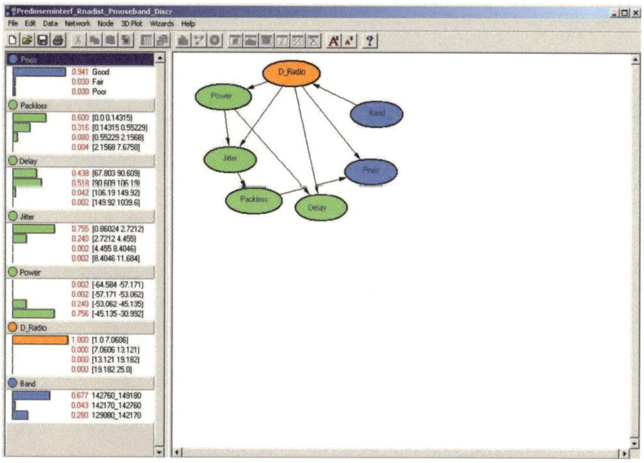

Fig. 9. Bayesian networks with best distance inference applied to ground floor

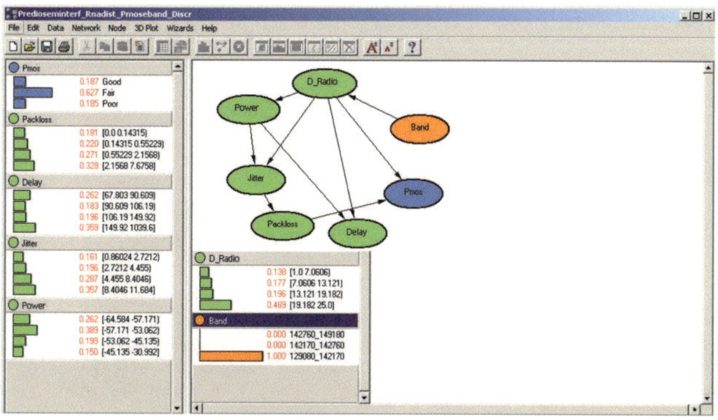

Fig. 10. Bayesian networks with worst throughput inference applied to ground floor

Another inference performed was the selection of the worst throughput, as shown in Fig. 12. The packet loss probability value was 47.8% for lying within 1.255% to 3.899%. Considering now the jitter, its probability for lying within 7.9975 ms to 12.43 ms was 51.7%. PMOS had the probability value of 43% for lying within 0.2645 and 3.3536. Finally, the delay has a 71.4% probability of lying within 150.87 ms and 3229.9 ms.

4.1.3 Outdoor Environment - Marituba's Square

Fig. 13 presents the inference results related to throughput with the best value. In the case of packet loss, the probability of loss lying within 0 to 0.85% is 44.3%. Considering now the jitter, its probability for lying within 4.66 to 7.33 ms is 33.2%. Finally, the PMOS has the probability values for lying within 3.8(Fair) to 4.02(Good) is 48.1% (the values of PMOS were codified in agreement with ITU-T Recommendation P.800 (ITU-TP800, 1996)).

Another inference performed is the selection of the worst distance in Fig. 14. The packet loss for the network with inference of worst distance is 51.8% lies within 2.04% to 6.81%. The jitter probability to be greater than 16.5 ms is 47.8%. The PMOS has the probability value of 51.8% for lying within 2.45 and 2.92 (Poor). Finally, the throughput metric has probability lying within 68496 bps and 70774 bps is 51.8%.

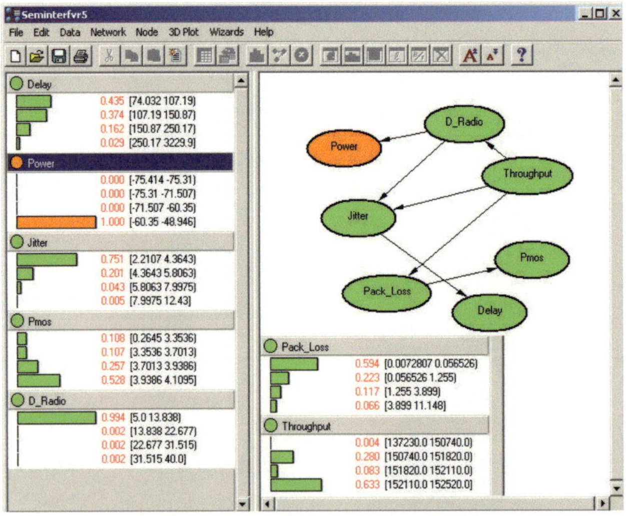

Fig. 11. Bayesian networks with best power inference applied to classroom's building

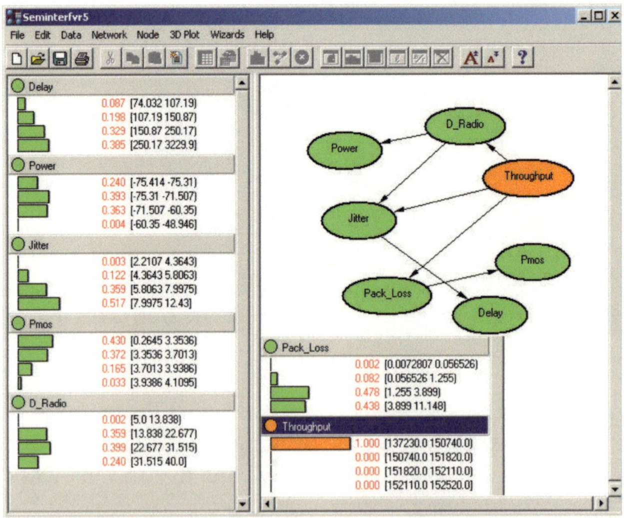

Fig. 12. Bayesian networks with worst throughput inference applied to classroom's building.

Through the use of Bayesian networks can be noticed that the QoS parameters applied to outdoor environment were degraded even in the best situation of the network, i.e. the best throughput. Referring to the worst case collected, can be seen that the parameters degraded, but the achieved distances are bigger than the measures at the indoor environment, i.e. 155 meters and 19 meters for the indoor environment . The difference between the equipments used can be the reason of this, but the densely arboreous environment contributes either. The use of this computational intelligence aids the decision maker to decide which is the best point to locate the access point, and how the Qos parameters will behave.

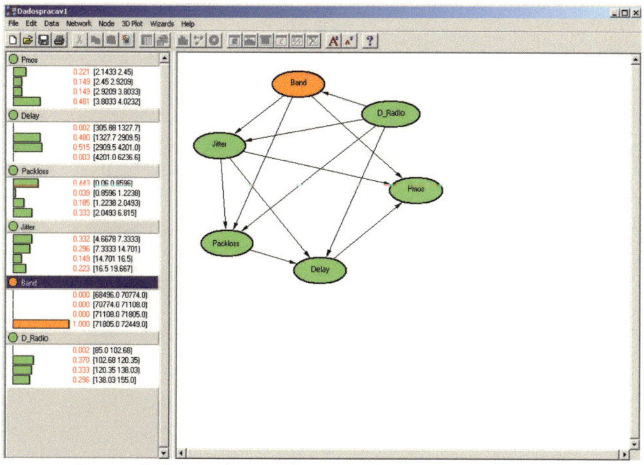

Fig. 13. Bayesian networks with best throughput inference applied to Marituba's square

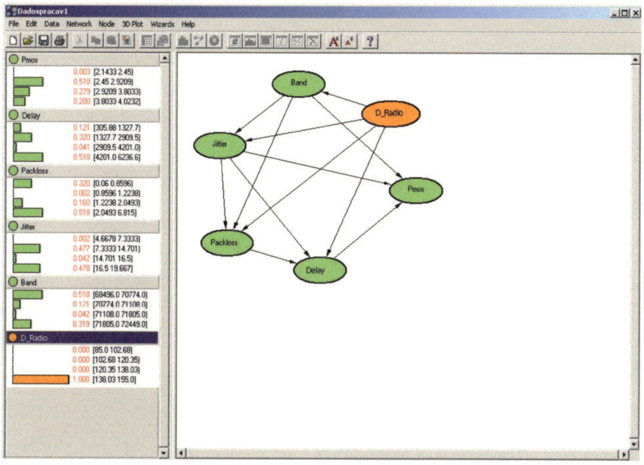

Fig. 14. Bayesian networks with worst distance inference applied to Marituba's square.

4.1.4 Outdoor Environment - University Parking lot

According to Fig. 15, the inference results related to best distance. In the case of packet loss, the probability of loss being equal to zero was 87.1%. Considering now the jitter, its probability for lying within 2.66 to 3.33 ms was 61.5%. Finally, the probability that PMOS values would lie within 3.89 (Fair) to 4.02 (Good) was 39.3% (the values of PMOS were codified in agreement with ITU-T Recommendation of P.800 (ITU-TP800, 1996))

Another inference performed was the selection of the worst distance, as shown in Fig. 16. The packet loss for the network with inference of worst distance had a 61.4% probability of being greater than 5%. The jitter probability value was 77.5% for lying within 15.95 and 34.66 ms. PMOS had the probability value of 76.9% for lying within 0 and 2.48 (Poor). Finally, the throughput metric has a 75.5% probability of lying within 57297 and 69527 bps.

The use of this computational intelligence aids the decision maker to decide which is the best point to locate the access point, and how the Qos parameters will behave.

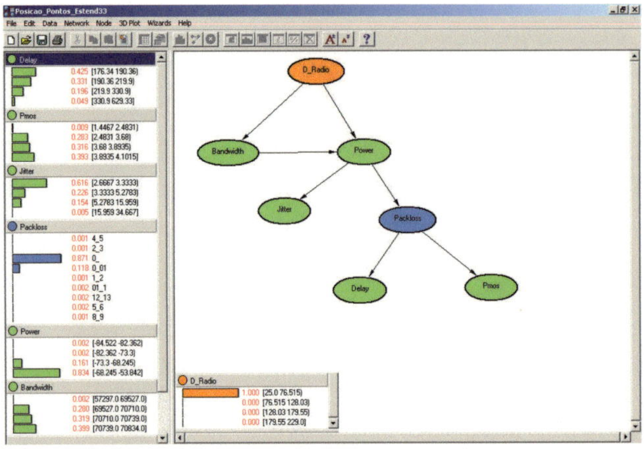

Fig. 15. Bayesian networks with best distance inference applied to parking lot

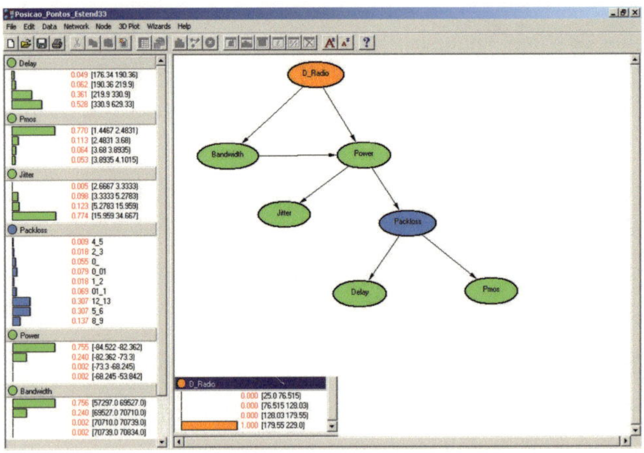

Fig. 16. Bayesian networks with worst distance inference applied to parking lot

5. Conclusion

In this chapter a novel WLAN planning and performance evaluation strategy with computational intelligence approach, i.e., bayesian networks was presented. The measuring technique was used through an empirical study concerning the behavior of the QoS parameters of an VoIP application in 802.11g network. This study was performed in two different indoors and outdoors environments, characterizing two types of scenarios. This was done to establish the correlation between the behavior of the QoS parameters and the distance.

The main contribution of the Bayesian strategy is the use of computational intelligence to differentiate the two environments and to validate the robustness of the methodology proposed here. Another contribution is about the installation of new hot spots at digital cities, that must be installed at State of Pará. In addition, due to the large application of wireless LAN, this strategy can be applied in real engineering design. This methodology will aid the decision makers where to locate the access point to better attend the public offices, according on the applications that must be used. It is important to mention that Bayesian network offers an

approach to select several scenarios of QoS. Therefore, it is possible to guarantee a minimum distance to the AP for VoIP application in an indoor or outdoor WLAN environment.
Finally, in real building there is a very strong trend to find similar scenarios to the presented ones in this paper, where different networks cohabit and where it is desirable that applications with rigid parameters of QoS carry out.

6. References

Amaldi, E., Capone, A., Cesana, M., Fratta, L. & Malucelli, F. (2005). Algorithms for wlan coverage planning, *in* G. Kotsis & O. Spaniol (eds), *Wireless Systems and Mobility in Next Generation Internet*, Springer Berlin / Heidelberg, New York, pp. 52–65.

Araújo, J., Rodrigues, J., Fraiha, S., Gomes, H., Reis, J., Vijaykumar, N., Cavalcante, G. & Francês, C. (2007). The influence of interference networks in qos parameters in a wlan 802.11g: a bayesian approach, *Proceedings of Conference Broadband Access Communication Technologies II*, Society of Photo-Optical Instrumentation Engineers, Boston, pp. 677604–1–677604–12.

Bianchi, G. (year 2000). Performance analysis of the ieee 802.11 dcf, *Journal of Selected Areas in Communications* **Vol. 18**(No. 3): 535–547.

Bosio, S., Capone, A. & Cesana, M. (year 2007). Radio planning of wireless local area networks, *IEEE/ACM Transactions on Networking* **Vol. 15**(No. 6): 1414–1427.

Callgen, S. (2010). `www.openh323.org`.

Discoverer, B. (2010). `www.bayesware.com`.

Heusse, M., Rousseau, F., Berger-Sabbatel, G. & Duda, A. (2003). Performance anomaly of 802.11b, *Proceedings of the Twenty-Second Annual Joint Conference of the IEEE Computer and Communications Societies - IEEE INFOCOM 2003*, IEEE, San Francisco, pp. 836–843.

Iperf, S. (2010). `http://dast.nlanr.net/Projects/Iperf/`.

ITU-TP800, R. (1996). *Methods for subjective determination of transmission quality* .

Jaffres-Runser, K., Gorce, J.-M. & Ubeda, S. (year 2007). Mono- and multiobjective formulations for the indoor wireless lan planning problem, *Computers and Operations Research* **Vol. 35**(No. 12): 3885–3901.

Kamenetsky, M. & Unbehaun, M. (2002). Coverage planning for outdoor wireless lan systems, *Proceedings of International Zurich Seminar on Broadband Communications Access, Transmission, Networking 2002*, IEEE, Zurich, pp. 491–496.

Lee, Y., Kim, K. & Choi, Y. (2002). Optimization of ap placement and channel assignment in wireless lans, *Proceedings of IEEE Conference on Local Computer Networks 2002*, IEEE, Florida, pp. 831–836.

Lu, J.-L. & Valois, F. (2006). Performance evaluation of 802.11 wlan in a real indoor environment, *Proceedings of IEEE Wireless Mobility*, IEEE, Montreal, pp. 140–147.

Mateus, G., Loureiro, A. & Rodrigues, R. (year 2001). Optimal network design for wireless local area network, *Annals of Operations Research* **Vol. 106**(No. 1-4): 331–345.

Medepalli, K., Gopalakrishnan, P., Famolari, D. & Kodama, T. (year 2004). Voice capacity of ieee 80211b 80211a and 80211g wireless lans, *IEEE Communications Society* **Vol. 3**(No.): 1549–1553. Globecomm.

Molina, O. & Alonso, I. (2004). Automatic planning optimal quality-cost wireless networks, the indoor pareto oriented abspad approach, *Proceedings of IEEE Personal, Indoor and Mobile Radio Communications*, IEEE, Barcelona, pp. 2545–2550.

NAVEGAPARA, P. (2010). `www.prodepa.gov.br`.

Netstumbler (2010). `www.netstumbler.com`.

Prommak, C., Kabara, J., Tipper, D. & Charnsripinyo, C. (2002). Next generation wireless lan system design, *Proceedings of IEEE Military Conference 2002*, IEEE, Florida, pp. 473–477.

Rodrigues, R., Mateus, G. & Loureiro, A. (2000). On the design and capacity planning of a wireless local area network, *Proceedings of IEEE/IFIP Network Operations and Management Symposium 2000*, IEEE, Honolulu, pp. 335–348.

Santana, A., Francês, C., Rocha, C., Carvalho, S., Vijaykumar, N., Rego, L. & Costa, J. (year 2007). Strategies for improving the modeling and interpretability of bayesian networks, *Data and Knowledge Engineering* **Vol. 63**(No. 1): 91–107.

Unbehaun, M. & Kamenetsky, M. (year 2003). On the deployment of picocellular wireless infrastructure, *IEEE Wireless Communications* **Vol. 10**(No. 6): 70–80.

Zhai, H., Wang, J. & Fang, Y. (year 2006). Providing statistical qos guarantee for voice over ip in the ieee 802.11 wireless lans, *IEEE Wireless Communications (Special Issue on Voice over Wireless Local Area Network)* **Vol. 13**(No. 1): 36–43.

18

Causal modelling based on bayesian networks for preliminary design of buildings

Berardo Naticchia and Alessandro Carbonari
Università Politecnica delle Marche, DACS department, Division of Building Construction (www.dacs-bc.univpm.it)
via delle Brecce Bianche, 60131 Ancona, Italy

1. Introduction

The adoption of innovative technologies in construction is sometimes difficult, due to the lack of adequate knowledge to properly estimate and size such systems in the professional environment. Moreover, the lack of proper simulation programs for the preliminary design of buildings which integrate the new technologies prevents the application of these systems to the contemporary construction market, often producing higher costs and less efficient buildings.

Despite the recognized validity of several new technological solutions through extended experimentation, and the numerous advances that are being obtained each year, only a small percentage of this technology is being applied to the erection of buildings. This can be explained by the fact that professional architects prefer to adopt standard techniques that they can control rather than try to apply new systems with a high risk of failure, which require assistance from technology experts in order to help architects arrive at their design choices.

The best way to overcome these limitations, while fostering wide and fast spread of recently developed technologies on the market, would be to provide professional designers with friendly and reliable simulation tools to help architects discern the best configuration during the conceptual phase of buildings which are to be equipped with these new solutions. In particular, Bayesian Networks will be shown to be a suitable tool for developing multi-criteria decision software programs, given their ease of use and flexibility. In fact, they are able to deal with the difficulty underlying even complex phenomena, by means of an explicit causal framework that links the variables affecting the system. In addition, they can be learned from the same raw data that researchers collect from experiments or advanced simulation tools (e.g. finite difference or finite element methods), automatically giving back accurate estimations to professionals, who in this way do not need to become involved in the use of complex and time consuming simulation programs, like the ones adopted by technology developers. In addition, Bayesian Networks based models can implement decisional functions which are more suitable and quicker than parametric analyses for rough sizing purposes.

Even though the Bayesian approach is very powerful, the best methodology to be moulded for its implementation needs to be carefully evaluated, because it must take into account several variables, mainly related to:

- how to build the probabilistic framework relative to complex phenomena involving hundreds of variables linked by non-linear relationships;
- how to use raw data coming from experiments or advanced simulation results to learn conditional probability tables among variables;
- how to validate the model under development.

In this chapter a methodology to build a reliable Bayesian model integrating both experimental data and prior knowledge is shown. It is expected to act as a preliminary simulation tool that is a lean and fast way to perform rough sizing, leaving the task of more accurate and time consuming forecasts to the following design stages.

Finally, its application to a practical case study for the design of glazed saddlebacked roofpond equipped buildings is taken as an example to show how this multi-criteria decision Bayesian model may be used to assist designers in the problems dealt with by architects during the preliminary stage of design.

2. State of the art

Despite the great potential and flexibility offered by the use of Bayesian Networks, as detailed in the following section 2.1, their application to building design must respond to some basic methodological precautions, which will be indicated in subsection 2.2.

2.1 Scientific background on Bayesian Networks

Bayesian networks can be extracted from the knowledge of experts, using a method called causal mapping: it is applied in the context of an information technology outsourcing decision (Nadkarni and Shenoy, 2004). Mathematical models can also be translated into qualitative patterns (Lucas, 2005), in order to infer conditional relations and the graphical structure of the network. Their application has been tested in many areas.

Bayesian Networks are used for the management of areas affected by salinity, and they offer the possibility to trade off different kinds of knowledge, like observed data, expert knowledge and results from simulations (Sadoddin et al., 2005). It has been demonstrated that they are able to evaluate the influence of management actions on different aspects of the model framework, such as biophysical, social and economic issues. Bayesian Networks are also applied to study the impact of design, manufacturing and operational decisions relative to oil drill platforms and to the external environment (Zhu et al., 2003). Other applications are known in the field of process monitoring and root cause analysis of complex industrial systems (Weidl et al., 2005). A methodology to be applied in the field of software architectural design, to obtain decisions regarding the adoption or rejection of the best alternative from a web of complex and often uncertain information, has also been proposed (Zhang et al., 2005).

The high flexibility of Bayesian Networks has also been shown by (Van Truong et al., 2009), where subjective knowledge, collected by means of questionnaire surveys with experts, was collected to build a network quantifying the most likely causes for delays in construction. Other research is also being carried out in the field of automatic parameter learning in the difficult case of incomplete datasets or sparse data (Wenhui et al., 2009). Bayesian models also have important properties including the possibility to arrive at decisions, which is critical in many fields, like maintenance processes (Zhiqiang et al., 2008): the networks can be developed from past data about failures and can then be used to obtain decisions, based on the probability of occurrence of future damaging events.

Many attempts have also been made in the field of automatic learning Bayesian Networks, whose final purpose would be to provide a machine learning process that finds the network's structure and its associated parameters, which best fit any available dataset (Lauría et al., 2007). However this cannot work properly when data of different kinds are available and they must be put together to develop the final model.

2.2 Advances obtained with respect to the state of the art

To the authors' knowledge, there are no systematic analyses concerning the applicability of Bayesian Networks to the preliminary design stage of innovative buildings, although it is well known that architects involved in this task must cope with a multi-criteria decision making process in order to reason about environmental, cost analysis, structural, aesthetic and other issues (Brouchlaghem, 2000). The software programs which are currently available on the market are mainly based on the numerical solution of complex analytical models. Although accurate and sometimes time-saving, they leave the final choice for the optimization of performance to the designer's intuition. In fact, an interesting first advance in this direction was pursued by testing Object Oriented models in the housing construction process: the opportunity to visualize and manage together many aspects of this process was appreciated (Harish et al., 2008). Indeed, the possibility to reason from uncertain inputs and to include long-term consequences for each scenario, makes Bayesian tools suitable for use in the early stage of preliminary design, when there is not a complete knowledge of the system and its boundary conditions.

The procedure proposed in this chapter is mainly intended to show how to use Bayesian models for building reliable and easy to use simulation tools, which can integrate several types of knowledge coming from different sources into a single probabilistic framework.

In addition, this methodology exploits the tool of Object Oriented Bayesian Networks, shortened to OOBNs (Koller et al., 1997), which also helps deal with new technologies which have intrinsic complexity (e.g. many variables interacting according to non-linear relationships) that is a well known challenge for those involved in modelling. Furthermore, they provide an explicit representation of the causal framework that links the variables affecting the system, through which a designer can analyze, criticize and then improve the preliminary project; in order to apply it, he/she needs only know the performance to be obtained and the input data.

As regards the specific case of roofponds, presented as a demonstration at the end of this chapter, current approaches proposed by researchers are suitable for executing parametric studies or for verifying thermal performance when boundary conditions are known. Instead, the model developed in the following is able to automatically predict the thermal behaviour of roofpond buildings using only rough input data, which is typical of the preliminary stage of design. This model reasons in a way similar to that adopted by expert designers when detailed data about the new construction are not available, and a heuristic method must be used to describe the system from a functional point of view, inferring the best choice for future design.

3. Developing complex Bayesian Network models

3.1 Brief overview on Bayesian Networks

The main asset of Bayesian Networks lays in the integration of qualitative physical patterns (Boborow, 1984) and computational algorithms elaborated in the field of artificial intelligence (Jensen, 2000) in order to create an intelligent support tool. The main utility of Bayesian Networks consists in the possibility to combine typical results from macroscopic and microscopic analyses (Naticchia et al., 2001). Combining the two approaches, designers have the possibility to perform a trial approach also considering very detailed numerical results in order to reach a higher reliability.

Over the last decade, Bayesian Networks (also called belief bayesian networks or causal probabilistic networks) have dominated the field of reasoning under uncertainty, thanks to the ability of such expert models to deal with incomplete or uncertain information (Pearl, 1988; Korb and Nicholson, 2004).

Bayesian Networks consist of two parts: a graphical model and an underlying conditional probability distribution. The graphical model is represented by a directed acyclic graph (DAG), whose nodes represent random variables, which are linked by arcs, corresponding to causal relationships with the previous ones. Each variable may take two or more possible states, of both numerical and label types. An arc from a variable A to another variable B denotes, in the general case, that A causes B. Using the standard terminology, A is said to be a parent of B (which is its child). The strength of that relationship is quantified by conditional probability tables (Wonnacott and Wonnacott, 1990), where the probability to observe each state of any child variable is given with respect to all combinations of its parents' states; in our example it would be generally billed $P(b|a)$, where A is conditionally independent of any variable of the domain that is not its parent, and "a" defines a generic state for variable A. The same holds for variable B. Thus we can obtain a conditional probability distribution over every domain, where the state of each variable can be determined by the knowledge only of the state of its parents, and the joint probability of a set of variables E can be computed applying the "chain rule" (Pearl, 1988):

$$P(E) = P(E_1,...,E_n) = P(E_n \mid parents(E_n)) \cdot ... \cdot P(E_2 \mid E_1) \cdot P(E_1) \tag{1}$$

Eq. (1) simplifies the computational process considerably, and it is also the first main feature of Bayesian Networks. In other words, the joint probability of any combination of variables E is given by the product between the variable E_n, given any sub-set of variables that includes only the parents of E_n, and any sub-set of variables that are simply ancestors of E_n, given the conditional probabilities of their parents. Thus the complete specification of any joint probability distribution does not require an absurdly huge database as is the case when every variable is considered to be dependent on the others (Charniak, 1991).

Secondly, the Bayesian explicit graphical representation also provides a clear understanding of the qualitative relationships among variables, allowing the user to reason about their causal correlations.

In addition, every node of a Bayesian Network can be conditioned with new information via a flow of information through the network. The probability of a set of "query" nodes is computed given the evidence on other nodes for which observations are already available. Furthermore, parameter updating is supported for any direction of reasoning: from causes to consequences ("predictive" reasoning) or from consequences to causes ("diagnostic" reasoning). This advantage derives from the application of the "Bayes Theorem":

$$P(H \mid e) = \frac{P(e \mid H) \cdot P(H)}{P(e)} \tag{2}$$

where H is the variable with unknown probability distribution; e is the set of variables for which evidence has been obtained.

Finally, Bayesian Networks have the important capability to update it from new evidence: this can be formulated by gradually substituting the prior probability distribution $P(H)$ with $P(H \mid e_n)$, that is the probability distribution of H conditioned upon a set of old evidence e_n. Similarly $P(e \mid H)$ becomes $P(e \mid e_n, H)$, and $P(e)$ becomes $P(e \mid e_n)$:

$$P(H \mid e_n, e) = \frac{P(e \mid e_n, H) \cdot P(H \mid e_n)}{P(e \mid e_n)} \tag{3}$$

3.2 Building the graphical structure

The three basic reference modules of elementary graphical structures are provided in Fig. 1 (Pearl, 1988): given the case of Fig. 1-a, the probability of C, given B, is exactly the same as the probability of C, given A and B. Therefore A and C are conditionally independent: that structure is called a *causal chain*. The *common causes* structure in Fig. 1-b is slightly more complex: if there is no evidence or information about B, then learning the probability distribution of A or C will change the probability distribution of the unknown variable between A or C; in the opposite case, when B is given, the knowledge of A or C will not change the probability distribution of the other. The last *common effects* structure in Fig. 1-c, represents the situation where an effect has two causes: the parents are marginally independent, but become dependent given information about the common effect.

While building any causal structure to develop a probabilistic model before validation, this must be compared with the elementary networks in Fig. 1, in order to verify that any conditional independence stated by the causal model really corresponds to the meaning assigned by the corresponding basic reference structure.

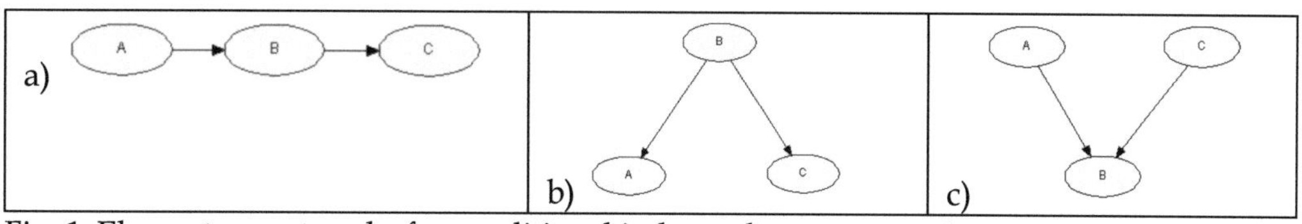

Fig. 1. Elementary networks for conditional independence assumptions.

3.3 Object Oriented Bayesian Networks

Probabilistic causal networks to model complex physical phenomena are expected to be made up of several elementary networks (each of them devoted to modelling a part of the whole process), and assembled through the use of Object Oriented Bayesian Networks (OOBNs). This functionality is particularly useful to provide a hierarchical description of complex technology systems, because it breaks down the whole domain into single units or fragments or elementary networks or more generally "objects". An object is the fundamental unit of an OOBN (Koller et al., 1997), representing either a node or an instantiation of a fragment network, which is an abstract description of a network containing both input and output nodes. Input nodes are depicted as ellipses with shadow dashed line borders, and output nodes are ellipses with shadow bold line borders, that can be shared by several networks. Fig. 2 depicts an example of a very simple OOBN, which is not intended to have a meaning but must be considered as an example, where the main elements are depicted: instances, input and output nodes linking the previous ones, standard nodes. "Node2" and "Node3" are output nodes which can transfer information to input nodes (like "node1") and to and from intermediate nodes.

In practice, input nodes are used to insert information (or evidence) from the user or from results of other elementary networks; intermediate nodes are used to perform computations; output nodes contain information that can be used directly for design purposes or is sent as input to another elementary network performing one of the next tasks.

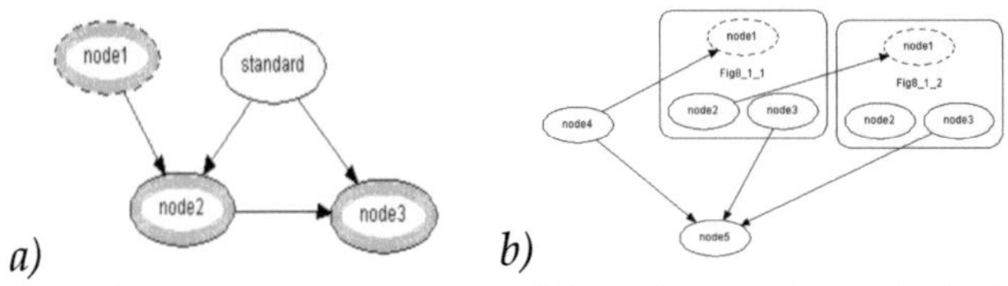

Fig. 2. Example of an object with interface variables (a) and of an OOBN (b).

3.4 Conditional probability estimation

In general there are two different ways of learning probabilities from data: with a known structure (where only probability parameters need to be estimated) and with an unknown structure (where the probabilistic framework must also be estimated). In the case of technology development, qualitative relationships among variables are learnt from expert aids, therefore only the conditional probabilities remain unknown. From a mathematical point of view, we deal with a domain $U=\{E_1, \ldots E_n\}$ made up of discrete variables, that is quantified by a finite collection of discrete physical probabilities, whose structure will be called B_s, as in (Heckerman, 1996). Considering the case of learning from a dataset with no missing data, that is to say, for each set of observations of the random sample $D=\{C_1, .., C_m\}$ the states of each variable belonging to U are given, the following theory holds if it is assumed that all the parameters are independent. Let us define B_{sh} as any random sample generated by a Bayesian network B_s, and r_i the number of states of a generic variable x_i; we will define the combination of states of a set of variables:

$$q_i = \prod\nolimits_{x_l \in \Pi_i} r_l \tag{4}$$

where $\prod i$ is the chosen set of variables. Let θ_{ijk} denote the probability that the generic variable is observed to assume one of its states k (x_i=k), given $\prod i$ = j for i limited between 1 and n, while j is limited between 1 and q_i and k between 1 and r_i. In addition we call:

$$\theta_{ij} = \cup_{k=1}^{r_i} \{\vartheta_{ijk}\},\ \theta_{B_s} = \cup_{i=1}^{n} \cup_{j=1}^{q_i} \{\vartheta_{ij}\}$$

and we suppose that each variable set θ_{ij} has a Dirichlet distribution:

$$p\left(\theta_{ij} \mid B_s^h, \xi\right) = c \cdot \prod_{k=1}^{r_i} \vartheta_{ijk}^{N'_{ijk}-1} \tag{5}$$

where c is a normalization constant, N'_{ijk} are the multinomial parameters of that distribution, limited between 0 and 1, finally ξ is the observed evidence. Eq. (5) can also be expressed in its explicit form using the gamma function Γ (Evans et al., 1993):

$$p\left(\theta_{ij} \mid B_s^h, \xi\right) = \frac{\Gamma\left(\sum_{k=1}^{r} N'_k\right)}{\prod_{k=1}^{r} \Gamma(N'_k)} \cdot \prod_{k=1}^{r_i} \vartheta_{ijk}^{N'_{ijk}-1} \tag{6}$$

Thus, if N_{ijk} is the number of observations in the database D in which $x_i=k$ and $\prod i=j$ we are able to update that distribution:

$$p\left(\theta_{ij} \mid D, B_s^h, \xi\right) = c \cdot \prod_{k=1}^{r_i} \vartheta_{ijk}^{N'_{ijk}+N_{ijk}-1} \tag{7}$$

Eq. (7) is applied to each case belonging to that database. Exploiting the properties of Dirichlet distributions, we can compute the probability that $x_i=k$ and $\prod i=j$ in the next case to be seen in the database C_{m+1} (but not observed yet) as:

$$p\left(C_{m+1} \mid D, B_s^h, \xi\right) = \prod_{i=1}^{n} \prod_{j=1}^{q_i} \frac{N'_{ijk} + N_{ijk}}{N'_{ij} + N_{ij}} \tag{8}$$

where:

$$N'_{ij} = \sum_{k=1}^{r_i} N'_{ijk}, \quad N_{ij} = \sum_{k=1}^{r_i} N_{ijk}$$

In the case of missing data in the database D, the "EM learning" algorithm can be applied (Lauritzen, 1995).

After learning the probabilities from a database D, it could be necessary to add other information from further empirical data. This can be carried out using the "sequential updating" method, that is a procedure to modify the network parameters over time in order to improve its performance. This method works by modifying the multinomial parameters of the Dirichlet distributions under the assumption of parameter independence. With the term "experience", we mean quantitative memory which can be based both on quantitative expert judgment and past cases (Spiegelalther and Laurintzen, 1990).

For the purpose of learning models relative to the preliminary design of buildings, the prior parameters in the first version of the network can be set with a particular equivalent sample size (by tuning the values N'_{ijk}), after which more data are added using the same procedure, starting from the new equivalent sample size and Dirichlet parameters, independently from the numerousness of the first dataset. As regards the equivalent sample size relative to the first learning procedure, the larger its size, the greater is the confidence in the previous parameter estimates and the slower the change due to adapting to new data. This technique is also valid with missing data.

The procedure described in the next paragraph is intended to provide a generally valid method, to find the optimum ratio between the equivalent sample size and the added empirical database. As further detailed in 4.1, this procedure requires an iterative adaptation of the parameters, until two quality indices of the network are optimized: sensitivity and case-based reasoning.

As an alternative, probabilities can be derived by deterministic relations. As required by the subdivision of variables into discrete intervals, this kind of algebra deals with real intervals (Alefeld and Herberg, 1983). In such a case, assuming that $[a_1, b_1)$, $[a_2, b_2)$ … $[a_n, b_n]$ are a set of contiguous intervals in the real field, the variables domain will be defined as:

$$X_i = \{[a_1, b_1) : p_1, [a_2, b_2) : p_2, .., [a_n, b_n] : p_n\} \tag{9}$$

In other words the probability distribution of a generic variable of the Bayesian model will be defined as:

$$\{P(x_i \in [a_1, b_1) = p_1); (x_i \in [a_2, b_2) = p_2), .., (x_i \in [a_n, b_n] = p_n)\} \tag{10}$$

This kind of interval subdivision will be assigned to each variable, both of the "parent" and "child" type. However, only the distribution of the parent variables is known and not that of the child variables. Subsequently a mathematical expression linking the state of each child to that of their parent variables can be used to compute the probability distribution of child variables (Hugin, 2008): at this juncture, a number of samples within each (bounded) interval of the parents are generated (generally according to the Monte Carlo Simulation method). Each of these samples will result in a "degenerated distribution" for the child node with each distribution corresponding to a given state for the parents. The final distribution assigned to the child node is the average over all the generated distributions. This amounts to counting the number of times a given child state appears when applying deterministic relationships to the generated sample.

4. Causal modelling for the preliminary design of buildings

4.1 Description of the general procedure

The general procedure suggested (and applied to a real case in the following) for setting up Bayesian models for the preliminary design of buildings, involves the following steps (Fig. 3):

1. decomposition of all the physical phenomena governing the technology into a set of several simpler processes, which describe their qualitative features through graphical structure representations (linked nodes) and their validation from a semantic point of view;
2. combination of the sub-networks developed on item no. 1 through interface variables into Object Oriented Networks in order to obtain only one whole model;
3. verification of the semantics of this whole model by technology experts, checking that the arrangement aggregation has not determined a lack of meaning;

4. formulation of a simplified release of available analytical methods to work out a first estimation of conditional probabilities;
5. preliminary updating of these conditional probability tables with raw data collected from simulations or field tests;
6. first evaluation of the quality of the network in step no. 5, through sensitivity analyses and case-based validations;
7. iterative refinement of parameters, adding further empirical information, while repeating again items no. 5 and 6.

The application of this 7-step procedure provides the following benefits:

1. explicit representation of all the complex phenomena involved in the building's behaviour through the graphical part of a Bayesian Network;
2. exploitation of both simplified relationships and experimental data to work out reliable (and validated) conditional probability networks;
3. production of a friendly simulation tool, which can act as an expert system in support of professional architects.

Validations in steps 1 and 3 can be performed by comparing the meaning of the qualitative causal relationships represented by the elementary networks in Fig. 1 with the real role played by each variable affecting the technology under development. The first probability learning from deterministic relations in step no. 4, is useful to estimate the parameter N'_{ijk} mentioned in paragraph 3.4 (assessing theoretical knowledge), while the raw data in step no. 5 add further knowledge to estimate the parameters N_{ijk}. Finally, network quality evaluation in steps no. 6 and 7 requires the use of sensitivity analysis and case-based reasoning. In general, any Bayesian network contains a high level of information if it is sensitive to changes in parameters and if it does not produce even probability distributions.
For the purpose of model development, the selection among several networks with different values of conditional probabilities is required, each generated from a different probability elicitation, corresponding to a different ratio between theoretical and empirical knowledge. For that purpose, sensitivity to changes in parameters is applied. This method can be exploited in order to find the best ratio between different amounts of experience and theoretical knowledge. The best solution is supposed to be the one giving back the sharpest probability distribution on each variable of interest. Entropy is the metric used to measure the variables' level of information (in the rest of this paper shortened to LOI): its lowest extreme is zero and corresponds to the maximum level of certainty; therefore the final aim is to minimize entropy. This is defined as (Korb and Nicholson, 2004):

$$H(x) = -\sum_{k \in x} P(k) \cdot \log_2(P(k)) \tag{11}$$

being the summation on k carried out for each possible state of the query node.

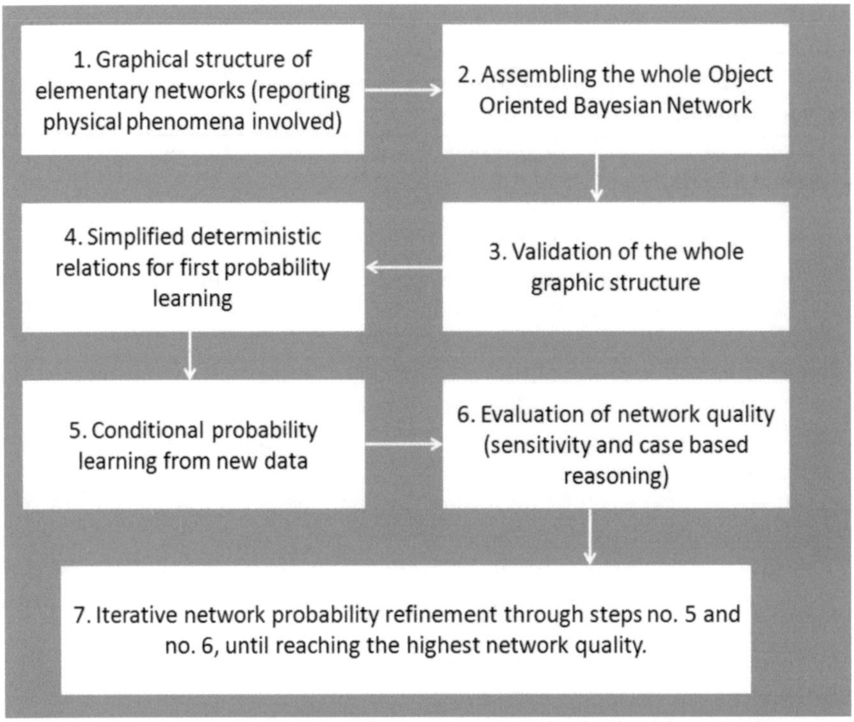

Fig. 3. Overall procedure for developing Bayesian networks for building design.

Entropy can also be computed for probability conditioned to some kind of evidence, having in this case P(x | E), where E is the evidence.

These metrics easily indicate the best solution, because if the varying of parameters through adding new data does not produce improvements in the network, it means that a flat point has been reached, which is also the best that can be obtained with such a configuration and no further improvements are allowed.

Accuracy will be evaluated by running the produced Bayesian networks (with different ratios between theoretical and empirical knowledge) on a set of test cases in order to find which one gives back the highest number of correct predictions or inferences (Korb and Nicholson, 2004). Input variables will be set both on average and extreme values, in order to generate meaningful test case studies, thereby performing a case-based reasoning on a set of observations different from the ones previously used for probability learning.

	Case a: RTE = 10/1	Case b: RTE = 1/1
Probability Distribution	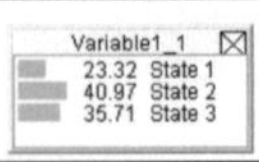	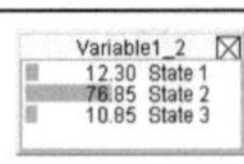
LOI	1.54	1.01

Table 1. Computation of LOI for two variables with different RTE.

4.2 The case study: glazed saddlebacked roofponds

As an example of application of the methodology described in subsection 4.1, a model for the rough sizing of glazed saddlebacked roofponds will be developed. Roofponds are mainly targeted to one and two-storey buildings located at average latitude climates, to provide cooling and heating loads necessary for air-conditioning (Marlatt et al, 1984). Throughout a normal year and in these specific types of dwellings, with outdoor temperatures ranging from 0°C to 46°C, roofponds allow inside temperatures to be maintained at between 20°C and 28°C with no conditioning.

"Roofponds" are a form of high-mass construction systems (Stein and Reynolds, 2000): as they require only the roof to be massive, they allow for considerable design freedom below, both in walls and fenestration. This strategy uses sliding panels of insulation over bags of water; panels slide open on winter days to collect sunlight and open again on summer nights to radiate heat to the sky when the ponds are used for cooling. The first roofponds to be tested were flat, usually used in warmer, less humid areas.

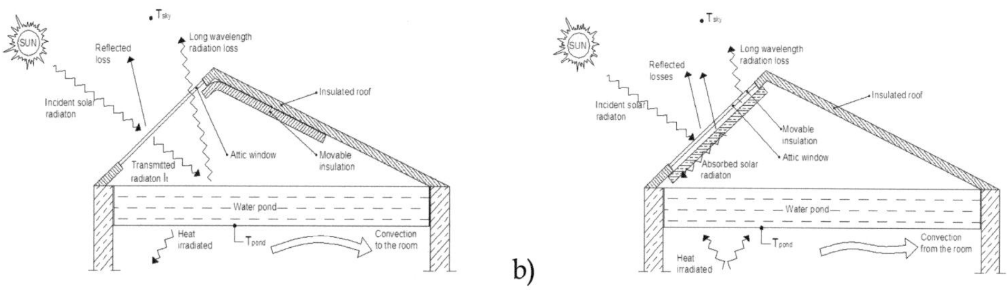

Fig. 4. Daytime glazed saddlebacked roofpond behaviour in winter (a) and summer (b).

Recently, a branch of research has concerned experiments on glazed saddlebacked roofponds, which are specifically designed for cooler climates (Fernández-González, 2003). This system consists of a "ceiling pond" under a pitched roof (to resist snowfalls), conventionally insulated on the north side, and with clear insulated glass on the south slope to collect solar energy (Fig. 4). For summer periods a movable insulating device is used to cover the glazed window and prevent solar gains inside the attic.

Glazed saddlebacked roofponds have even been tested in Muncie, Indiana, in order to show that they are able to shift average internal temperatures closer to the comfort range, increasing them during the winter and decreasing them during the summer. The smoothing of temperature swings during both seasons is useful not only for reducing HVAC average consumption, but also for increasing internal comfort. The most surprising effect is registered for the increment in the minimum extreme temperatures during the coldest month, which are responsible for the greatest amount of fuel required by the HVAC system (Fernández-González, 2003).

4.3 Analytical models for preliminary probability quantification

Carrying out step no. 4 of Fig. 3 requires the availability of deterministic relations towards a first probability learning. In this paragraph we show how technology developers can work out simplified relations even when starting from complex simulation models.

A finite difference approach was used for accurate transient thermal simulations of roofponds (Lord, 1999; Fernandez-Gonzalez, 2004; Fernandez-Gonzalez, 2003). It predicts temperature courses inside roofpond equipped buildings, given external climate and occupancy schedules as boundary conditions. The finite difference method solves the one-dimensional unsteady equation of conduction:

$$\frac{\partial T}{\partial t} = \alpha \cdot \frac{\partial^2 T}{\partial x^2} \tag{12}$$

where α is the diffusivity and T is temperature varying with time t and position x. The first step is the domain subdivision in small elements connected through nodes. Subsequently the energy balance given by (Athienitis and Santanouris, 2002) must be solved at all nodes:

$$C_i \cdot \frac{\partial T_i}{\partial t} = Q_i + \sum_j U_{ij} \cdot (T_j - T_i) \tag{13}$$

where i is the node of interest and j is any other node linked in some way to the previous one through a mean having thermal conductance equal to U_{ij}. Q_i is the heat generated at the level of the node of interest.

Approximate solutions of the finite difference model above were worked out for the preliminary learning of some elementary Bayesian networks. For instance, writing Eq. (13) for the nodes representing internal air and roofpond and solving that system of two differential equations, the general solutions for internal air and roofpond temperature courses are written in the form (Naticchia et al., 2007):

$$T = C_1 + C_2 \cdot e^{-At} \tag{14}$$

where C_1, C_2 and A are constant terms. Neglecting the time dependent term (the second term of the sum), but considering only the long-term behaviour of pond and internal air temperatures and eventually rearranging those equations in order to explicitly express the two temperatures, the average long-term expected values of internal air and roofpond temperatures are obtained. This equation can easily be used for preliminary probability learning, which means neglecting the building's transient behaviour and approximating it with its long-term forecast.

Once the average values are known, the temperature swings must be computed, according to (Balcomb et al., 1980). The basic equation for the computation of swings is given by:

$$\Delta T(swing) = \frac{0.733 \cdot A_{tot} \cdot q_s}{DHC} \tag{15}$$

where ΔT is the temperature gradient, DHC the total diurnal heat capacity, A_{tot} the sum of the size of all collection surfaces and q_s is the total amount of solar heat gains through south oriented windows. These equations have been implemented in the model to estimate average temperatures and corresponding swings in both seasons following any choice of input parameters.

4.4 Example of model development

The Bayesian model for glazed saddlebacked roofponds was built by combining, through the OOBNs tool, several networks which simulate the thermal behaviour of roofpond equipped buildings with other networks for the estimation of the parameters acting as inputs to the previous ones, leading, finally, to one decision network. The model was based on the platform provided by the program software Hugin Expert™.

In particular, and referring to Fig. 5, the whole model was organized according to three different levels:

- the first level includes seven elementary networks to compute solar heat gains in both seasons, split into attic and main room contributions, besides the needed climatic inputs;
- the eight second level networks compute the internal average air temperatures and swings for both the roofpond building and its benchmark, when operating in heating and cooling modes;
- the third level is made up of one decision model, solving the problem of choosing the best combination of input variables to optimize the project, finding the best trade-off between benefits pursued in the cold and warm season.

4.4.1 Development of the first level

Elementary networks no. 4, 5, 6 and 7 estimate solar gains through the attic and south oriented windows respectively, in the form of mean values in both seasons for the roofpond equipped building and its benchmark. The basic relation implemented is as follows:

$$I_{\text{int}} = I_{ext} \cdot SHGC \tag{16}$$

where I represents irradiation and SHGC is the solar heat gain coefficient (Athienitis and Santanouris, 2002). Networks no. 1, 2 and 3 estimate the input parameters necessary for the computation above, such as average sky temperature and emissivity, irradiation and its angle of incidence on external windows, according to methods suggested by available literature (Balcomb et al, 1980; ASHRAE, 2001), also neglecting non-south oriented window contributions for solar gains. These parameters vary according to climate and building features.

4.4.2 Development of the second level

Four elementary networks of level no. 2 were devoted to computing the average internal temperatures in the attic and internal rooms of the roofpond building and its benchmark in both seasons. In particular, networks no. 8 and 9 (the second depicted in Fig. 6) estimate pond and internal temperatures in summer and winter respectively for the roofpond equipped building, based on outputs from the first level networks and other user input parameters. Winter average long-term internal temperatures were computed according to analytical relations put in the form of eq. (14), which has the advantage of being arranged in explicit form. It is a function of heat gains (pond, solar and internal) and losses, mainly due to envelopes and air ventilation (Lord, 1999).

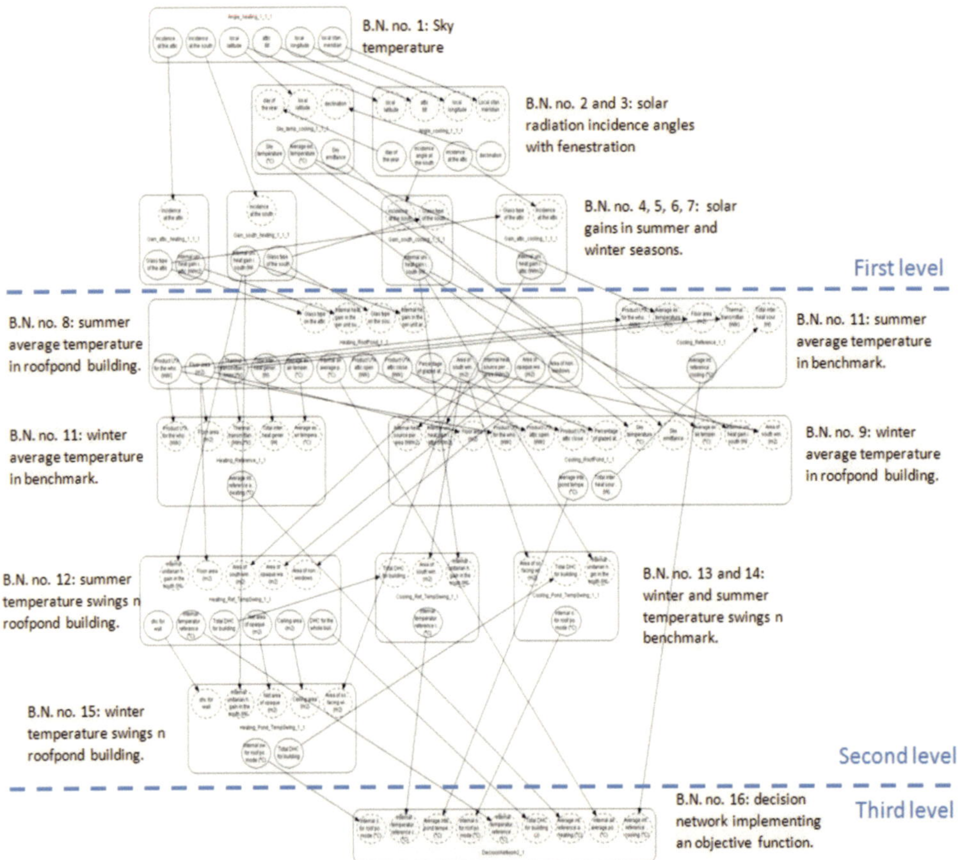

Fig. 5. OOBN representation the whole Bayesian model (a total of 16 networks made up of 219 variable nodes), where only interface nodes are visible.

The case of network no. 8 regarding summer behaviour was slightly more complicated, because the application of thermal balance equations to the main room and attic of the roofpond equipped buildings leads to a system with no explicit variables: in this case pond temperature is affected not only by its exchange with the interior but also with the sky. Hence, an explicit equation built on many statistical observations generated by a system including both exchanges between the pond and the interior and between the pond and the exterior. This statistical empirical equation was used to estimate pond temperatures in function of the sky and external conditions. Thermal exchanges with the sky and the interior were inferred from this equation. Validation showed that the model is accurate with an error never exceeding 10%, which was considered as acceptable for preliminary learning. Similar approaches were used to build networks no. 10 and 11, relative to benchmarks, which are simpler given the absence of the roofpond. The other networks are relative to temperature oscillation estimation, in accordance with the theory related to eq. (15) (Balcomb et al., 1980).

4.4.3 Development of the third level

Considering that optimizing design choices for winter periods does not guarantee that the same holds for the summer, at this level one large network implementing an objective function to be maximized was set up. It considers two contributions: economic savings deriving from winter benefits that the roofpond determines with respect to its benchmark and from summer benefits. The general form of the objective function is given by:

$$EES = EES_h + 2.5 \cdot EES_c \tag{17}$$

where expected energy savings (EES) in the cooling mode (EES_c) are more important than those in the heating mode (EES_h), because of the difference in fuel and electricity prices. Each term includes energy saving derived from shifting the average temperatures closer to the comfort value and reducing the temperature oscillations around the mean; in addition climate influence and the whole thermal inertia of the building under development are considered. Further details about the model can be found in (Naticchia et al., 2007)

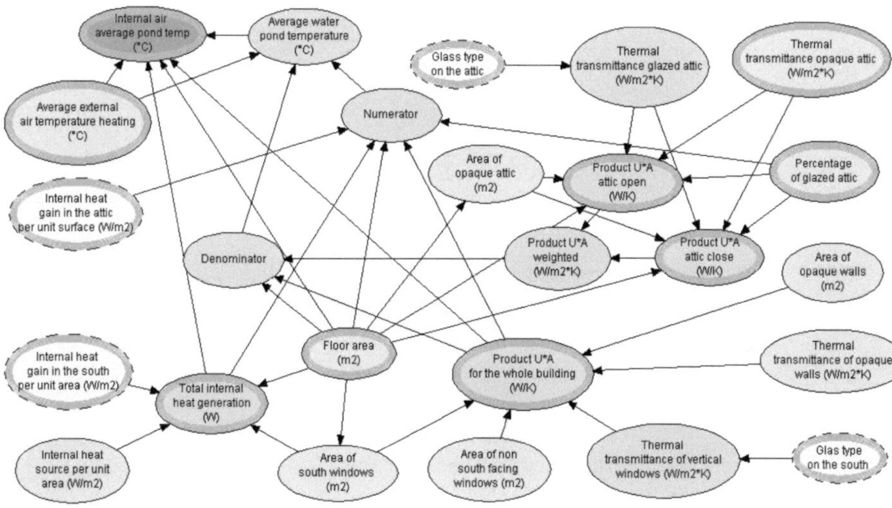

Fig. 6. Elementary network no. 9.

4.5 Model refinement and validation

After having implemented the analytical relations into the networks for the approximate relationships in paragraph 4.4, they were subsequently refined using empirical data and by monitoring this process through sensitivity analysis and case-based reasoning, according to the procedure suggested in paragraph 4.1. This paragraph will show some examples of how this could be performed, as it can be applied to any model under development. It has the practical advantage of using all the observations which derive from experiments and simulations worked out by complex software tools.

In the particular case of roofponds, the finite difference model described in paragraph 4.3 and based on eq. (13) was implemented on a wide set of real cases to build numerous databases used to implement the sequential updating (paragraph 4.1) on the preliminary conditional probability tables, derived from the equations in paragraph 4.4. This sequential updating refines, at each step, the parameters of the Dirichlet distributions underlying the networks, until it optimizes the two quality indices described in paragraph 4.1. Defining as "theoretical experience" the *a priori* information inserted through simplified analytical laws and "experimental experience" the information deriving from further observations, this method allows the optimum ratio to be found between the importance given to the first and second samples with the aim of learning the Dirichlet parameters.

This procedure must be applied to each elementary network included in the model. For instance, let us consider the Bayesian Network no. 6 of level 1 (Fig. 7), relative to the computation of the average hourly SHGC value for the attic window of the roofpond building in winter for direct and diffuse radiation coming from the exterior.

After the first learning, based on the use of approximate analytical relationships, a database was generated through accurate simulation. Three values of RTE were then tried: RTE = 10/1; RTE = 3/1; RTE = 1/1. Tab 2 shows LOI values computed in function both of each RTE value chosen and of evidence imposed to parent variables. It can be noticed that RTE = 1/1 gives back the LOI values lower than the initial theoretical model, meaning that entropy is getting closer to zero and that probability distributions are more expressive.

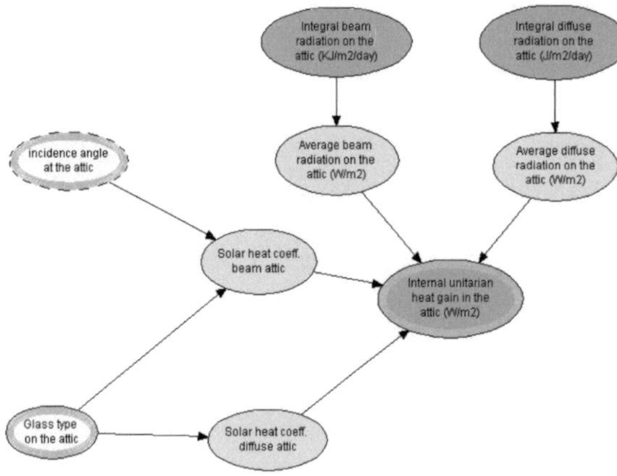

Fig. 7. Graphical structure of elementary network no. 6.

In addition, Tab. 3 shows that adding empirical evidence with RTE = 1/1 redirects the values computed by the network towards the real values. In order to perform this case-based reasoning it is necessary to divide any dataset into two parts: the first (and bigger) is generally used for model learning and the second (smaller) is generally used to compare the data provided by the network with the ones recorded by testing or simulation. In this case 90% of the data were used for model learning and 10% for model validation according to case-based reasoning.

It was evident that adding further experience to this elementary network did not improve the output of the network, which means that it reached a flat point, beyond which no improvements can be obtained at LOI level. Case-based reasoning must be evaluated on the network's capability to estimate the correct value for output variables: to that end the interval or contiguous intervals with the highest probability values must be checked. Fig. 8 shows how probability propagation algorithms update output variables probability according to evidence on input variables: black bars are relative to evidence assignment, while the other probability values are timely and automatically recomputed by the network according to those inputs.

Query node	*Evidence*	*LOI for theoretical model*	*LOI for RTE = 1/10*	*LOI for RTE = 1/3*	*LOI for RTE = 1/1*
Q_{auh}	None	2.4	2.4	2.42	2.43
Q_{auh}	T_a="0 to 50", BIA = 9000 to 12000	1.08	0.63	0.88	0.9
Q_{auh}	T_a="0 to 50", BIA = 3000 to 6000"	0.99	1.06	1.06	0.87

Table 2. Sensitivity analysis.

Query	*Evidence*	*Results from theoretical model*	*Results from RTE = 10/1*	*Results from RTE = 1/1*	*Real value*
Q_{auh}	T_a="0 to 50", BIA = 3000 to 6000"	70 to 90 (91.6%)	70 to 90 (92.9%)	70 to 90 (97.04%)	83.8
Q_{auh}	T_a="0 to 50", BIA = 3000 to 6000"	70 to 90 (49.9%) 90 to 110 (49.9%)	70 to 90 (49.9%) 90 to 110 (45.4%)	50 to 70 (25.2%) 70 to 90 (49.09%)	69.4

Table 3. Case-based reasoning for accuracy survey.

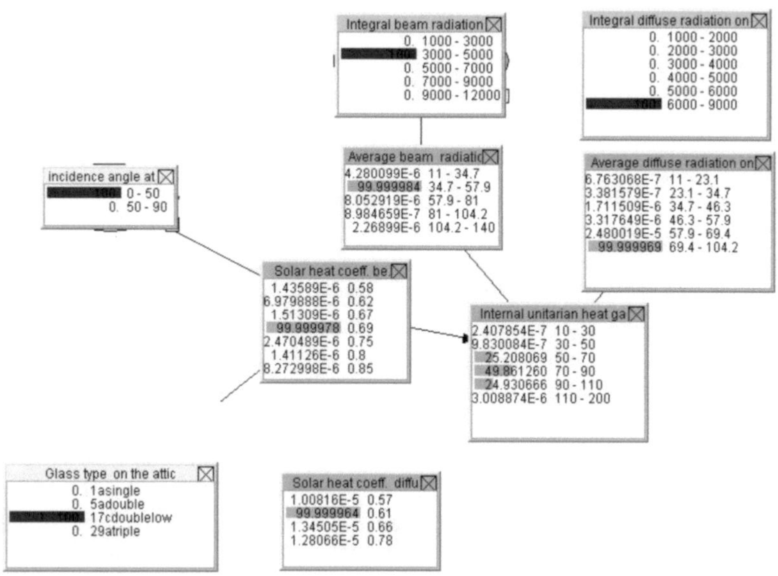

Fig. 8. Example of probability updating for elementary network no. 6.

The same procedure was applied for the other networks of the first and second levels. For instance, network no. 11 is optimized by assigning RTE = 1/1; networks no. 8 and 10 were optimized by assigning RTE = 1/5. In general, this method gives back the amount of experimental data which is sufficient to guarantee good estimations and it can be applied every time technology developers are able to perform controlled tests or software simulations.

5. Applications of the final model

5.1 Overview of the model's applicability

At this level, each sub-network constituting the overall explicit whole network has successfully undergone a validation procedure. Three basic practical applications of Bayesian reasoning within the architectural design profession are (Naticchia et al., 2007):

- determination of the best design solution among several possibilities;
- optimal sizing of building parameters;
- approximate sizing under conditions of uncertainty.

In the first case, designers could be supported in the process of discerning the best choice among several likely building configurations using the model's energy efficiency-based objective function.

The second aspect is typical of a rough-sizing process: often, before a designer may size any building parameter, the viable choices on the market need to be investigated in terms of various issues. This probabilistic model allows bottom-up reasoning, that is to say, querying the objective function in order to derive the proper values that yield the highest utility for the particular issue being considered.

Finally, when there is no certain knowledge about some given parameters (e.g. type of glazing for south facing windows), the designer should be able to make inferences in the case of uncertain distribution over several values and give back a probability distribution for the objective function that is useful for carrying out an exploration of the two previously mentioned design aspects.

5.2 First application case: choice of the best design solution

Suppose a designer specifies a glazed, saddleback roofpond application for a one-storey 100 m^2 residence located in Salt Lake City, Utah. All the architectural features have been determined except for the total area of the solar attic window. Two available options consist in an area equal to either 50% or 35% of the total floor. All the input values were inserted into the model: climatic parameters in accordance with Salt Lake City characteristics; "5a double clear" type of glass; south facing window area equal to 5 m^2; total area of the other windows equal to 35 m^2. Thermal transmittance of walls and attic respectively equal to 0.035 $W/(m^2 \cdot K)$ and 0.02 $W/(m^2 \cdot K)$; "5/8 in gypsum panel" installed as a ceiling; "4 in thick brick" used for walls; 0.3 m deep pond. Fig. 9 depicts some of the results obtained from the two assumed cases. It can be noticed from temperature diagrams that in the second case (attic window area equal to 35% of the floor) the temperature difference between the roofpond building and the benchmark working in heating mode is higher than in the second, meaning that it brings higher positive benefits in winter. On the contrary, the second case is less advantageous in the summer. Similar remarks could be confirmed for temperature swings. Given this ambiguous situation, the use of the objective function (in BN no. 16) with the computation of the expected utility, makes us conclude that the second option is better. This is likely to be due to the lower winter losses caused by the smaller area of glazing.

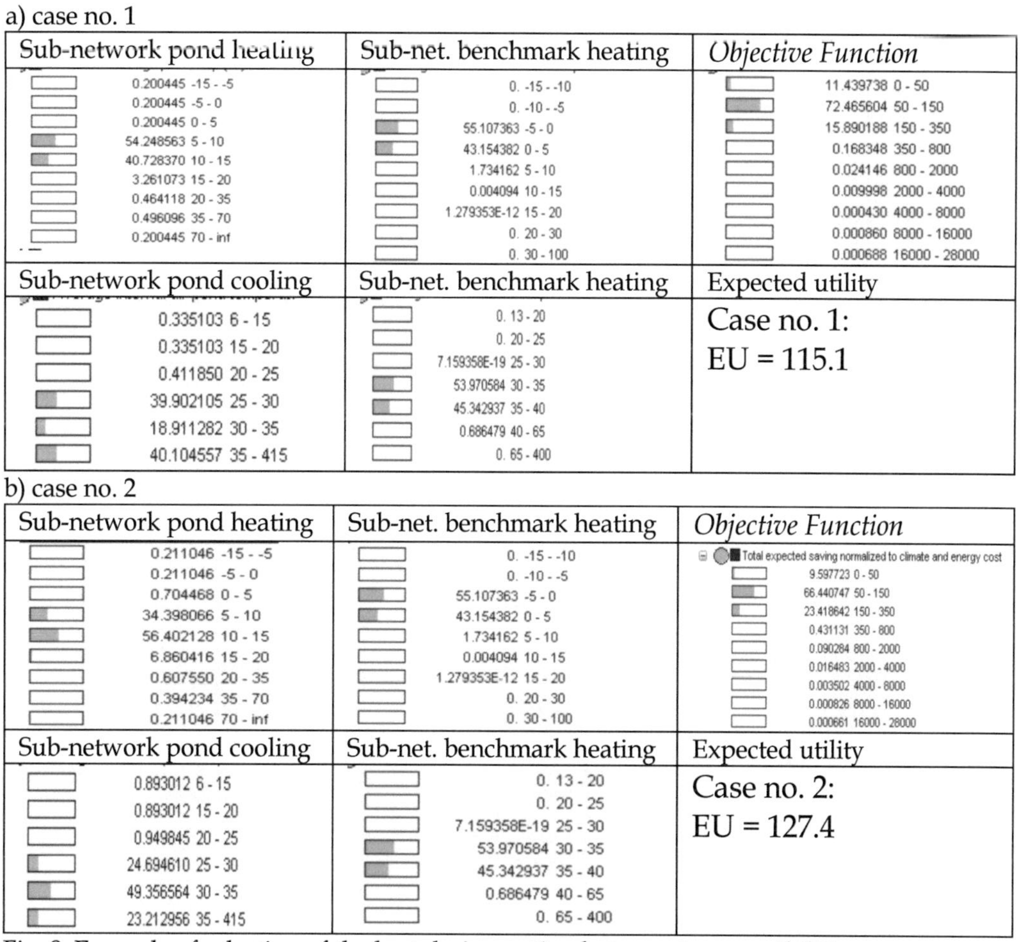

Fig. 9. Example of selection of the best design option between two possibilities.

5.3 Second application case: optimal sizing of building parameters

As a second scenario let us assume that a designer is employing a roofpond strategy in a passively conditioned building made up of a detached single-family dwelling of 200 m^2 in Seattle, Washington. The designer is free to determine the area of south-facing glazing. Some input parameters include: thermal transmittance of walls equal to 0.04 $W/(m^2 \cdot K)$, while for the attic equal to 0.02 $W/(m^2 \cdot K)$; area of non-south facing windows equal to 20 m^2; the glazed window attic area being 0.35 times the floor area; ceiling of "1/2 in gypsum panel" type, wall of "4 in thick hollow brick fired clay" type and 0.3 m deep pond on the roof. The first case in Fig. 10 depicts the results obtained at the level of the Objective Function and the south facing window area, that was left free to vary. The probability distribution relative to the area of south windows considers the first interval as the only one not to be chosen, while the others cannot be excluded at this level.

If the objective function value is maximized by the introduction of evidence in its highest interval, the south facing probability distribution changes accordingly: it prompts an optimum value between two intervals having approximately the same probability, that can be interpreted as being on average 7 m^2. Moreover, in order to show how this model is sensitive to the choice of the decision variable, the possibility that the total area of non-south facing glazing is changed from 20 to 35 m^2 is considered, leaving all the other parameters unchanged (case no. 2 in Fig. 10). The probabilistic model has the ability to adjust itself: the objective function value now suggests that it would be more opportune to increase the south facing glazing to between 7 and 12 m^2.

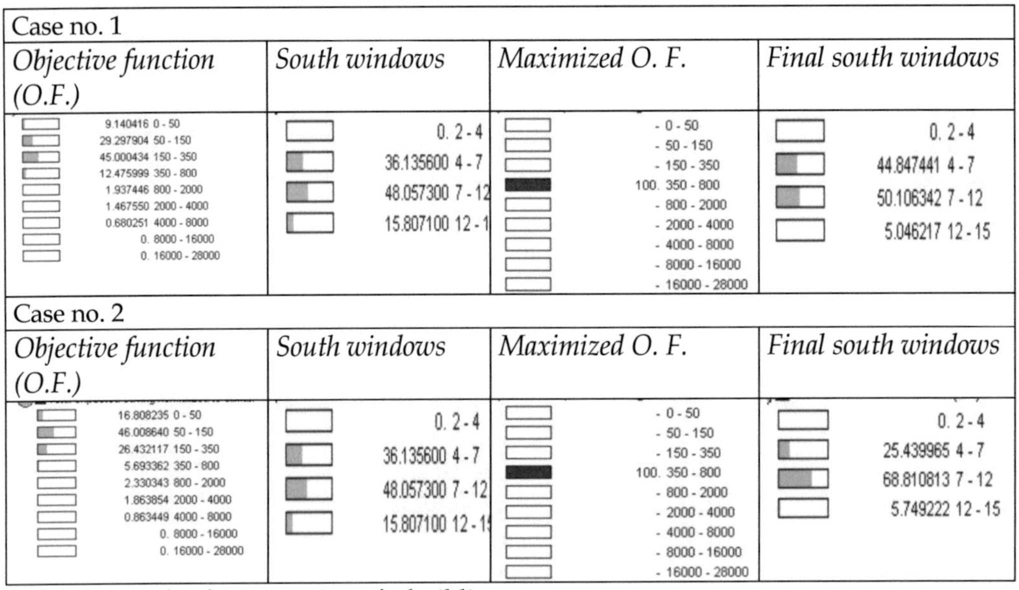

Fig. 10. Example of optimization of a building parameter.

5.4 Third application case: approximate sizing under conditions of uncertainty

In the third scenario a designer for a one-storey residence in Seattle, Washington is implementing the glazed saddleback roofpond application and needs to determine the most appropriate type of glazing for all the building windows, while the one relative to skylights of the solar collection space above the roofpond was chosen to be "5a double clear". For the other windows the designer must choose between "5a double clear" and "17c double low-emission" glazing. In addition to this choice, there is uncertainty regarding the optimum area of the south facing glazing, which will be determined after the preliminary design in accordance with other needs.

At this juncture, it will suffice to approximate that there is a 30% probability of choosing a total area of 5.5 m^2 and a 70% chance of choosing a total area of 9.5 m^2. Fig. 11 depicts the likelihood distribution inserted as the variable "area of south windows", which lets the network reason under uncertainty.

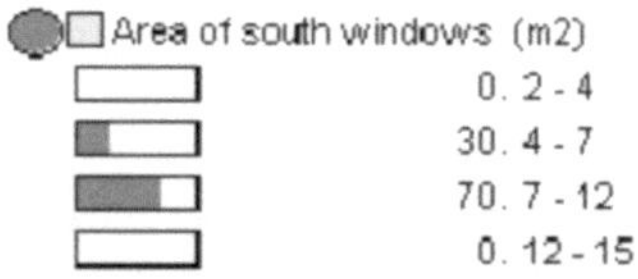

Fig. 11. Likelihood distribution inserted in the network

Other decision variables are: floor area of 200 m^2; area of non-south facing windows of 20 m^2; ratio of glazed attic surface out of floor equal to 0.35.

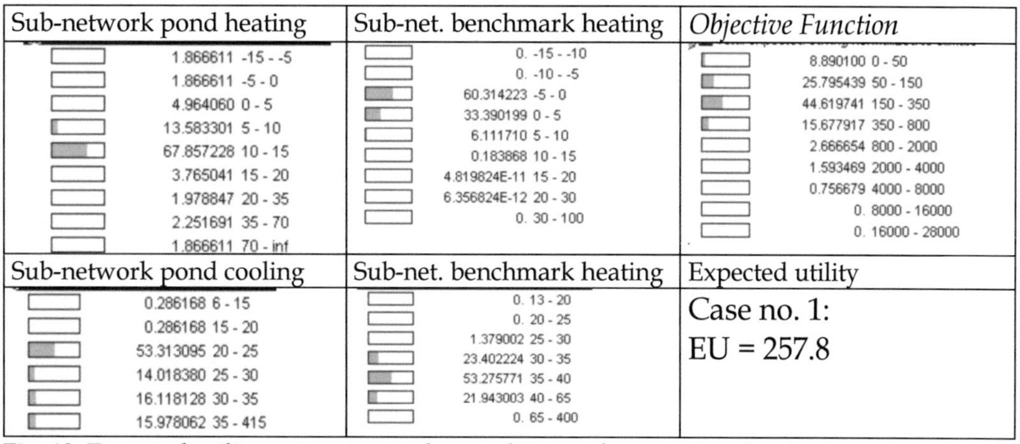

Fig. 12. Expected utility estimation under conditions of uncertainty for the 1st option

In the case of Fig. 12 all the windows are assigned as the "5a double clear" type. The consequent result is an expected utility of 257.8. In the second case (implemented in a way similar to the first case) all the building windows were assigned a "17c low emission" glass type, which gave back an expected utility of 293.25. The results suggest that the "17c double low emission" glazing would work better in terms of thermal performance. This is probably due to the thermal transmittance of the glazed windows utilized in the second case, which is lower than that found in the first case (2.89 versus 3.25 $W/(m^2 \cdot K)$).

6. Conclusion

Bayesian Networks are a powerful tool to build accurate models for supporting professional architects in the preliminary design phase of buildings. Their friendly graphical structure is easy to interpret and their learning algorithms are capable of reproducing even non-linear relationships relative to complex phenomena, which involve a number of different sub-processes. This chapter deals with a procedure that research scientists can adopt to develop Bayesian networks for modelling the behaviour of new technologies and favouring their fast spread into the market, simplifying the task of designers who have to make design choices based on estimated building parameters. Developing such models involves discerning which kind of knowledge and data must be inserted in order to obtain a reliable network. For that purpose, this study suggests breaking down the whole process into sub-processes, each simulated by one elementary network, according to the OOBNs approach. Causal

relationships among variables in every elementary network are then learnt, starting from both simplified analytical relationships and from experimental data or those deriving from numerical simulations: the procedure proposed in this chapter suggests how to measure the level of quality of the network, which can gradually be tuned by changing the importance of subjective relationships, with respect to the data.
Once the model is built, it has the benefit of performing both predictive and diagnostic reasoning, as well as reasoning under conditions of uncertainty. Therefore it is capable of supporting architects in several single or combined basic tasks: the determination of the best design solution among different available choices; the optimal sizing of building parameters; the approximate sizing under conditions of uncertainty. These applications include what could realistically be of interest for professional designers, who would use the Bayesian models as an expert system to drive them towards fast and accurate designing.

7. References

Alefeld, G. and Herzberg, J. (1983) Introduction to interval computation, Academic Press, New York, NY.

Athienitis A.K., Santanouris M., (2002) Thermal Analysis and Design of Passive Solar Buildings, James and James Science Publisher, London, 2002

Balcomb J.D., Barley D., McFarland R., Perry J., Wray W., Noll S., Passive Solar Design Handbook, vol. 2, Passive Solar Design Analysis, US Department of Energy, Washington, DC, 1980.

Boborow, D. G., (1984) Qualitative reasoning about physical systems, MIT Press, Cambridge, UK.

Charniak, E., Bayesian networks without tears, Artificial Intelligence magazine, vol. 12, no. 4, pp. 50-63, 1991.

Evans, M., Hastings, N., Peacock, B., (1993) Statistical distributions, 2nd edition, John Wiley and sons, New York.

Fernández-González, A. Characterizing Thermal Comfort in Five Different Passive Solar Heating Strategies. Proceedings of the Architectural Research Centers Consortium 2003 Conference, April 10-12, Tempe, AZ.

Fernandez-Gonzalez A., RP_performance: a design tool to simulate the thermal performance of skytherm North roofpond systems, in: Proceedings of the 33rd American Solar Energy Society National Conference, Portland, OR, July 9–14, 2004.

Heckerman, D. (1996) Bayesian Networks for knowledge discovery, in Fayyad, U.M., Piatetsky-Shapiro, G., Smyth, P., Uthurusamy, R., Advances in knowledge discovery and data mining, AAAI Press/MIT Press.

Hugin Expert, (2008) Hugin API – Reference Manual, version 7.0, Hugin Expert A/S.

Jensen, F. V., (2000) Bayesian Networks and decision graphs, Springer-Verlag, New York.

Koller, D., and Pfeffer, A., Object-oriented Bayesian networks, in 13th Conference on Uncertainty in Artificial Intelligence (Morgan Kaufmann, San Francisco, 1997), D. Geiger and P. Shenoy, Eds., pp. 302-313.

Korb, K. B., Nicholson, A. E., (2004) Bayesian artificial intelligence, Chapman & Hall/CRC edition.

Lauría, Eitel J.M.; Duchessi, Peter J.; A methodology for developing Bayesian networks: An application to information technology(IT)implementation, *European Journal of Operational Research*, v 179, n 1, p 234-252, May 16, 2007.

Lauritzen, S. L., The EM algorithm for graphical association models with missing data, Computational Statistics & Data Analysis, vol.19, pp.191-201, (1995).

Lord D., An interactive,Web-based computer program for thermal design of roofponds, Ph.D. Thesis, California Polytechnic State University, College of Architecture and Environmental Design, Architecture Department, 1999.

Lucas, P.J.F., Bayesian network modeling with qualitative pattern, Artificial Intelligence, vol. 163, pp. 233-263, (2005).

Marlatt, W. P., Murray, K. A., Squier, S. E., (1984) Roofpond systems, Energy Technology Engineering center for the U. S. Department of Energy (contract DE-AM03-76SF00700), Canoga Park, California.

Nadkarni, S., Shenoy, P. P., A causal mapping approach to constructing Bayesian networks, Decision support systems, vol. 38, pp. 259-281.

Naticchia, B., De Grassi, M., (2001) Modelling environmental complexity for sustainable design practice (Ch, 9 pp. 135-160), in Towards Sustainable Buildings, N. Maiellaro, Kluwer Academic Publisher, pp. 135-160, Netherlands.

Naticchia B., Fernandez-Gonzalez A., Carbonari A., "Bayesian Network model for the design of roofpond equipped buildings", Energy and Buildings, vol. 39, 2007, pp. 258-272.

Stein, B., Reynolds, J. S., (2000) Mechanical and electrical equipments for buildings, John Wiley and sons, New York, 9th ed.

Pearl, J., (1988) Probabilistic reasoning in Intelligent Systems: networks of plausible inference, Morgan Kauffman ed., San Mateo, California.

Sadoddin, A., Letcher, R.A., Jakeman, A.J., Newham, L.T.H., A Bayesian decision network approach for assessing the ecological impacts of salinity management, Mathematics and computers in simulation, vol. 69, pp. 162-176.

Spiegelalther, D., Laurintzen, S. L., Sequential updating of conditional probabilities on directed graphical structures, Networks, vol. 20, pp. 579-605.

Van Truong, Luu; Soo-Yong, Kim; Nguyen Van, Tuan; Stephen O., Ogunlana "Quantifying schedule risk in construction projects using Bayesian belief networks", International Journal of Project Management, v 27, n 1, p 39-50, January 2009.

Weidl, G., Madsen, A.L., Israelson, S., Applications of object-oriented Bayesian networks for condition monitoring, root cause analysis and decision support on operation of complex continuous processes, Computers and Chemical Engineering, vol. 29, pp. 1996-2009.

Wenhui, Liao; Qiang, Ji; Learning Bayesian network parameters under incomplete data with domain knowledge, Pattern Recognition, v 42, n 11, p 3046-3056, November 2009.

Wonnacott, T. H. and Wonnacott, R. J., (1990) Introductory statistics - 5th ed., John Wiley and sons, New York, pp. 582-615.

Zhang, H., Jarzabek, S., A Bayesian Network approach to rational architectural design, ,International Journal of Software Engineering and Knowledge Engineering, v 15, n 4, August, 2005, p 695-717.

Zhiqiang, Cai; Shudong, Sun; Shubin, Si; Bernard, Yannou; Maintenance management system based on Bayesian networks, *2008 International Seminar on Business and Information Management*, ISBIM 2008, v 2, p 42-45, 2009.

Zhu, J.Y., Deshmukh, A., Application of Bayesian decision networks to life cycle engineering in Green design and manufacturing, Engineering applications of Artificial Intelligence, vol. 16, pp. 91-103.

19

Bayesian networks methods for traffic flow prediction

Enrique Castillo
University of Cantabria
Spain

Santos Sánchez-Cambronero and José María Menéndez
University of Castilla-La Mancha
Spain

During the last decades, "Transport Demand" and "Mobility" has been a continuously developing branch in the transport literature. This is reflected in the great amount of research papers published in scientific magazines dealing with trip matrix estimation (see (Doblas & Benítez, 2005)) and traffic assignment problems (see (Praskher & Bekhor, 2004)). Current traffic models reproduce the mobility using several data inputs, in particular prior trip matrix, link counts, etc. (see (Yang & Zhou, 1998)) which are data only from a subset of the problem variables, and its size will depend on the available budget for the study being carried on. Among the problems faced for solving these models we can emphasize the high number of possible solutions which is usually solved by choosing the solution where the model results best fits real data. Nevertheless a model does not have to reproduce only the real data, but also must reproduce accurately all the variables. To this end, the aim of this paper consists of presenting two Bayesian network models for traffic estimation, trying to bring a new tool to the transportation field. The first one deals with the problem of link flows, trip matrix estimation and traffic counting location and in the second one we propose a Bayesian network for route flow estimation (and hence link flows and $\mathcal{OD}$ flows) using data from plate scanning technique together with a model for optimal plate scanning device location. Since a Gaussian Bayesian network is used, these models allow us to update the predictions from a small subset of real data and probability intervals or regions are obtained to get an idea of the associated uncertainties. In addition dealing with data from the plate scanning approach we improve the under-specification level of the traffic flow estimation problem.

1. Some background on Bayesian network models

A Bayesian network (see, (Castillo et al., 1999)) is a pair $(\mathcal{G}, \mathcal{P})$, where $\mathcal{G}$ is a directed acyclic graph (DAG) defined on a set of nodes $\mathbf{X}$ (the random variables), and $\mathcal{P} = \{p(x_1|\pi_1), \ldots, p(x_n|\pi_n)\}$ is a set of n conditional probability densities (CPD), one for each variable, and Π_i is the set of parents of node X_i in $\mathcal{G}$. The set P defines the associated joint probability density of all nodes as

$$p(\mathbf{x}) = \prod_{i=1}^{n} p(x_i|\pi_i). \tag{1}$$

The graph $\mathcal{G}$ contains all the qualitative information about the relationships among the variables, no matter which probability values are assigned to them[1]. Complementary, the probabilities in $\mathcal{P}$ contain the quantitative information, i.e., they complement the qualitative properties revealed by the graphical structure.

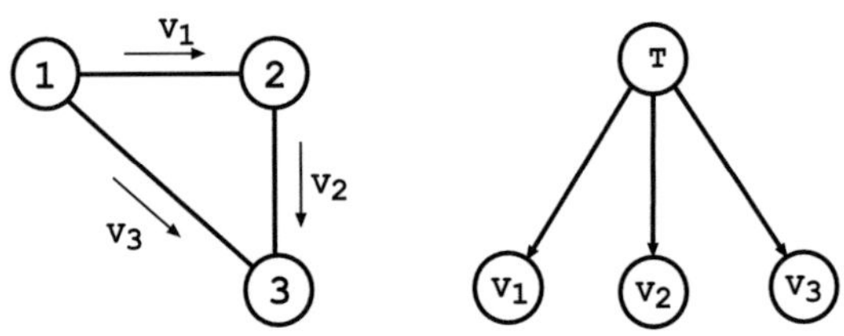

Fig. 1. A traffic network and its associated Bayesian network.

As an example of how a traffic network can be represented by means of $\mathcal{G}$ let us consider the simple traffic network in Figure 1. Assume that we have only the $\mathcal{OD}$ pair $(1,3)$ and two routes $\{(1,2),(3)\}$. Then, it is clear that the link flows v_1, v_2 and v_3 depend on the $\mathcal{OD}$ flow t, leading to the Bayesian network in the right part of Figure 1, where the arrows go from parents to sons. Note that link flow v_a has the t $\mathcal{OD}$ flow as a parent, and the t $\mathcal{OD}$ flow has v_a as son, if link a is contained in at least one path of such a $\mathcal{OD}$ pair.
Since a normal model is going to be used, the particular case of Gaussian Bayesian networks is presented next. A Bayesian network $(\mathcal{G},\mathcal{P})$ is said to be a Gaussian Bayesian network (see (Castillo et al., 1997a;b)) if and only if the joint probability distribution (JPD) associated with its variables X is a multivariate normal distribution, $N(\mu,\Sigma)$, i.e., with joint probability density function:

$$f(x) = (2\pi)^{-n/2}|\Sigma|^{-1/2}\exp\left\{-1/2(x-\mu)^T\Sigma^{-1}(x-\mu)\right\}, \tag{2}$$

where μ is the n-dimensional mean vector, Σ is the $n \times n$ covariance matrix, $|\Sigma|$ is the determinant of Σ, and μ^T denotes the transpose of μ.
In transportation problems, when some variables are observed, one needs to consider the other variables conditioned on the observations, and then the remaining variables change expected values and covariances. The following equations permit updating the mean and the covariance matrix of the variables when some of them are observed. They illustrate the basic concepts underlying exact propagation in Gaussian network models (see (Anderson, 1984)). These updating equations are:

$$\mu_{Y|Z=z} = \mu_Y + \Sigma_{YZ}\Sigma_{ZZ}^{-1}(z-\mu_Z), \tag{3}$$

$$\Sigma_{Y|Z=z} = \Sigma_{YY} - \Sigma_{YZ}\Sigma_{ZZ}^{-1}\Sigma_{ZY}, \tag{4}$$

[1] This allows determining which information is relevant to given variables when the knowledge of other variables becomes available. As we will see in section 2.3 this fact is the basis of the traffic counts location problem.

where **Y** and **Z** are the set of unobserved and observed variables, respectively, which allow calculating the conditional means and variances given the actual evidence, i.e. they are the updating equations for the means and variance-covariance matrix of the unobserved variables and the already deterministic values of observed variables, when the later have been observed.
Note that instead of using a single process with all the evidences, one can incorporate the evidences one by one. In this way, one avoids inverting the matrix $\mathbf{\Sigma_{ZZ}}$, which for some solvers can be a problem because of its size. Note also that the conditional mean $\mu_{Y|Z=z}$ depends on z but the conditional variance $\Sigma_{Y|Z=z}$ does not.

2. Traffic link count based method for traffic flow prediction

In this section a method for traffic flow prediction using Bayesian networks with data from traffic counts is presented (see (Castillo, Menéndez & Sánchez-Cambronero, 2008a)).

2.1 Model assumptions

In our Gaussian Bayesian network for traffic flow prediction using data from traffic counts, we make the following assumptions:

Assumption 1: The vector **T** with elements t_{ks} of $\mathcal{OD}$ flows from origin k to destination s, are multivariate normal $N(\boldsymbol{\mu}_T, \mathbf{\Sigma}_T)$ random variables with mean $\boldsymbol{\mu}_T$ and variance-covariance matrix $\mathbf{\Sigma}_T$.
At this point we have to note that the **T** random variables are correlated. In particular, at the beginning and end of vacation periods the traffic increases for all $\mathcal{OD}$ pairs and strong weather conditions reduce traffic flows in all $\mathcal{OD}$ pairs. This fact can be formulated as follows:

$$t_{ks} = \zeta_{ks} U + \eta_{ks}, \tag{5}$$

where ζ_{ks} are positive real constants, U is a normal random variable $N(\mu_U, \sigma_U^2)$, and η_{ks} are independent normal $N(0, \gamma_{ks}^2)$ random variables. The meanings of these variables are as follows:

- U : A random positive variable that measures the level of total mean flow. This means that flow varies randomly and deterministically in situations similar to those being analyzed (weekend period, labor day, beginning or end of a general vacation period, etc.).
- ζ : A column matrix which element ζ_{ks} measures the relative weight of the traffic flow between origin k to destination s with respect to the total traffic flow (including all $\mathcal{OD}$ pairs).
- η : A vector of independent random variables with null mean such that its ks element measures the variability of the $\mathcal{OD}$ pair ks flow with respect to its mean.

Assumption 2: The conditional distribution of each link flow **V** given the $\mathcal{OD}$ flows is the following normal distribution

$$v_{ij}|\mathbf{T} \sim N\left(\mu_{v_{ij}} + \sum_{k,s \in \Pi_{ij}} \beta_{ijks}(t_{ks} - \mu_{t_{ks}}), \psi_{ij}^2\right),$$

where v_{ij} is the traffic flow in link l_{ij}, β_{ijks} is the regression coefficient of v_{ij} on t_{ks}, which is zero if the link l_{ij} does not belong to any path of the $\mathcal{OD}$ pair ks, ψ_{ij}^2 is its variance and Π_{ij} is the set of parents of link l_{ij}.

Note that this forces our model to satisfy the flow conservation laws. If there are no errors in measurements, that is, $\psi_{ij}^2 = 0; \forall l_{ij} \in \mathcal{A}$, where $\mathcal{A}$ is the set of links, the conservation laws hold exactly. If errors are allowed ($\psi_{ij}^2 \neq 0$) they are statistically satisfied.
Note that this regression relationship, comes from the well known flow equilibrium equation:

$$v_{ij} = \sum_{ks} t_{ks} \left(\sum_{r} p_r^{ks} \delta_{ijr}^{ks} \right), \tag{6}$$

where t_{ks} and are the flows of the $\mathcal{OD}$ ks, p_r^{ks} is the probability of the user travelling from k to s to choose the path r, and δ_{ijr}^{ks} is the incidence matrix, i.e., it takes value 1 if link l_{ij} belongs to path r of the $\mathcal{OD}$ pair ks, and 0, otherwise.
In fact, Equation (6) can be written as

$$v_{ij} = \sum_{ks} t_{ks} \left(\sum_{r} p_r^{ks} \delta_{ijr}^{ks} \right) = \sum_{ks} \beta_{ijks} \mu_{t_{ks}} + \sum_{ks} \beta_{ijks} (t_{ks} - \mu_{t_{ks}}), \tag{7}$$

and then, it becomes apparent that

$$E[v_{ij}] = \mu_{v_{ij}} = \sum_{ks} \beta_{ijks} \mu_{t_{ks}} \tag{8}$$

$$\beta_{ijks} = \sum_{r} p_r^{ks} \delta_{ijr}^{ks}. \tag{9}$$

Note that the p_r^{ks} depend on the intensities of the traffic flow. Thus, this model is to be assumed conditional on the p_r^{ks} values. If one desires to combine this model with traffic assignment models, one can obtain the p_r^{ks} values from the predicted $\mathcal{OD}$ pair flows and iterate until convergence (this is a particular example of a bi-level method that will be explained in chapter 2.2.2 combined with the WMV assignment model).
Therefore, we can assume that the link flows are given by

$$\mathbf{V} = \boldsymbol{\beta}\mathbf{T} + \boldsymbol{\varepsilon}, \tag{10}$$

where $\boldsymbol{\varepsilon} = (\varepsilon_1, \varepsilon_2, \ldots, \varepsilon_n)$ are mutually independent normal random variables, independent of de random variables in $\mathbf{T}$, and ε_ℓ has mean $E[\varepsilon_\ell]$ and variance $\psi_\ell^2; \ell = 1, 2, \ldots, n$. These variables represent the traffic flow that enters or exits the link ℓ_{ij} apart from that going from the origin to the destination node of such a link. In particular, they can be assumed to be null.

Note also that assumption 1 is reasonable because when there is a general increase or decrease of flows (the U value), this affects to all ODs, and this effect can be assumed to be proportional, and the random variables η_{ks} account for random variations over these proportional distributions of flows.
Therefore, from (5), we have

$$\mathbf{T} = (\zeta \mid \mathbf{I}) \begin{pmatrix} U \\ -- \\ \boldsymbol{\eta}^T \end{pmatrix}$$

and the variance-covariance matrix $\boldsymbol{\Sigma}_\mathbf{T}$ of the $\mathbf{T}$ variables becomes

$$\boldsymbol{\Sigma}_\mathbf{T} = (\zeta \mid \mathbf{I}) \, \boldsymbol{\Sigma}_{(\mathbf{U}, \boldsymbol{\eta})} \begin{pmatrix} \zeta^T \\ -- \\ \mathbf{I} \end{pmatrix} = \sigma_U^2 \zeta \zeta^T + \mathbf{D}_{\boldsymbol{\eta}}, \tag{11}$$

where the matrices $\mathbf{\Sigma}_{(\mathbf{U},\boldsymbol{\eta})}$ and $\mathbf{D}$ are diagonal.
Then, we have

$$\begin{pmatrix} \mathbf{T} \\ \mathbf{V} \end{pmatrix} = \begin{pmatrix} \mathbf{I} & | & \mathbf{0} \\ - & + & - \\ \boldsymbol{\beta} & | & \mathbf{I} \end{pmatrix} \begin{pmatrix} \mathbf{T} \\ -- \\ \boldsymbol{\varepsilon} \end{pmatrix}$$

which implies that the mean $E[(\mathbf{T},\mathbf{V})]$ is

$$E[(\mathbf{T},\mathbf{V})] = \begin{pmatrix} E(U)\boldsymbol{\zeta} \\ -------- \\ E(U)\boldsymbol{\beta}\boldsymbol{\zeta} + E(\boldsymbol{\varepsilon}) \end{pmatrix}, \tag{12}$$

and since the variance-covariance matrix of $(\mathbf{T},\boldsymbol{\varepsilon})$ is

$$\mathbf{\Sigma}_{(\mathbf{T},\boldsymbol{\varepsilon})} = \begin{pmatrix} \mathbf{\Sigma}_{\mathbf{T}} & | & \mathbf{0} \\ -- & + & -- \\ \mathbf{0} & | & \mathbf{D}_{\varepsilon} \end{pmatrix},$$

the variance-covariance matrix of $(\mathbf{T},\mathbf{V})$ becomes

$$\begin{aligned} \mathbf{\Sigma}_{(\mathbf{T},\mathbf{V})} &= \begin{pmatrix} \mathbf{I} & | & \mathbf{0} \\ - & + & - \\ \boldsymbol{\beta} & | & \mathbf{I} \end{pmatrix} \begin{pmatrix} \mathbf{\Sigma}_{\mathbf{T}} & | & \mathbf{0} \\ -- & + & -- \\ \mathbf{0} & | & \mathbf{D}_{\varepsilon} \end{pmatrix} \begin{pmatrix} \mathbf{I} & | & \mathbf{0} \\ - & + & - \\ \boldsymbol{\beta} & | & \mathbf{I} \end{pmatrix}^T \\ & \begin{pmatrix} \mathbf{\Sigma}_{\mathbf{T}} & | & \mathbf{\Sigma}_{\mathbf{T}}\boldsymbol{\beta}^T \\ --- & + & ------- \\ \boldsymbol{\beta}\mathbf{\Sigma}_{\mathbf{T}} & | & \boldsymbol{\beta}\mathbf{\Sigma}_{\mathbf{T}}\boldsymbol{\beta}^T + \mathbf{D}_{\varepsilon} \end{pmatrix}. \end{aligned} \tag{13}$$

All these assumptions imply that the joint PDF of $(t_{12}, t_{13}, \dots t_{ks}, v_{12}, v_{23}, \dots, v_{ij})$ can be written as

$$f(t_{12}, t_{13}, \dots t_{ks}, v_{12}, v_{23}, \dots, v_{ij}) = f_{N(\mu_{\mathbf{T}}, \mathbf{\Sigma}_{\mathbf{T}})}(t_{12}, t_{13}, \dots t_{ks}) \prod_{ks} f_{N(\mu_{v_{ij}} + \sum\limits_{ks \in \Pi_{ij}} \beta_{ijks}(t_{ks} - \mu_{t_{ks}}), \psi_{ij}^2)}(V_{ks}), \tag{14}$$

and can be used to predict traffic flows when information from traffic counts becomes available. The idea consists of using the joint distribution of $\mathcal{OD}$ pairs and link traffic flows conditioned on the available information. In fact, since the remaining variables (those not known) are random, the most informative item we can get is its conditional joint distribution, and this is what the Bayesian network methodology supplies.

Now the most convenient graph for this problem (from our point of view), is going to be described: the $\mathcal{OD}$ flows t_{ks} should be the parents of all link flows v_{ij} used by the corresponding travelers, and the error variables should be the parents of the corresponding flows, that is, the ε_{ij} must be parents of the v_{ij}, and the η_{ks} must be parents of the t_{ks}. Finally, the $\mathbf{U}$ variable must by on top (parent) of all $\mathcal{OD}$ flows, because it gives the level of them (high, intermediate or low). This solves the problem of "parent" being well defined, without the need for recursion in general graphs. One could seemingly have a "deadlock" situation in which it is not clear what node is the parent of which other node (see (Sumalee, 2004))

2.2 The Bayesian network model in a bi-level approach

Up to now the Bayesian network trip matrix estimation model (BNME) has been considered as a ME model, i.e., a model able to predict the $\mathcal{OD}$ and link flows from a given probability matrix with elements p_r^{ks} or β_{ijks}. The main difference with other methods is that it gives the joint conditional distributions of all not observed variables and that makes no difference between $\mathcal{OD}$ flows and link flows, in the sense that information of any one of them gives information about the others indistinctly. In particular the marginal distributions of any flow ($\mathcal{OD}$ or link) are supplied by the method, so that not only predictions can be obtained but probability intervals or regions.

In this chapter we show that the BNME can be easily combined with some assignment method to obtain the equilibrium solution of the traffic problem and therefore obtain a more realistic solution, for example in the congested case, using a bi-level approach. In particular, the proposed model is combined with an assignment model which identifies the origin and destination of the travelers who drive on a link (see (Castillo, Menéndez & Sánchez-Cambronero, 2008b)). Among its advantages, we can emphasize that the β_{ijks} coefficients are easily calculated and the most important, the route enumeration is avoided. This method, called the "*Wardrop minimum variance (WMV) method*", is next, combined with the BNME proposed method, but first let us give a detailed explanation of it.

2.2.1 The WMV assignment model

In this section a User Equilibrium based optimization problem is presented that, given the t_{ks} $\mathcal{OD}$ flows, deals with the link ℓ_{ij} flows x_{ijks} coming from node k (origin) and going to node s (destination). The balance of all these flows particularized by origins and destinations, allows us classifying the link flows by ODs.

This important information can be used, not only to have a better knowledge of the user behavior and the traffic in the network but to make decisions, for example, when some network events take place. In addition, this method avoids the route enumeration problem which is a very important issue because including a sub routine which deals with this problem is always a thorny issue.

The problem is formulated as follows:

$$\underset{\mathbf{x}}{\text{Minimize}}\, Z = \sum_{\ell_{ij}\in A} \int_0^{\sum\limits_{k,s} x_{ijks}} c_{ij}(x)dx + \frac{\lambda}{m} \sum_{\ell_{ij}\in A} \sum_{k,s} (x_{ijks} - \mu)^2 \tag{15}$$

subject to

$$t_{ks}(\delta_{ik} - \delta_{is}) = \sum_{\ell_{ij}\in \mathcal{A}} x_{ijks} - \sum_{\ell_{ji}\in \mathcal{A}} x_{jiks} \quad \forall i; \quad \forall k,s; \quad k \neq s, \tag{16}$$

$$\mu = \frac{1}{m} \sum_{\ell_{ij}\in A} \sum_{k,s} x_{ijks}, \tag{17}$$

$$x_{ijks} \geq 0 \quad \forall i,j,k,s, \tag{18}$$

where $c_{ij}(\cdot)$ is the cost function for link ℓ_{ij}, x_{ijks} is the flow through link ℓ_{ij} with origin node k and destination node s, $\lambda > 0$ is a weighting factor, δ_{ik} are the Dirac deltas ($\delta_{ik} = 0$, if $i \neq k$, $\delta_{ii} = 1$), μ is the mean of the x_{ijks} variables, and m is its cardinal. We have also assumed that the cost on a link depends only on the flow on that link.

Note that equation (16) represents the flow balance associated with the OD-pair (k,s), for all nodes, and that the problem (15)-(18) for $\lambda = 0$ is a statement of the Beckmann et al. formulation of the Wardrop UE equilibrium problem, but stated for each OD pair. As the cost function we have selected the Bureau of Public Roads (BPR) type cost functions, because it is generally accepted and has nice regularity properties, but other alternative cost functions with the same regularity properties (increasing with flow, monotonic and continuously differentiable) can be used instead. This function is as follows:

$$c_{ij}\left(\sum_{k,s} x_{ijks}\right) = c_{ij}\left[1+\alpha_{ij}\left(\frac{\sum_{k,s} x_{ijks}}{q_{ij}}\right)^{\gamma_{ij}}\right], \tag{19}$$

where for a given link ℓ_{ij}, c_{ij} is the cost associated with free flow conditions, q_{ij} is a constant measuring the flow producing congestion, and α_{ij} and γ_{ij} are constants defining how the cost increases with traffic flow. So the total flow v_{ij} through link ℓ_{ij} is:

$$v_{ij} = \sum_{k,s} x_{ijks}. \tag{20}$$

The problem (15)-(18) for $\lambda = 0$ becomes a pure Wardrop problem and has unique solution in terms of total link flows, but it can have infinitely many solutions in terms of x_{ijks}, though they are equivalent in terms of link costs (they have the same link costs). Note that any exchange of users between equal cost sub-paths does not alter the link flows nor the corresponding costs. So, given an optimal solution to the problem, exchanging different OD users from one sub-path to the other leads to another optimal solution with different x_{ijks} values, though the same link flows v_{ij}. To solve this problem one can choose a very small values of λ. In this case, the problem has a unique solution. Note also that since for $\lambda > 0$, (15) is strictly convex, and the system (16)-(18) is compatible and convex, the problem (15)-(18) has a unique solution, which is a global optimum.

2.2.2 Combining the BN model and the WMV equilibrium model

In this section the Bayesian network model is combined with the new WMV assignment model described in section 2.2.1 using a bi-level algorithm. The aim of proposing this assignment method instead of, for example, an standard SUE assignment model is twofold. First this method avoids the route enumeration which is a very important issue. Second, once the flows x_{ijks} are known, the β_{ijks} coefficients can be easily calculated as:

$$\beta_{ijks} = \frac{x_{ijks}}{t_{ks}}. \tag{21}$$

which is a very important data for the BNME proposed method and allow us an easily implementation of it.

Algorithm 1 (Bi-level algorithm for the BN and WMV models).

INPUT. *$E[U]$, the ζ matrix of relative weight of each OD-pair, the cost coefficients $c_{ij}, \alpha_{ij}, q_{ij}$ and γ_{ij}, $\forall l_{ij} \in A$, and the observed link flows, are the data needed by the algorithm.*

OUTPUT. *The predictions of the $\mathcal{OD}$ and link flows given the observed flows.*

Step 0: Initialization. *Initialize the $\mathcal{OD}$ flows to the initial guess for $E[\mathbf{T}]$:*

$$\mathbf{T}_0 = E[\mathbf{T}] = E[U]\zeta. \tag{22}$$

Step 1: Master problem solution. *The WMV optimization problem (15)-(18) is solved.*

Step 2: Calculate the $\boldsymbol{\beta}$ matrix. *The $\boldsymbol{\beta}$ matrix, of regression coefficients of the $\mathbf{V}$ variables given $\mathbf{T}_0$, is calculated using Equation (21).*

Step 3: Subproblem: Update the $\mathcal{OD}$ and link flow predictions using the Bayesian network. *The new OD-pair $\mathbf{T}$ and link $\mathbf{V}$ flows are predicted using equations shown in this section, which are:*

$$E[\mathbf{V}] = E[U]\boldsymbol{\beta}\zeta + E[\varepsilon] \tag{23}$$
$$\mathbf{D}_{\boldsymbol{\eta}} = Diag\,(vE[\mathbf{T}]) \tag{24}$$
$$\boldsymbol{\Sigma}_{\mathbf{TT}} = \sigma_U^2 \zeta\zeta^T + \mathbf{D}_{\boldsymbol{\eta}} \tag{25}$$
$$\boldsymbol{\Sigma}_{\mathbf{TV}} = \boldsymbol{\Sigma}_{\mathbf{TT}}\boldsymbol{\beta}^T \tag{26}$$
$$\boldsymbol{\Sigma}_{\mathbf{VT}} = \boldsymbol{\Sigma}_{\mathbf{TV}} \tag{27}$$
$$\boldsymbol{\Sigma}_{\mathbf{VV}} = \boldsymbol{\beta}\boldsymbol{\Sigma}_{\mathbf{TT}}\boldsymbol{\beta}^T + \mathbf{D}_{\boldsymbol{\varepsilon}} \tag{28}$$
$$E[\mathbf{Y}|\mathbf{Z}=\mathbf{z}] = E[\mathbf{Y}] + \boldsymbol{\Sigma}_{\mathbf{YZ}}\boldsymbol{\Sigma}_{\mathbf{ZZ}}^{-1}(\mathbf{z} - E[\mathbf{Z}]) \tag{29}$$
$$\boldsymbol{\Sigma}_{\mathbf{Y}|\mathbf{Z}=\mathbf{z}} = \boldsymbol{\Sigma}_{\mathbf{YY}} - \boldsymbol{\Sigma}_{\mathbf{YZ}}\boldsymbol{\Sigma}_{\mathbf{ZZ}}^{-1}\boldsymbol{\Sigma}_{\mathbf{ZY}} \tag{30}$$
$$E[\mathbf{Z}|\mathbf{Z}=\mathbf{z}] = \mathbf{z} \tag{31}$$
$$\boldsymbol{\Sigma}_{\mathbf{Z}|\mathbf{Z}=\mathbf{z}} = \mathbf{0} \tag{32}$$
$$\mathbf{T} = E[\mathbf{Y}|\,\mathbf{Z}=\mathbf{z}]|_{(\mathbf{Y},\mathbf{Z})=T} \tag{33}$$

Step 4: Convergence checking. *Compute actual error by means of*

$$error = (\mathbf{T}_0 - \mathbf{T})^T(\mathbf{T}_0 - \mathbf{T}). \tag{34}$$

If the error is less than the tolerance, stop and return the values of $\mathbf{T}$ and $\mathbf{V}$. Otherwise, let $\mathbf{T}_0 = \mathbf{T}$ and continue with Step 1.

Equation (22) is the initial $\mathcal{OD}$ flow matrix calculated using the random variable $\mathbf{U}$, which gives an estimation of the global flow in the system, and the relative weight vector $\boldsymbol{\zeta}$, which gives the relative importance of the different $\mathcal{OD}$ flows. This $\mathbf{T}_0$ matrix with elements t^0_{ks}, is initially the input data for the problem (15)-(18) and is the initial guess for the $\mathcal{OD}$ matrix with which the calculations are started.

As it has been indicated, for the non-observed data ($\mathcal{OD}$ or/and link flows), one can supply a probability interval, obtained from the resulting conditional probabilities given the evidence. The relevance of the proposed method consists of using the covariance structure of all the variables involved. The importance of this information has been pointed out by (Hazelton, 2003), who shows how the indeterminacy of the system of equations relating link and $\mathcal{OD}$ flows, due to the larger number of the latter, can be compensated by the covariance structure.

2.3 Optimal counting location method using Bayesian networks

This section describes how the Bayesian network model can be also used to select the optimal number and locations of the links counters based on maximum correlation (see (Castillo, Menéndez & Sánchez-Cambronero, 2008b)). To deal with this problem, a simple procedure based on the correlation matrix is described below.

Algorithm 2 (Optimal traffic counting locations.).

INPUT. *The set of target variables to be predicted (normally $\mathcal{OD}$ flows), a variance tolerance, and the initial variance-covariance matrix $\mathbf{\Sigma_{ZZ}}$, or alternatively, σ_{U} and the matrices ζ, $\mathbf{D}_{\eta}$, $\mathbf{D}_{\varepsilon}$ and $\boldsymbol{\beta}$.*

OUTPUT. *The set of variables to be observed (normally link flows).*

Step 0: Initialization. *If the initial variance-covariance matrix $\mathbf{\Sigma_{TV}}$ is not given, calculate it using (23)-(28).*

Step 1: Calculate the correlation matrix. *The correlation matrix* ***Corr*** *with elements*

$$Corr_{ab} = \frac{Cov(X_a X_b)}{\sqrt{\sigma_{x_a}\sigma_{x_b}}} \tag{35}$$

is calculated from the variance-covariance matrix $\mathbf{\Sigma_{TV}}$.

Step 2: Select the target and observable variables. *Select the target variable (normally among the OD-flows) and the observable variable (normally among the link flows), by choosing the largest absolute value of the correlations in matrix* ***Corr****. Note that a value of $Corr_{ab}$ close to 1 means that variables a and b are highly correlated. Therefore if the knowledge of a certain target variable is desired, it is more convenient to observe a variable with greater correlation coefficient because it has more information than other variables on the target variable.*

Step 3: Update the variance-covariance matrix $\mathbf{\Sigma_{TV}}$. *Use formulas (30) and (32) to update the variance-covariance matrix.*

Step 4: End of algorithm checking. *Check residual variances of the target variables and determine if they are below the given threshold. If they are, stop the process and return the list of observable variables. If there are still variables to be observed, continue with Step 1. Otherwise, stop and inform that there is no solution with the given tolerance and provide the largest correlation in order to have a solution.*

Note that equation (30), which updates the variance-covariance matrix, does not need the value of the evidence, but only the evidence variable. Thus, the algorithm can be run without knowledge of the evidences. Note also that this algorithm always ends, either with the list of optimal counting locations or with a threshold[2] value for the correlation coefficient for the problem to have a solution.

[2] Because the model determines the links to be observed, this selection is done with a given error level, therefore the quality of the results depend on it.

2.4 Example of applications: The Nguyen-Dupuis network

In this section, we illustrate the previous models using the well known Nguyen-Dupuis network. It consists of 13 nodes and 19 links, as shown in Figure 2.

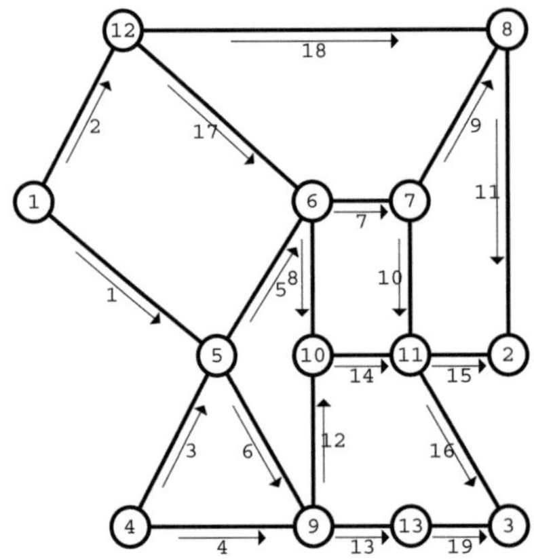

Fig. 2. The Nguyen-Dupuis network.

2.4.1 Selecting an optimal subset of links to be observed

Because the selection procedure is based on the covariance matrix of the link and $\mathcal{OD}$ flows, we need this matrix. To simplify, in this example, the matrix $\mathbf{D}_{\mathcal{E}}$ is assumed to be diagonal with diagonal elements equal to 0.1 (a very small number), that is, we assume that there is practically no measurement error in the link flows. The matrix $\mathbf{D}_{\eta}$ is also assumed to be diagonal with variances equal to the $(0.1E[t_{ks}])^2$.
The mean and standard deviation of U are assumed to be $E[U] = 100$ and $\sigma_U = 20$, respectively, and the assumed elements of the ζ matrix are given in left part of Table 1.

$\mathcal{OD}$	ζ	Prior $\mathbf{T}_0$
1-2	0.4	40
1-3	0.8	80
4-2	0.6	60
4-3	0.2	20

link	c_{ij}	q_{ij}	α_{ij}	γ_{ij}
1 -5	7	70	1	4
1 -12	9	56	1	4
4 -5	9	56	1	4
4 -9	12	70	1	4
5 -6	3	42	1	4
5 -9	9	42	1	4
6 -7	5	70	1	4
6 -10	5	28	1	4
7 -8	5	70	1	4
7 -11	9	70	1	4

link	c_{ij}	q_{ij}	α_{ij}	γ_{ij}
8 -2	9	70	1	4
9 -10	10	56	1	4
9 -13	9	56	1	4
10-11	6	70	1	4
11-2	9	56	1	4
11-3	8	56	1	4
12-6	7	14	1	4
12-8	14	56	1	4
13-3	11	56	1	4

Table 1. Data needed for solving the example :Prior $\mathcal{OD}$ flow, ζ matrices and link parameters.

Because we have no information about the beta matrix $\boldsymbol{\beta}$, we have used a prior $\mathcal{OD}$ trip matrix $\mathbf{T_0}$ and solved the problem (15)-(18) with the cost coefficients $c_{ij}, \alpha_{ij}, q_{ij}$ and γ_{ij} $\forall \ell \in \mathcal{A}$, shown in right part of Table 1, to obtain one. The method has been used, and Table 2 shows this initial matrix.

	Link proportions (β matrix)																		
OD	1 -5	1 -12	4 -5	4 -9	5 -6	5 -9	6 -7	6 -10	7 -8	7 -11	8 -2	9 -10	9 -13	10-11	11-2	11-3	12-6	12-8	13-3
1 -2	**0.00**	1.00	0.00	0.00	0.00	0.00	0.00	0.00	0.00	0.00	1.00	**0.00**	**0.00**	0.00	0.00	0.00	0.00	**1.00**	0.00
1 -3	**0.84**	0.16	0.00	0.00	0.47	0.37	0.37	0.26	0.00	0.37	0.00	**0.00**	**0.37**	0.26	0.00	0.63	0.16	**0.00**	0.37
4 -2	**0.00**	0.00	0.36	0.64	0.36	0.00	0.36	0.00	0.36	0.00	0.36	**0.64**	**0.00**	0.64	0.64	0.00	0.00	**0.00**	0.00
4 -3	**0.00**	0.00	0.00	1.00	0.00	0.00	0.00	0.00	0.00	0.00	0.00	**0.00**	**1.00**	0.00	0.00	0.00	0.00	**0.00**	1.00

Table 2. Initial β matrix.

With this information and considering a threshold value of 1 for the variances of the $\mathcal{OD}$ flows, one can use Algorithm 2 where the initial variance-covariance matrix $\mathbf{\Sigma_{TV}}$ has been calculated using (25)-(28).

After solving this problem, one knows that to predict the $\mathcal{OD}$ matrix at the given quality level it is necessary to observe only the following links:

$$1-5, \quad 12-8, \quad 9-10, \quad 9-13. \tag{36}$$

It is interesting to observe the boldfaced columns associated with these links in the beta matrix in Table 2 to understand why these link flows are the most adequate to predict all the $\mathcal{OD}$ flows.

	Variances in each iteration				
Variable	0	1	2	3	Final
OD-pair 1-2	80.00	28.82	0.10	0.10	0.10
OD-pair 1-3	320.00	0.14	0.14	0.14	0.14
OD-pair 4-2	180.00	64.95	52.06	0.24	0.24
OD-pair 4-3	20.00	7.20	5.78	5.23	0.11
Link 1-5	226.48	**0.00**	**0.00**	**0.00**	**0.00**
Link 1-12	128.92	28.94	0.20	0.20	0.20
Link 4-5	23.73	8.62	6.93	0.13	0.13
Link 4-9	154.50	45.93	33.87	5.46	0.31
Link 5-6	159.32	8.68	6.98	0.16	0.16
Link 5-9	44.49	0.12	0.12	0.12	0.12
Link 6-7	117.21	8.68	6.96	0.15	0.15

	Variances in each iteration				
Variable	0	1	2	3	Final
Link 6-10	22.20	0.11	0.11	0.11	0.11
Link 7-8	23.73	8.62	6.93	0.13	0.13
Link 7-11	42.77	0.12	0.12	0.12	0.12
Link 8-2	173.30	51.48	7.08	0.23	0.23
Link 9-10	73.39	26.47	21.26	**0.00**	**0.00**
Link 9-13	112.01	7.34	5.91	5.36	**0.00**
Link 10-11	159.74	26.47	21.29	0.21	0.21
Link 11-2	73.38	26.51	21.27	0.20	0.20
Link 11-3	126.29	0.16	0.16	0.16	0.16
Link 12-6	8.20	0.10	0.10	0.10	0.10
Link 12-8	80.10	28.92	**0.00**	**0.00**	**0.00**
Link 13-3	112.01	7.35	5.92	5.36	0.20

Table 3. Variance of all variables initially and after updating the evidences in each step

Table 3 shows the variance of each variable at each stage, i.e., after the observable variables are being observed. In the Iteration 0 column, the variances when the information is not available are shown. In Iteration 1, the variances after observing the first observable variable $w_{1,5}$ are shown, and so on. It is interesting to see that the variances decrease with the knowledge of new evidences. At the end of the process all variances are very small (smaller than the selected threshold value). Note that the unobserved link flows can also be estimated with a small precision. Table 4 shows the correlation sub matrices at each iteration, together with the associated largest absolute value (boldfaced) used to choose the target and observable variable.

2.4.2 $\mathcal{OD}$ matrix estimation

Once the list of links to be observed have been obtained, one can observe them. The observed flows corresponding to these links have been simulated assuming that they are normal random variables with their corresponding means and standard deviations. The resulting flows were:

$$\hat{v}_{1,5} = 59.73; \quad \hat{v}_{12,8} = 36.12; \quad \hat{v}_{9,10} = 39.68; \quad \hat{v}_{9,13} = 49.87;$$

To estimate the $\mathcal{OD}$ flows with Algorithm 1 one needs some more data. We have assumed $E[\varepsilon] = \mathbf{0.1}$ and a tolerance value to check convergence of 0.00001. The initial guess for **T**:

$$\mathbf{T_0} = E[\mathbf{T}] = E[U]\zeta. \tag{37}$$

Iteration 1				
	OD-pair			
Link	1-2	1-3	4-2	4-3
1-5	0.800	**1.000**	0.800	0.800
1-12	0.988	0.881	0.831	0.831
4-5	0.798	0.798	0.998	0.798
4-9	0.838	0.838	0.976	0.910
5-6	0.840	0.973	0.917	0.840
5-9	0.799	0.999	0.799	0.799
6-7	0.842	0.963	0.932	0.842
6-10	0.798	0.998	0.798	0.798
7-8	0.798	0.798	0.998	0.798
7-11	0.799	0.999	0.799	0.799
8-2	0.975	0.839	0.913	0.839
9-10	0.799	0.799	0.999	0.799
9-13	0.841	0.967	0.841	0.926
10-11	0.839	0.913	0.974	0.839
11-2	0.799	0.799	0.999	0.799
11-3	0.800	1.000	0.800	0.800
12-6	0.795	0.994	0.795	0.795
12-8	0.999	0.800	0.800	0.800
13-3	0.841	0.967	0.841	0.926
Target variable=OD-pair 1-3				
Observed variable=link 1-5.				

Iteration 2			
	OD-pair		
Link	1-2	4-2	4-3
1-12	0.998	0.444	0.444
4-5	0.442	0.994	0.442
4-9	0.513	0.934	0.733
5-6	0.442	0.992	0.442
6-7	0.443	0.993	0.443
7-8	0.442	0.994	0.442
8-2	0.930	0.740	0.514
9-10	0.444	0.998	0.444
9-13	0.442	0.442	0.992
10-11	0.444	0.998	0.444
11-2	0.444	0.998	0.444
12-8	**0.998**	0.444	0.444
13-3	0.442	0.442	0.992
Target variable=OD-pair 1-2			
Observed variable=link 12-8.			

Iteration 3		
	OD-pair	
Link	4-2	4-3
4-5	0.993	0.306
4-9	0.918	0.657
5-6	0.991	0.306
6-7	0.991	0.306
7-8	0.993	0.306
8-2	0.986	0.307
9-10	**0.998**	0.308
9-13	0.306	0.990
10-11	0.997	0.308
11-2	0.998	0.308
13-3	0.306	0.990
Target variable=OD-pair 4-2		
Observed variable=link 9-10.		

Iteration 4	
	OD-pair
Link	4-3
4-9	0.982
9-13	**0.989**
13-3	0.989
Target variable=OD-pair 4-3	
Observed variable=link 9-13.	

Table 4. Correlation sub matrices

The initial values of v_{ij} and x_{ijks} variables, using (37), are shown in table 5, and the resulting $\mathcal{OD}$ and link flows after convergence of the process are shown in Table 6 column 2, and the final $\boldsymbol{\beta}$ matrix is shown in Table 7.

Thus, the resulting $\mathcal{OD}$ matrix estimates and the prediction for the link flow variables are shown in Table 6. In addition, for the sake of comparison, in algorithm 1 the WMV has been replaced by a Logit SUE assignment, and the results are shown in column 4 of Table 6. Note that the results are very similar.

Link	Cost	v_{ij}	1-2	1-3	4-2	4-3
1 -5	13.0	67.3	0.0	67.3	0.0	0.0
1 -12	16.1	52.7	40.0	12.7	0.0	0.0
4 -5	9.2	21.7	0.0	0.0	21.7	0.0
4 -9	17.8	58.3	0.0	0.0	38.3	20.0
5 -6	14.9	59.3	0.0	37.5	21.7	0.0
5 -9	11.3	29.8	0.0	29.8	0.0	0.0
6 -7	6.4	51.0	0.0	29.2	21.7	0.0
6 -10	6.6	21.0	0.0	21.0	0.0	0.0
7 -8	5.0	21.7	0.0	0.0	21.7	0.0
7 -11	9.3	29.2	0.0	29.2	0.0	0.0

Link	Cost	v_{ij}	1-2	1-3	4-2	4-3
8 -2	14.4	61.7	40.0	0.0	21.7	0.0
9 -10	12.2	38.3	0.0	0.0	38.3	0.0
9 -13	14.6	49.8	0.0	29.8	0.0	20.0
10-11	9.1	59.3	0.0	21.0	38.3	0.0
11-2	11.0	38.3	0.0	0.0	38.3	0.0
11-3	13.2	50.2	0.0	50.2	0.0	0.0
12-6	11.8	12.7	0.0	12.7	0.0	0.0
12-8	17.6	40.0	40.0	0.0	0.0	0.0
13-3	17.9	49.8	0.0	29.8	0.0	20.0

Table 5. Link cost, link total flows, and link flows after using WMV assignment.

OD or	Link flows		
link	BN-WMV	Prior	BN-SUE
1-2	36.15	40.00	38.07
1-3	72.81	80.00	71.81
4-2	67.72	60.00	61.15
4-3	22.45	20.00	29.19
1 -5	59.73	67.27	59.73
1 -12	49.19	52.73	50.02
4 -5	28.07	21.74	34.33
4 -9	62.10	58.26	56.00
5 -6	60.48	59.25	60.87
5 -9	27.36	29.76	33.33
6 -7	52.88	50.95	52.63

OD or	Link flows		
link	BN-WMV	Prior	BN-SUE
6 -10	20.64	21.03	22.10
7 -8	27.92	21.74	21.94
7 -11	24.97	29.21	30.70
8 -2	64.07	61.74	58.09
9 -10	39.68	38.26	39.68
9 -13	49.87	49.76	49.87
10-11	60.28	59.28	61.80
11-2	39.80	38.26	41.14
11-3	45.45	50.24	51.36
12-6	13.04	12.73	13.87
12-8	36.12	40.00	36.12
13-3	49.82	49.76	49.63

Table 6. $\mathcal{OD}$ and link flows resulting from algorithm 1, and replacing the WMV by a Logit SUE method.

	Link proportions ($\boldsymbol{\beta}$ matrix)																		
OD	1 -5	1 -12	4 -5	4 -9	5 -6	5 -9	6 -7	6 -10	7 -8	7 -11	8 -2	9 -10	9 -13	10-11	11-2	11-3	12-6	12-8	13-3
1 -2	0.00	1.00	0.00	0.00	0.00	0.00	0.00	0.00	0.00	0.00	1.00	0.00	0.00	0.00	0.00	0.00	0.00	1.00	0.00
1 -3	0.82	0.18	0.00	0.00	0.45	0.38	0.34	0.28	0.00	0.34	0.00	0.00	0.38	0.28	0.00	0.62	0.18	0.00	0.38
4 -2	0.00	0.00	0.41	0.59	0.41	0.00	0.41	0.00	0.41	0.00	0.41	0.59	0.00	0.59	0.59	0.00	0.00	0.00	0.00
4 -3	0.00	0.00	0.00	1.00	0.00	0.00	0.00	0.00	0.00	0.00	0.00	0.00	1.00	0.00	0.00	0.00	0.00	0.00	1.00

Table 7. Final $\boldsymbol{\beta}$ matrix.

Before analyzing the previous results, one must realize that we are using two sources of information: that contained in the Bayesian network (the joint normal distribution of links and $\mathcal{OD}$ flows), and the observed link or $\mathcal{OD}$ flows. We must note that the first one can be very informative. In fact, when no observations are available it is the only one supplying information about flows, but when observations become available it can still be more informative than that contained in the observations, if the number of observed links is small.

3. Plate scanning based method for traffic prediction.

This section shows how the Bayesian network tool can be also used with data from the plate scanning technique (see (Sánchez-Cambronero et al., 2010). Therefore, first the plate scanning approach will be introduced, and then the model will be described (see (Castillo, Menéndez & Jiménez.P, 2008)).

3.1 Dealing with the information contained in the data from the plate scanning technique.

The idea of plate scanning consists of registering plate numbers and the corresponding times of the circulating vehicles when they travel on some subsets of links. This information is then used to reconstruct vehicle routes by identifying identical plate numbers at different locations and times. In order to clarify the concepts, let us consider a traffic network $(\mathcal{N}, \mathcal{A})$ where $\mathcal{N}$ is

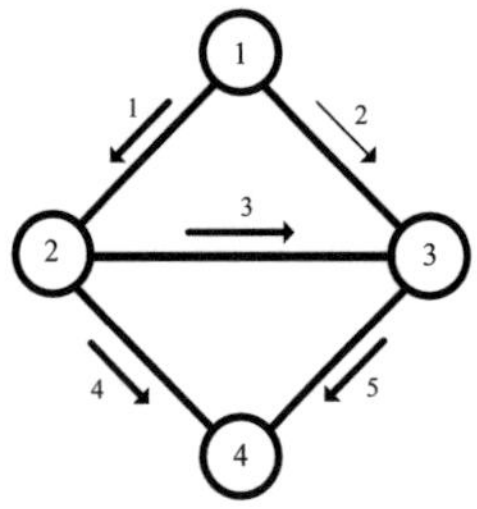

Fig. 3. The elementary example network used for illustrative purposes.

a set of nodes and $\mathcal{A}$ is a set of links. We have used the simple network in Figure 3 of 4 nodes and 5 links. Table 8 shows the 4 OD-pairs considered and the corresponding 7 paths used in this example.

$\mathcal{OD}$	path code (r)	Links
1-4	1	1 3 5
1-4	2	1 4
1-4	3	2 5
2-4	4	3 5
2-4	5	4
1-2	6	1
2-3	7	3

Table 8. Set of 4 OD-pairs and 7 paths considered in the elementary example.

We assume that we have selected a nonempty subset $\mathcal{SC} \subset \mathcal{A}$ of $n_{sc} \neq 0$ links to be scanned. To illustrate, consider the scanned subset $\mathcal{SC}$ of 4 links[3]

$$\mathcal{SC} \equiv \{1,3,4,5\}. \tag{38}$$

In the scanned links the plate numbers and the passing times[4] of the users are registered, i.e., the initially gathered information $\mathcal{I}$ consists of the set

$$\mathcal{I} \equiv \{(I_k, \ell_k, \tau_k);\ k = 1,2,\ldots,m;\ \ell_k \in \mathcal{SC}\}, \tag{39}$$

[3] This subset is not arbitrary, but has been carefully selected as we will see.

[4] Passing times are used only to identify the scanned user routes.

where I_k is the identification number (plate number) of the k-th observed user, $\ell_k \in \mathcal{SC}$ is the link where the observation took place, τ_k is the corresponding pass time through link ℓ_k, and m is the number of observations.

For illustration purposes, a simple example with 28 registered items is shown in left part of Table 9, where the plate numbers of the registered cars and the corresponding links and passing times, in the format second-day-month-year, are given.

item k	# Plate I_k	Link ℓ_k	Time τ_k
1	1256 ADL	1	00001 19-12-2009
2	3789 BQP	3	00022 19-12-2009
3	7382 BCD	2	00045 19-12-2009
4	9367 CDF	1	00084 19-12-2009
5	9737 AHH	1	00123 19-12-2009
6	3789 BQP	5	00145 19-12-2009
7	7382 BCD	5	00187 19-12-2009
8	6453 DGJ	4	00245 19-12-2009
9	9737 AHH	3	00297 19-12-2009
10	9367 CDF	4	00309 19-12-2009
11	3581 AAB	1	00389 19-12-2009
12	6299 HPQ	4	00478 19-12-2009
13	9737 AHH	5	00536 19-12-2009
14	3581 AAB	3	00612 19-12-2009
15	1243 RTV	3	00834 19-12-2009
16	7215 ABC	1	00893 19-12-2009
17	8651 PPT	3	01200 19-12-2009
18	3581 AAB	5	01345 19-12-2009
19	1974 PZS	1	01356 19-12-2009
20	1256 ADL	4	01438 19-12-2009
21	2572 AZP	1	01502 19-12-2009
22	6143 BBA	3	01588 19-12-2009
23	7614 CAB	1	01670 19-12-2009
24	6143 BBA	5	01711 19-12-2009
25	1897 DEP	2	01798 19-12-2009
26	1897 DEP	5	01849 19-12-2009
27	2572 AZP	4	01903 19-12-2009
28	7614 CAB	4	01945 19-12-2009
⋮	⋮	⋮	⋮

item z	# Plate I_z	Scanned links C_{s_z}	Code s
1	1256 ADL	{1,4}	2
2	3789 BQP	{3,5}	4
3	7382 BCD	{5}	3
4	9367 CDF	{1,4}	2
5	9737 AHH	{1,3,5}	1
6	6453 DGJ	{4}	5
7	3581 AAB	{1,3,5}	1
8	4769 CCQ	{3}	7
9	2572 AZP	{1,4}	2
10	6143 BBA	{3,5}	4
11	7614 CAB	{1,4}	2
12	1897 DEP	{5}	3
13	6299 HPQ	{5}	3
14	7215 ABC	{1}	6
15	1974 PZS	{1}	6
16	1243 RTV	{3}	7
⋮	⋮	⋮	⋮

			Scanned links				
OD	r	s	1	3	4	5	$\hat{w}_s$
1-4	1	1	X	X		X	2
1-4	2	2	X		X		4
1-4	3	3				X	3
2-4	4	4		X		X	2
2-4	5	5			X		1
1-2	6	6	X				2
2-3	7	7		X			2

Table 9. Example of registered data by scanned links and the data after been processed.

Note that a single car user supplies one or more elements (I_k, ℓ_k, τ_k), in fact as many as the number of times the corresponding user passes through an scanned link. For example, the user with plate number 9737 AHH appears three times, which means he/she has been registered when passing through three scanned links (1, 3 and 5).

A cross search of plate numbers contained in the different (I_k, ℓ_k, τ_k) items of information and check of the corresponding passing times allows one determining the path or partial paths followed by the scanned users. This allows building the set

$$\{(I_z, C_{s_z})|\ z = 1, 2, \ldots, n;\ C_{s_z} \in \mathcal{P}(\mathcal{SC})\}, \tag{40}$$

where C_{s_z} is the subset of links associated with the I_z user, which includes all links in which the user has been scanned (scanned partial path of that user), n is the number of registered users, and $\mathcal{P}(\mathcal{SC})$ is the set of parts of $\mathcal{SC}$, which contains $2^{n_{sc}}$ elements. A registered user

has an associated C_{s_z} subset only if the corresponding scanned links belong to its route. Of course, a non-registered user appears in no registered links, which corresponds to $C_{s_z} = \emptyset$. We associate with each user the subset C_{s_z} of scanned links contained in his/her route, and call a subset C_{s_z} of scanned links feasible if there exists a user which associated subset is C_{s_z}. Note that not all subsets of scanned links are feasible and each route must leads to a feasible subset. Upper right part of Table 9, shows the plate numbers and the associated scanned subpath (C_{s_z} set of registered links for given users) of the registered users in columns two and three. For example, the user with plate number $3581AAB$ appears as registered in links 1,3 and 5, thus leading to the set $C_{s_z} \equiv \{1,3,5\}$.
To obtain the feasible C sets one needs only to go through each possible path and determine which scanned links are contained in it. The two first columns of bottom part of Table 9 shows all paths, defined by the $\mathcal{OD}$ and r (the order of the path within the OD) values. The third column corresponds with the set of scanned link code s which, in this case, is the same as the route code because we have full route observability. Finally, the last columns corresponds with the scanned links and its associated sets (each indicated by an X).
An important point to note is that since all combinations of scanned links are different for all paths, and this happens because the set of links to be scanned has been adequately selected, the scanning process allows identifying the path of any scanned user. Therefore, using this information, one obtains the observed number of users $\hat{w}_s$ with associated s-values and C_s sets (see Table 9). This allows us to summarize the scanned observations as

$$\{\hat{w}_s : s \in \mathcal{S}\}, \tag{41}$$

where $\mathcal{S}$ is the set $\mathcal{S} \equiv \{1,2,\ldots,n\}$ and n the number of different C_s sets in $\mathcal{S}$, which is the information used by the proposed model to estimate the traffic flows. Note also that standard models are unable to deal with this problem, i.e., to handle the information in the form (41).
To control this type of information, the traffic flow must be disaggregated in terms of the new variables $\hat{w}_s$, which refer to the flow registered by the scanned links in C_s. Then, one needs to write the conservation laws as follows:

$$\hat{w}_s = \sum_{r \in \mathcal{R}} \delta_{sr} f_r; \quad r \in \mathcal{R}; \quad s \in \mathcal{S}, \tag{42}$$

where f_r is the flow of route r, δ_{sr} is one if the route r contains all and only the links in C_s.

3.2 Model assumptions

In this section, the model assumptions for the BN-PLATE (see (Sánchez-Cambronero et al., 2010) model are introduced. Note that there are important differences with the model described in section 2 in which the model was built considering OD-pair and link flows, instead of route and scanned links flows, respectively.
Therefore assuming the route and subsets of scanned link flows are multivariate random variables, we build a Gaussian Bayesian network using the special characteristics of traffic flow variables. To this end, we consider the route flows as parents and the subsets of scanned link flows as children and reproduce the conservation law constraints defined in (42) in an exact or statistical (i.e., with random errors) form. In our Gaussian Bayesian network model we make the following assumptions:

Assumption 1: The vector $\mathbf{F}$ of route flows is a multivariate normal $N(\boldsymbol{\mu}_F, \boldsymbol{\Sigma}_F)$ random variable with mean $\boldsymbol{\mu}_F$ and variance-covariance matrix $\boldsymbol{\Sigma}_F$.

For the same reason than in the BN-WMV model, it is clear that the **F** random variables are correlated. Therefore:

$$f_r = k_r U + \eta_r, \tag{43}$$

where $k_r, r = 1, \ldots, m$ are positive real constants, one for each route r, U is a normal random variable $N(\mu_U, \sigma_U^2)$, and η_r are independent normal $N(0, \gamma_r^2)$ random variables. The meanings of these variables are as follows:

- U : A random positive variable that measures the level of total mean flow.
- $\mathbf{K}$: A column matrix whose element k_r measures the relative weight of the route r flow with respect to the total traffic flow (including all routes).
- $\boldsymbol{\eta}$: A vector of independent random variables with null mean such that its r element measures the variability of the route r flow with respect to its mean.

Assumption 2: The flows associated with the combinations of scanned link flows and counted link flows can be written as

$$\mathbf{W} = \boldsymbol{\Delta}\mathbf{F} + \varepsilon, \tag{44}$$

where w_s, f_r and δ_{sr} have the same meaning than before, and $\boldsymbol{\varepsilon} = (\varepsilon_1, \varepsilon_2, \ldots, \varepsilon_n)$ are mutually independent normal random variables, independent of the random variables in **F**, and ε_s has mean $E(\varepsilon_s)$ and variance $\psi_s^2; s = 1, 2, \ldots, n$. The ε_s represents the error in the corresponding subset of scanned links. In particular, they can be assumed to be null i.e. the plate data is got error free.

Then, following these assumptions, we have

$$\mathbf{F} = \left(\mathbf{K} \mid \mathbf{I} \right) \begin{pmatrix} U \\ -- \\ \boldsymbol{\eta}^T \end{pmatrix} \tag{45}$$

and the variance-covariance matrix $\boldsymbol{\Sigma}_{\mathbf{F}}$ of the **F** variables becomes

$$\boldsymbol{\Sigma}_{\mathbf{F}} = \left(\mathbf{K} \mid \mathbf{I} \right) \boldsymbol{\Sigma}_{(U,\boldsymbol{\eta})} \begin{pmatrix} \mathbf{K}^T \\ -- \\ \mathbf{I} \end{pmatrix} = \sigma_U^2 \mathbf{K}\mathbf{K}^T + \mathbf{D}_{\boldsymbol{\eta}}, \tag{46}$$

where the matrices $\boldsymbol{\Sigma}_{(U,\boldsymbol{\eta})}$ and $\mathbf{D}_{\boldsymbol{\eta}}$ are diagonal.
From (44) and (45)

$$\begin{pmatrix} \mathbf{F} \\ \mathbf{W} \end{pmatrix} = \begin{pmatrix} \mathbf{I} & | & \mathbf{0} \\ - & + & - \\ \boldsymbol{\Delta} & | & \mathbf{I} \end{pmatrix} \begin{pmatrix} \mathbf{F} \\ -- \\ \varepsilon \end{pmatrix},$$

which implies that the mean $E[(\mathbf{F}, \mathbf{W})]$ is

$$E[(\mathbf{F}, \mathbf{W})] = \begin{pmatrix} E(U)\mathbf{K} \\ ------- \\ E(U)\boldsymbol{\Delta}\mathbf{K} + E(\varepsilon) \end{pmatrix}, \tag{47}$$

and the variance-covariance matrix of $(\mathbf{F}, \mathbf{W})$ becomes

$$\boldsymbol{\Sigma}_{(\mathbf{F},\mathbf{W})} = \begin{pmatrix} \boldsymbol{\Sigma}_{\mathbf{F}} & | & \boldsymbol{\Sigma}_{\mathbf{F}}\boldsymbol{\Delta}^T \\ -- & + & ---- \\ \boldsymbol{\Delta}\boldsymbol{\Sigma}_{\mathbf{F}} & | & \boldsymbol{\Delta}\boldsymbol{\Sigma}_{\mathbf{F}}\boldsymbol{\Delta}^T + \mathbf{D}_{\varepsilon} \end{pmatrix}. \tag{48}$$

Now we have to define the graph: the route flows f_r are the parents of all link flow combinations w_s used by the corresponding travelers, and the error variables are the parents of the corresponding flows, that is, the ε_s are the parents of the w_s, and the η_r are the parents of the F_r. Finally, the U variable is on top (parent) of all route flows, because it gives the level of them (high, intermediate or low).
In this section we consider the simplest version of the proposed model, which considers only the route flows, and the scanned link flow combinations. Therefore, a further analysis requires that a model with all variables must be built i.e. including the mean and variance matrix of the all variables ($U, \eta_r; r = 1, 2, \ldots, m$ and $\varepsilon_s; s = 1, 2, \ldots, n$).

3.3 Using the model to predict traffic flows

Once we have built the model, we can use its JPD (similar to the one defined in (14)) to predict route flows (and therefore $\mathcal{OD}$ and link flows) when the information becomes available. In this section we propose an step by step method to implement the plate scanning-Bayesian network model:

Step 0: Initialization step. Assume an initial $\mathbf{K}$ matrix (for example, obtained from solving a SUE problem for a given out-of-date prior OD-pair flow data), the values of $E[U]$ and σ_U, and the matrices $\mathbf{D}_\varepsilon$ and $\mathbf{D}_\eta$.

Step 1: Select the set of links to be scanned. The set of links to be scanned must be selected. This chapter deals with this problem in Section 3.4 providing several methods to select the best set of links to be scanned.

Step 2: Observe the plate scanning data. The plate scanning data $\hat{w}_s$ are extracted.

Step 3: Estimate the route flows. The route matrix $\mathbf{F}$ with elements f_r are estimated using the Bayesian network method, i.e., using the following formulas (see (47), (48),(3) and (4)):

$$E[\mathbf{F}] = E[U]\mathbf{K} \tag{49}$$

$$E[\mathbf{W}] = E[U]\mathbf{\Delta K} + E[\varepsilon] \tag{50}$$

$$\mathbf{D}_{\boldsymbol{\eta}} = \text{Diag}\left(\nu E[\mathbf{F}]\right), \tag{51}$$

$$\mathbf{\Sigma_{FF}} = \sigma_U^2 \mathbf{K}\mathbf{K}^T + \mathbf{D}_{\boldsymbol{\eta}} \tag{52}$$

$$\mathbf{\Sigma_{FW}} = \mathbf{\Sigma_{FF}}\mathbf{\Delta}^T \tag{53}$$

$$\mathbf{\Sigma_{WF}} = \mathbf{\Sigma_{FW}} \tag{54}$$

$$\mathbf{\Sigma_{WW}} = \mathbf{\Delta}\mathbf{\Sigma_{FF}}\mathbf{\Delta}^T + \mathbf{D}_{\boldsymbol{\varepsilon}} \tag{55}$$

$$E[\mathbf{F}|\mathbf{W} = \mathbf{w}] = E[\mathbf{F}] + \mathbf{\Sigma_{FW}}\mathbf{\Sigma_{WW}^{-1}}(\mathbf{w} - E[\mathbf{W}]) \tag{56}$$

$$\mathbf{\Sigma_{F|W=w}} = \mathbf{\Sigma_{FF}} - \mathbf{\Sigma_{FW}}\mathbf{\Sigma_{WW}^{-1}}\mathbf{\Sigma_{WF}} \tag{57}$$

$$E[\mathbf{W}|\mathbf{W} = \mathbf{w}] = \mathbf{w} \tag{58}$$

$$\mathbf{\Sigma_{W|W=w}} = \mathbf{0} \tag{59}$$

$$\mathbf{F} = E[\mathbf{F}|\, \mathbf{W} = \mathbf{w}]|_{(\mathbf{F},\mathbf{W})=F} \tag{60}$$

where ν is the coefficient of variation selected for the η variables, and we note that **F** and **W** are the unobserved and observed components, respectively.

Step 4. Obtain the F vector. Return the the f_r route flows as the result of the model. Note that from **F** vector, the rest of traffic flows (link flows and $\mathcal{OD}$ pair flows) can be easily obtained.

3.4 The plate scanning device location problem

Due to the importance of the traffic count locations to obtain good traffic flow predictions, this section deals with the problem of determining the optimal number and allocation of plate scanning devices(see (Mínguez et al., 2010)).

3.4.1 Location rules

In real life, the true error or reliability of an estimated $\mathcal{OD}$ matrix is unknown. Based on the concept of maximal possible relative error (MPRE), (Yang & Zhou, 1998) proposed several location rules. We have derived analogous rules based on prior link and flow values and the following measure (RMSRE, root mean squared relative error):

$$\text{RMSRE} = \sqrt{\frac{1}{m}\sum_{i\in\mathcal{I}}\left(\frac{t_i^0 - t_i}{t_i^0}\right)^2}, \tag{61}$$

where[5] t_i^0 and t_i are the prior and estimated flow of $\mathcal{OD}$-pair i, respectively, and m is the number of $\mathcal{OD}$-pairs belonging to the set $\mathcal{I}$. Since the prior $\mathcal{OD}$ pair flows t_i^0 are known and there are the best available information, they are used to calculate the relative error.
Given the set R of all possible routes, any of them corresponding to a unique $\mathcal{OD}$ pair, if R_i is the set of routes belonging to $\mathcal{OD}$-pair i, we have $t_i^0 = \sum_{r\in R_i} f_r^0$, and then the RMSRE can be expressed as:

$$\text{RMSRE} = \sqrt{\frac{1}{m}\sum_{i\in I}\left(\frac{t_i^0 - \sum_{r\in R_i} f_r^0 y_r}{t_i^0}\right)^2}, \tag{62}$$

where y_r is a binary variable equal to one if route r is identified uniquely (observed) through the scanned links, and zero otherwise. Note that the minimum possible RMSRE-value corresponds to $y_r = 1;\ \forall r \in R$, where $t_i = t_i^0$ and RMSRE=0.
However, if $n_{\text{sc}} = \sum_{\forall r\in R} y_r \le n_r$ then RMSRE> 0, and then, one interesting question is: how do we select the routes to be observed so that the RMSRE is minimized? From (62) we obtain

$$m \times \text{RMSRE}^2 = \sum_{i\in I}\left(1 - \sum_{r\in R_i}\frac{f_r^0}{t_i^0} y_r\right)^2, \tag{63}$$

where it can be deduced that the bigger the value of $\sum_{r\in R_i}\frac{f_r^0}{t_i^0} y_r$ the lower the RMSRE. If the set of routes is partitioned into observed ($\mathcal{OR}$) and unobserved ($\mathcal{UR}$) routes associated with $y_r = 1$ or $y_r = 0$, respectively, (63) can be reformulated as follows

$$m \times \text{RMSRE}^2 = \sum_{i\in I}\left(1 - \sum_{r\in(R_i\cap\mathcal{OR})}\frac{f_r^0}{t_i^0}\right)^2 = \sum_{i\in I}\left(\sum_{i\in(R_i\cap\mathcal{UR})}\frac{f_r^0}{t_i^0}\right)^2, \tag{64}$$

[5] from now on, and for simplicity, we denoted each $\mathcal{OD}$ pair as i instead of ks and each link as a instead of ℓ_{ij}

so that routes to be observed ($y_r = 1$) should be chosen minimizing (64).
The main shortcoming of equations (63) or (64) is their quadratic character which makes the RMSRE minimization problem to be nonlinear. Alternatively, the following RMARE (root mean absolute value relative error) based on the mean absolute relative error norm can be defined:

$$\text{RMARE} = \frac{1}{m}\sum_{i\in I}\left|\frac{t_i^0 - t_i}{t_i^0}\right| = \frac{1}{m}\sum_{i\in I}\left|\frac{t_i^0 - \sum_{r\in R_i} f_r^0 y_r}{t_i^0}\right|, \tag{65}$$

and since the numerator is always positive for error free scanners ($0 \le \sum_{r\in R_i} f_r^0 y_r \le T_i^0;\ \forall i \in I$), the absolute value can be dropped, so that the RMARE as a function of the observed and unobserved routes is equal to

$$\text{RMARE} = 1 - \frac{1}{m}\left(\sum_{i\in I}\sum_{r\in(R_i\cap\mathcal{OR})}\frac{f_r^0}{t_i^0}\right) = \frac{1}{m}\left(\sum_{i\in I}\sum_{r\in(R_i\cap\mathcal{UR})}\frac{f_r^0}{t_i^0}\right), \tag{66}$$

which implies that minimizing the RMARE is equivalent to minimizing the sum of relative route flows of unobserved routes, or equivalently, maximize the sum of relative route flows of observed routes. Note that this result derives in a rule that can be denominated the **Maximum Relative Route Flow** rule.
The above location rule has been derived by supposing that the prior trip distribution matrix is reasonably reliable and close to the actual true value, because the accuracy of the prior matrix has a great impact on the estimates of the true $\mathcal{OD}$ matrix. Note that even though the knowledge of prior $\mathcal{OD}$ pair flows could be difficult in practical cases, the aim of the proposed formulation is determining which $\mathcal{OD}$ flows are more important than others in order to prioritize their real knowledge.
Since the proper identifiability of routes must be made through plate scanner devices in links, an additional rule related to links should be considered, which states that scanned links must allow us to identify uniquely the routes to be observed ($y_r = 1$) from all possible routes being considered. This rule can be denominated the **Full Identifiability of Observed Path Flows** rule.

3.4.2 Location models

The first location model to be proposed in this chapter considers full route observability, i.e. RMSRE$= 0$, but including budget considerations. In the transport literature, each link, is considered independently of the number of lanes it has. Obviously, when trying to scan plate numbers links with higher number of lanes are more expensive. Then:

$$\text{M}_1 = \underset{z}{\text{Minimize}} \sum_{a\in\mathcal{A}} \mathcal{P}_a z_a \tag{67}$$

subject to

$$\sum_{a\in\{\mathcal{A}\}} (\delta_a^r + \delta_a^{r_1})(1 - \delta_a^r\delta_a^{r_1}) z_a \ge 1 \left\{ \begin{array}{l} \forall (r, r_1)|r < r_1 \\ \sum_{a\in\mathcal{A}} \delta_a^r\delta_a^{r_1} > 0 \end{array}\right. \tag{68}$$

$$\sum_{a\in\mathcal{A}} z_a\delta_a^r \ge 1; \forall r, \tag{69}$$

where z_a is a binary variable taking value 1 if the link a is scanned, and 0, otherwise, r and r_1 are paths, $\mathbf{\Delta}$ is the route incidence matrix with elements δ_a^r.

Note that constraint (68) forces to select the scanned links so that every route is uniquely defined by a given set of scanned links (every row in the incidence matrix $\boldsymbol{\Delta}$ is different from the others) and (69) ensures that at least one link for every route is scanned (every row in the incidence matrix $\boldsymbol{\Delta}$ contains at least one element different from zero). Both constraints force the *maximum relative route flow* and *full identifiability of observed path flows* rules to hold. Note also that all $\mathcal{OD}$ pairs are totally covered. In addition, this model allows the estimation of the required budget resources $\mathcal{B}^* = \sum_{a\in\mathcal{A}} \mathcal{P}_a z_a^*$ for covering all $\mathcal{OD}$ pairs in the network. However, budget is limited in practice, meaning that some $\mathcal{OD}$ pairs or even some routes may remain uncovered, consequently based on (66) the following model is proposed in order to observe the maximum relative route flow:

$$\mathrm{M}_2 = \underset{\mathbf{y},\mathbf{z}}{\text{Maximize}} \quad \sum_{\forall i\in I}\sum_{r\in R_i} \frac{f_r^0}{t_i^0} y_r \tag{70}$$

subject to

$$\sum_{a\in\{\mathcal{A}\}} (\delta_a^r + \delta_a^{r_1})(1-\delta_a^r\delta_a^{r_1}) z_a \geq y_r \begin{cases} \forall (r,r_1)|r<r_1 \\ \sum\limits_{a\in\mathcal{A}} \delta_a^r\delta_a^{r_1} > 0 \end{cases} \tag{71}$$

$$\sum_{a\in\mathcal{A}} z_a\delta_a^r \geq y_r; \quad \forall r, \tag{72}$$

$$\sum_{a\in\mathcal{A}} \mathcal{P}_a z_a \leq \mathcal{B}, \tag{73}$$

where y_r is a binary variable equal to 1 if route r can be distinguished from others and 0 otherwise, z_a is a binary variable which is 1 if link a is scanned and 0 otherwise, and $\mathcal{B}$ is the available budget.

Constraint (71) guarantees that the route r is able to be distinguished from the others if the binary variable y_r is equal to 1. Constraint (72) ensures that the route which is able to be distinguished contains at least one scanned link. Both constraints (71) and (72) ensure that all routes such that $y_r = 1$ can be uniquely identified using the scanned links. It is worthwhile mentioning that using y_r instead of 1 in the right hand side of constraints (71) and (72) immediately converts into inactive the constraint (69) for those routes the flow of which are not fully identified.

Note that the full identifiability of observed path flows is included in the optimization itself and it will be ensured or not depending on the available budget $\mathcal{B}$. Note also that previous models can be easily modified in order to include some practical considerations as for example the fact that some detectors are already installed and additional budget is available. To do that one only need to include the following constraint to models M_1 or M_2

$$z_a = 1; \quad \forall a \in \mathcal{OL}. \tag{74}$$

where $\mathcal{OL}$ is the set of already observed links (links with scanning devices already installed).

3.5 Example of application

In this section we illustrate the proposed methods by their application to a simple example. Consider the network in Figure 4 with the routes and OD-pairs in Table 10, which shows the feasible combination of scanned links after solving the M_1 model ($\mathcal{SL} = \{1,2,3,4,7,8\}$). Next, the proposed method in Section 3.3 is applied.

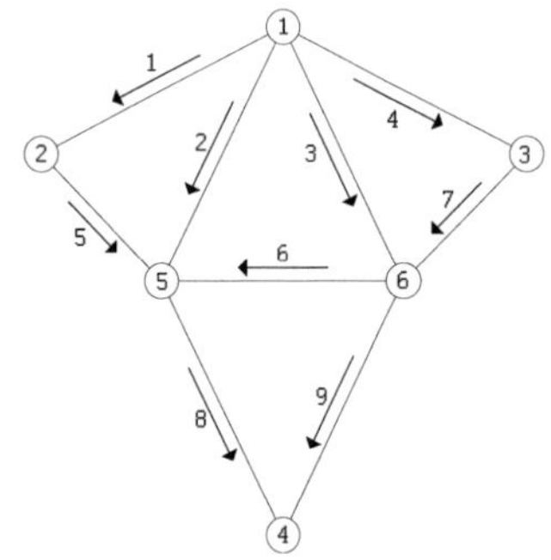

Fig. 4. The elementary example network

$\mathcal{OD}$	path code (r)	Links	set code (s)	Scanned links 1	2	3	4	7	8
1-4	1	1 5 8	1	X					X
1-4	2	2 8	2		X				X
1-4	3	3 9	3			X			
1-4	4	3 6 8	4			X			X
1-4	5	4 7 9	5				X	X	
1-4	6	4 7 6 8	6				X	X	X
2-4	7	5 8	7						X
2-4	8	7 6 8	8					X	X
3-4	9	7 9	9					X	

Table 10. Required data for the simple example.

Step 0: Initialization step. To have a reference flow, we have considered that the true route flows are those shown in the second column of Table 11. The assumed mean value was $E[U] = 10$ and the value of σ_U was 8. The initial matrix $\mathbf{K}$ is obtained by multiplying each true route flow by an independent random uniform $U(0.4, 1.3)/10$ number. The $\mathbf{D}_\varepsilon$ is assumed diagonal matrix, the diagonal of which are almost null (0.000001) because we have assumed error free in the plate scanning process. $\mathbf{D}_\eta$ is also a diagonal matrix which values are associated with a variation coefficient of 0.4.

Step 1: Select the set of links to be scanned. The set of links to be scanned have been selected using the M_2 model for different available budget, i.e. using the necessary budget for the devices needed to be installed in the following links:

$$\mathcal{SL} \equiv \{1,2,3,4,7,8\}; \mathcal{SL} \equiv \{1,4,5,7,9\}; \mathcal{SL} \equiv \{1,4,7,9\};$$

$$\mathcal{SL} \equiv \{4,7,9\}; \mathcal{SL} \equiv \{1,5\};\ \mathcal{SL} \equiv \{2\}.$$

Step 2: Observe the plate scanning data. The plate scanning data w_s is obtained by scanning the selected links as was explained in Section 3.1.

Step 3: Estimate the route flows. The route flows $\mathbf{F}$ with elements f_r are estimated using the Bayesian network method and the plate scanning data, i.e., using the formulas (49)-(60)

Route	True flow	Method	Scanned links 0	1	2	3	4	5	6
1	5.00	BN	4.26	4.35	5.00	4.91	5.00	5.00	5.00
		LS	4.26	4.26	5.00	4.26	5.00	5.00	5.00
2	7.00	BN	6.84	7.00	7.76	7.89	7.91	7.85	7.00
		LS	6.84	7.00	6.84	6.84	6.84	6.84	7.00
3	3.00	BN	3.45	3.52	3.91	3.00	3.00	3.00	3.00
		LS	3.45	3.45	3.45	3.00	3.00	3.00	3.00
4	5.00	BN	3.00	3.07	3.41	3.46	3.47	3.45	5.00
		LS	3.00	3.00	3.00	3.00	3.00	3.00	5.00
5	6.00	BN	5.36	5.47	6.08	6.00	6.00	6.00	6.00
		LS	5.36	5.36	5.36	6.00	6.00	6.00	6.00
6	4.00	BN	3.37	3.45	3.82	4.00	4.00	4.00	4.00
		LS	3.38	3.38	3.38	4.00	4.00	4.00	4.00
7	10.00	BN	8.90	9.08	10.00	10.25	10.28	10.00	10.00
		LS	8.90	8.90	10.00	8.90	8.90	10.00	10.00
8	7.00	BN	3.97	4.06	4.50	7.00	7.00	7.00	7.00
		LS	3.97	3.97	3.97	7.00	7.00	7.00	7.00
9	5.00	BN	5.45	5.57	6.18	5.00	5.00	5.00	5.00
		LS	5.45	5.45	5.45	5.00	5.00	5.00	5.00

Table 11. Route flow estimates using BN and LS approaches

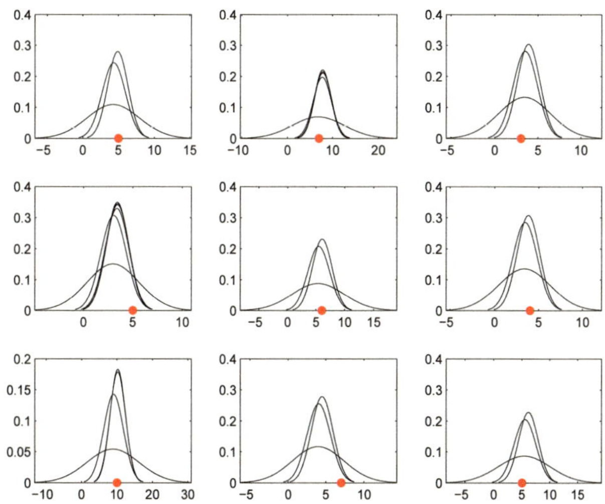

Fig. 5. Conditional distribution of the route flows

The method has been repeated for different subsets of scanned links shown in step 2 of the process. The resulting predicted route flows are shown in Table 11. The first rows correspond to the route predictions using the proposed model. With the aim of illustrating the improvement resulting from the plate scanning technique using Bayesian networks, we have compared the results with the standard method of Least Squares (LS) using the same data. The results appear in the second rows in Table 11. The results confirm that the plate scanning method using Bayesian networks outperforms the standard method of Least Squares for several reasons:

- The BN tool provides the random dependence among all variables. This fact allows us to improve the route flow predictions even though when we have no scanned link belonged to this particular route. Note that using the LS approach the prediction is the prior flow (the fourth column in Table 11, i.e with 0 scanned links in the network).
- The BN tool provides not only the variable prediction but also the probability intervals for these predictions using the JPD function. Fig. 5 shows the conditional distributions of the route flows the different items of accumulated evidence. From left to right and from top to bottom $f_1, f_2 \ldots$ predictions are shown. In each subgraph the dot represents the real route flow in order to analyze the predictions.

4. References

Anderson, T. W. (1984). *An introduction to multivariate statistical analysis*, John Wiley and Sons, New York.

Castillo, E., Gutiérrez, J. M. & Hadi, A. S. (1997a). *Expert systems and probabilistic network models*, Springer Verlag, New York.

Castillo, E., Gutiérrez, J. M. & Hadi, A. S. (1997b). Sensitivity analysis in discrete Bayesian networks, *IEEE Transactions on Systems, Man and Cybernetics* **26(7)**: 412–423.

Castillo, E., Menéndez, J. M. & Jiménez.P (2008). Trip matrix and path flow reconstruction and estimation based on plate scanning and link observations, *Transportation Research B* **42**: 455–481.

Castillo, E., Menéndez, J. M. & Sánchez-Cambronero, S. (2008a). Predicting traffic flow using Bayesian networks, *Transportation Research B* **42**: 482–509.

Castillo, E., Menéndez, J. M. & Sánchez-Cambronero, S. (2008b). Traffic estimation and optimal counting location without path enumeration using Bayesian networks, *Computer Aided Civil and Infraestructure Engineering* **23**: 189–207.

Castillo, E., Sarabia, J. M., Solares, C. & Gómez, P. (1999). Uncertainty analyses in fault trees and Bayesian networks using form/sorm methods, *Reliability Engineering and System Safety* **65**: 29–40.

Doblas, J. & Benítez, F. G. (2005). An approach to estimating and updating origin-destination matrices based upon traffic counts preserving the prior structure of a survey matrix, *Transportation Research* **39B**: 565–591.

Hazelton, M. L. (2003). Some comments on origin-destination matrix estimation, *Transportation Research* **37A**: 811–822.

Mínguez, R., Sánchez-Cambronero, S., Castillo, E. & Jiménez, P. (2010). Optimal traffic plate scanning location for od trip matrix and route estimation in road networks, *Transportation Research B* **44**: 282–298.

Praskher, J. N. & Bekhor, S. (2004). Route choice models used in the stochastic user equilibrium problem: a review, *Transportation Reviews* **24**: 437–463.

Sánchez-Cambronero, S., Rivas, A., Gallego, I. & Menéndez, J. (2010). Predicting traffic flow in road networks using bayesian networks and data from an optimal plate scanning device location., *in* A. F. J.Filipe & B. Sharp. (eds), *Proceedings of 2nd Internationa Conference on Agents and Artificial Intelligence.*, INSTICC-Springer., Valencia, Spain., pp. 552–559.

Sumalee, A. (2004). Optimal road user charging cordon design: a heuristic optimization approach, *Journal of Computer Aided Civil and Infrastructure Engineering* **19**: 377–392.

Yang, H. & Zhou, J. (1998). Optimal traffic counting locations for origin-destination matrix extimation, *Transportation Research* **32B**: 109–126.

20

Accommodating uncertainty in grazing land condition assessment using Bayesian Belief Networks

H. Bashari[A,C] and C.S. Smith[b]
[a] *Department of Natural Resources, Isfahan University of Technology, Isfahan, 84156-83111, Iran*
[b] *The University of Queensland, School of Agriculture and Food Sciences, St Lucia QLD 4072, Australia*

1. Introduction

Rangelands are semi-natural landscapes, an important global resource that covers more than 47 percent of the land area of Earth (332 million hectares) (Tueller, 1998). They have been used for many purposes (e.g. grazing, bee industry, hunting, mining and tourism). Rangeland ecosystems are highly variable in terms of their biophysical components such as rainfall and soil type (Gross et al., 2003 & 2006). The primary production of grasses can vary up to 10 times from year to year (Kelly & Walker, 1976). In addition, there are often clear conflicts in the multiple objectives of rangeland use and management (e.g. production and conservation).

Land managers and technical assistance specialists require a system for assessing rangeland condition in order to know where to focus management efforts and for a better understanding of ecosystem processes (Karfs et al., 2009). The assessment of the present condition of the land and monitoring of relevant and meaningful changes are essential for preventing land degradation (Liu, 2009). Range assessment is also essential to evaluate the effectiveness of implemented management practices and to identify the ecological problems in rangelands before its condition becomes seriously degraded (Manske, 2004).

Several key obstacles emerge when considering rangeland condition, namely, 1) no single entity can handle all aspects of rangeland condition and 2) rangeland condition varies in time and space (Bellamy & Lowes, 1999). This introduces uncertainty into rangeland management and therefore assessment tools. Although researchers have developed sophisticated methods of assessing rangeland condition, it is not easy to accommodate the uncertainty associated with the indicators used. Almost all of the rangeland condition assessment tools available at present use deterministic or 'hard' criteria to assess condition against a set of indicators, which does not represent the true variability or uncertainty associated with condition assessment. We believe Bayesian belief Networks (BBNs) (Jensen, 2001) provide a tool that can help solve this problem.

2. Bayesian Belief Networks

In the late 1980s, BBNs were introduced to accommodate uncertainty in the modeling of complex systems (Pearl, 1988). BBNs provide a probabilistic and dynamic representation of the relationships between variables using conditional probability (Jensen, 1996). They consist of qualitative and associated quantitative parts. The qualitative part is a directed graph (cause and effect diagram) with a set of nodes representing relationships between the variables under study. The quantitative part is a set of conditional probabilities that explain the strength of the dependences between variables represented.

BBNs have two main functions that make them valuable assessment tools. The first is a scenario, or what if, analysis where particular states of input nodes are selected to reveal the probability of outcomes occurring (Figure 1a). The second is diagnostic analysis where particular states of outcomes are selected to reveal the probability of inputs occurring (Figure 1b).

Some other key aspects of BBNs that make them attractive assessment tools are:

- They are graphical, which facilitates communication about systems behavior among managers;
- They are updatable, meaning that their conditional probabilities can be updated over time using monitoring records. Thus, nodes, states and relationships can be modified as new knowledge about the system becomes available;
- They provide an integrative framework that combines qualitative and quantitative knowledge, plus probabilities obtained from monitoring, experiential knowledge and outputs from other models.

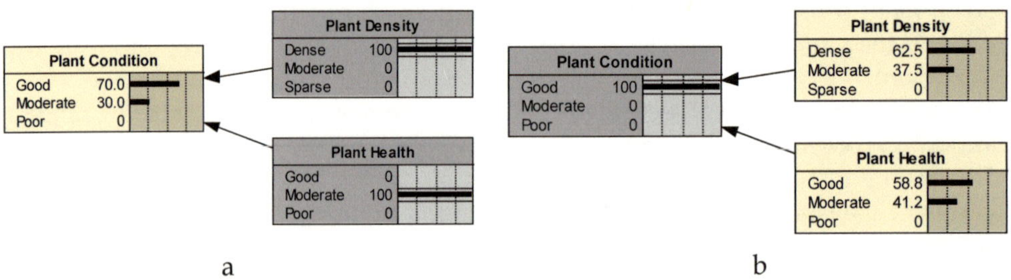

Fig. 1. A BBN used in (a) Predictive mode, (b) Diagnostic mode.

3. Stocktake

Stocktake is a Decision Support System (DSS) for paddock-scale grazing land condition monitoring and management (Department of Primary Industries and Fisheries, 2004). It has been developed and used recently in Australian pastures to assist land managers to assess grazing land condition, long term carrying capacity and calculate short-term forage budgets. It is designed to be applied to different land types and can be used in the broad scale assessment of grazing land condition. The simplicity and repeatability of Stocktake assists managers to assess grazing land condition based on a ABCD grazing land condition scoring framework by using indicators such as pasture composition, tree density, weeds and soil erosion (Chilcott et al., 2003) and relates grazing land condition to grass growth potential for different land types. This enables land managers to evaluate the effect that suboptimal grazing land condition will have on long-term carrying capacity. The forage budgeting component of this DSS provides a tool for land managers to regulate stock numbers according to seasonal forage supply.

Grazing land condition in Stocktake is comprised of three components including pasture condition, soil condition and woodland condition. Grazing land condition directly influences the ecosystem functioning, biodiversity and long term carrying capacity. It is affected by long term paddock management and its rate of change is slow over a number of seasons or years. The pasture condition component of grazing land condition indicates the capacity of the pasture to capture and transfer solar energy into edible components for livestock, capture rainfall and to preserve soil condition and nutrient cycling. Pasture condition depends on the presence of 3P grasses (perennial, palatable and productive grasses), crown cover and health of 3P grasses, species diversity and weed infestation. Soil condition indicates the soil capacity to capture rainfall, cycle and store nutrients, habitat for seed germination, support for the growth of seedlings and to resist erosion. Woodland condition indicates the ability of vegetation to regulate ground water and cycle nutrients (Department of Primary Industries and Fisheries, 2004).

The data requirements for assessing grazing land condition using Stocktake are limited to qualitative data that is very easy to collect. Large amounts of these data, including photos are recorded and stored for future reference. Although this DSS can assist land managers to plan, implement and monitor a grazing land management strategy for the whole property, it lacks the capability of assessing the effect of different grazing management plans on grazing land condition. It also does not incorporate uncertainty inherent in rangeland ecosystems in the assessment of grazing land condition or carrying capacity. In the following section, we demonstrate how BBNs can be used to change the Stocktake monitoring procedure into a predictive DSS.

4. The Grazing Land Condition Model

The development of a grazing land condition model consisted of two main steps: (a) conceptual model development, and (b) converting the conceptual model into a predictive grazing land condition model (Figure 2).

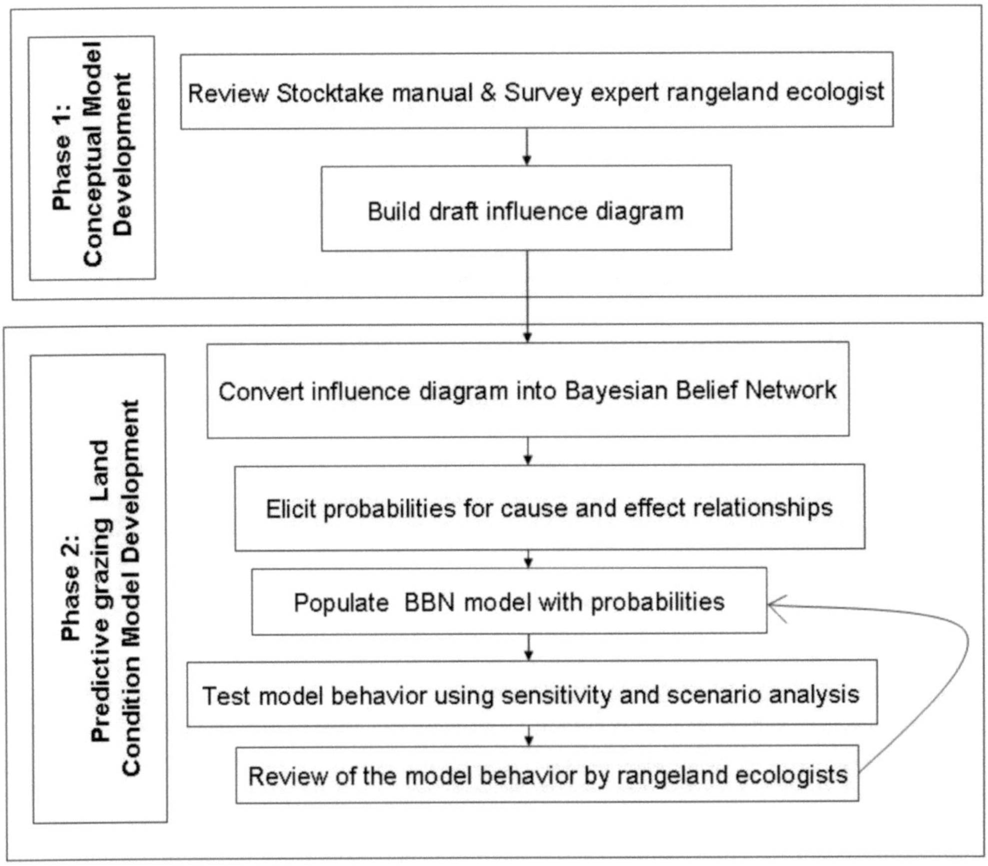

Fig. 2. Steps used to build a predictive grazing land condition model.

4.1. Conceptual Model Development

The purpose of conceptual model development was to build an influence diagram capturing the key variables believed to influence grazing land condition within the Stocktake land condition assessment approach. First, we reviewed the Stocktake manual, followed by a meeting held with an expert of the CSIRO who had expertise in grazing land condition assessment. We used the information from the manual and the meeting to build a draft influence diagram. The influence diagram (Fig.3) contained: (a) key environmental variables believed to influence pasture condition (b) key environmental variables believed to influence soil condition, and (c) key woodland variables believed to influence woodland condition. The draft influence diagram was reviewed by the grazing land condition expert and the influence diagram altered based on the feedback received.

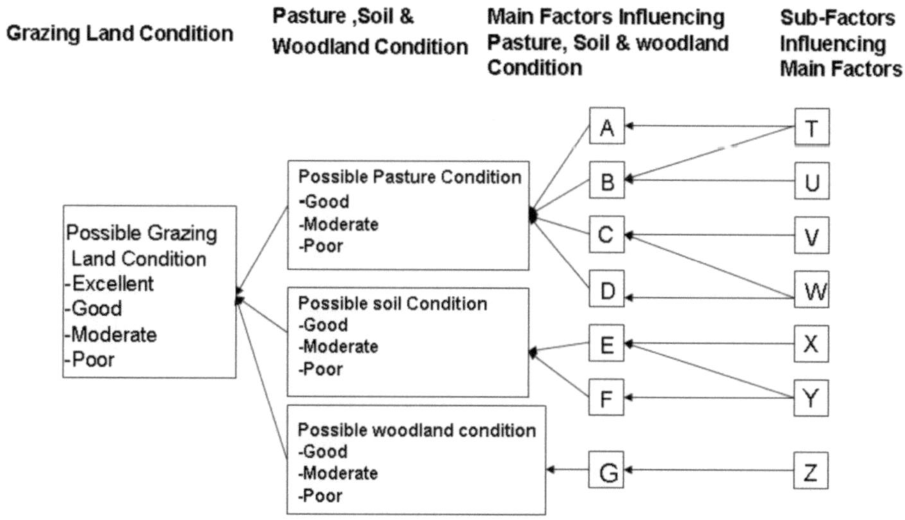

Fig. 3. Framework used to construct an influence diagram for grazing land condition model.

Next, states were defined for each node in the influence diagram. Figure 4 shows the completed influence diagram for Ironbark-Spotted Gum Woodland in south-east Queensland, Australia. Table 1 lists the states and the definitions for each node in the influence diagram.

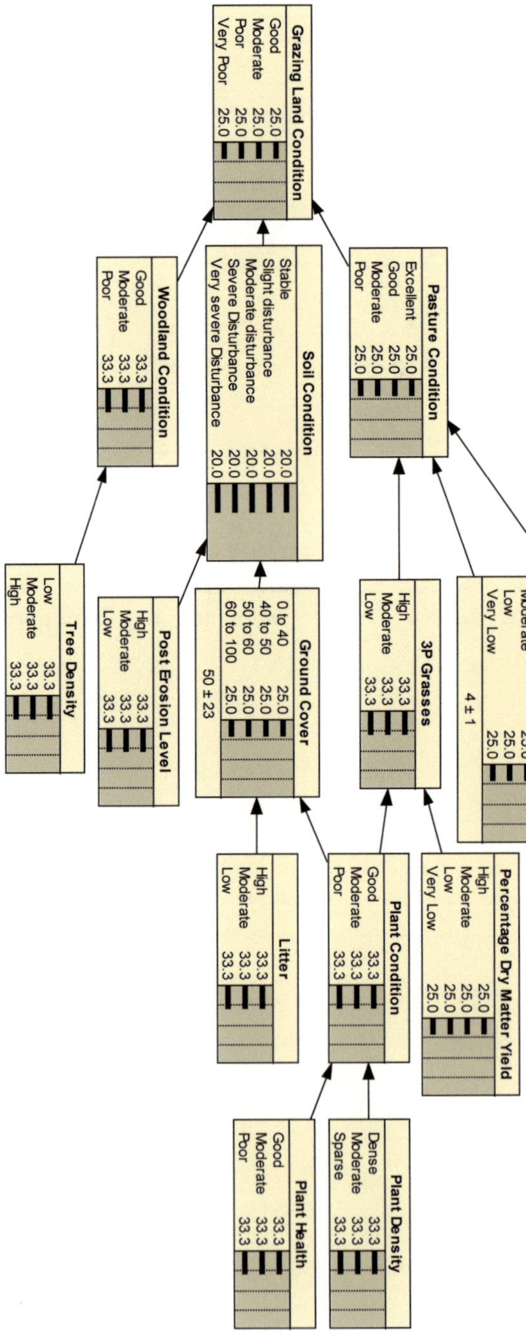

a

Fig. 4. Influence diagram for Bayesian grazing condition model.

Node	Definition and classes
Grazing Land Condition	This node represents the overall pasture, soil and woodland condition and represents the efficiency of ecosystem functioning. Grazing land condition is slow to change and it is an indicator of long term safe carrying capacity. Good: good coverage of perennial grasses, little bare ground, few weeds, no woodland thickening. Moderate: some decline of perennial grasses, soil condition and thickening in density of woody plants. Poor: general decline of perennial grasses and past erosion and obvious thickening in density of woody plants. Very Poor: general lack of perennial grasses, severe erosion, thickets of woody plants cover most of the area.
Pasture Condition	This node represents the status of perennial, palatable and productive grasses (3P grasses) , species diversity and weed infestation. Excellent: good coverage of 3P grasses, few weeds Good: some decline of 3P grasses and increase in less favoured species. Moderate: general decline in 3P grasses, large amounts of less favoured species Poor: general lack of 3P grasses.
Soil Condition	This node represents the soil health in terms of capacity of soil to absorb and store rainfall, resist erosion, nutrient cycling, and habitat for seed germination. Stable: good soil surface condition and no signe of erosion and soil movement. Slight Disturbance: some decline in soil condition, some signe of post erosion and increased surface runoff. Moderate Disturbance: obvious signe of past erosion (sheet or rill erosion), and high current erosion susceptibility. Plant pedestalling occurring, gravel and stone pavements common. Severe Disturbance: high erosion level (rill or gully erosion), bedrock at the surface. Very Severe Disturbance: severe erosion or scalding, gully erosion more than 15 mm deep.
Three P Grasses	This node represents the status of perennial, palatable and productive grasses. High: good coverage of 3P grasses. Moderate: some decline of 3P grasses. Low: general decline or lack of 3P grasses
3P Grasses Diversity Desirable	This node represents if a variety of the native species with high grazing value are dominant. High: pasture consists of more than 5 desirable species. Moderate: pasture consists of 3 to 5 desirable species. Low: pasture consists of 2 to 3 desirable species. Very Low: pasture consists of one or less desirable species.
Weeds	This node represents how much of the pasture are covered by unpalatable, invader and in some cases poisonous species. Weeds have direct influence on the paddock productivity. None: there are not any weeds in the pasture. Low: few weeds and no significant infestations. Moderate: there are large amounts of weeds. High: weeds are dominant and infested heavily.

Node	Definition and classes
Ground Cover	This node represents how much of the soil surface is protected against rain. It includes vegetation cover, leaf litter, dung, sticks or rocks. 0 to 4o: this percentage of ground is covered by vegetation cover, leaf litter, dung, sticks or rocks. 40 to 50: this percentage of ground is covered by vegetation cover, leaf litter, dung, sticks or rocks. 50 to 60: this percentage of ground is covered by vegetation cover, leaf litter, dung, sticks or rocks. 60 to 100: this percentage of ground is covered by vegetation cover, leaf litter, dung, sticks or rocks.
Post Erosion Level	This node represents the status of erosion in the past. High: there are obvious and severe erosion signs. Moderate: there are some erosion signs. Low: there are few erosion signs.
Percentage Dry Matter Yield	This node represents how much of the pasture yield is comprised by 3P grasses. High: 3P grasses comprise 80% or more of pasture yield. Moderate: 3P grasses comprise 60 to 80% of pasture yield. Low: 3P grasses comprise 10 to 60% of pasture yield. Very Low: 3P grasses comprise less than 10% of pasture yield.
Plant Condition	This node represents the status of plants in terms of their crown cover and healthiness. Good: plants are dense and healthy. Moderate: moderate density and some plants dead. Poor: sparse and many plants dead.
Plant Density	This node represents the number of individual plants in a given area. Dense: the crowns of 3P grasses are not sparse and there is not bare ground in-between. Moderate: the crowns of 3P grasses are not dense and there is some bare ground in-between. Sparse: the crowns of 3P grasses are sparse and there is much bare ground in-between.
Plant Health	This node represents the healthiness status of 3P grasses. Good: the 3P grasses are healthy and they are not diseased, discoloured or poor growth. Moderate: some of the 3P grasses are healthy and some are diseased, discoloured or dead. Poor: many of 3P grasses are unhealthy, diseased, discoloured or dead.
Litter	This node indicates if there are litter between the tussocks of grasses. High: little bare ground and much litter in-between. Moderate: some bare ground and litter in-between. Poor: much bare ground and low litter in-between.

Table 1. Definition for nodes and their classes in the land condition model, adapted from Department of Primary Industries and Fisheries, 2004)

4.2. Eliciting probabilities for the model

Conditional Probability Tables (CPTs) characterize the relationships between nodes within a BBN (Bashari et al., 2009). To produce a predictive model, the CPTs in the grazing land condition model influence diagram were populated using subjective probability estimates obtained from the expert who participated in building the influence diagram. It was necessary to elicit subjective probability estimates because measured probabilities were not available and can only be obtained from long-term studies.

A CPT calculator developed by Cain (2001) was used in the probability elicitation process to maintain logical consistency in the estimated probabilities. It also reduced the number of probabilities that had to be elicited from the expert to populate the BBN. The CPT calculator works by reducing a CPT to the minimum number of scenarios for which probabilities need to be estimated. These scenarios allow the CPT calculator to determine the relative influence of each factor on the probability of outcomes. Once probabilities for these scenarios are elicited, the calculator checks for logical consistency and then interpolates probabilities for all scenarios in the CPT.

To illustrate, the shaded lines in Table 2 represent the reduced CPT for the node "Plant Condition", which has two input nodes; plant density and plant health. In the reduced CPT, (a) the first line represents the best-case scenario where all of the parent nodes of "plant condition" are in the best state, (b) the last line represents the worst-case scenario where all of the parent nodes of "plant condition" are in the worst state, and (c) the remaining shaded lines represent scenarios where only one parent node is not in the best state. Probabilities for the shaded lines are elicited from an expert, after which the CPT calculator interpolates probabilities for the full CPT (Table2). For parentless nodes in the grazing land condition influence diagram (for example, the "plant density", "plant health" "litter" and "Tree density" and " Post erosion level") uniform probability distributions were specified for their CPTs (each state was given equal probability).

Factors influencing Plant Condition		Probability of Plant Condition (%)		
Plant Density	Plant Health	Good	Moderate	Poor
Dense	Good	100	0	0
Dense	Moderate	70	30	0
Dense	Poor	0	50	50
Moderate	Good	60	40	0
Moderate	Moderate	42	58	0
Moderate	Poor	0	50	50
Sparse	Good	0	40	60
Sparse	Moderate	0	30	70
Sparse	Poor	0	0	100

Table 2. The full probability table for "plant condition" interpolated using the CPT calculator (the scenarios for which probabilities were elicited are highlighted).

4.3.Testing Model Behavior

To test the behavior of the completed grazing land condition model, and to highlight any inconsistencies, a sensitivity analysis was performed and the results compared with the expectations of rangeland scientists (Table 3). The measure of sensitivity used was entropy reduction (Marcot, 2006)

Table 3. Sensitivity of grazing land condition to the key environmental variables (variables are listed in order of influence on grazing land condition from most to least influential)

Node	Entropy reduction
Pasture condition	0.923
3P grasses	0.5061
Soil condition	0.2622
Plant condition	0.2522
Ground cover	0.2272
Plant density	0.09224
Percentage dry matter	0.08194
Plant health	0.06302
Weeds and annuals	0.02942
Post erosion level	0.01361
3P grasses diversity desirable	0.005579
Litter	0.00366
Woodland condition	0.0007306
Tree density	0.0006092

Sensitivity is calculated as the degree of entropy reduction I, which is the expected difference in information bits H between variable Q with q states and findings variable F with f states, after (Marcot, 2006):

$$I = H(Q) - H(Q \mid F) = \sum_q \sum_f \frac{P(q,f)\log_2[P(q,f)]}{P(q)P(f)}$$

The sensitivity analysis revealed that pasture condition was the most influential factor on grazing land condition, followed by 3P grasses (which directly influences pasture condition). Species composition in most grassland ecosystems has proved to be a good indicator of ecosystem processes (Heady, 1975). Soil and plant condition had similar influence on grazing land condition.

BBN models have the ability to provide rangeland managers with decision support through their analytic capabilities. As mentioned before, two main types of analysis can be performed using a BBN, (a) prediction, and (b) diagnosis. Predictive analysis can be used to answer "what if" questions and diagnostic analysis can be used to answer "how" questions. Figure 5 is an example of the grazing land condition model used for predictions. Here, the selected states of input nodes (outer boxes) represent a scenario for a site. In Figure 5a, the model shows that, under the selected scenario, the chance of this site being in good condition is high (85.6%). Also there is 75% chance of the site having excellent pasture condition. The model also indicates the probable causes for this condition, that is, good plant condition (100%) and high 3P grasses (100%). These causes were also highlighted by sensitivity analysis as being influential on grazing land condition (Table 3).

Besides answering "what if" questions, the BBN grazing land condition model can also help to answer "how" questions. For example, how might grazing land condition fall in a poor state? Figure 5b is an example of the grazing condition model being used to answer this question using diagnosis. The model shows that it is most likely if pasture condition is poor and soil condition is severely disturbed, and in turn, low abundance of 3P grasses.

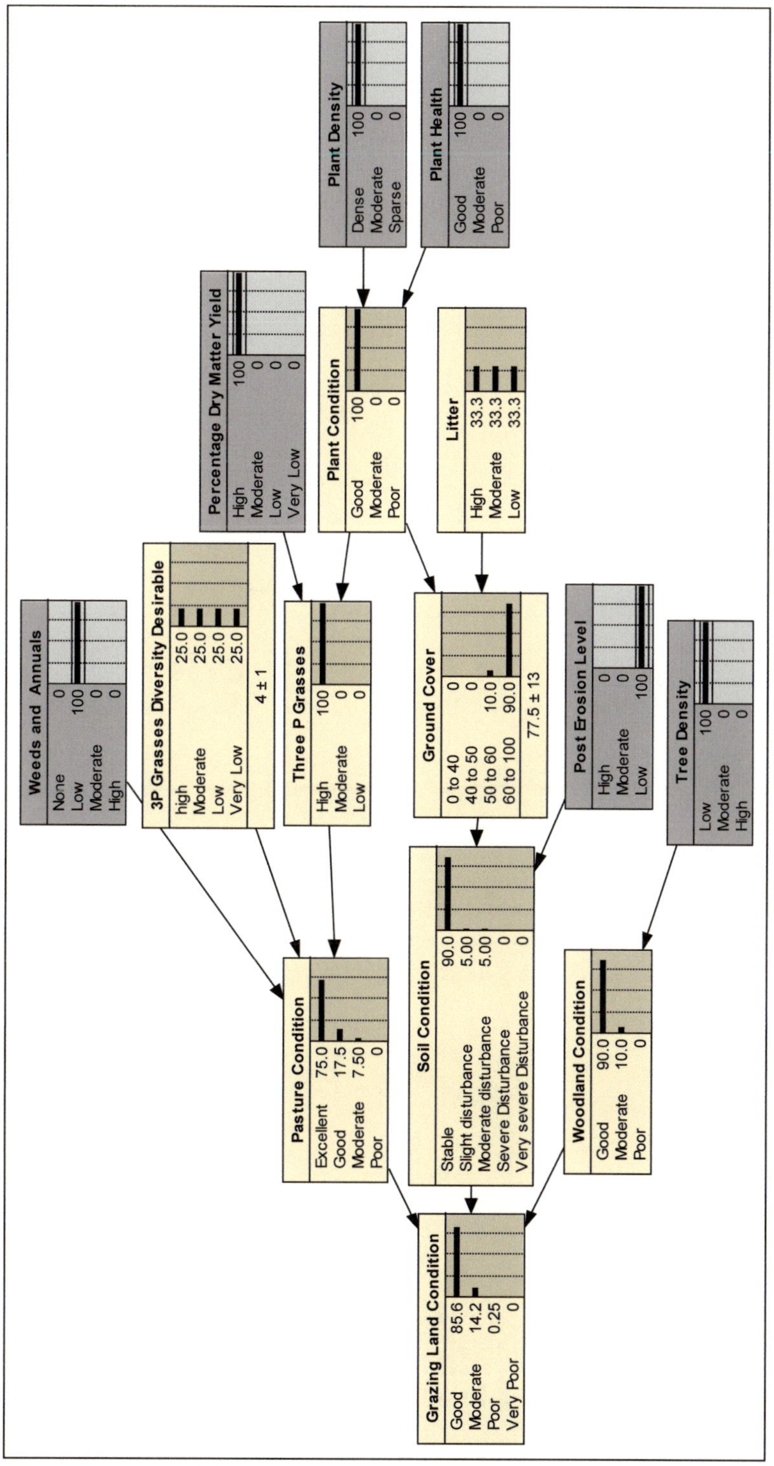
Plant Density
Dense 100
Moderate 0
Sparse 0
Plant Health
Good 100
Moderate 0
Poor 0
Percentage Dry Matter Yield
High 100
Moderate 0
Low 0
Very Low 0
Plant Condition
Good 100
Moderate 0
Poor 0
Litter
High 33.3
Moderate 33.3
Low 33.3
Weeds and Annuals
None 0
Low 100
Moderate 0
High 0
3P Grasses Diversity Desirable
high 25.0
Moderate 25.0
Low 25.0
Very Low 25.0
4 ± 1
Three P Grasses
High 100
Moderate 0
Low 0
Ground Cover
0 to 40 0
40 to 50 0
50 to 60 10.0
60 to 100 90.0
77.5 ± 13
Post Erosion Level
High 0
Moderate 0
Low 100
Tree Density
Low 100
Moderate 0
High 0
Pasture Condition
Excellent 75.0
Good 17.5
Moderate 7.50
Poor 0
Soil Condition
Stable 90.0
Slight disturbance 5.00
Moderate disturbance 5.00
Severe Disturbance 0
Very severe Disturbance 0
Woodland Condition
Good 90.0
Moderate 10.0
Poor 0
Grazing Land Condition
Good 85.6
Moderate 14.2
Poor 0.25
Very Poor 0

a

b

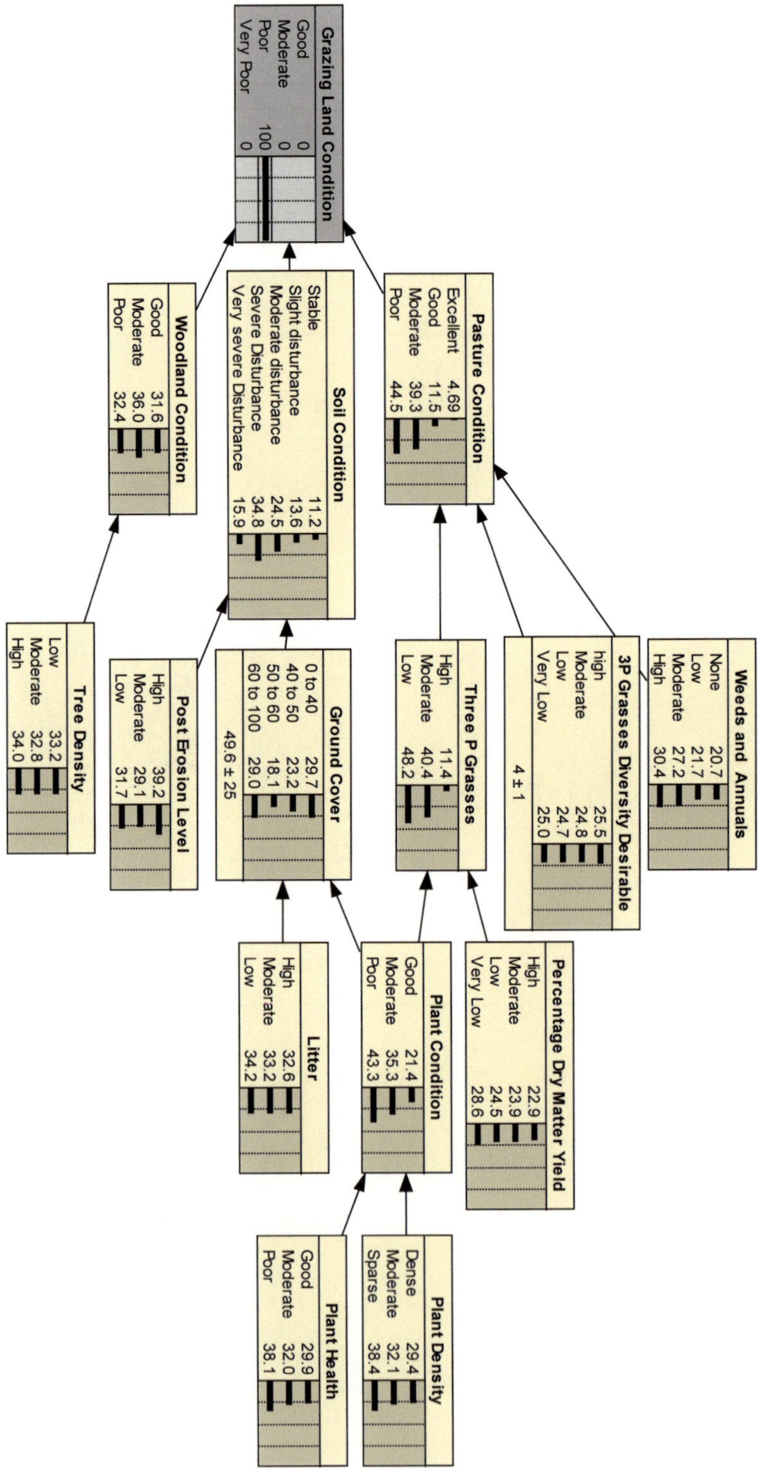

Fig. 5. Using the Bayesian belief network for diagnosis.

5. Conclusion

Most methods for assessing rangeland condition are deterministic. Stocktake is a local-level monitoring tool that is flexible, adaptive and easy to use by local land users for monitoring and documenting changes in grazing land condition in order to guide and support management responses accordingly. Integration of a condition assessment tool, such as Stocktake, with BBN allows for the construction of cause and effect models and allows uncertainty to be explicitly incorporated into condition assessment. The predictive and diagnostic capabilities of BBNs have the potential to provide valuable information to rangeland managers by allowing them to conduct scenario analysis. The simplicity of the approach, the graphical nature of the models and their scenario analysis capabilities also facilitates the communication of rangeland condition dynamics with land managers.

6. Acknowledgment

The authors are grateful to Mr Bruce Alchin from the University of Queensland for his assistance in this study and to Dr. John McIvor from CSIRO Sustainable Ecosystems, for his participation in the workshops associated with this study and the sharing of his expert knowledge.

7. References

Bellamy, A.J & Lowes, D., (1999). Modeling change in state of complex ecological systems in space and time: an application to sustainable grazing management, *Environment International*, 25, 701-712

Boyd, C.S. & Syeicar, T.J. (2009). Managing complex problems in rangeland ecosystems. *Rangeland ecology & management*, 62, 491-499.

Cain, J.(2001). *Planning improvements in natural resources management: Guidelines for using Bayesian networks to support the planning and management of development programs in the water sector and beyond*, Center for Ecology and Hydrology, Wallingford, UK.

Department of Primary Industries and Fisheries (2004).*Stocktake balancing supply and demand*, DPIF, Brisbane.

Gross, J., McAllister, R., Abel, N., Stafford-Smith, D.& Maru, Y. (2003). Australian rangelands as complex adaptive systems: A conceptual model and preliminary results. *Paper presented to modeling and simulation society of Australian and Newzealand Inc (MODISM)*, Tonsville, Australia.

Gross, J.E., McAllister, R.R.J., Abel, N., Smith, D.M.S. & Maru, Y. (2006) Australian rangelands as complex adaptive systems: A conceptual model and preliminary results, *Environmental modelling & software*, 21, 1264-1272.

Heady, H. F. (1975). Range condition and trend, *Paper presented to Evaluation and mapping of tropical African rangelands*, Addis Ababa, International Livestock Center for Africa.

Jensen, F.V. (1996). An introduction to Bayesian Networks, University College London Press.

Jensen, F.V. (2001). Bayesian network and decision graphs, Springer, New York.

Karfs, R.A., Abbott, B.N., Scarth, P.F. & Wallace, J.F. (2009). Land condition monitoring information for reef catchments: a new era. *Rangeland Journal*, 31, 69-86.

Kellner, K., & Moussa, A.S. (2009). A conceptual tool for improving rangeland management decision-making at grassroots level: the local-level monitoring approach. *African journal of range & forage science*, 26, 139-147.

Kelly, R.D. & Walker, B.H. (1976). The effects of different forms of land use on ecology of a semi-arid region in south-eastern Rhodesia. *Ecology*, 64, 535-76.

Liu, A. (2009). Monitoring and evaluation as tools for rangeland management. Edited by: Squires, V.R., Xinshi, L., Tao, W. & Youlin, Y., Rangeland degradation and recovery in China's pastoral lands.

Manske, L.L. (2004). Simplified assessment of range condition. *Grassland report in Dickinson Research extension Center.*

Marcot, B.G., Steventon, J.D., Sutherland, G.D.& McCann, R.K., (2006). Guidelines for developing and updating Bayesian belief networks applied to ecological modeling and conservation1. *Canadian Journal of Forest Research* 36,, 3063–3074.

Pearl, J. (1988). *Probabilistic Reasoning in Intelligent Systems: Networks of Plausible Inference,* Morgan Kaufmann, San Francisco.

Tueller, P. (1988). *Vegetation science applications for rangeland analysis and management*, Kluwer Academic

21

Classification of categorical and numerical data on selected subset of features

Zaher Al Aghbari
Department of Computer Science
University of Sharjah
UAE

1. Abstract

Many Data Mining techniques use the whole features space in the classification process. This feature space might contain irrelevant, or redundant, features that could reduce the accuracy of classification. This paper presents an approach to select a subset of features that are most relevant to the classification application. We use a wrapper approach to search for relevant subset of features, which will be used in the classification of two datasets: categorical teachers' dataset and numerical image dataset. Naïve Bayesian algorithm and *K*-Nearest Neighbor algorithm are used to classify and estimate the accuracy of the categorical data and numerical data, respectively. The experimental results for both categorical and numerical datasets indicate that classification accuracy is improved by removing the irrelevant features and using only the relevant subset of the feature space.

Key-Words: - Data mining, classification, feature selection, wrapper approach, image classification, and categorical data classification.

2. Introduction

Numerous data mining and machine learning applications employ feature selection techniques to improve their performance accuracy and efficiency. Instead of performing the task on the whole input feature set, these applications optimize the solution by selecting only the relevant subset of features, discarding the irrelevant ones, and perform the task on the selected subset of features. As a result, the running time cost of the system is reduced. However, minimizing the set of features may lead to degradation of classification accuracy. Thus, it is essential to include sufficient number features to achieve comparable, or better, classification accuracy as compared to including the whole input set of features.

If we assume that the whole set of features is S, then for some applications it is necessary to develop an algorithm that can reduce the set of features S to a subset of relevant features F, where F < S. Those eliminated features are irrelevant, or redundant, features and can negatively contribute to the classification accuracy. Therefore, before performing the classification task, the relevant subset of features should be searched for. There are two methods to search for the relevant features. In the first method, the search can be performed based on prior knowledge of the feature space and the targeted results; however, this method is subjective and based on the user's intuition and it difficult to apply the same method to different applications (John et. al, 1994). In the second method, a heuristic algorithm is developed to automatically select a subset of features, F, from the whole set of features, S, that will be sufficient to improve accuracy. However, with a moderate size of S, the number of subsets to be considered grows exponentially with the number of features S (Guyon & Elisseeff, 2003). There are two heuristic approaches in the literature to select the relevant subset of features: filter approach and wrapper approach.

The filter approach tries to find a subset of features independently of the inductive algorithm that will use this subset in classification. This is achieved by applying some statistics to select strong relevant features and filter out the weak relevant ones before executing the classification algorithm. In contrast, wrapper approach searches for subsets of features using cross-validation and compares the performance of the classification algorithm with each tested subset in order to select the optimal one. Although the wrapper approach achieves better classification performance compared to filter approach, it requires more time for computations (Guyon & Elisseeff, 2003). The filter approach emphasizes the discovery of relevant features that maximizes the classificaiton accuracy, while the wrapper approach searches for relevant features that minimizes the classification error (Lui & Kender, 2003).

Some scientific applications, such as fusion physics and remote sensing, necessitate the use of feature selection algorithms (Cantu-Paz et al., 2004). In fusion physics, the goal of scientists is not to build a predictor but to identify which features are related with an interesting state of the plasma. In remote sensing, feature selection algorithms are used to automate the identification of human settlements in satellite imagery, which is an essential step in the production of maps of human settlements that are used in studies of urbanization, population movement, etc.

In this paper, we present an approach to select a subset of features that are most relevant to the classification application. We use the Sequential Forward Selection algorithm (SFS) in a wrapper approach to search for relevant subset of features. The selected subset of features will be used in the classification of two datasets: categorical teachers' dataset and numerical image dataset. Naïve Bayesian algorithm and K-Nearest Neighbor algorithm are used to classify and estimate the accuracy of the categorical data and numerical data, respectively.

In Section 2, we survey the related work to classification based on selected subset of features. Then, in Section 3, we present the algorithms for searching a subset of features and the classification algorithms. In Section 4, we present our experimental results. Finally, we conclude the paper in Section 5.

3. Related Work

Developing heuristic algorithms that efficiently searches the space of features and selects the best subset that maintains the same or better performance was a field of research for the past

4 decades. One of the most common feature selection algorithms is genetic algorithms (GA) (Holland, 1975; Laanaya et al., 2005; Vafaie & Imam, 1994; Hao et al., 2003). In GA, a population of candidate solutions of selected subsets of features is always maintained. Candidate solutions are sometimes named as individuals, chromosomes, etc. Each individual is an encoded representation of features of the problems at hand. Each feature in an individual is termed as Gene. The evolution starts from a population of completely random individuals and happens in generations. In each generation, the fitness of the whole population is evaluated; multiple individuals are stochastically selected from the current population based on their fitness, mutated or recombined to form a new population, which becomes current in the next iteration of the algorithm (Laanaya et al., 2005). This generalization process is repeated until a termination condition is achieved such as a solution that satisfies minimum criteria is found, which could be fixed number of generations is reached.

Another feature selection algorithm is called importance score, which is based on greedy-like search (Vafaie & Imam, 1994). The algorithm is based on determining the importance score of each feature using a fitness function and then it performs a greedy-like search to obtain the minimum set of features that maximizes the recognition of some learned rules.

Secquential backwork elimination (SBE) (Marill & Green, 1963) and Sequential forward selection (SFS) (Whitney, 1971) are greedy wrappers used to select the relevant subset of features. SBE start the search with a full set and in each iteration it examines all subsets by removing one feature and retains the subset the gives the highest accuracy as a basis for the next iteration. On the other hand, SFS starts with an empty set and in every iteration it adds one feature to the subset. The search terminates after the accuracy of the current subset cannot be improved by removing (in case of SBE), or adding (in case of SFS), any other feature. However, the drawback of SFS is that once a feature is selected it cannot be removed even if its removal will increase performance accuracy. Similary, in SBE, once a feature is removed it cannot be included even if its inclusion will increase performance accuracy.

A recent algorithm called Basic Sort-Merge Tree (BSMT) (Lui & Kender, 2003) is proposed to choose a very small subset of features. BSMT can be divided into two parts: the creation of a tree of feature subsets, and the manipulation of the tree to create a feature subset of desired cardinality or accuracy. Each part uses a heuristic greedy method. The algorithm reduces the cardinality of the input data by sorting the individual features by their effectiveness in categorization, and then merging these features pairwise into feature sets of cardinality two. Repeating this Sort-Merge process several times results in a subset of features that is efficient and accurate, which is then used in the classification process.

A memory-based algorithm, called leave-one-out cross validation (LOOCV) (Moore & Lee, 1994) employs backward and forward hill-climbing techniques to search for the best subset of features without having to exhaustively evaluate all possible subsets.

In this paper, we present SFS based search algorithm that avoids evaluating all possible subsets of features in a wrapper approach and then the selected subsets of the categorical dataset are classified by a Naïve Bayes algorithm and the selected subsets of the numerical datasets are classified by a K-Nearest Neighbor algorithm.

4. Feature Selection

There are two major approaches, namely the filter approach and wrapper approach, to select the relevant subset of features that will improve system performance in terms of cost and accuracy. As compared to the filter approach, the wrapper approach improves the system performance by reducing the classification error. However, the wrapper approach requires more computations (Guyon & Elisseeff, 2003).

4.1 Feature Subset Selection

Selecting a subset of features has many potential benefits for classification applications:

- Reduces dimensionality to improve classification.
- Reduces compuatational cost and storage requirements.
- Reduces training time.
- Facilitates data understanding.

A simple greedy algorithm called Sequential Forward Selection SFS was proposed by Whitney in 1971 (Whitney, 1971) to search for the best subset of features. SFS (see below) starts with an empty feature subset (see line 1). In each iteration, one feature is added to the feature subset. To determine which feature to add, the algorithm tentatively adds to the candidate feature subset one feature that is not already selected and tests the accuracy of a classifier built on the tentative feature subset. The feature that results in the highest accuracy is added to the feature subset (lines 3-8). If we have added all the features or there is no improvement accrued from adding any further features, the search stops and returns the current set of features (line 9). This algorithm returns a single solution which contains the same selected subset of features on a given problem at every run. As shown in Fig. 1, the SFS algorithm takes as input the whole set of input features and returns the relevant subset of features.

Algorithm: SFS

Input: *whole set of input features, S*

Output: *best subset of features, F*

1) Let current subset, $F = \phi$
2) While size of $F < \tau$, where τ is the maximum allowed size of F.
3) for each $f \in S$
4) set $F' \leftarrow f \cup F$
5) evaluate F' and keep result
6) set $F \leftarrow F'$ of best result
7) set $S \leftarrow S - f$ of best result
8) keep evaluation result of current F
9) Return F

4.2 Subset Selection for Categorical Data

Bayes theory is a statistical method that measures the probability of a record in belonging to different classes. A method called Naïve Bayesian classifier (NB-Classifier), which is based on Bayes theory (Tan et al., 2006), is used to measure the accuracy of classification of our categorical teachers dataset into three classes: assistant professor, associate professor and full professor. Each record in the staff dataset consists of six features: name, age, nationality, salary, number of research works, and number of advisees.

The NB-Classifier is particularly suited when the dimensionality of the inputs is high. Despite its simplicity, NB-Classifier can often outperform more sophisticated classification methods. The NB-Classifier requires only one scan of the training data. Furthermore, it can easily handle missing values by simply omitting their probabilities when calculating the likelihoods of membership in each class. This method handles discrete values; however, if an attribute has continuous data, such as salary, these continuous values are divided into ranges. The ranges we used in our experiment are presented in Section 4. Table 1 summarizes the major notation used in this Section and subsequent sections.

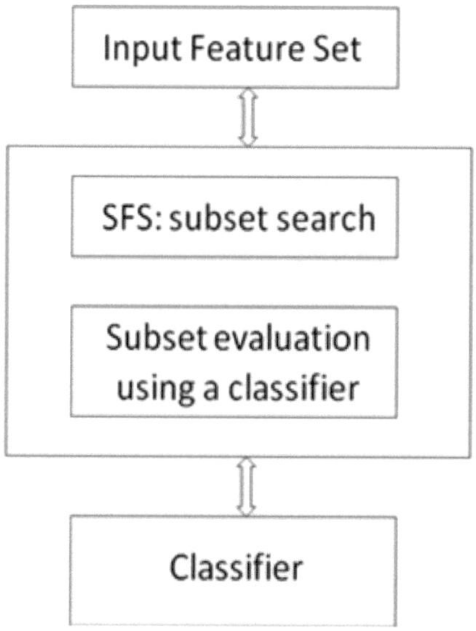

Fig. 1. feature subset search and evaluation using a wrapper approach

The *NB*-Classifier can be viewed as a specialized form of a Bayesian network, termed naïve because it relies on two important simplifying assumptions: independence and normality. That is it assumes that the predictive features x_{ik} of an observed staff record x_i are conditionally independent given the class c_j. These assumptions support very efficient algorithms for both learning and classification. An *NB*-classifier is often represented graphically as shown in Fig. 2, where the direction of the arrows state that the predictive attributes x_{i1}, x_{i2}, …, x_{in} are conditionally independent given the class c_j.

symbol	description
X	A set of s observed records $X = x_1, x_2, \ldots, x_s$
x_{ij}	feature j of the observed record x_i
C	A set of m classes $C = c_1, c_2, \ldots, c_m$
$P(c_i)$	Prior probability associated with class c_i
$P(x_i)$	Probability of occurrence of record x_i
$P(x_i \mid c_j)$	Conditional probability that given class c_j the record x_i satisfies it
$P(c_j \mid x_i)$	Posterior probability that estimates the probability of c_j given x_i

Table 1. summary of notation used in this paper

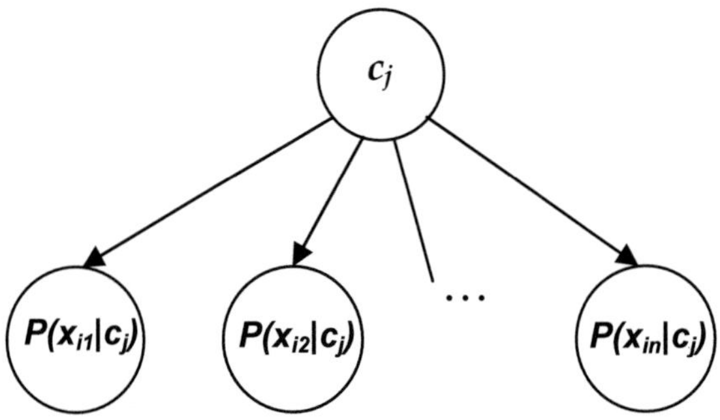

Fig. 2. A Bayesian network that represent the NB-classifier.

Let a set of classes $C = c_1, c_2, \ldots, c_m$ denote the classes of the observed staff records (training set) $X = x_1, x_2, \ldots, x_s$. Consider each observed record x_i as a vector of random variables denoting the predictive feature values $x_{i1}, x_{i2}, \ldots, x_{in}$. Then, given a test instance x to be classified, first, using Bayes rule (Eq. 1) we compute the posterior probabilities of each class and then predict the class with the highest probability as the class of x.

$$P(c_j \mid x_i) = \frac{P(x_i \mid c_j)P(c_j)}{P(x_i)} \tag{1}$$

From the training set, $P(c_j)$ is computed by counting the number of occurrences of c_j. For each feature x_{ik}, the number of occurrences is counted to determine $P(x_i)$. Similarly, assuming categorical features, the probability $P(x_i \mid c_j)$ can be estimated by counting how often each value x_{ik} occurs in the class in the training set.

Since a staff record has n independent features, we compute $P(x_{ik} \mid c_j)$ for every feature and then estimate $P(x_i \mid c_j)$ by the conjunction of all conditional probabilities of the features as shown in Eq. 2.

$$P(x_i \mid c_j) = \prod_{k=1}^{n} P(x_{ik} \mid c_j) \tag{2}$$

The posterior probability, Eq. 1, is estimated for every class and then predict the class with the highest probability as the class of the test instance x. The *NB*-classifier is simple and efficient approach to classify new staff record instances.

4.3 Subset Selection for Numerical Data

We use the K-Nearest Neighbor (*KNN*) algorithm to classify the numerical dataset (image dataset) using the selected subset of features. Each image is represented by a feature vector of size 64 and an image may belong to one of the following 12 classes: beach, garden, desert, snow, sunset, rose, banana, tomato, copper, tiger, wood, and gorilla.

KNN algorithm measures the classification accuracy of the selected subset of feature based on a distance function, $d(q, p)$, Eq. 3, where $p: p_1, p_2, \ldots, p_d$ and $q: q_1, q_2, \ldots, q_d$ are two vectors representing two images.

$$\mathrm{D}(q, p) = \left(\sum_{i=1}^{d} |q_i - p_i|^2 \right)^{0.5} \tag{3}$$

Algorithm: K-Nearest Neighbor

Input: t_{DB}, K, I_Q

Output: *Class to which* I_Q *is assigned*

(1) $L_K = 0$
(2) *for each* $t \in t_{DB}$ *do*
(3) *compute* $d(t, I_Q)$ *using Eq. 3*
(4) *if* L_K *contains* $< K$ *items*
(5) $L_K = L_K \cup t$
(6) *else if* $d(t, I_Q) < d(I_Q, K^{th})$
(7) $L_K = L_K - K^{th}$
(8) $L_K = L_K \cup t$
(9) *Assign* t *to the majority class in* L_K

Generally, the *KNN* algorithm works as follows:

- A number of images are prepared to be the training dataset. We performed stratified sampling to build the training dataset, which are representative images from all the pre-defined classes. These representative images in the training set include the class information.

- The algorithm maintains an ascending order list LK that keeps the K nearest neighbors found so far.

- Each image in the database IQ is then compared with each image t in the training set tDB by computing their Euclidean distance (Eq. 3).

- If the list LK contains less than K images from the training dataset, add image t into the list, otherwise, if the distance between IQ and image t is less than the distance between IQ and the Kth neighbor in LK, remove the Kth neighbor from the LK and add t to LK.

- Finally, the query image IQ is classified to the majority class in the retrieved K nearest images in LK.

5. Experimental Results

We performed our experiments on two datasets: categorical teachers' dataset and numerical image dataset. The first part measures the accuracy of classification of our categorical teachers dataset into three classes: assistant professor, associate professor and full professor. Each record in the staff dataset consists of six features: name, age, nationality, salary, number of research works, and number of advisees.

The *NB*-Classifier technique handles discrete values. If a feature value type is continuous, such as salary, these continuous values are discretized by dividing it into ranges. Each of the three classes is given a value that is assistant professor class is assigned 1, associate professor is assigned 2 and full professor is assigned 3. Before the *NB*-Classifier tests features, these features need to be encoded first. In our experiment, we encoded the ages between 20 and 30 as 1, those between 31 and 40 as 2, those between 41 and 50 as 3, and the ages above 51 are encoded as 4. Salary is divided into ranges as follows: 10,000-20,000 as 1, above 20,000-30,000 as 2, above 30,000-40,000 as 3, above 40,000-50,000 as 4, above 50,000-60,000 as 5, above 60,000-70,000 as 6, and above 70,000 is encoded as 7. Number of research is encoded as 1 for 0-25, 2 for 26-50, and 3 for 51 researches and above. Similarly, number of advisees was divided into ranges as follows: 1-5 as 1, 6-10 as 2, 11-15 as 3, and so on. The name and nationality features were not considered in our experiment as they are clearly irrelevant.

A training staff dataset of size 50 records was prepared and used to search for the subset of features and evaluate them. As seen in Table 2, in the first iteration, the number of research works gave the best classification accuracy among all other features and thus it was used as a basis for second iteration. In the second iteration, the subset of number of research works and salary gave the best classification accuracy. In the third iteration, the technique evaluated all possible subsets by adding another unselected to the basis from the second iteration; however, the best subset in this iteration did not give classification accuracy higher than the subset found in the second iteration. Therefore, the search was stopped and the best subset of features found in the second iteration was returned.

Iteration Number	Best Subset of Features	Classification Accuracy
1	{number of research works}	0.67
2	{number of research works, salary}	0.87
3	{number of research works, salary, age}	0.85

Table 2. subset of features of categorical data

In the second part, several experiments were performed using 419 images. The accuracies of all possible combinations of 64 features, which represent the images, are found by SFS and measured by *KNN* classifier.

First, using stratified sampling, a sample of 60 images was selected, five different images from each class were chosen. *KNN* algorithm estimates the class of each image by selecting ten nearset neighbors. All subset of features were evaluated and their accuracies were measured by SFS when training images are included in evaluation and when they are excluded.

Dataset Size	Training Dataset Included	Size of selected subset	Best Classification Acurracy
60	Yes	51	0.35
	No	36	0.67
139	Yes	52	0.67
	No	37	0.79
419	Yes	56	0.74
	No	39	0.80

Table 3. subset of features of image data

Similar experiments were performed on different image dataset sizes; that is using 139 images from the set of 419. The training dataset in each experiment was chosen to be 60% of the dataset size. The number of training images selected from each class is propotional to the number of images in the class and these training images are selected randomly. Table 3 summaries the results of the three experiments. Notice that as we include the training images in the testing phase, the classification accuracy increase, which is expected because the system will be able to correctly classify those images used in the training. Also, note that as the training image dataset is included in the testing phase the size of the subset of features is reduced.

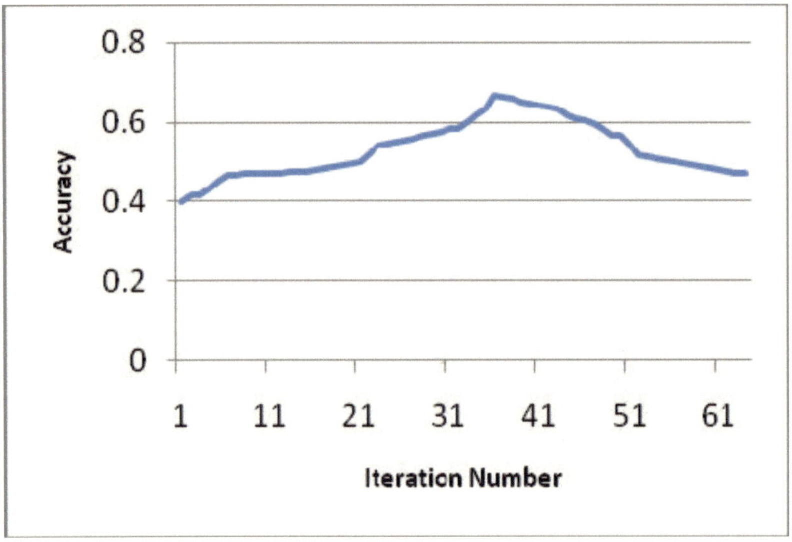

Fig. 3. Classification accuracy of the tested subsets of features.

The highest classification accuracy of the subset of features (with the training image dataset included in the test) at each iteration is shown in Fig. 2 for the image dataset of size 60. The X-axis represents the iteration number and the Y-axis represents the classification accuracy. Note that in the classification accuracy increases as more features are added to the basis of previous iteration till it reaches a peak at which the system had found the best subset of features, then as more feauters are added the accuracy degrades. The other datasets depict the same trend. The highest classification accuracy in Fig. 3 was reached for the following subset of feature numbers: {6, 3, 23, 9, 24, 32, 33, 12, 13, 14, 19, 20, 29, 30, 35, 36, 40, 41, 45, 50, 51, 56, 25, 49, 34, 61, 37, 52, 53, 57, 59, 17, 21, 42, 55}. The order of the features in the subset depicts the order of their inclusion in subset, which is based on their contributions to classification accuracy.

6. Conclusion

In this paper, we presented a wrapper approach to select the best subset of features that result in the highest classification accuracy. We use an SFS approach to search for the best subset of features. The Naïve Bayes algorithm and *K*-Nearest Neighbor algorithm are used to classify and estimate the accuracy of the categorical data and image data, respectively. This approach is evaluated using two datasets: categorical teachers' dataset and image dataset. The experimental results for both categorical and image datasets show the feasibility of the presented techniques in classifying categorical and numerical data. Such techniques are useful in many applications to decrease the performance cost and increase the classification accuracy.

7. References

John, G. H.; Kohavi, R. & Pfleger, K. (1994). Irrelevant Features and the Subset Selection Problem, In *Proceedings of the International Conference on Machine Learning,* pp. 121-129.

Guyon, I. & Elisseeff, A. (2003). Overfitting in Making Comparisons Between Variable Selection Methods, *Journal of Machine Learning Research* , vol. 3, pp. 1371-1382.

Cantu-Paz, E.; Newsam, S. & Kamath, C. (2004). Feature Selection in Scientific Applications, *Proceedings of International Conference on Knowledge Discovery and Data Mining,* Seattle, WA, United States.

Liu, Y. & Kender, J. R. (2003). Sort-Merge Feature Selection for Video Data, *SIAM Data Mining Conference (SDM),* San Francisco, USA.

Holland, J. (1975). Adaptation in Natural and Artificial Systems, *University of Michigan Press,* Ann Arbor, MI., USA.

Laanaya, H.; Martin, A.; Khenchaf, A. & Aboutajdine, D. (2005). Feature Selection Using Genetic Algorithms For Sonar Images Classification With Support Vector, *ECPS* Conference, 15-18 March, Brest, France.

Vafaie, H. & Imam, I.F. (1994). Feature selection methods: Genetic algorithms vs. greedy-like search, *Proceedings of International Conference on Fuzzy and Intelligent Control Systems.*

Hao, H.; Liu, C. & Sako, H. (2003). Comparison of Genetic Algorithm and Sequential Search Methods for Classifier Subset Selection, *Proceedings of the Seventh International Conference on Document Analysis and Recognition (ICDAR'03),* vol. 2, pp. 765-769.

Marill, T. & Green, D.M. (1963). On the Effectiveness of Receptors in Recognition Systems, *IEEE Trans. on Information Theory,* vol. 9, pp.11-17.

Whitney, A.W. (1971). A Direct Method of Nonparametric Measurement Selection, *IEEE Trans. on Computers,* vol. 20, no. 9, pp. 1100-1103.

Moore, A.W. & Lee, M.S. (1994). Efficient Algorithms for Minimizing Cross Validation Error, *in Cohen, W. and Hirsh, H. eds., Machine Learning: Proceedings of the 11th International Conference, Morgan Kaufmann,* 1994.

Tan, P-N.; Steinbach, M. & Kumar, V. (2006). Introductin to Data Mining, *Addison Wesley.*

22

Learning self-similarities for action recognition using conditional random fields

Imran N. Junejo
University of Sharjah
U.A.E.

Abstract

Human action recognition is a complex process due to many factors, such as variation in speeds, postures, camera motions etc. Therefore an extensive amount of research is being undertaken to gracefully solve this problem. To this end, in this paper, we introduce the application of *self-similarity surfaces* for human action recognition. These surfaces were introduced by Shechtman & Irani (CVPR'07) in the context of matching similarities between images or videos. These surfaces are obtained by matching a small patch, centered at a pixel, to its larger surroundings, aiming to capture *similarities* of a patch to its neighborhood. Once these surfaces are computed, we propose to transform these surfaces into Histograms of Oriented Gradients (HoG), which are then used to train Conditional Random Fields (CRFs). Our *novelty* lies in recognizing the utility of these self-similarity surfaces for human action recognition. In addition, in contrast to Shechtman & Irani (CVPR'07), we compute only a few of these surfaces (two per frame) for our task. The proposed method does not rely on the structure recovery nor on the correspondence estimation, but makes only mild assumptions about the rough localization of a person in the frame. We demonstrate good results on a publicly available dataset and show that our results are comparable to other well-known works in this area.

1. Introduction

Visual recognition and understanding of human actions has attracted much of the attention over the past three decades Moeslund et al. (2006); Wang et al. (2003); Turaga et al. (2008) and still remains an active research area of computer vision. A good solution to the problem holds a huge potential for many applications such as the search and the structuring of large video archives, video surveillance, human-computer interaction, gesture recognition and video editing. Recent work has demonstrated the difficulty of the problem associated with the large variation of human action data due to the individual variations of people in expression, posture, motion and clothing; perspective effects and camera motions; illumination variations; occlusions and disocclusions; and distracting effects of scenes surroundings. In addition, actions frequently involve and depend on manipulated objects adding another layer of variability.

Various approaches using different constructs have been proposed over the years for action recognition. These approaches can be roughly categorized on the basis of representation used by the researchers. Time evolution of human silhouettes was frequently used as action description. For example, Bobick & Davis (2001) proposed to capture the history of shape changes using temporal templates and Weinland et al. (2006) extend these 2D templates to 3D action templates. Similarly, based on silhouettes, notions of *action cylinders* Syeda-Mahmood et al. (2001), and *space-time shapes*Yilmaz & Shah (2005a); Gorelick et al. (2007) have also been introduced. Recently, researchers have started analyzing video sequences as space-time volumes, built by various *local features*, such as intensities, gradients, optical flow etc Fathi & Mori (2008); Jhuang et al. (2007); Filipovych & Ribeiro (2008). Original work in this area is that of Laptev & Lindeberg (2003), and Niebles et al. (2006); Liu & Shah (2008); Jingen et al. (2008); Mikolajczyk & Uemura (2008); Bregonzio et al. (2009); Rapantzikos et al. (2009) and Gilbert et al. (2008) represent some of the recent work in this area. Using these space-time or other local image features, researchers have also attempted at modeling the complex dynamic human motion by adopting various machine learning approaches Ali et al. (2007); Weinland & Boyer (2008); Jia & Yeung (2008): Hidden Markov Models (HMMs) Brand et al. (1997); Wilson & Bobick (1995); Ikizler & Forsyth (2007); Support Vector Machines (SVMs) Ikizler et al. (2008); Yeffet & Wolf (2009); Prototype Trees Lin et al. (2009); or Conditional Random Fields (CRF) and its variants Sminchisescu et al. (2005); Natarajan & Nevatia (2008), using features such as histograms of combined shape context and edge features, fast-fourier transforms of angular velocities, and blocked-based features of silhouettes etc. Some other works based on multiview geometry or pose learning includes that of Syeda-Mahmood et al. (2001); Yilmaz & Shah (2005b); Carlsson (2003); Rao et al. (2002); Shen & Foroosh (2008); Parameswaran & Chellappa (2006); Ogale et al. (2006); Ahmad & Lee (2006); Li et al. (2007); Lv & Nevatia (2007); Shen & Foroosh (2008), requiring either identification of body parts or the estimation of corresponding points between video sequences.

Our approach builds upon the concept self-similarities as introduced by Shechtman & Irani (2007). For a given action sequence, the approach consists of computing *similarities* of the pose to itself in each frame. This we call as the *self-similarity surface*. This surface has been introduced in context of image and video matching previously by the Shechtman & Irani (2007). They build on the assumption that for a phenomenon or a pattern captured in different forms, even though different representation and their corresponding measures vary significantly, there exists a common underlying visual property of patterns, which is captured in terms of the local intensity properties. However, in their work Shechtman & Irani (2007), these surfaces are computed very densely in an image, whereas we perform very sparse sampling, i.e. we compute only two self-similarity surface for an entire image. Also, in this paper, we introduce the usage of these surfaces for the human action recognition. We believe that this *novel* application of these surfaces is very significant for understanding the human actions, and provides acceptable accuracy compared to other well-known methods.

In the rest of the paper we operationalize self-similarity surface for human action sequences. The rest of the paper is organized as follows. In the next section we review related work. Section 3 gives a formal definition of self-similarity surface using image color features. Section 4 describes the representation and training of action sequences based on HoG descriptors, constructed from the self-similarity surfaces. In Section 5 we test the method on public dataset and demonstrate the practicality and the potential of the proposed method. Section 6 concludes the paper.

2. Related Work

The methods most closely related to our approach are that of Shechtman & Irani (2007); Benabdelkader et al. (2004); Cutler & Davis (2000); Carlsson (2000). Recently for image and video matching, Shechtman & Irani (2007) explored *local* self-similarity descriptors. The descriptors are constructed by correlating the image (or video) patch centered at a pixel to its surrounding area by the sum of squared differences. The correlation surface is transformed into a binned log-polar representation to form a local descriptor used for image and video matching. Differently to this method, we explore the structure of similarities between *all* pairs of time-frames in a sequence. The main focus of our work is on the use of self-similarities for action recognition which was not addressed in Shechtman & Irani (2007). The Figure 3 shows the self-similarity descriptor extracted from two separate images. The figure on the top left has three marked points: 1,2 and 3. The right three images in the first row shows the computed self-similarity descriptor for each of these marked points, respectively. In row two, the leftmost image also has three points marked at almost the same location as the one in row one above. The corresponding self-similarity descriptors are shown in the second row as well. This image demonstrates that even when we have difference images containing same phenomenon, (even in the presence of some perspective distortion), the computed self-similarity descriptors, as can be seen above, bear great similarities. Consequently, the Figure 2 shows the self-similarity descriptors at work. Figure 2(a) shows an input image. A self-similarity descriptor of this image is extracted which is then matched to the descriptors extracted from a database of a large number of images. In the figure, red corresponds to the highest similarity values.
Our approach has a closer relation to the notion of video self-similarity used by Benabdelkader et al. (2004); Cutler & Davis (2000). In the domain of periodic motion detection, Cutler and Davis Cutler & Davis (2000) track moving objects and extract silhouettes (or their bounding boxes). This is followed by building a 2D matrix for the given video sequence, where each entry of the matrix contains the absolute correlation score between the two frames i and j. Their observation is that for a periodic motion, this similarity matrix will also be periodic. To detect and characterize the periodic motion, they use the Time-Frequency analysis. Following this, Benabdelkader et al. (2004) use the same construct of the self-similarity matrix for gait recognition in videos of walking people. The periodicity of the gait creates diagonals in the matrix and the temporal symmetry of the gait cycles are represented by the cross-diagonals. In order to compare sequences of different length, the self-similarity matrix is subdivided into small units. Both of these works focus primarily on videos of walking people for periodic motion detection and gait analysis. The method in Carlsson (2000) also concerns gait recognition using temporal similarities between frames of different image sequences. None of the methods above explores the notion of self-similarity for action recognition. In addition, we perform very sparse sampling of the foreground space.s

2.1 Overview of our approach

The overview of the proposed approach is as shown in Fig. 1. Whereas, Shechtman & Irani (2007) compute the self-similarity based descriptor densely, we divide the foreground into just two portions, the top and the bottom, as shown in the figure. What we do is basically match the center of the top patch with its surroundings, within a certain radius. And we repeat the same process for the bottom part of the foreground. This results in two self-similarity surfaces, explained below, which are then converted into HoG based descriptors Dalal & Triggs (2005). Once we have these pose descriptors for all action sequences of all classes, we train a Condi-

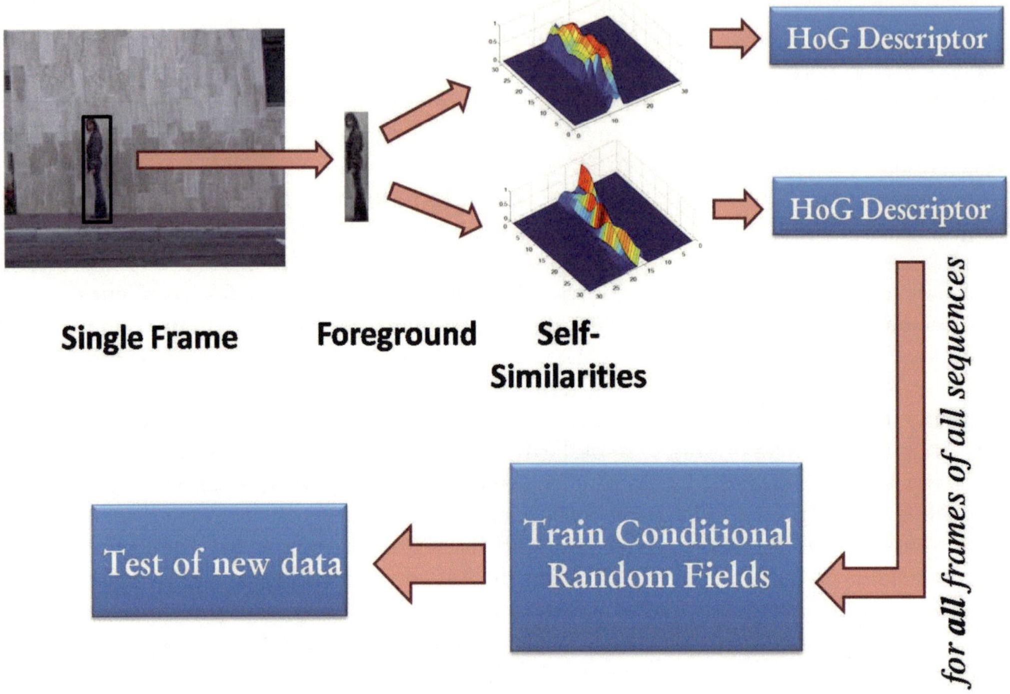

Fig. 1. Overview of the proposed solution. Initially, the foreground is extracted from an action sequence. Once extracted, the foreground is divided into the upper part and the lower part. Each part is used to compute the self-similarity descriptor, as shown in right above. Each surface is then transformed into HoG features. These features are computed for all sequences of each action class and a probabilistic model i.e. Conditional Random Fields, is learned for performing accurate action recognition.

tion Random Field Lafferty et al. (2001), during our training phase. During the testing phase, a test sequence also goes through the same process, and is assigned a probable sequence label that maximizes our conditional model.

3. Self-Similarity Surfaces

In this section we define self-similarities computed from an action sequence. The main contribution of the paper is the introduction of this self-similarity descriptor for action recognition, with the rationale that poses from different action sequences produce self-similarities of distinct patterns or structures, thus allowing us to perform action recognition.
Originally introduced by Shechtman & Irani Shechtman & Irani (2007), the descriptor captures internal geometric layout of local self-similarities within images by using only the color information. Essentially, what it does is capture self-similarity of edges, color, repetitive patterns and complex textures in a simple and unified way Shechtman & Irani (2007). The notion of

Fig. 2. The figure above shows the matching capabilities of the self-similarity descriptors. (a) shows the input (or the test) image. A self-similarity descriptor of this image is extracted. Once this is done, the descriptor is efficiently matched to the descriptors extracted from a database of a large number of images. In the figure above, red corresponds to the highest similarity values. (image courtesy of Shechtman & Irani (2007)

self-similarity is closely related to the notion of statistical co-occurrence, which is captured by the Mutual Information (MI). An example of this is shown in Figure 1.

The self-similarity descriptor is computed as follows: First, the object (or the actor) is extracted from the action sequence. This can be done by simple application of any background subtraction method (we use the extracted foregrounds provided by Shechtman & Irani (2007)). Once such a foreground is obtained, we divide it into two equal parts (the upper and the lower part). The center p of each patch, generally represented by a 5×5 patch, is compared to the surrounding patches within a radius (generally of size 15 or 30, depending on the size of the foreground object). The comparison of the patches is made by a simple application of *sum of square differences* (SSD). The result surface $Y_p(x,y)$, is then normalized into a correlation surface:

$$S_p = \exp\left(-\frac{Y_p(x,y)}{\sigma_{auto}}\right) \tag{1}$$

where σ_{auto} is a constant that takes into account noise, and common variations in color, illumination etc (for our experiments, we set its value to 2.5).

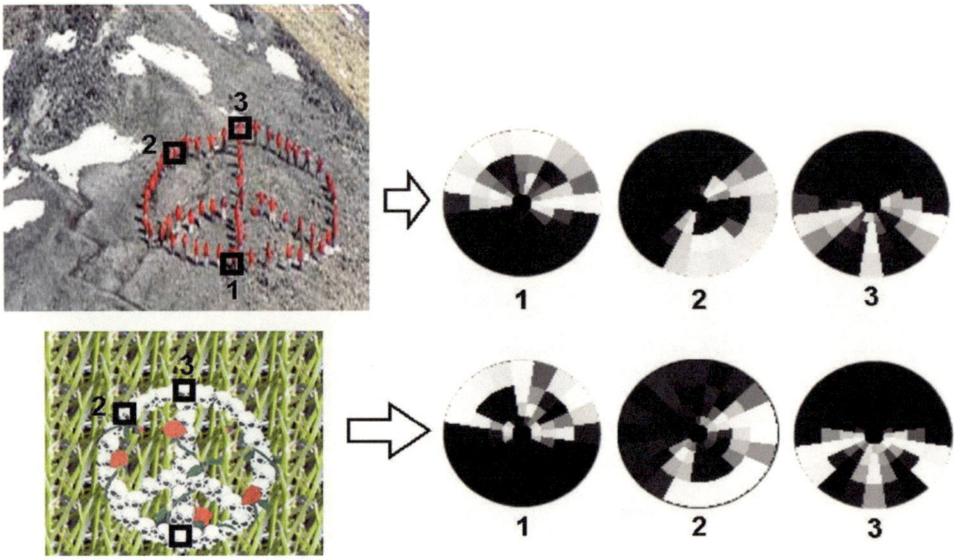

Fig. 3. The figure above shows the self-similarity descriptor extracted from two separate images. The figure on the top left has three marked points: 1,2 and 3. The right three images in the first row shows the computed self-similarity descriptor for each of these marked points, respectively. In row two, the leftmost image also has three points marked at almost the same location as the one in row one above. The corresponding self-similarity descriptors are shown in the second row as well. This image demonstrates that even when we have difference images containing same phenomenon, (even in the presence of some perspective distortion), the computed self-similarity descriptors, as can be seen above, bear great similarities. (image courtesy of Shechtman & Irani (2007)

The surface S_p is then transformed to better distinguish the spatial appearances. To this end, for describing the spatial appearance of a person at each image frame, we compute Histograms of Oriented Gradients (HoG) Dalal & Triggs (2005). This feature, originally used to perform human detection, characterizes the local shape by capturing the gradient structure. In our implementation, we use 8 bin histograms for each of 5×7 blocks defined on (the upper and the lower parts of the) bounding box around the person in each frame. The final self-similarity descriptor $\mathbf{D}^i$, for an image i, is then a concatenation of HoG features obtained from the upper and the lower part of the bounding box.
As shown in the 5, in addition to the SSDs based on the color information, we also test on self-similarity surfaces computed from optical flows. The optical flow is computed by Lucas and Kanade method Lucas & Kanade (1981) on person-centered bounding boxes using two consecutive frames.

4. Modeling & Recognition

At this stage, for each action sequence, we have obtained self-similarity surfaces, two for each frame. As described above, this self-similarity surface is then converted into HoG features. In this section, we aim to learn these features for each action class to perform action recognition. For this purpose, we chose Conditional Random Fields (CRFs) Lafferty et al. (2001)(cf. Fig. 4a). To the best of our knowledge, no work exists that learns these self-similarity surfaces for action recognition.

4.1 Conditional Random Fields

CRFs are a probabilistic framework for segmenting and labeling sequence data. Exhibiting many advantages over the traditional Hidden Markov Models (HMMs), CRFs provide a great flexibility by relaxing the conditional independence assumption generally made for the observation data.

The general framework for CRFs is as follows: Let $\mathbf{X}$ be a random variable over data sequence to be labeled and let $\mathbf{R}$ be a random variable over our corresponding label sequences. All components of $\mathbf{R}_i$ of $\mathbf{R}$ are assume to range over a finite label sequence $\mathcal{R}$, which in our case can be action sequences like `bend, wave, jump, hop, run, walk` etc. Generally, in training dataset, the random variables $\mathbf{R}$ and $\mathbf{X}$ are jointly distributed, but in the case of CRFs we construct the conditional model $p(\mathbf{R}|\mathbf{X})$, rather than explicitly modeling the marginal $p(\mathbf{X})$: Let $\mathcal{G} = (V,E)$ be an undirected graph over our set of random variables $\mathbf{R}$ and $\mathbf{X}$ (cf. Fig. 4b). Then $(\mathbf{R},\mathbf{X})$ is a conditional random field in case, when conditioned on X, the random variable $\mathbf{R}_i$ obey the Markov property with respect to the graph: $p(\mathbf{R}_i|\mathbf{R}_j,\mathbf{X},i \neq j) = p(\mathbf{R}_i|\mathbf{R}_j,\mathbf{X},i \sim j)$, where $\sim$ means i and j are neighbors in $\mathcal{G}$ Lafferty et al. (2001). Let $\mathcal{C}(\mathbf{X},\mathbf{R})$ be the set of maximal clique in $\mathcal{G}$, then the CRFs define the conditional probability of the label sequence given the observed sequence as

$$p_\theta(\mathbf{R}|\mathbf{X}) = \frac{1}{Z(\mathbf{X})} \prod_{c \in \mathcal{C}(\mathbf{R},\mathbf{X})} \phi_\theta^c(\mathbf{R}_c,\mathbf{X}_c) \tag{2}$$

where $Z(\mathbf{X})$ is the normalization factor over all states of the sequences, and is given as:

$$Z(\mathbf{X}) = \sum_{\mathbf{R}} \prod_{c \in \mathcal{C}(\mathbf{R},\mathbf{X})} \phi_\theta^c(\mathbf{R}_c,\mathbf{X}_c) \tag{3}$$

and ϕ_θ^c is the potential function of the clique c, and characterizes according to the set of selected features f_θ so that:

$$\phi_\theta^c(\mathbf{R}_c,\mathbf{X}_c) = \exp\left(\sum_{t=1}^{T}\sum_n \lambda_n f_\theta(\mathbf{R}_c,\mathbf{X}_c,t)\right) \tag{4}$$

where the model parameters $\psi = \{\lambda_n\}$ are a set of real weights, per feature. Each feature function $f_\theta(\mathbf{R}_c,\mathbf{X}_c,t)$ is either a state function $s_k(\mathbf{r}_t,\mathbf{x}_t,t)$ or a transition function $g_k(\mathbf{r}_{t-1},\mathbf{r}_t,\mathbf{x}_t,t)$. State functions depend on a single hidden variable in the model, while the transition function can depend on a pair of hidden variables Lafferty et al. (2001).

Linear-chain CRFs, as shown in Fig. 4b, are widely used in many applications. Accordingly, the cliques of such a conditional model include pair of neighboring sates $(\mathbf{r}_{t-1},\mathbf{r}_t)$, whereas the connectivity among the observation is unrestricted. Therefore, arbitrary complex observation dependencies can be added to the model without out affected complicating the inferences, as these observations are known and fixed.

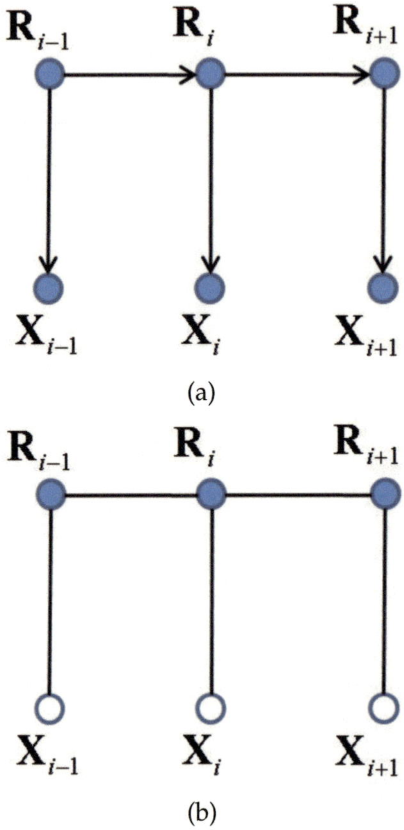

Fig. 4. (a) Graphical structures of simple HMMs. (b) Graphical Structure in the case of CRFs, where the open circle indicates that the data is not generated by the model.

4.2 Parameter Learning

Given a fully labeled training data set $\mathcal{D} = \{\mathbf{R}^i, \mathbf{X}^i\}_{i=1}^N$, the CRF parameters can be obtained by optimizing the conditional log-likelihood function

$$\mathcal{O}(\theta) = \sum_{i=1}^{N} \log p_\theta(\mathbf{R}^i | \mathbf{X}^i) \tag{5}$$

The derivative of (5) with respect to the λ_k associated with clique c is given as:

$$\begin{aligned} \frac{\partial \mathcal{O}(\theta)}{\partial \lambda_k} &= \sum_i \sum_t f_\theta(\mathbf{R}^i_{t,c}, \mathbf{X}^i, t) \\ &\quad - \sum_i \sum_t \sum_{c \in C} \sum_{\mathbf{R}_c} p_\theta(\mathbf{R}_c | \mathbf{X}^i_t) f_\theta(\mathbf{R}_{t,c}, \mathbf{X}^i, t) \end{aligned} \tag{6}$$

where $\mathbf{R}_{t,c}$ denotes the variable $\mathbf{R}$ at time stamp t in clique c of the CRF, and $\mathbf{R}_c$ ranges over assignment to c.

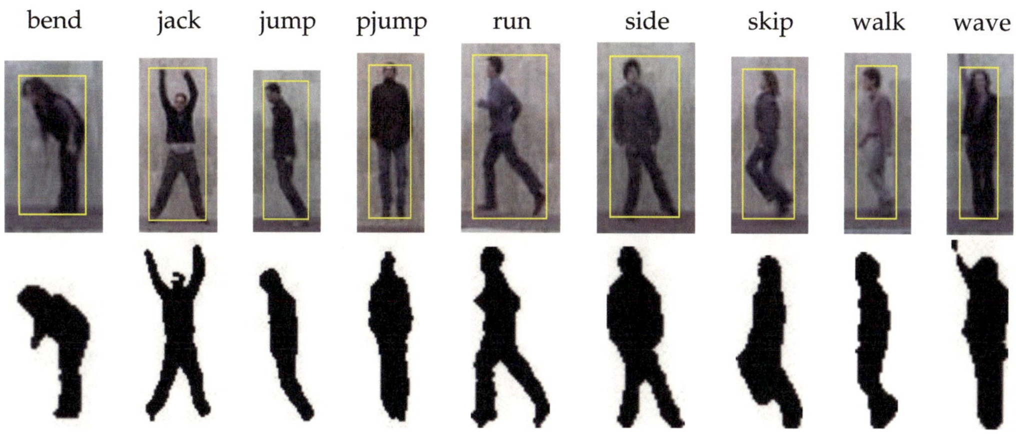

Fig. 5. Weizman dataset: The top row shows instances from the nine different action sequences. The bottom row depicts the extracted silhouettes. The dataset contains ninety-three action sequences videos of varying lengths, each having a resolution of 180×144. The nine difference action are: `bending down`, `jumping back`, `jumping`, `jumping in place`, `galloping sideways`, `running`, `walking`, `waving one hand` and `waving two hands`.

To reduce over-fitting, generally a penalized likelihood is used for training the parameters: $\mathcal{O}(\theta) + \frac{1}{2\sigma^2}\|\theta\|^2$ where the second term is the log of a Gaussian prior with variance σ^2, i.e. $P(\theta) = \exp(\frac{1}{2\sigma^2}\|\theta\|^2)$

This convex function can be optimized by a number of techniques such as the Quasi-Newton optimization methods. Specially for discrete-valued chain models, the observation dependent normalization can be efficiently computed by tensor/matrix multiplication Sminchisescu et al. (2005).

4.3 Action Recognition

So far what we have is a labeled data sequence $\mathcal{D} = \{\mathbf{R}^i, \mathbf{X}^i\}_{i=1}^N$, and we have computed the CRF model parameters θ^*. Now, once we have a new test sequence $\mathbf{x}$, we perform the same task as defined above in Section 3: we extract the foreground, divide it into two parts and compute the self-similarities. We then convert these self-similarities to HoG descriptors. Once we have these descriptors, we are ready to determine the test sequence's correct class assignment. What we want to do is to to estimate the most probable sequence label $\mathbf{r}^*$ that maximizes the conditional model.

$$\mathbf{r}^* = \arg\max_{\mathbf{r}} P(\mathbf{r}|\mathbf{x}, \theta^*) \tag{7}$$

where the parameters θ^* are learned from the training samples. While another option could be to use Viterbi path, in our experiments we the above maximal marginal probabilities for training, and the Viterbi path for labeling a new sequence for performing action recognition.

	bend	wave1	wave2	pjump	skip	jack	jump	run	walk	side
bend	**66.7**	11.1	0.0	0.0	0.0	22.2	0.0	0.0	0.0	0.0
wave1	22.2	**55.6**	22.2	0.0	0.0	0.0	0.0	0.0	0.0	0.0
wave2	0.0	0.0	**66.7**	22.2	0.0	0.0	0.0	0.0	0.0	11.1
pjump	0.0	0.0	0.0	**66.7**	0.0	0.0	0.0	0.0	0.0	33.3
skip	0.0	0.0	0.0	10.0	**20.0**	0.0	20.0	30.0	20.0	0.0
jack	11.1	0.0	0.0	0.0	0.0	**88.9**	0.0	0.0	0.0	0.0
jump	0.0	0.0	0.0	11.1	11.1	0.0	**66.7**	0.0	11.1	0.0
run	0.0	0.0	0.0	0.0	10.0	10.0	0.0	**60.0**	20.0	0.0
walk	0.0	0.0	0.0	0.0	0.0	0.0	0.0	30.0	**70.0**	0.0
side	11.1	0.0	0.0	0.0	0.0	0.0	0.0	0.0	0.0	**88.9**

Fig. 6. Cross-validation results for action recognition of the Weizman dataset when the whole foreground patch is used for computing the self-similarity surface. The dataset contains ninety-three action sequences videos of varying lengths, each having a resolution of 180×144. The first row shows label for each action type, the second row show a sample frame from an action sequence, while the last row shows the extracted silhouette from the action sequence.

For all recognition experiments in the next section, we report results for n-fold cross-validation and make sure the actions of the same person do not appear in the training and in the test sets simultaneously.

5. Experimental results

In this section, we put the proposed method of using the self-similarities for action recognition to test. For this reason, we validate our approach on the publicly available Weizman dataset and compare our results with some of the other significant work done in the area. Some instances from the data set are shown in Fig. 5.

5.1 Experiments with Weizman actions dataset

To asses the discriminative power of our method on real video sequences we apply it to the standard single-view video dataset with nine classes of human actions performed by nine subjects Gorelick et al. (2007)(see Fig. 5(top)). The dataset contains ninety-three action sequences videos of varying lengths, each having a resolution of 180×144. The nine difference action are: `bending down`, `jumping back`, `jumping`, `jumping in place`, `galloping sideways`, `running`, `walking`, `waving one hand` and `waving two hands`. Using

	bend	wave1	wave2	pjump	skip	jack	jump	run	walk	side
bend	**77.8**	0.0	22.2	0.0	0.0	0.0	0.0	0.0	0.0	0.0
wave1	11.1	**55.6**	33.3	0.0	0.0	0.0	0.0	0.0	0.0	0.0
wave2	11.1	11.1	**55.6**	0.0	0.0	22.2	0.0	0.0	0.0	0.0
pjump	11.1	0.0	0.0	**77.8**	0.0	0.0	0.0	0.0	0.0	11.1
skip	0.0	0.0	0.0	10.0	**30.0**	0.0	0.0	30.0	20.0	10.0
jack	11.1	0.0	0.0	0.0	0.0	**88.9**	0.0	0.0	0.0	0.0
jump	0.0	0.0	0.0	0.0	11.1	0.0	**66.7**	11.1	0.0	11.1
run	0.0	0.0	0.0	0.0	30.0	10.0	0.0	**50.0**	10.0	0.0
walk	0.0	0.0	0.0	0.0	10.0	0.0	0.0	0.0	**90.0**	0.0
side	0.0	0.0	0.0	11.1	0.0	0.0	0.0	0.0	22.2	**66.7**

Fig. 7. Cross-validation results for action recognition of the Weizman dataset when the self-similarity surfaces are created based on the optical flows computed between consecutive frames of the action sequences.

the extracted silhouettes, we compute the self-similarity surfaces of the foreground and then transform them into HoG features for action learning.

We have described above that the foreground is divided into two different parts and then we compute the self-similarity surfaces. However, we performed experiment with dividing the foreground into different parts rather than just two. For example, the results for the case when the whole foreground object is used for computing the self-similarity surface is shown in Fig. 6. As can be seen in the figure, the action recognition results for n-fold cross-validation is almost 66%. In addition, we also test on the self-similarity surfaces that are computed based on the optical flow computed between consecutive frame of the action sequence. The confusion matrix for this testing is shown in Fig. 7. The accuracy for this approach reaches 66%. Tests were also performed on dividing the foreground into three and six parts, but no improvement in the accuracy was observed.

However, our experiments show that the best results are obtained when the foreground is divided into the top and the bottom part and using only the color information. Results for the n-fold cross-validation for this case are depicted in Fig. 8. As the confusion matrix shows, the accuracy reached for our method is 70%.

These accuracy results are very encouraging, specially since we are using very sparse descriptors for the pose (just two per frame). Although higher accuracy results have been reported Ikizler & Duygulu (2007), accuracy of the proposed method is comparable to the well known

	bend	wave1	wave2	pjump	skip	jack	jump	run	walk	side
bend	**100.0**	0.0	0.0	0.0	0.0	0.0	0.0	0.0	0.0	0.0
wave1	12.0	**67.0**	0.0	0.0	0.0	12.0	0.0	0.0	12.0	0.0
wave2	0.0	23.0	**67.0**	0.0	0.0	12.0	0.0	0.0	0.0	0.0
pjump	12.0	0.0	0.0	**34.0**	12.0	23.0	0.0	0.0	0.0	23.0
skip	10.0	0.0	0.0	0.0	**40.0**	0.0	10.0	40.0	0.0	0.0
jack	0.0	0.0	0.0	0.0	0.0	**100.0**	0.0	0.0	0.0	0.0
jump	0.0	0.0	0.0	0.0	12.0	0.0	**67.0**	0.0	23.0	0.0
run	0.0	0.0	0.0	0.0	30.0	0.0	0.0	**60.0**	10.0	0.0
walk	0.0	0.0	0.0	0.0	0.0	0.0	10.0	0.0	**90.0**	0.0
side	0.0	0.0	0.0	0.0	0.0	34.0	0.0	0.0	0.0	**67.0**

Fig. 8. Cross-validation results for action recognition of the Weizman dataset when the foreground patch is divided into an upper and a lower part for computing the self-similarity surface.

work of Niebles et al. (2006) and also with the recently reported results in Resendiz & Ahuja (2008) posted for the same dataset.

6. Conclusion

We propose a *novel* usage self-similarity surfaces for action recognition. These surfaces are computed on an extracted foreground of a person performing an action. In contrast to Shechtman & Irani (2007); Niebles et al. (2006); Resendiz & Ahuja (2008), we compute only a few of these surfaces per frame, in fact just two features per frame. Experimental validation on Weizman datasets confirms the stability and utility of our approach. The proposed method does not rely on the structure recovery nor on the correspondence estimation, but makes only mild assumptions about the rough localization of a person in the frame. This lack of strong assumptions is likely to make our method applicable to action recognition beyond controlled datasets when combined with the modern techniques for person detection and tracking.

7. References

Ahmad, M. & Lee, S. (2006). HMM-based human action recognition using multiview image sequences, *Proc. ICPR*, pp. I:263–266.

Ali, S., Basharat, A. & Shah, M. (2007). Chaotic invariants for human action recognition, *Proc. ICCV*.

Benabdelkader, C., Cutler, R. & Davis, L. (2004). Gait recognition using image self-similarity, *EURASIP J. Appl. Signal Process.* **2004**(1): 572–585.

Bobick, A. & Davis, J. (2001). The recognition of human movement using temporal templates, *PAMI* **23**(3): 257–267.

Brand, M., Nuria, O. & Pentland, A. (1997). Coupled hidden markov models for complex action recognition, *Proc. CVPR*.

Bregonzio, M., Gong, S. & Xiang, T. (2009). Recognising action as clouds of space-time interest points, *Proc. CVPR*, pp. 1948–1955.

Carlsson, S. (2000). Recognizing walking people, *Proc. ECCV*, pp. I:472–486.

Carlsson, S. (2003). Recognizing walking people, *I. J. Robotic Res.* **22**(6): 359–370.

Cutler, R. & Davis, L. (2000). Robust real-time periodic motion detection, analysis, and applications, *PAMI* **22**(8): 781–796.

Dalal, N. & Triggs, B. (2005). Histograms of oriented gradients for human detection, *Proc. CVPR*, pp. I:886–893.

Fathi, A. & Mori, G. (2008). Action recognition by learning mid-level motion features, *Proc. CVPR*.

Filipovych, R. & Ribeiro, E. (2008). Learning human motion models from unsegmented videos, *Proc. CVPR*.

Gilbert, A., Illingworth, J. & Bowden, R. (2008). Scale invariant action recognition using compound features mined from dense spatio-temporal corners, *Proc. ECCV*, pp. I: 222–233.

Gorelick, L., Blank, M., Shechtman, E., Irani, M. & Basri, R. (2007). Actions as space-time shapes, *PAMI* **29**(12): 2247–2253.

Ikizler, N., Cinbis, R. G. & Duygulu, P. (2008). Human action recognition with line and flow histograms, *Proc. ICPR*.

Ikizler, N. & Duygulu, P. (2007). Human action recognition using distribution of oriented rectangular patches, *Workshop on Human Motion*, pp. 271–284.

Ikizler, N. & Forsyth, D. (2007). Searching video for complex activities with finite state models, *Proc. CVPR*.

Jhuang, H., Serre, T., Wolf, L. & Poggio, T. (2007). A biologically inspired system for action recognition, *Proc. ICCV*.

Jia, K. & Yeung, D.-Y. (2008). Human action recognition using local spatio-temporal discriminant embedding, *Proc. CVPR*.

Jingen, L., Saad, A. & Shah, M. (2008). Recognizing human actions using multiple features, *Proc. CVPR*.

Lafferty, J., McCallum, A. & Pereira, F. (2001). Conditional random fields: Probabilistic models for segmenting and labeling sequence data, *In Proceedings of the Eighteenth International Conference on Machine Learning (ICML)*.

Laptev, I. & Lindeberg, T. (2003). Space-time interest points, *Proc. ICCV*, pp. 432–439.

Li, R., Tian, T. & Sclaroff, S. (2007). Simultaneous learning of nonlinear manifold and dynamical models for high-dimensional time series, *Proc. ICCV*.

Lin, Z., Jiang, Z. & Davis, L. S. (2009). Recognizing actions by shape-motion prototype trees, *Proc. ICCV*.

Liu, J. & Shah, M. (2008). Learning human actions via information maximization, *Proc. CVPR*.

Lucas, B. & Kanade, T. (1981). An iterative image registration technique with an application to stereo vision, *Image Understanding Workshop*, pp. 121–130.

Lv, F. & Nevatia, R. (2007). Single view human action recognition using key pose matching and viterbi path searching, *Proc. CVPR*.
Mikolajczyk, K. & Uemura, H. (2008). Action recognition with motion-appearance vocabulary forest, *Proc. CVPR*, pp. 1–8.
Moeslund, T., Hilton, A. & Krüger, V. (2006). A survey of advances in vision-based human motion capture and analysis, *CVIU* **103**(2-3): 90–126.
Natarajan, P. & Nevatia, R. (2008). View and scale invariant action recognition using multiview shape-flow models, *Proc. CVPR*, pp. 1–8.
Niebles, J., Wang, H. & Li, F. (2006). Unsupervised learning of human action categories using spatial-temporal words, *Proc. BMVC*.
Ogale, A., Karapurkar, A. & Aloimonos, Y. (2006). View-invariant modeling and recognition of human actions using grammars, *Proc. Workshop on Dynamic Vision*, pp. 115–126.
Parameswaran, V. & Chellappa, R. (2006). View invariance for human action recognition, *IJCV* **66**(1): 83–101.
Rao, C., Yilmaz, A. & Shah, M. (2002). View-invariant representation and recognition of actions, *IJCV* **50**(2): 203–226.
Rapantzikos, K., Avrithis, Y. & Kollias, S. (2009). Dense saliency-based spatiotemporal feature points for action recognition, In Proc. CVPR.
Resendiz, E. & Ahuja, N. (2008). A unified model for activity recognition from video sequences, *Proc. ICPR*, pp. 1–4.
Shechtman, E. & Irani, M. (2007). Matching local self-similarities across images and videos, *Proc. CVPR*.
Shen, Y. & Foroosh, H. (2008). View invariant action recognition using fundamental ratios, *Proc. CVPR*.
Sminchisescu, C., Kanaujia, A., Li, Z. & Metaxas, D. (2005). Conditional models for contextual human motion recognition, *Proc. CVPR*.
Syeda-Mahmood, T., Vasilescu, M. & Sethi, S. (2001). Recognizing action events from multiple viewpoints, *Proc. EventVideo*, pp. 64–72.
Turaga, P. K., Chellappa, R., Subrahmanian, V. S. & Udrea, O. (2008). Machine recognition of human activities: A survey, *IEEE Trans. Circuits Syst. Video Techn.* **18**(11): 1473–1488.
Wang, L., Hu, W. & Tan, T. (2003). Recent developments in human motion analysis, *Pattern Recognition* **36**(3): 585–601.
Weinland, D. & Boyer, E. (2008). Action recognition using exemplar-based embedding, *Proc. CVPR*.
Weinland, D., Ronfard, R. & Boyer, E. (2006). Free viewpoint action recognition using motion history volumes, *CVIU* **103**(2-3): 249–257.
Wilson, A. & Bobick, A. (1995). Learning visual behavior for gesture analysis, *IEEE Symp. on Comp. Vision*.
Yeffet, L. & Wolf, L. (2009). Local trinary patterns for human action recognition, *Proc. ICCV*.
Yilmaz, A. & Shah, M. (2005a). Actions sketch: A novel action representation, *Proc. CVPR*, pp. I:984–989.
Yilmaz, A. & Shah, M. (2005b). Recognizing human actions in videos acquired by uncalibrated moving cameras, *Proc. ICCV*, pp. I:150–157.

Probabilistic modelling and recursive bayesian estimation of trust in wireless sensor networks

Mohammad Momani[1] and Subhash Challa[2]
[1]University of Technology Sydney, Australia mmomani@eng.uts.edu.au
[2]NICTA, VRL, University of Melbourne, Australia, Subhash.Challa@nicta.com.au

1. Introduction

Wireless Sensor Networks closely resemble a human behaviour model, in which a number of nodes that have just met are able to communicate with each other based on mutual trust levels developed over a period of time. WSNs are characterised by their performance of an additional function to the traditional functions of an ad-hoc network, which is monitoring events and reporting data and, as such, the sensed data represent the core component of trust-modelling in this research.

The trust-modelling problem in wireless networks is characterised by uncertainty. It is a decision problem under uncertainty and the only coherent way to deal with uncertainty is through probability. There are several frameworks for reasoning under uncertainty, but it is well accepted that the probabilistic paradigm is the theoretically sound framework for solving a decision problem involving uncertainty. Some of the trust models introduced for sensor networks employ probabilistic solutions mixed with ad-hoc approaches. None of them produces a full probabilistic answer to the problem. Each node's reliability is an unknown quantity. The ensuing decision problems concern is which nodes are to be trusted. It is these decision problems; regarding when to terminate nodes, that motivate research in trust models.

We look at applying trust evaluation to WSNs, providing continuous data in the form of a new reputation system we call GTRSSN: *Gaussian Trust and Reputation System for Sensor Networks*. It has been argued that previous studies on WSNs focused on the trust associated with the routing and the successful performance of a sensor node in some predetermined task. This resulted in looking at binary events. The trustworthiness and reliability of the nodes of a WSN, when the sensed data are continuous, has not been addressed. Our main contribution is therefore the introduction of a statistical approach; a theoretically sound Bayesian probabilistic approach for modelling trust in WSNs in the case of continuous sensor data; that is, we derive a Bayesian probabilistic reputation system and trust model for WSNs, as presented in our work in (Momani et al., 2007a) and (Momani et al., 2007b).

2. Node Misbehaviour Classification

The main idea behind reputation and trust-based systems is to discover and exclude misbehaving nodes and to minimise the damage caused by inside attackers. Node misbehaviour can be classified in two categories: communication misbehaviour and data misinforming. Most of the researchers classify node misbehaviour in the same way they model trust: from the communication point of view. However, as discussed so far, WSNs are deployed to sense events and report data, so the node misbehaviour diagram presented in (Srinivasan et al., 2007) is extended by introducing a new branch addressing sensor data misbehaviour; misinforming, as a second category of nodes' misbehaviour classification in WSNs, as illustrated below in Figure 1, to reflect the way trust is being modelled here.

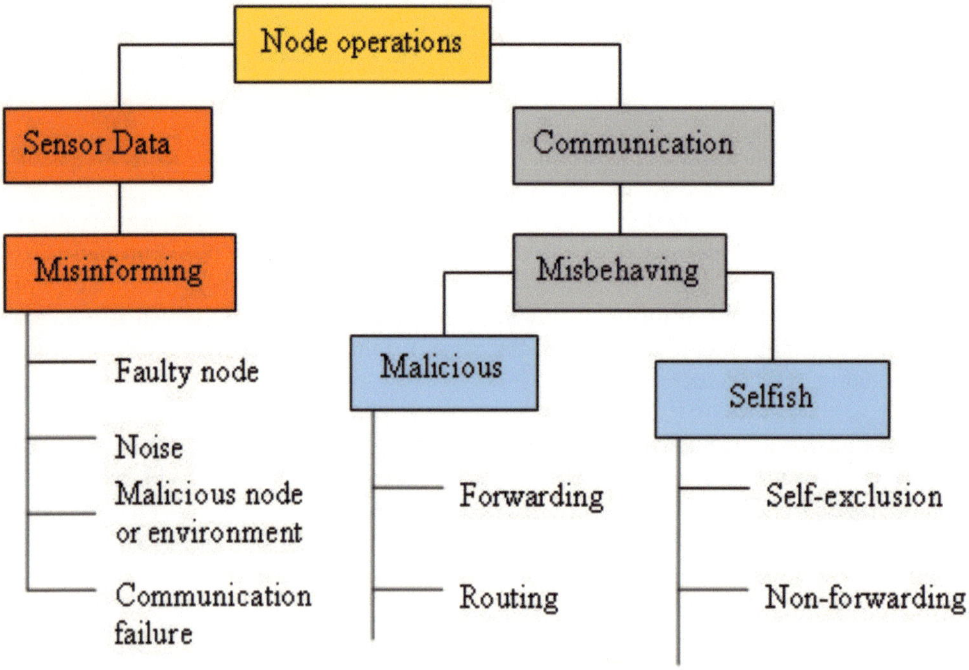

Fig. 1. Node misbehaviour classification

As can be seen from the diagram in Figure 1, the new branch dealing with sensor data includes the misinforming behaviour of a sensor node. This can be caused due to a faulty node, a node that is damaged or has expired, or due to a noise, as sensor data are not without noise, a malicious node or environment. The node might have been captured or the environment is malfunctioning or there might have been a communication failure, or there has been interference or the communication between nodes is cut off for some reason. The communication misbehaviour classification is due to the node being malicious, an intruder attacking and damaging the network, or the node is selfish, trying to save resources for later usage. Further detailed information regarding the node misbehaviour communication branch is provided in (Srinivasan et al., 2006).

3. Modelling Trust

Initially, the primary focus of the research on trust in WSNs was on whether a node will detect appropriately, will or will not report the detected event(s), and will route information. The uncertainty in these actions warranted the development of reputation systems and corresponding trust models. Modelling trust in general is the process of representing the trustworthiness of one node in the opinion of another node, that is, how much one node trusts every other node in the surrounding area, and it has been the focus of many researchers from different domains. In other words, trust-modelling is simply the mathematical representation of a node's opinion of another node in a network. Figure 2 below shows the two main sources for trust formation in WSNs: the observation of the behaviour of the surrounding nodes, direct trust and the recommendation from other nodes, indirect trust.

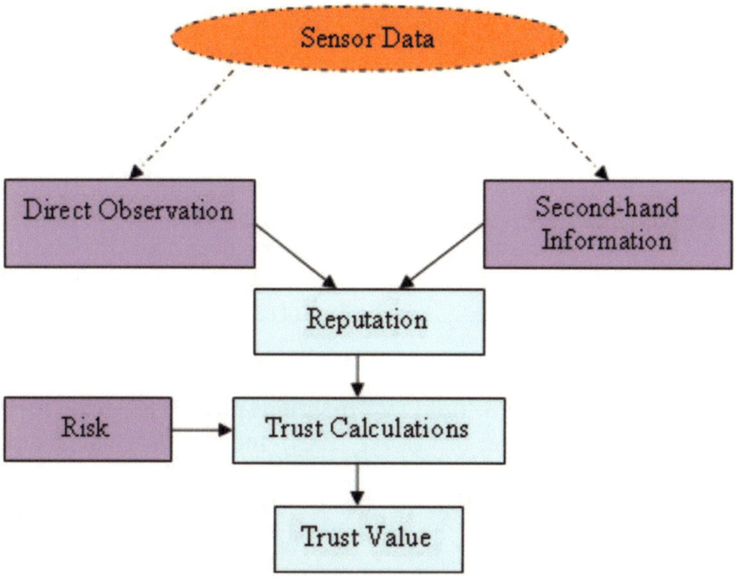

Fig. 2. Trust computational model for WSN

3.1. Direct Observations

A node will observe a neighbouring node's behaviour and build a reputation for that node based on the observed data. The neighbouring node's transactions data are direct observations referred to as *first-hand information*. By their nature, the considered events are binary, and the mathematical trust models developed for WSNs are for binary transactions. We argue that the problem of assessing a reputation based on observed data is a statistical problem. Some trust models make use of this observation and introduce a probabilistic modelling. For example, the reputation-based framework for high integrity sensor networks (RFSN) trust model presented in (Ganeriwal & Srivastava, 2004) by Ganeriwal and Srivastava uses a Bayesian updating scheme known as the *Beta Reputation System* for assessing and updating the nodes' reputations. The Beta reputation system was introduced by Josang and Ismail (Jøsang & Ismail, 2002), who used the Beta distribution to model binary statistical events.

3.2. Second-hand Information

A second source of information in trust-modelling is information provided by other nodes. This source of information is referred to as *second-hand information*. It consists of information gathered by nodes as first-hand information and converted into an assessment. Due to the limitations of a WSN, the second-hand information is summarised before being shared. For example, the RFSN in (Ganeriwal & Srivastava, 2004) uses the Beta probability model and share the values of the parameters of the probability distributions as second-hand information. This shared information is not hard data for the node receiving the information. A proper way is required to incorporate this new information into the trust model by combining it with observed data. While some trust models build reputations purely on the basis of observations, most of them attempt to use the second-hand information. The reasons are obvious from a statistical point of view. But the interest is also motivated by the desire to speed up the assessment of reputations. Due to the asymmetric transactions in a network, some nodes may not have enough observations about all neighbouring nodes.

Using shared information improves the efficiency and speed of reputation assessment, however, combining the two sources of information is handled differently by different trust models. For example, the RFSN uses the Dempster-Shafer Belief Theory. The Belief Theory is a framework for reasoning under uncertainty that differs from the probabilistic framework. The discussion of the fundamental differences between these two theories is beyond the scope of this research. Although the two approaches can be joined in some cases, they differ in their philosophies on how to treat uncertainty. The RFSN uses both of them in the same problem. We propose a probabilistic treatment of trust, and apply it to the case of continuous sensor data.

Although a reputation system is designed to reduce the harmful effect of an unreliable or malicious node, such a system can be used by a malicious node to harm the network. Systems such as the RFSN in (Ganeriwal & Srivastava, 2004) and the distributed reputation-based beacon trust system (DRBTS) in (Srinivasan et al., 2006) are confronted with the issue of what second-hand information is allowed to be shared. For example, some prohibit negative second-hand information to be shared, in order to reduce the risk of a negative campaign by malicious nodes. Our proposed model incorporates all of the second-hand information. To resolve the issue of the validity of the information source, the information is modulated using the reputation of the source. This probabilistic approach rigorously answers the question of how to combine the two types of data in the exercise of assessing reputations in a sensor network. It is based on work undertaken in modelling *Expert Opinion* (Lindley & Singpurwalla, 1986; Morris, 1971; West, 1984). Expert opinions are used whenever few data are observed. The expert opinion is second-hand information that is merged with hard data according to the laws of probability. Information provided by knowledgeable sources is known as "expert opinion" in the statistical literature. These opinions are modulated by existing knowledge about the experts themselves, to provide a calibrated answer.

4. The Beta Reputation System

The Beta Reputation System was proposed by Josang and Ismail in (Jøsang & Ismail, 2002) to derive reputation ratings in the context of e-commerce. It was presented as a flexible system with foundations in the theory of statistics, and is based on the Beta probability density function. The Beta distribution can be used in the probability modelling of binary events. Let θ be a random variable representing a binary event, $\theta = 0; 1$, and p the probability that the event occurs, $\theta = 1$. Then the Beta-family of probability distributions, a continuous family of functions indexed by two parameters a and β, can be used to represent the probability density distribution of p, noted as $Beta(a, \beta)$, as shown in equation (1):

$$f(p \mid \alpha, \beta) = \frac{\Gamma(\alpha+\beta)}{\Gamma(\alpha)\Gamma(\beta)} p^{\alpha-1}(1-p)^{\beta-1} \tag{1}$$

where $0 \leq p \leq 1$; $a > 0$; $\beta > 0$. If the number of outcomes where there are r occurrences and s non-occurrences of the event is observed, then using a Bayesian probabilistic argument, the probability density function of p can be expressed as a Beta distribution, where $a = r + 1$ and $\beta = s + 1$. This probabilistic mechanism is applied to model the reputation of an entity using events of completion of a task by the assessed entity. The reputation system counts the number r of successful transactions, and the number s of failed transactions, and applies the Beta probability model. This provides for an easily updatable system, since it is easy to update both r and s in the model. Each new transaction results either in r or s being augmented by 1.
For the RFSN (Ganeriwal & Srivastava, 2004) Ganeriwal and Srivastava used the work of Josang and Ismail presented in (Jøsang & Ismail, 2002), in their trust model for WSNs. For each node n_j, a reputation R_{ij} can be carried by a neighbouring node n_i. The reputation is embodied in the Beta model and carried by two parameters a_{ij} and β_{ij}. a_{ij} represents the number of successful transactions node n_i had with n_j, and β_{ij} represents the number of unsuccessful transactions. The reputation of node n_j maintained by node n_i is $R_{ij} = Beta(a_{ij} + 1, \beta_{ij} + 1)$. The trust is defined as the expected value of the reputation, as shown in equation (2):

$$T_{ij} = E(R_{ij}) = E(Beta(\alpha_{ij}+1, \beta_{ij}+1)) = \frac{\alpha_{ij}+1}{\alpha_{ij}+\beta_{ij}+2} \tag{2}$$

Second-hand information is presented to node n_i by another neighbouring node n_k. Node n_i receives the reputation of node n_j by node n_k, R_{kj}, in the form of the two parameters a_{kj} and β_{kj}. Using this new information, node n_i combines it with its current assessment R_{ij} to obtain a new reputation R_{ij}^{new}, as given in equation (3):

$$R_{ij}^{new} = Beta(\alpha_{ij}^{new}, \beta_{ij}^{new}) \tag{3}$$

where

$$\alpha_{ij}^{new} = \alpha_{ij} + \frac{2\alpha_{ik}\alpha_{kj}}{(\beta_{ik}+2)(\alpha_{kj}+\beta_{kj}+2)(2\alpha_{ik})} \tag{4}$$

$$\beta_{ij}^{new} = \beta_{ij} + \frac{2\alpha_{ik}\beta_{kj}}{(\beta_{ik}+2)(\alpha_{kj}+\beta_{kj}+2)(2\alpha_{ik})} \tag{5}$$

Note that node n_i uses its reputation of node n_k in the combination process. The authors of the RFSN defined how their trust model can be used in practice. They brought out some important points concerning the way information is to be used to avoid two major problems: (i) data incest, and (ii) a game theoretic set-up. Some researchers (Agah et al., 2004; Liu et al., 2004) have looked into the game theory aspect, which is no doubt inherent in a problem with malicious nodes in the network. However, a game theory solution might be difficult to obtain, in view of the large number of nodes. The RFSN forces the WSNs protocols into an exchange of information that limits any game aspect. The effectiveness of the notion of reputation and trust resides in the assumption that the majority of nodes in any neighbourhood is trustworthy, therefore creating a resilience of the system. Trust assessment is used to flush out the bad nodes. In combining information, the authors of the RFSN followed the approach of (Jøsang & Ismail, 2002), by mapping the problem into a Dempster-Shafer belief theory model (Shafer, 1976), solving it using the concept of belief discounting, and conducting a reverse mapping from belief theory to probability. In our work we find it unnecessary to use the Belief theory. Rather, probability theory, and the ensuing work on expert opinion provide a way to combine the two types of information.

5. Expert Opinion Theory

The use of expert opinion has received much attention in the statistical literature. It allows for the formal incorporation of informed knowledge into a statistical analysis. Expert opinion, or informed judgement, is often available in the form of vendor information, engineering knowledge, manufacturer's knowledge, or simply an opinion formed over time. It is often a subjective opinion based on knowledge. Its main departure from hard data is that it cannot be claimed as objectively observed data. Nevertheless, it is often valuable information that has been formed over the course of time. In our case, reputation is offered to neighbouring nodes as an opinion. The node making the assessment has not observed that reputation, and therefore treats it as an opinion. Early work to formalise ad-hoc procedures for the use of expert opinion includes (Dalkey & Helmer, 1963; Morris, 1971). Morris (Morris, 1974) recognised the importance of treating the expert opinion as data, stating the general principle on which subsequent work was based. The topic was further enlarged by the Bayesian statistical community to the problem of reconciliation prior information from different sources (Dawid, 1987; French, 1980; Genest & Schervish, 1985; Lindley et al., 1979), a topic that dated back to Winkler (Winkler, 1968). Lindley (Lindley, 1983) highlighted the theory in the statistical arena, with others following with work on reliability (Aboura & Robinson, 1995; Mazzuchi & Soyer, 1993; Singpurwalla, 1988), on maintenance optimization (Aboura, 1995; Mazzuchi & Soyer, 1996; Van Nortwijk et al., 1992) and on nuclear safety (Cooke, 1994).

The probabilistic approach adopted in the elicitation and use of expert opinion considers the opinion given by the expert as data and treats it according to the laws of probability. If θ is a random variable, and μ represents an opinion from an expert about θ, then $P(\theta \mid \mu)$ obtains, using Bayes' theorem as discussed in appendix A, the following formula, as shown in equation (6):

$$P(\theta \mid \mu) = \frac{P(\mu \mid \theta)P(\theta)}{P(\mu)} \tag{6}$$

$$P(\mu) = \int_{\theta} P(\mu \mid \theta)P(\theta)d\theta \tag{7}$$

- $P(\mu \mid \theta)$ is the likelihood function, and represents the analyst model of the expert's input
- $P(\theta)$ is the distribution that represents any prior knowledge the analyst may have about the quantity of interest
- $P(\mu)$ is the normalising constant

Bayes' theorem inverses the probability, so that the evidence μ highlights the value of θ that is most likely. The likelihood function $L(\theta) = P(\mu \mid \theta)$ refers to where the expert opinion is modelled. As an example, consider the reliability scenario of (Aboura & Robinson, 1995). In it, an expert provides reliability estimates for a device or machine. The work was undertaken in the context of maintenance optimisation.

Figure 3 shows the expert's input along the unknown reliability curve that the analyst wants to estimate. Each assessment by the expert is about the reliability as a time t_i, in the form of a value $0 < r_i < 1$. If the expert was perfect, and assuming that the reliability at time t_i is $e^{-\lambda t_i^{\beta}}$, then

$$r_i = e^{-\lambda t_i^{\beta}} \quad (8)$$

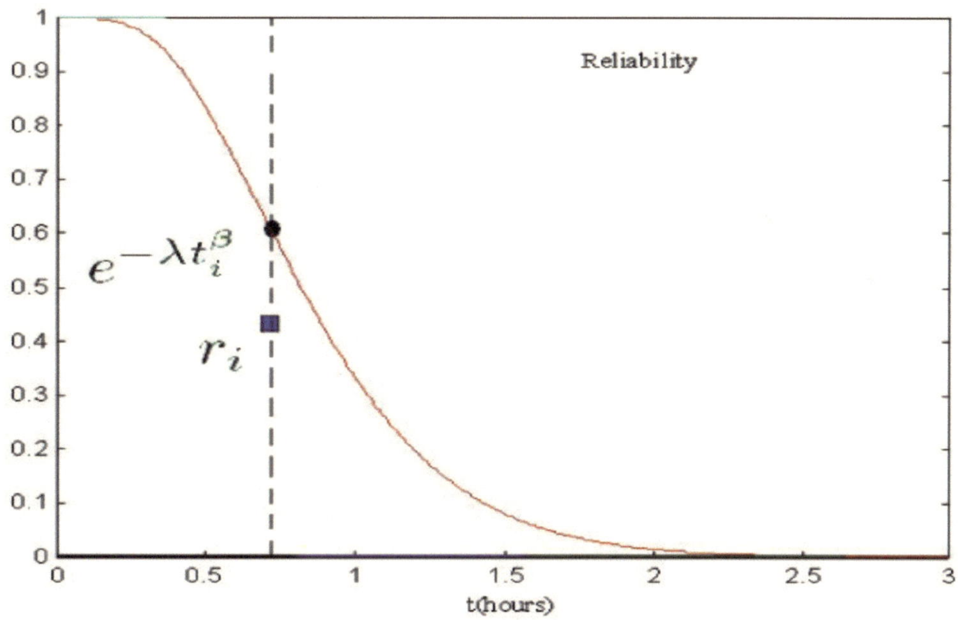

Fig. 3. Expert opinion r_i for reliability at time t_i

However, it will not necessarily be the case, and a probability distribution is needed to model the input. That probability distribution is the likelihood function, in this case

$$L(\lambda, \beta) = P(r_i \mid \lambda, \beta) \quad (9)$$

The authors of (Aboura & Robinson, 1995) modelled it using a Beta distribution, such that

$$E(r_i \mid \lambda, \beta) = \alpha_i + \sigma_i e^{-\lambda t_i^{\beta}} \tag{10}$$

where σ_i and α_i are inflation and bias, respectively, carried by the expert about the reliability at time t_i. These two values reflect the analyst's modulation of the expert opinion. To model several correlated inputs, a Dirichlet model is used. Once the likelihood function is built, then it can be used to combine the actual expert opinion with any existing knowledge about the random variable of interest. The analyst may not only have prior knowledge but also some observed data y about a random variable of interest, θ. Bayes' theorem is applied to combine the three sources of information, as shown in equation (11):

$$P(\theta \mid y, \mu) = \frac{P(y \mid \theta, \mu) P(\mu \mid \theta) P(\theta)}{P(y, \mu)} \tag{11}$$

One often writes, $P(\theta \mid y,\mu) \propto P(y \mid \theta,\mu)\ P(\mu \mid \theta)\ P(\theta)$, the denominator being a normalising constant that does not affect the combination occurring in the numerator. This seemingly simple operation can effectively combine many sources of information. We use it to model the reputation of a node when opinions about that node are provided by other nodes.

6. GTRSSN: Gaussian Trust and Reputation System for Wireless Sensor Networks

Taking into consideration the above discussion, let us assume that the wireless sensor network shown in Figure 4 consists of N nodes $(n_1, n_2, ..., n_N)$, and the corresponding matrix $\Gamma = [\Gamma_{i,j}]$ is given as follows:

$$\Gamma = [\Gamma_{i,j}] = \begin{pmatrix} 1 & 0 & 0 & 1 \\ 0 & 1 & 1 & 0 \\ 0 & 1 & 1 & 0 \\ 1 & 0 & 0 & 1 \end{pmatrix}$$

If node n_i is connected to node n_j, then $\Gamma_{i,j} = \Gamma_{j,i} = 1$, otherwise it is equal to (0). Let (X) be a field variable monitored in the environment where the WSN is deployed. This variable, might represent temperature, chemical component or atmospheric value, is detected and estimated by the sensor nodes and it is assumed to be of a continuous nature. The nodes are synchronised and can report at discrete times $t = 0, 1, 2, ..., k$.

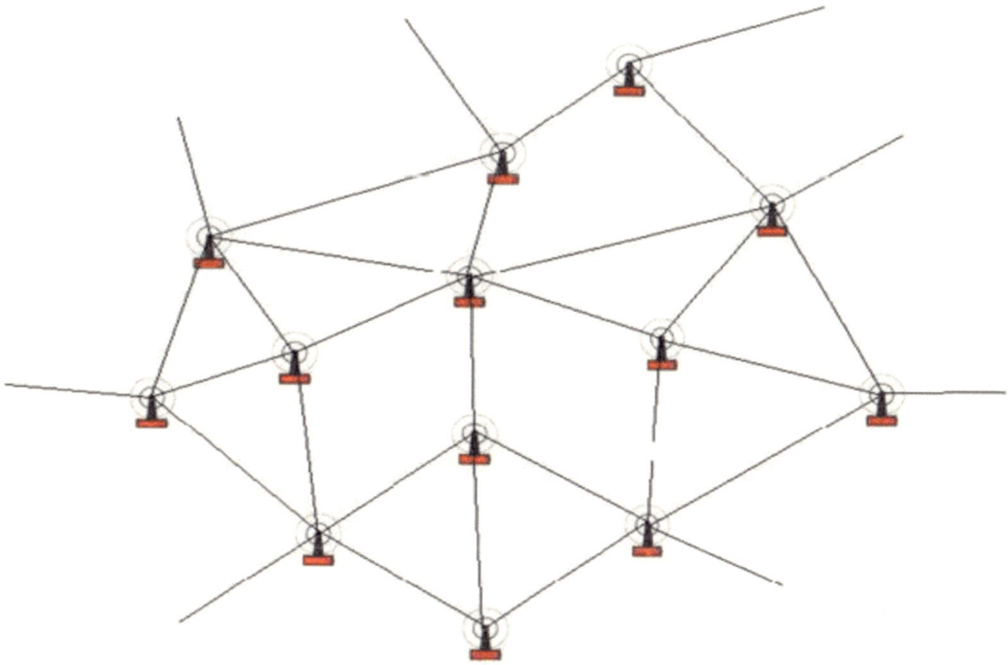

Fig. 4. Network of wireless sensor nodes

The random variable $(X_{n_i} = X_i)$ is the sensed value reported by node n_i, $i = 1, \ldots, N$. $x_i(t)$ is the realisation of that random variable at time t. Each node n_i, $i = 1, \ldots, N$ has a time series $(x_i(t))$. These time series are most likely different, as nodes are requested to provide readings at different times, depending on the sources of the requests. It could also be that the nodes provide such readings when triggered by particular events. We assume that each time a node provides a reading, its one-hop neighbours that route its report see that report, and can evaluate the reported value. For example, if node n_j reports $x_j(t)$ at some time t, then node n_i, such that $\Gamma_{i,j} = 1$, obtains a copy of that report for routing purposes, and has its own assessment $x_i(t)$ of the sensed variable. Let $y_{i,j}(t) = x_j(t)-x_i(t)$. From the node n_i perspective, $X_i(t)$ is known, and $Y_{i,j}(t) = X_j(t) - X_i(t)$ represents the error that node n_j commits in reporting the sensed field value $X_j(t)$ at time t. $Y_{i,j}(t)$ is a random variable modelled as a Normal (Gaussian), as shown in equation (12):

$$Y_{i,j}(t) \sim N(\theta_{i,j}, \tau^2) \tag{12}$$

where τ is assumed to be known (error variance), and it is the same for all nodes. If we let $\bar{y}_{i,j}$ be the mean of the observed error, as observed by n_i about n_j reporting, as in equation (13):

$$\bar{y}_{i,j} = \sum_{t=1}^{k} y_{i,j}(t)/k \tag{13}$$

then

$$(\theta_{i,j} \mid y_{i,j}) \sim N(\overline{y}_{i,j}, \tau^2 / k) \tag{14}$$

where $y_{i,j} = \{y_{i,j}(t)$, for all t values at which a report is issued by n_j and routed through $n_i\}$. This is a well-known straightforward Bayesian updating where a diffuse prior is used.

We let $\mu_{i,j} = \overline{y}_{i,j}$ and $\sigma_{i,j}^2 = \tau^2 / k$. Recall that k is node-dependent. It is the number of reports issued by node n_j and routed through n_i, and differs from node to node. We define the reputation as the probability density function, as in equation (15):

$$R_{i,j} = N(\mu_{i,j}, \sigma_{i,j}^2) \tag{15}$$

where $\mu_{i,j} = \overline{y}_{i,j}$ and $\sigma_{i,j}^2 = \tau^2 / k$ are the equivalent of a_{ij} and β_{ij} in RFSN (Ganeriwal & Srivastava, 2004).

Trust is defined differently, since we want it to remain between (0) and (1), a convention that seems to be unanimous among researchers, except for the occasional translation to the scale [-1, 1]. In our trust model, we define the trust to be the probability, as shown in equations (16) and (17):

$$T_{i,j} = \text{Prob}\{|\theta_{i,j}| < \varepsilon\} \tag{16}$$

$$\begin{aligned} T_{i,j} &= \text{Prob}\{-\varepsilon < \theta_{i,j} < +\varepsilon\} \\ &= \phi\left(\frac{\varepsilon - \overline{y}_{i,j}}{\tau / \sqrt{k}}\right) - \phi\left(\frac{-\varepsilon - \overline{y}_{i,j}}{\tau / \sqrt{k}}\right) \\ &= \phi\left(\frac{\varepsilon - \mu_{i,j}}{\sigma}\right) - \phi\left(\frac{-\varepsilon - \mu_{i,j}}{\sigma}\right) \end{aligned} \tag{17}$$

where ϕ is the cumulative probability distribution (cdf) of the Normal $N(0, 1)$. As shown in Figure 5, the area under the *Gaussian* curve $N(\mu_{i,j}, \sigma_{i,j}^2)$ within the interval $[-\varepsilon, +\varepsilon]$ is the trust value. The bigger the error θ_{ij} is, meaning its mean shifting to the right or left of 0, and the more spread that error is, the lower the trust value is. Each node n_i maintains a line of reputation assessments composed of T_{ij} for each j, such that $\Gamma_{i,j} \neq 0$ (one-hop connection). T_{ij} is updated for each time period t for which data is received from some connecting node j. The filled areas in Figure 5 represent the Gaussian Trust T_{ij} in two cases.

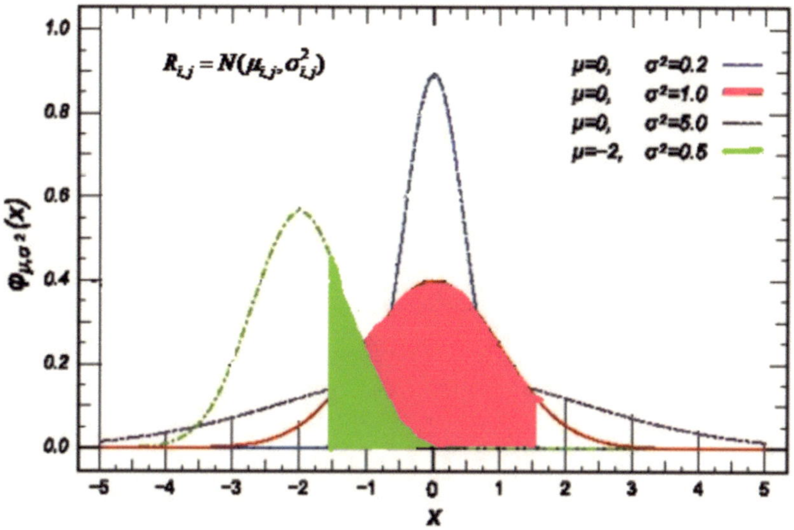

Fig. 5. Normal (Gaussian) distribution example

In addition to data observed in form of $y_{i,j} = \{y_{i,j}(t)\}$, for all t values at which a report is issued by n_j and routed through n_i}, node n_i uses *second-hand information* in the form of $(\mu_{l_s,j}, \sigma_{l_s,j})$, s = 1, …, m, from the m nodes connected to n_j and n_i, as shown in Figure 6, below. This is an "expert opinion", that is, soft information from external sources. Each of these m nodes has observed node n_j reports and produced assessments of its error in the form of $(\mu_{l_s,j}, \sigma_{l_s,j})$, s = 1,…, m, and consequently $T_{ls,j}$, s = 1, …, m. In using the expert opinion theory, one needs to modulate it. Node n_i uses its own assessment of the nodes $n_{l_1}, \ldots, n_{l_m}$, in the form of $(\mu_{i,l_s}, \sigma_{i,l_s})$, s = 1, … , m, and consequently $T_{i,ls}$, s = 1, …, m.

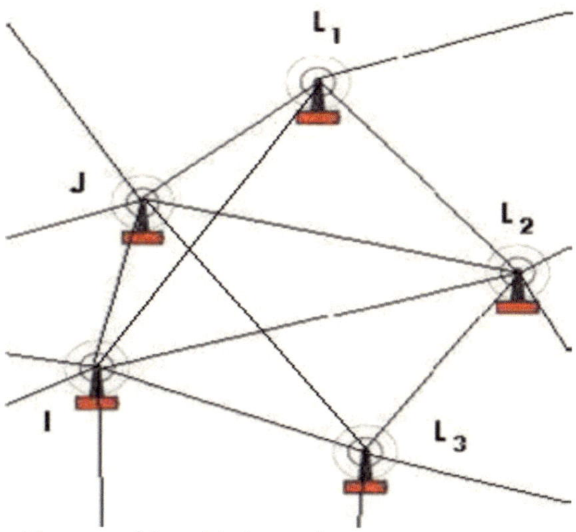

Fig. 6. Nodes that provide second-hand information

Using Bayes' theorem, the probability distribution of $\theta_{i,j}$ is obtained using the observed data along with the second-hand modulated information, as shown in equation (18):

$$P(\theta_{i,j} \mid y_{i,j},(\mu_{l_1,j},\sigma_{l_1,j}),\ldots,(\mu_{l_m,j},\sigma_{l_m,j}) \\ ,(\mu_{i,l_1},\sigma_{i,l_1}),\ldots,(\mu_{i,l_m},\sigma_{i,l_m})) \tag{18}$$

and it is proportional to the product of three terms shown in equations (19), (20) and (21):

$$P(y_{i,j} \mid \theta_{i,j},(\mu_{l_1,j},\sigma_{l_1,j}),\ldots,(\mu_{l_m,j},\sigma_{l_m,j}), \\ (\mu_{i,l_1},\sigma_{i,l_1}),\ldots,(\mu_{i,l_m},\sigma_{i,l_m})) \tag{19}$$

$$P((\mu_{l_1,j},\sigma_{l_1,j}),\ldots,(\mu_{l_m,j},\sigma_{l_m,j}) \mid \theta_{i,j},(\mu_{i,l_1},\sigma_{i,l_1}),\ldots,(\mu_{i,l_m},\sigma_{i,l_m})) \tag{20}$$

and

$$P(\theta_{i,j} \mid (\mu_{i,l_1},\sigma_{i,l_1}),\ldots,(\mu_{i,l_m},\sigma_{i,l_m})) \tag{21}$$

The first term, equation (19), reduces to $P(y_{i,j} \mid \theta_{i,j})$ through conditional independence, and is equal to the product of the likelihoods

$$\prod_{t=1}^{k} N(\theta_{i,j},\tau^2) \tag{22}$$

The third term, equation (21), also reduces to $P(\theta_{i,j})$, due to the conditional independence of $\theta_{i,j}$ from $(\mu_{i,l_1},\sigma_{i,l_1}),\ldots,(\mu_{i,l_m},\sigma_{i,l_m})$, and it represents the prior distribution of $\theta_{i,j}$ which we model as a diffuse prior $N(0,\infty)$.

The second term, equation (20), models the use of the second-hand information. This term requires some elaboration and can be reduced to the product of equation (23) through conditional independence arguments.

$$\prod_{s=1}^{m} P((\mu_{l_s,j},\sigma_{l_s,j}) \mid \theta_{i,j},(\mu_{i,l_s},\sigma_{i,l_s})) \tag{23}$$

To derive $P((\mu_{l_s,j},\sigma_{l_s,j}) \mid \theta_{i,j},(\mu_{i,l_s},\sigma_{i,l_s}))$ for each $s = 1, \ldots, m$, we observe the following: for some t's,

$$\theta_{i,j} = x_j(t) - x_i(t) \tag{24}$$

and for some t's

$$\theta_{l,j} = x_j(t) - x_l(t) \tag{25}$$

and, if all t's were the same, then

$$\theta_{i,j} = x_j(t) - x_i(t) = (x_j - x_l) + (x_l - x_i) = \theta_{l,j} + \theta_{i,l} \tag{26}$$

But not all *t*'s are the same, so all data are not used for all assessments. We inspire ourselves from this relationship to model the expert opinion likelihood. We assume that

$$\theta_{l,j} \sim \theta_{i,j} - \theta_{i,l} \tag{27}$$

$$\mu_{l,j} \sim \theta_{i,j} - \mu_{i,l} \tag{28}$$

and we model

$$\mu_{l,j} \sim N(\theta_{i,j} - \mu_{i,l}, var) \tag{29}$$

where we choose *var* to be inversely related to node's n_i assessment of the reputation of node n_l, that is

$$var = \left(\frac{1}{T_{i,l}} - 1\right)\psi \tag{30}$$

where ψ is a model parameter.

$$\mu_{l,j} \sim N(\theta_{i,j} - \mu_{i,l}, \left(\frac{1}{T_{i,l}} - 1\right)\psi) \tag{31}$$

leads to equation (32):

$$\prod_{s=1}^{m} P((\mu_{l_s,j}, \sigma_{l_s,j}) \mid \theta_{i,j}, (\mu_{i,l_s}, \sigma_{i,l_s})) = \prod_{s=1}^{m} N(\theta_{i,j} - \mu_{i,l_s}, \left(\frac{1}{T_{i,l_s}} - 1\right)\psi) \tag{32}$$

and consequently proves that equation (33)

$$P(\theta_{i,j} \mid y_{i,j}, (\mu_{l_1,j}, \sigma_{l_1,j}), \ldots, (\mu_{l_m,j}, \sigma_{l_m,j}), (\mu_{i,l_1}, \sigma_{i,l_1}), \ldots, (\mu_{i,l_m}, \sigma_{i,l_m})) \tag{33}$$

is a Normal (Gaussian) distribution with mean and variance as shown in equations (34) and (35) respectively:

$$\mu_{i,j}^{new} = \frac{\sum_{s=1}^{m} \frac{(\mu_{l_s,j} + \mu_{i,l_s})}{\left(\frac{1}{T_{i,l_s}} - 1\right)\psi} + (k\overline{y} / \tau^2)}{\sum_{s=1}^{m} \frac{1}{\left(\frac{1}{T_{i,l_s}} - 1\right)\psi} + (k / \tau^2)} \tag{34}$$

$$\sigma_{i,j}^{2\ new} = \frac{1}{\sum_{s=1}^{m} \frac{1}{\left(\frac{1}{T_{i,l_s}} - 1\right)\psi} + (k/\tau^2)} \tag{35}$$

These values $(\mu_{i,j}^{new}, \sigma_{i,j}^{new})$, along with $(\mu_{i,j}, \sigma_{i,j})$, are easily updatable values that represent the continuous Gaussian version of the $(\alpha_{i,j}, \beta_{i,j})$ and $(\alpha_{i,j}^{new}, \beta_{i,j}^{new})$ of the binary approach in (Ganeriwal & Srivastava, 2004), as derived from the approach in (Jøsang & Ismail, 2002). The solution presented is simple and easily computed, keeping in mind that the solution applies to networks with limited computational power. In the binary work, $(\alpha_{i,j}, \beta_{i,j})$ are obtained through a Bayesian approach, while $(\alpha_{i,j}^{new}, \beta_{i,j}^{new})$ are obtained through the combination approach of Belief functions. The Gaussian solution provides a full probabilistic approach in the case of continuous sensor data.

Some would object to the use of a diffuse prior, which, in effect, forces a null prior trust value, regardless of the ε value. A way to remedy to this is to start with a $N(\mu_0, \sigma_0^2)$ prior distribution for all θ_{ij}, such that the prior trust is (1/2). This choice not only answers the diffuse prior issue, but also allows the choice of the parameters involved. ε can be determined: given μ_0 and σ_0, μ_0 is most likely to be set to (0). Therefore, σ_0 and ε determine each other. Once one is set, the other is automatically deducted. Note that the prior is really node-dependent, making our definition of trust, and therefore ε, node-dependent. In practice, it is most likely that all priors are tuned to the same values so that the prior trusts are started at some level, say (1/2), with a proper prior $\theta_{i,j}$, as shown in equation (36):

$$\theta_{i,j} \sim N(\mu_0, \sigma_0^2) \tag{36}$$

The reputation parameters $\mu_{i,j}$ and $\sigma_{i,j}^2$ are presented in equations (37) and (38):

$$\mu_{i,j} = \frac{(\mu_0/\sigma_0^2) + (k\bar{y}_{i,j}/\tau^2)}{(1/\sigma_0^2) + (k/\tau^2)} \tag{37}$$

$$\sigma_{i,j}^2 = \frac{1}{(1/\sigma_0^2) + (k/\tau^2)} \tag{38}$$

and the updated values are presented in equations (39) and (40) respectively:

$$\mu_{i,j}^{new} = \frac{(\mu_0/\sigma_0^2) + \sum_{s=1}^{m} \frac{(\mu_{l_s,j} + \mu_{i,l_s})}{\left(\frac{1}{T_{i,l_s}} - 1\right)\psi} + (k\bar{y}_{i,j}/\tau^2)}{(1/\sigma_0^2) + \sum_{s=1}^{m} \frac{1}{\left(\frac{1}{T_{i,l_s}} - 1\right)\psi} + (k/\tau^2)} \tag{39}$$

$$\sigma_{i,j}^{2\ new} = \frac{1}{(1/\sigma_0^2) + \sum_{s=1}^{m} \frac{1}{\left(\frac{1}{T_{i,l_s}} - 1\right)\psi} + (k/\tau^2)} \tag{40}$$

Once $\mu_{i,j}^{new}$ and $\sigma_{i,j}^{2\ new}$ are formulated, the new trust value $T_{i,j}^{new}$ will be presented as shown in equation (41):

$$T_{i,j}^{new} = \phi\left(\frac{\varepsilon - \mu_{i,j}^{new}}{\sigma_{i,j}^{new}}\right) - \phi\left(\frac{-\varepsilon - \mu_{i,j}^{new}}{\sigma_{i,j}^{new}}\right) \tag{41}$$

We call this trust and reputation system (GTRSSN), which stands for Gaussian Trust and Reputation System for Sensor Networks. It can be seen as an extension of the concepts of RFSN and DRBTS for sensor data and it introduces a full probabilistic approach to the combination of information in the reputation assessment.

7. Simulation Results

To verify the theory introduced in this chapter, several simulation experiments in different scenarios were developed. The results from the simulations conducted on the network shown in Figure 7, for one scenario, where only a random region from the network is selected to report data on every time series, are presented in this section. In all simulation experiments, the trust relationship between four nodes (1, 6, 7 and 13) in a sub-network of the fifteen-node network shown in Figure 7 is calculated.

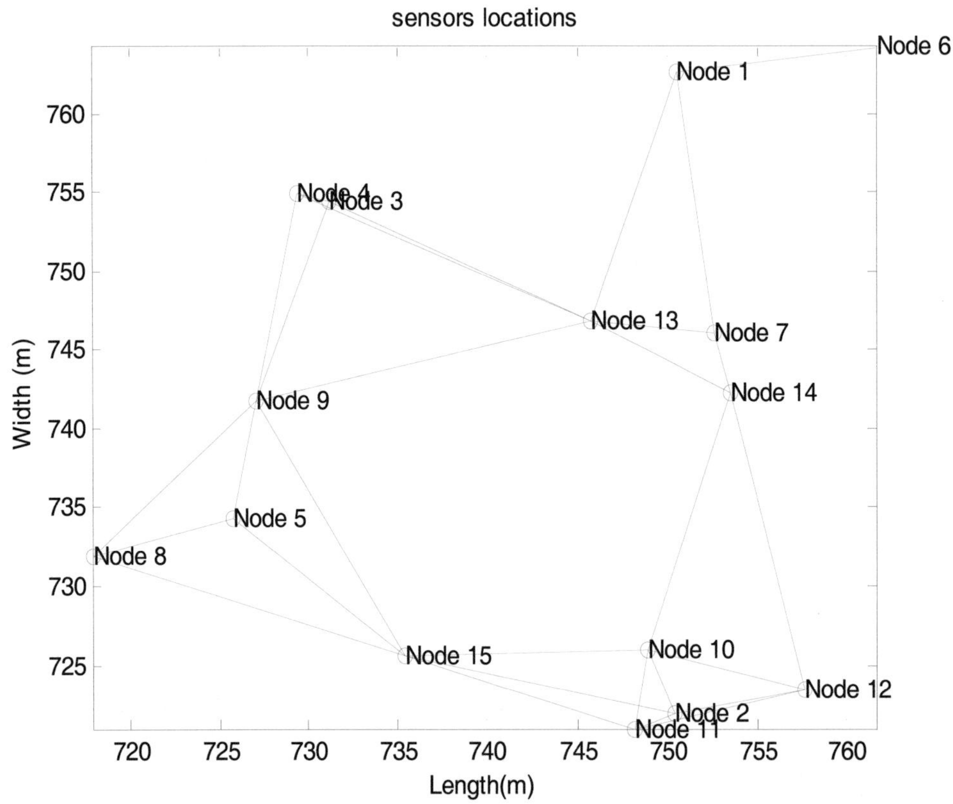

Fig. 7. Wireless Sensor Network Diagram

In this scenario and as stated before, it is assumed that, at each time slot a group of nodes are selected to report their sensed data, and when one node is sending its own reading to a specific node in the group, all the surrounding nodes connected to the sending node hear the reported value and start to send the output of that reading as a second-hand information to the receiving node regarding the sending node. The output of that reading between the sending and the receiving nodes is regarded as the direct observation, as discussed before. In other words, and in the case of selected sub-network, when node (7) is sending its reading to node (1), nodes (6) and (13) hear the reported data, use it to find the trust between them and node (7) and report that trust to node (1) as second-hand information about node (7). Node (1), at the same time, uses the reading reported directly from node (7) to calculate the direct trust between node (1) and node (7).

7.1. No faulty or malicious nodes are present in the network

At the beginning, it is assumed that all nodes are working properly, that no faulty or malicious nodes exist in the network, and report the sensed event (temperature) with minimum error. Figure 8 below presents the result of the simulation and shows the trust value between node (1) and the other nodes (6, 7 and 13). At first node (1) assesses node (13) based on the direct interactions only between the two nodes, without second-hand information, and then node (1) assesses node (13) based on the direct information between the two nodes and the second-hand information received from node (7) about node (13), with second-hand information. Node (1) performs the same assessment procedure for all nodes directly connected to it.

It can be seen from Figure 8 that trust values between node (1) and nodes (7) and (13) are slightly different but they eventually all converge to the value of one. The trust value between node (1) and node (6) is the same in both cases, with and without second-hand information as there is no second-hand information for node (6). Node (6) is not connected to any other node other than node (1).

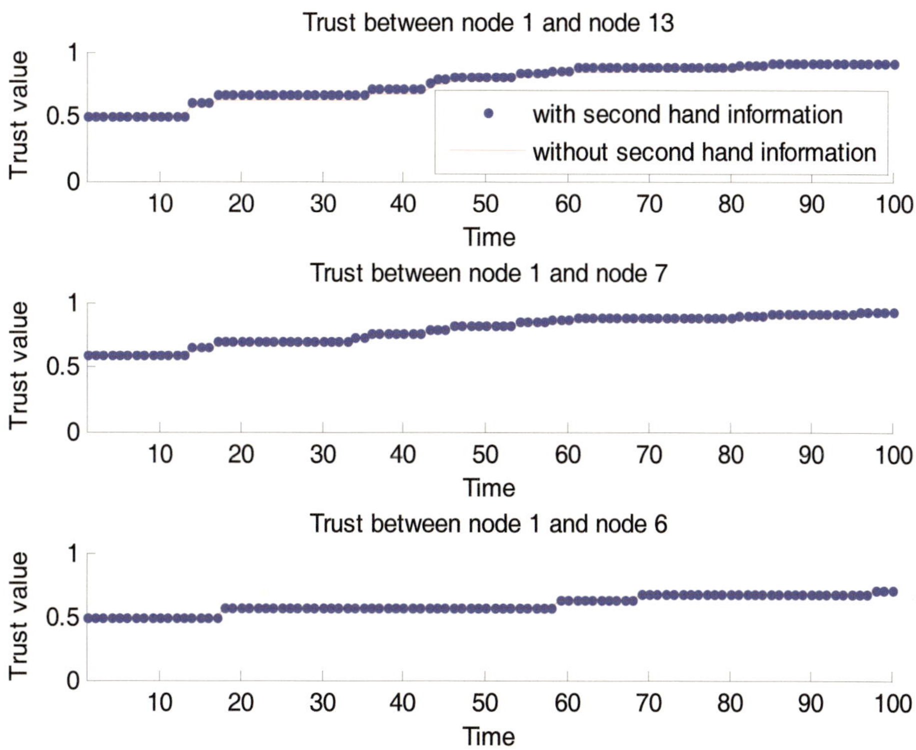

Fig. 8. All nodes are normal

7.2. Node (13) is Faulty or Malicious

In another experiment, the same network was simulated, but with the introduction of a significant error in node (13) readings, that is, node (13) is faulty or malicious. Simulation results are shown in Figure 9, below and, as can be seen from Figure 9, the trust value between node (1) and node (13) dropped to almost zero for both cases, with and without second-hand information, which means node (7) is assessing node (13) as a faulty or malicious node. The situation for node (6) is not affected, as there is no connection between node (6) and node (13). The interesting result here is that the trust value between node (1) and node (7) is not affected in either case even though there is a connection between node (7) and node (13). Node (13) is faulty, and one would think that it could harm the reputation of node (7), but that was not the case, which proves that the modulation in the approach makes the reputation system robust to bad-mouthing attacks.

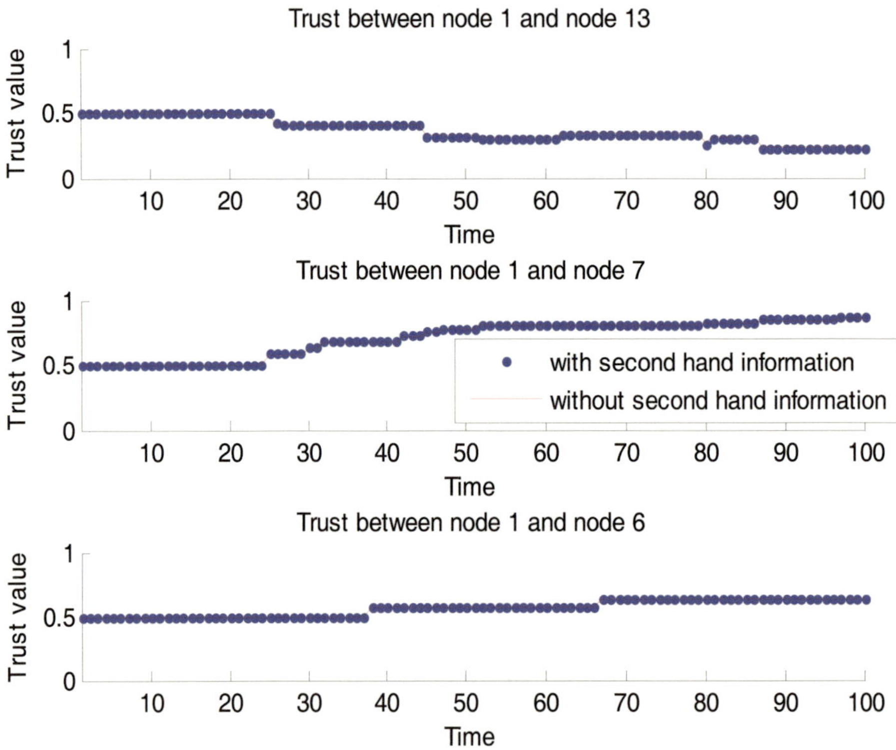

Fig. 9. Node (13) is faulty

7.3. Node (7) and Node (13) are Faulty

In this simulation experiment, it has been assumed that node (7) and node (13) are faulty. The results of the simulation are presented in Figure 5.10, showing that the trust values for both nodes (7) and (13) are dropping to zero in both cases. Node (6) is assumed reliable and the trust value associated with it is the same in both cases, as there is no connection between node (6) and the other faulty nodes, (7) or (13), to affect that trust value.

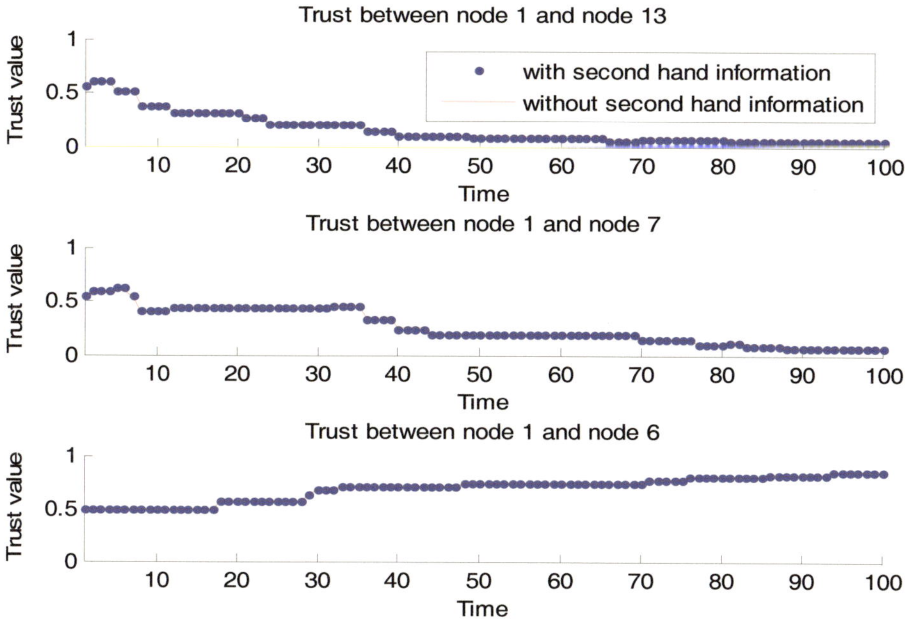

Fig. 10. Node (7) and node (13) are faulty

7.4. Node (6) is Faulty or Malicious

The simulation results presented in Figure 11 below show that when node (6) is faulty or malicious, nothing almost will change in the trust values between node (1) and either of nodes (7) and (13), as there is no direct or indirect connection between them. In other words, when node (6) is faulty, node (1) will discover that, as it has a direct connection with node (6) and the direct trust with node (6) will be affected. As there is no indirect trust for node (6), both trust values will stay on the initial trust value or will decrease to the value of zero.

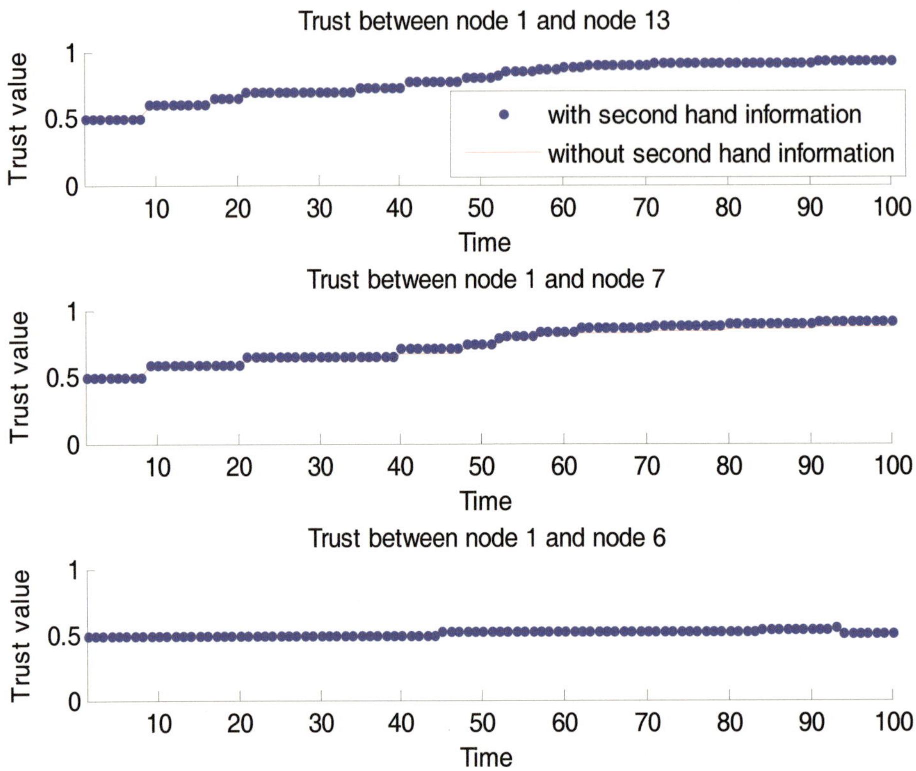

Fig. 11. Node (6) is faulty

7.5. Node (1) is Faulty or Malicious

It is assumed in this experiment that node (1) is faulty or malicious. Node (1) is the main node in the sub-network and is acting as the receiving node, and all the simulations show the trust relationship between node (1) and all the other nodes connected to it. As can be seen from Figure 5.12, the direct trust value for both nodes (7) and (13), is declining to the value of zero, as node 1 is faulty. That will leave the two nodes (7) and (13) to assess each other indirectly, which is a very interesting case again, as both nodes (7) and (13) are now assessing node (1) as a faulty node, so the indirect trust value for both nodes are slowly converging to the value of one. The trust value for node (6) is set to the initial value (0.5) and will decrease on both values to zero, as there is no second-hand information available to node (6) and node (1) is a faulty node.

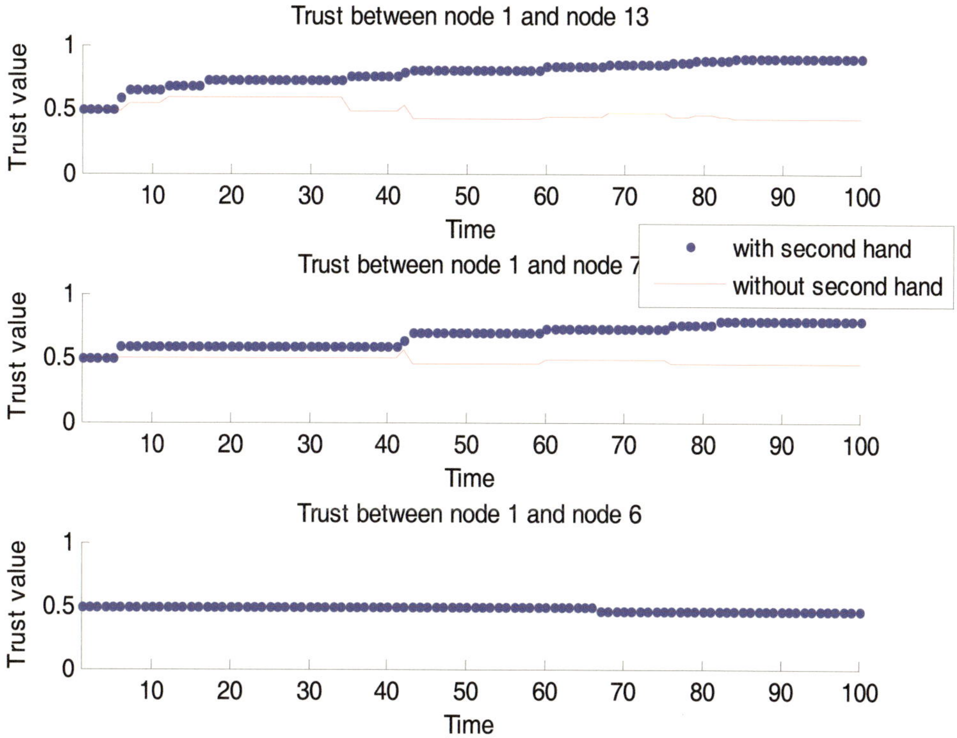

Fig. 12. Node (1) is a malicious node

The last example shows precisely the reason the trust system is instituted. It allows the classification of nodes into separate sets according to their trustworthiness. In the last example, it is known that node (1) is faulty, since it is a simulation exercise. The results should clearly indicate to the network that node (1) is faulty. However, it could also be the case that the nodes (7) and (13) are malicious. The trust system works on the assumption that a majority of nodes in a neighbourhood are reliable. This principle helps purge the system of bad elements. In this case, at this point, it is observed that the developed trust system is effective in distinguishing among nodes.

8. Conclusion

It has been argued that the trust-modelling problem is characterised by uncertainty, and the only coherent way to deal with uncertainty is through probability. Even though some of the trust models introduced for sensor networks employ probabilistic solutions mixed with ad-hoc approaches, none of them produces a full probabilistic answer to the problem. In this chapter we introduced a theoretically sound Bayesian probabilistic approach for calculating trust and reputation systems in WSNs. We introduced a new Gaussian Trust and Reputation System for Sensor Networks (GTRSSN), which we believe is a breakthrough in modelling trust in WSNs, as previous studies in WSNs focused on the trust associated with the routing and the successful performance of a sensor node in some predetermined task, that is, looking at binary events to model trust and the trustworthiness and reliability of the nodes of a WSN when the sensed data is continuous has not been addressed before. Having said that, introducing the sensor data as a major component of trust leads to the modification of node misbehaviour classification, the trust computational model and the way first-hand and second-hand information is formulated. These issues have been presented in this chapter. Also, a brief summary about the Beta reputation system and the expert opinion theory has been presented. A very detailed GTRSSN, which is the significant contribution of this research, has also been presented, with some simulation results. The simulation results show the implications of sensor data for the direct and indirect trust relationship between nodes, which helps to distinguish among nodes and purge the bad nodes from the network.

9. References

Aboura, K. & Robinson, N. I. (1995). Optimal Replacement in a Renewal Process, CSIRO.

Aboura, K. (1995). Bayesian Adaptive Maintenance Plans using Initial Expert Reliability Estimates, CSIRO.

Agah, A.; Das, S. K. & Basu, K. (2004). *A Game Theory based Approach for Security in Wireless Sensor Networks*. IEEE International Conference on Performance, Computing, and Communications, Phoenix, Arizona.

Cooke, R. (1994). Uncertainty in Dispersion and Deposition in Accident Consequence Modelling Assessed with Performance-based Expert judgement. *Reliability Engineering and System Safety* 45: 35-46.

Dalkey, N. C. & Helmer, O. (1963). An Experimental Application of Delphi Method to the Use of Experts. *Management Science* 9(3): 458-467.

Dawid, A. P. (1987). The Difficulty about Conjunction. *The Statistician* 36: 91-97.

French, S. (1980). Updating of Belief in the Light of Someone else's Opinion. *Journal of Royal Statistical Society* 143: 43-48.

Ganeriwal, S. & Srivastava, M. B. (2004). *Reputation-based Framework for High Integrity Sensor Networks*. The 2nd ACM Workshop on Security of Ad-hoc and Sensor Networks Washington DC, USA

Genest, C. & Schervish, M. J. (1985). Modelling Expert Judgments for Bayesian Updating. *The Annals of Statistics* 13(3): 1198-1212.

Jøsang, A. & Ismail, R. (2002). The Beta Reputation System. *The 15th Bled Electronic Commerce Conference*. Bled, Slovenia.

Lindley, D. V. & Singpurwalla, N. D. (1986). Reliability (and fault tree) Analysis using Expert Opinions. *Journal of the American Statistical Association* 81: 87-90.

Lindley, D. V.; Tversky, A. & Brown, R. V. (1979). On the Reconciliation of Probability Assessments. *Journal of Royal Statistical Society* 142: 146-180.

Lindley, D. V. (1983). Reconciliation of Probability Distributions. *Operations Research Journal* 31: 866-880.

Liu, Y.; Comaniciu, C. & Man, H. (2004). *A Bayesian game approach for intrusion detection in wireless ad hoc networks*. Proc. 2006 workshop on Game theory for communications and networks, ACM Int. Conf., Pissa, Italy.

Mazzuchi, T. A. & Soyer, R. (1993). A Bayes Method for Assessing Product Reliability during Development Testing. *IEEE Transactions on Reliability* 42(3): 503-510.

Mazzuchi, T. A. & Soyer, R. (1996). Adaptive Bayesian Replacement Strategies. *Bayesian Statistics* **5**: 667-674.

Momani, M.; Challa, S. & Aboura, K. (2007a). Modelling Trust in Wireless Sensor Networks from the Sensor Reliability Prospective. *Innovative Algorithms and Techniques in Automation, Industrial Electronics and Telecommunications.* Sobh, T., Elleithy, K., Mahmood, A. and Karim, M., Springer Netherlands.

Momani, M.; Aboura, K. & Challa, S. (2007b). *RBATMWSN: Recursive Bayesian Approach to Trust Management in Wireless Sensor Networks.* The Third International Conference on Intelligent Sensors, Sensor Networks and Information, Melbourne, Australia, IEEE.

Morris, P. A. (1971). Bayesian Expert Resolution. *Department of Engineering-Economic Systems,* Stanford University. Ph.D.

Morris, P. A. (1974). Decision Analysis Eexpert Use. *Management Science* 20: 1233-1241.

Shafer, G. (1976). *A Mathematical Theory of Evidence,* Princeton University Press.

Singpurwalla, N. D. (1988). An Interactive PC-Based Procedure for Reliability Assessment Incorporating Expert Opinion and Survival Data. *Journal of the American Statistical Association* **83**(401): 43-51.

Srinivasan, A.; Teitelbaum, J.; Liang, H.; Wu, J. & Cardei, M. (2007). Reputation and Trust-based Systems for Ad-hoc and Sensor Networks. *Algorithms and Protocols for Wireless Ad-hoc and Sensor Networks.* Boukerche, A., Wiley & Sons.

Srinivasan, A.; Teitelbaum, J. & Wu, J. (2006). DRBTS: Distributed Reputation-based Beacon Trust System. *The 2nd IEEE International Symposium on Dependable, Autonomic and Secure Computing (DASC '06).*

Van Nortwijk, J. M.; Dekker, R.; Cooke, R. & Mazzuchi, T. A. (1992). Expert Judgement in Maintenance Optimization. *IEEE Transactions on Reliability* 41(3): 427-432.

West, M. (1984). Bayesian aggregation. *Journal of the Royal Staistical Society* 147(4): 600-607.

Winkler, R. L. (1968). The Concensus of Subjective Probability Distributions. *Management Science* 15: B61-B75.

24

Time-frequency analysis using Bayesian regularized neural network model

Imran Shafi, Jamil Ahmad, Syed Ismail Shah and Ataul Aziz Ikram
Iqra University Islamabad Campus, Sector H-9
Pakistan

1. Introduction

During the last twenty years there has been spectacular growth in the volume of research on studying and processing the signals with time–dependant spectral content. For such signals we need techniques that can show the variation in the frequency of the signal over time. Although some of the methods may not result in a proper distribution, these techniqes are generally known as time–frequency distributions (TFDs) (1, Boashash 2003). The TFDs are two–dimensional (2–D) functions which provide simultaneously, the temporal and spectral information and thus are used to analyze the non–stationary signals. By distributing the signal energy over the time–frequency (TF) plane, the TFDs provide the analyst with information unavailable from the signal's time or frequency domain representation alone. This includes the number of components present in the signal, the time durations and frequency bands over which these components are defined, the components' relative amplitudes, phase information, and the instantaneous frequency (IF) laws that components follow in the TF plane.
There has been a great surge of activity in the past few years in the TF signal processing domain. The pioneering work in this area is performed by (2, Claasen & Mecklenbrauker 1980), (3, Janse & Kaizer 1983), and (4, Boashash 1978). They provided the initial impetus, demonstrated useful methods for implementation and developed ideas uniquely suited to the situation. Also, they innovatively and efficiently made use of the similarities and differences of signal processing fundamentals with quantum mechanics. Claasen and Mecklenbrauker devised many new ideas, procedures and developed a comprehensive approach for the study of joint TFDs. However Boashash is believed to be the first researcher, who used various TFDs for real world problems. He developed a number of new methods and particularly realized that a distribution may not behave properly in all respects or interpretations, but it could still be used if a particular property such as the IF is well defined. The research presented in (6, Flandrin & Escudie 1980) transcribed directly some of the early quantum mechanical results, particularly the work on the general class of distributions, into signal analysis. The work in (3, Janse & Kaizer 1983) developed innovative theoretical and practical techniques for the use of TFDs and introduced new methodologies remarkable in their scope.
Historically the spectrogram has been the most widely used tool for the analysis of time–varying spectra. The spectrogram is expressed mathematically as the magnitude–square of the short–time Fourier transform (STFT) of the signal, given by

$$S(t,\omega) = \left| \int_{-\infty}^{\infty} s(t)\, h(t-\tau) e^{-j\omega\tau} d\tau \right|^2 \quad (1)$$

where $s(t)$ is the signal and $h(t)$ is a window function. Nevertheless, the spectrogram has severe drawbacks, both theoretically, since it provides biased estimators of the signal IF and group delay, and practically, since the Gabor–Heisenberg inequality makes a tradeoff between temporal and spectral resolutions unavoidable. However STFT and its variation being simple and easy to manipulate, are still the primary methods for analysis of the signals with time varying spectral contents and most commonly used today.
There are other approaches with a motivation to improve upon the spectrogram, with an objective to clarify the physical and mathematical ideas needed to understand time–varying spectrum. These techniques generally aim at devising a joint function of time and frequency, a distribution that will be highly concentrated along the IFs present in a signal and cross terms (CTs) free thus exhibiting good resolution. One form of TFD can be formulated by the multiplicative comparison of a signal with itself, expanded in different directions about each point in time. Such formulations are known as quadratic TFDs (QTFDs) because the representation is quadratic in the signal. This formulation was first described by Eugene Wigner in quantum mechanics (7, Wigner 1932) and introduced in signal analysis by Ville (8, Ville 1946) to form what is now known as the Wigner–Ville distribution (WVD). The WVD is the prototype of distributions that are qualitatively different from the spectrogram, produces the ideal energy concentration along the IF for linear frequency modulated signals, given by

$$W(t,\omega) \triangleq \frac{1}{2\pi} \int_{-\infty}^{\infty} s^*(t - \frac{1}{2}\tau) s(t + \frac{1}{2}\tau) e^{-j\omega\tau} d\tau \quad (2)$$

It is found that the spectrogram results in a blurred version (5, Cohen 1995), which can be reduced to some degree by use of an adaptive window or by combination of spectrograms. On the other hand, the use of WVD in practical applications is limited by the presence of non-negligible CTs, resulting from interactions between signal components. These CTs may lead to an erroneous visual interpretation of the signal's TF structure, and are also a hindrance to pattern recognition, since they may overlap with the searched TF pattern. Moreover If the IF variations are non–linear, then the WVD cannot produce the ideal concentration. Such impediments, pose difficulties in the correct analysis of non–stationary signals, are dealt in various ways and historically many techniques are developed to remove them partially or completely. They were partly addressed by the development of the Choi–Williams distribution (9, Choi & Williams 1989), followed by numerous ideas proposed in literature with an aim to improve the TFDs' concentration and resolution for practical analysis (10, Shafi et al. 2009). Few other important non–stationary representations among the Cohen's class of bilinear TF energy distributions include the Margenau–Hill distribution (11, Margenau & Hill 1961), their smoothed versions (12, Hippenstiel & Oliveira 1990), and others with reduced CTs (13, Jeong & Williams 1992) are members of this class. Nearly at the same time, some authors also proposed other time–varying signal analysis tools based on a concept of scale rather than frequency, such as the scalogram (14, Daubechies 1990) (the squared modulus of the wavelet transform), the affine smoothed pseudo WVD (15, Rioul & Flandrin 1992) or the Bertrand distribution (16, Bertrand 1988). The theoretical properties and the application fields of this large variety of these existing methods are now well determined, and wide–spread. Although many other QTFDs have been proposed in the literature, no single QTFD can be effectively used in all possible applications. This is because different QTFDs suffer from one or more problems (5, Cohen 1995).

An ideal TFD function roughly requires the four properties namely (i) high clarity i.e high concentration along individual components, (ii) CTs' elimination, (iii) good mathematical properties, and (iv) lower computational complexity. These characteristics are necessary for an easy visual interpretation of their outcomes and a good discrimination between known patterns for non-stationary signal classification tasks. To analyze the signals well, choosing an appropriate TFD function is important. Which TFD function should be used depends on what application it applies on. On the other hand, the short comings make specific TFDs suited only for analyzing non–stationary signals with specific types of properties and TF structures. Half way in this decade, there has been an enormous amount of work towards achieving high concentration and good resolution along the individual components and to enhance the ease of identifying the closely spaced components in the TFDs. The aim has been to correctly interpret the fundamental nature of the non–stationary signals in the TF domain.

We shall present a novel Bayesian regularized artificial neural network (ANN) based method for computing highly informative TFDs. The proposed method provides a way to obtain a non-blurred and high resolution version of the TFDs of signals whose frequency components vary with time. The resulting TFDs do not have the CTs that appear in case of multicomponent signals in some distributions such as WVD, thus providing visual way to determine the IF of non-stationary signals. It is proved that Bayesian inference framework and ANN learning capabilities can be successfully applied in the TF field, where they have not been used before.

2. Bayesian regularized Neural Network based Framework for Computing De-blurred TFDs

This section presents the Bayesian regularized ANN model (BRNNM) based correlation vectored taxonomy algorithm to compute the TFDs that are highly concentrated in the TF plane (23, Shafi et al. 2008). The degree of regularization is automatically controlled in the Bayesian inference framework and produces networks with better generalized performance and lower susceptibility to over–fitting. The grayscale spectrograms and pre–processed WVD of known signals are vectored and clustered as per the elbow criterion to constitute the training data for multiple ANNs. The best trained networks are selected and made part of the localized neural networks (LNNs). Test TFDs of unknown signals are then processed through the algorithm and presented to LNNs. Experimental results demonstrate that appropriately vectored and clustered data and the regularization, with input training under Mackay's evidence framework, once processed through LNNs produce high resolution TFDs.

Bayesian regularization involves modifying the usually used objective function, such as the mean sum of squared network errors (24, MacKay 1992).

$$mse = \frac{1}{N}\sum_{k=1}^{K}(e_k)^2 \tag{3}$$

where mse, e_k, and N represent MSE, network error and network errors' taps for averaging respectively. It is possible to improve generalization if the performance function is modified by adding a term that consists of the mean of the sum of squares of the network weights and biases

$$msereg = \gamma mse + (1-\gamma)msw \tag{4}$$

where γ, $msereg$, and msw are the performance ratio, performance function and mean of the sum of squares of network weights and biases, respectively. msw is mathematically described as under:

$$msw = \frac{1}{n}\sum_{i=1}^{n}(w_i)^2 \tag{5}$$

using this performance function causes the network to have smaller weights and biases, and this forces the network response to be smoother and less likely to over fit. Moreover it helps in determining the optimal regularization parameters in an automated fashion.

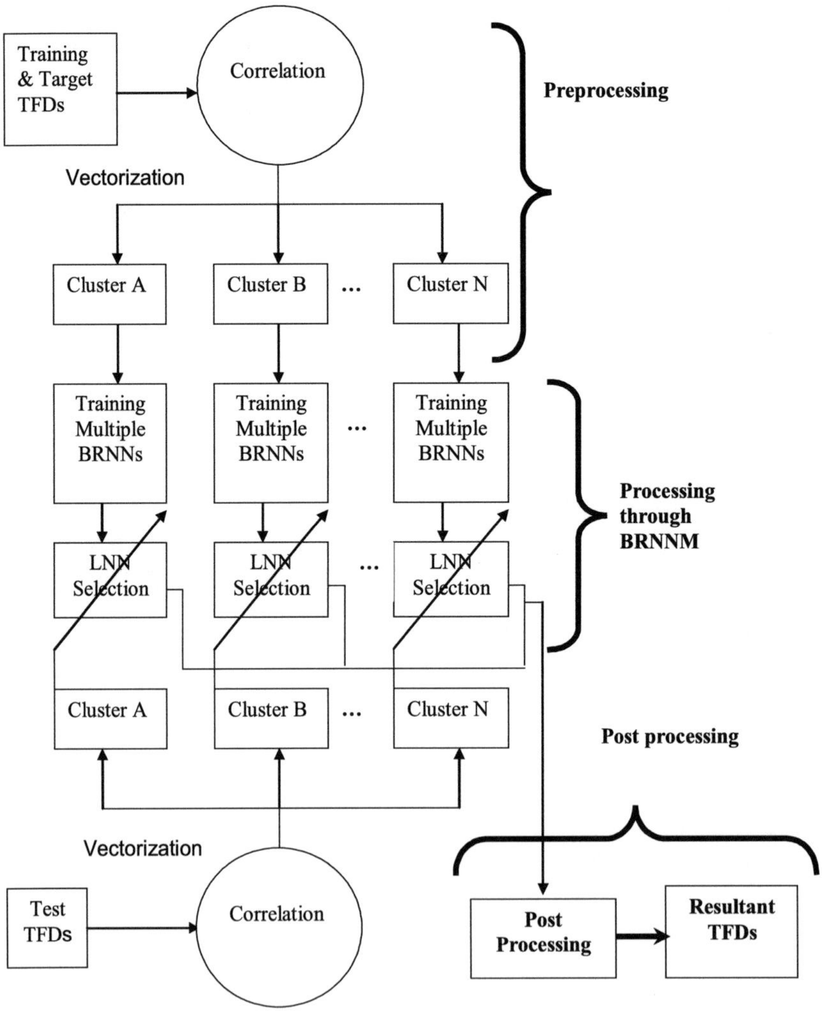

Fig. 1. Flow diagram of the method

Fig. 1 is the overall block representation of the proposed ANN based framework. This block diagram highlights three major modules of the method that include (i) pre–processing of training data, (ii) processing through the BRNNM and (iii) post–processing of output data. These modules and the rationale of the proposed method are described below:

2.1 Pre–processing of Training Data

Fig. 2 depicts the block diagram for this module. It consist of four major steps, namely (i) two–step pre–processing of target TFDs, (ii) vectorization, (iii) subspaces selection and direction vectors, and (iv) correlation and taxonomy. They are described as follows.

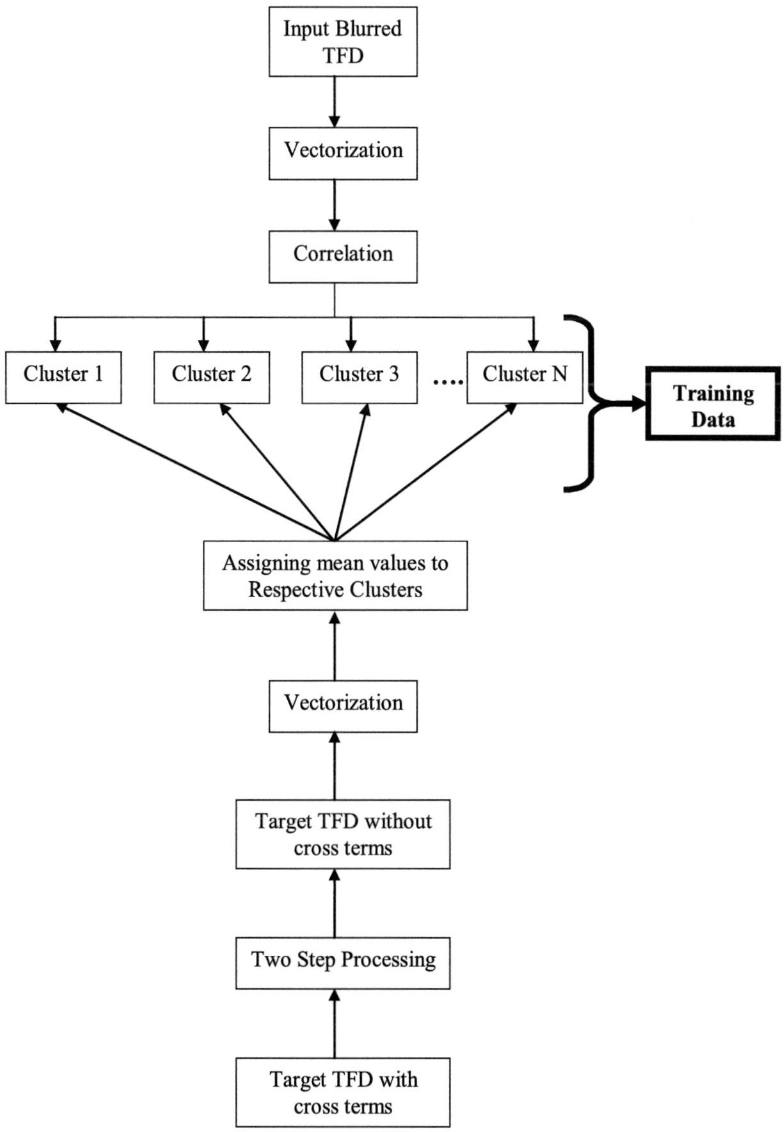

Fig. 2. Pre-processing of training data

2.1.1 Two–step pre–processing of target TFDs

The highly concentrated WVD of various known signals are used as the target TFDs. As will be shown in Fig. 4, the WVD suffers from CTs which make them unsuitable to be presented as targets to the ANNs (17, Hagan, Demuth & Beale 1996). The CTs are therefore eliminated before the WVD is fed to the ANN. This is achieved in two steps:

1. The WVD is multiplied point by point with spectrogram of the same signal obtained with a hamming window of reasonable size.
2. All values below a certain threshold are set to zero.

The resultant target TFDs are shown in Fig. 5, which are fed to the ANN after vectorization described as follows.

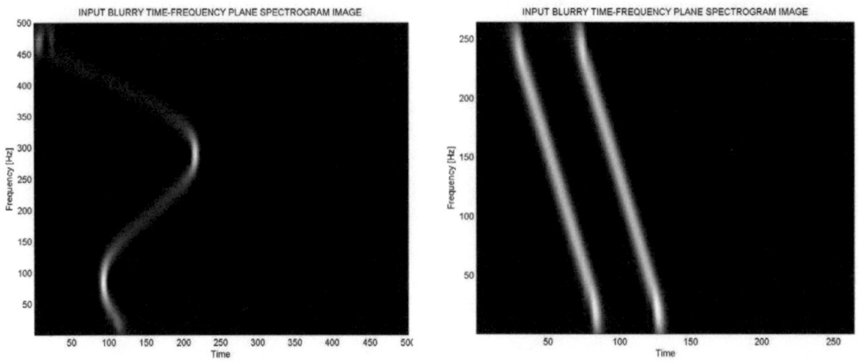

(a) (b)

Fig. 3. The spectrograms used as input training images of the (a) sinusoidal FM, and (b) parallel chirp signals.

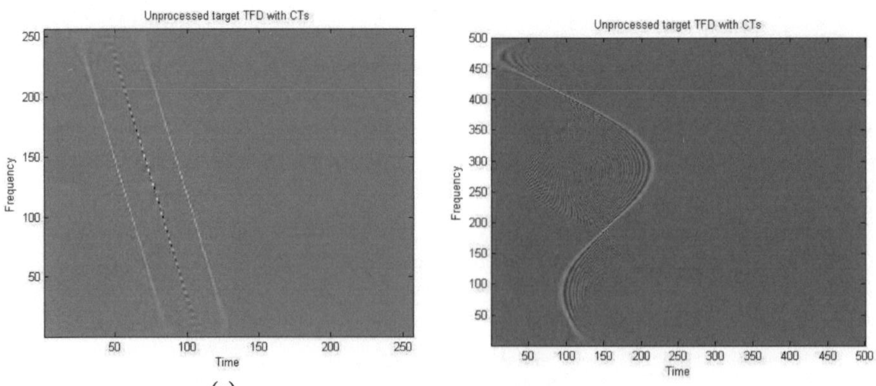

(a) (b)

Fig. 4. Target TFDs with CTs unsuitable for training ANN taking WVD of the, (a) parallel chirps' signal, and (b) sinusoidal FM signal.

2.1.2 Vectorization

(1) Input TFDs. Fig. 3 depicts input spectrograms. A TFD is considered as 2–D image which can be mathematically described as a matrix of pixels depicting grayscale values e.g.,

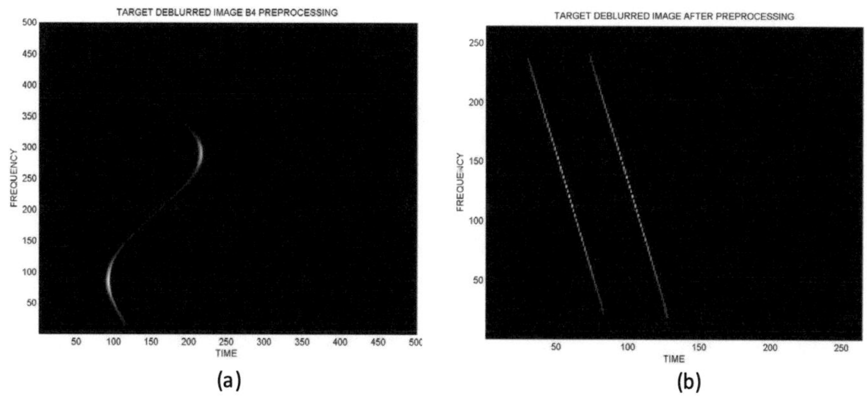

Fig. 5. Target TFDs without CTs suitable for training ANN after pre-processing WVD of the, (a) parallel chirps' signal, and (b) sinusoidal FM signal.

$\begin{pmatrix} a_{11} & \cdots & a_{1n} \\ \vdots & \ddots & \vdots \\ a_{m1} & \cdots & a_{mn} \end{pmatrix}$. These pixel values can be used to generate vectors, for example, a vector of length three will contain three pixel values of a row/column of TFD image. The suitable vector length is decided after experimenting with various vector lengths (3, 5, 7 and 9). The decision is made based on visual results. Each input TFD image is thus converted to vectors. These vectors are paired with the vectors obtained from target TFDs, to be subsequently used for training.

(2) Target TFDs. After CTs' removal from the target WVD, they are vectored. Next the mean values of these vectors are computed. For example, if $\langle a_{11}, a_{12}, a_{13} \rangle$ is a pixel vector of the input TFD and $\langle b_{11}, b_{12}, b_{13} \rangle$ is the vector representing the same region of the target TFD, then $\frac{(b_{11}+b_{12}+b_{13})}{3}$ will become the target numerical value for the input vector. Mean values are taken as targets with a view that the IF can be computed by averaging frequencies at each time instant, a definition suggested by many researchers (5, Cohen 1995).

2.1.3 Subspaces selection and direction vectors

1. *Elbow Criterion.* The elbow criterion is a common rule of thumb to determine what number of clusters should be chosen. It states that number of clusters be chosen so that adding another cluster does not add sufficient information (18). More precisely, if the percentage of variance explained by the clusters is plotted against the number of clusters, the first clusters will add much information (explain a lot of variance), but at some point the marginal gain will drop, giving an angle in the graph (the elbow). On the following graph (Fig. 6) which is drawn for the problem in hand, the elbow is indicated by the "goose egg". The number of clusters chosen is therefore three.

2. The number of subspaces N_s into which vectors are distributed is decided based on elbow criterion in relation to underlying image features like edges present in the data. The edge is considered because it is one of the important image underlying features and characteristics. Moreover it is well established fact that blurring mostly causes loss of edge information (19, Gonzalez & Wintz 1987). An edge could be ascending (1,2,3),

descending $(3,2,1)$, wedge $(1,3,2)$, flat $(1,1,1)$, triangular $(1,3,1)$ etc. Empirically it is found that going from three to four clusters does not add sufficient information, as the end result has no significant change in entropy values as indicated in Table 1 and evident from Fig. 6. The impact of clustering is noted for six different test images (TIs), described in section 3. As a result of this study, $N_s = 3$ is chosen considering the first three most general types of edges.

3. The sub space direction vectors v_n $(n = 1,2 \ldots N_s)$ are selected that will best represent the subspaces. As these subspaces are defined on the basis of edges, so three directional vectors v_h, v_c, v_l are computed in the following manner:

   (a) v_h is obtained by rearranging (any) 3 integers in descending order.

   (b) $v_c$is obtained by rearranging (any) 3 integers in a wedge shape where the highest value occurs in the middle and values on either side are in descending order.

   (c) v_l is obtained by rearranging (any) 3 integers in ascending order.

4. All the direction vectors v_h, v_c, v_l are normalized.

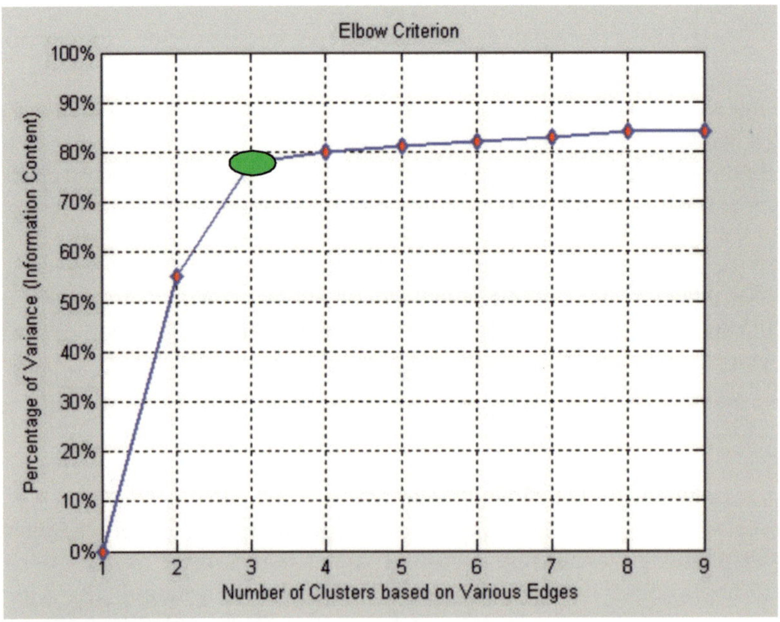

Fig. 6. Elbow criterion

2.1.4 Correlation & Taxonomy

1. An input vector x_i is chosen from input spectrogram. The correlation between each input vector x_i from input TFD and each direction vector v_h, v_c, v_l is calculated, i.e. $t_{ij} = x_i^T v_j$ where $j = h, c, l$.

2. There will be N_s product values obtained as a result of last step for each input vector x_i. To find the best match, if t_{ic} has the largest value then this indicates that the input x_i is most similar to the directional vector v_c, which implies that the vector is wedge type.

Description	E_Q (bits) for test TFDs					
	TI 1	*TI 2*	*TI 3*	*TI 4*	*TI 5*	*TI 6*
No cluster	20.539	18.113	18.323	19.975	21.548	17.910
2 clusters	13.523	12.294	12.421	11.131	14.049	11.940
3 clusters	**8.623**	**6.629**	**7.228**	**5.672**	**8.175**	**6.948**
4 clusters	8.101	6.300	7.202	5.193	8.025	6.733
5 clusters	7.998	6.187	7.111	5.012	7.939	6.678
6 clusters	7.877	6.015	7.019	5.995	7.883	6.661

Table 1. Entropy values vs clusters

3 Step (2) is repeated for all input vectors. Consequently all the vectors are classified based on the type of edge they represent and N_s clusters are obtained. The statistical data revealing various vector types in the two TFD images depicted in Fig. 3 is shown in Table 2.

4 The pairs of vectors from training and target TFDs are formed. These pairs are divided into training set and validation set for training phase and by observing error on these two sets, the aspect of overfitting is avoided.

These steps of vectorization, correlation and taxonomy are further elaborated in graphical form by Fig. 7.

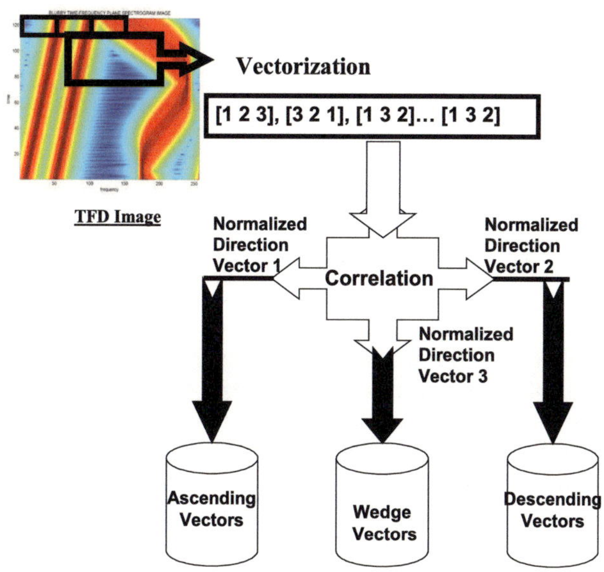

Fig. 7. Vectorization, correlation and taxonomy of TFD image.

2.2 Processing through Bayesian Regularized Neural Network Model

Fig. 8 represents this module. There are three steps in this module, namely (i) training of BRNNM, (ii) selecting the LNNs, and (iii) testing the LNNs. They are discussed as under.

Various parameters (input training TFDs)	*Cluster 1 ascending edge type vectors*	*Cluster 2 descending edge type vectors*	*Cluster 3 wedge edge type vectors*
Sinusoidal FM signal	19157	18531	112
Parallel chirps' signal	4817	4959	52
The best ANN	$ANN-3$	$ANN-2$	$ANN-1$
Time consumed for training	308 *seconds*	114 *seconds*	55 *seconds*
MSE converged	2.54×10^{-4}	3.56×10^{-4}	1.38×10^{-2}

Table 2. Cluster parameters

2.2.1 ANN Training

1. Since the ANN is being used in a data–rich environment to provide high resolution TFDs, it is important that it does well on data it has not seen before, i.e. that it can generalize. To make sure that the network does not become over trained. the error is monitored on a subset of the data that does not actually take part in the training. This subset is called the validation set other than the training set. If the error of the validation sets increases the training stops. For this purpose, alternate pairs of vectors from input and target TFDs are included in training and validation sets.
2. The input vectors represented by x_i and the mean values y_i of the pixel values, of the corresponding window from the target TFD image are used to train the multiple ANNs under Bayesian framework. There are three ANNs trained for each cluster, being the smallest numerical value to check the advantage of training multiple ANNs. This selection has no relation with the number of subspaces or direction vectors.
3. Step (2) is repeated until all pairs of input and corresponding target vectors are used for training.

2.2.1.1 Topology, Architecture and Training TFDs

To address the stated problem, Bayesian Regularized LMB training algorithm is used with feed forward back propagation ANN architecture and 40 neurons in the single hidden layer. This architecture is chosen after an empirical study (20; 21, Shafi et at. 2006, Ahmad et al. 2006). We experiment with various training algorithms using different parameters such as different activation functions between layers, number of hidden layers and number of neurons. Also the positive impact of localised processing by selecting the best trained ANN out of many is ascertained (22, Shah et al. 2007). The 'tansig' and 'poslin' transfer functions are used respectively representing the hidden layer of sigmoid neurons followed by an output layer of positive linear neurons. Multiple layers of neurons with nonlinear transfer functions allow the network to learn linear and nonlinear relationships between input and output vectors. The linear output layer lets the network produce values outside the range $[-1.+1]$.
The spectrograms and WVD of the two signals are used as input and target TFDs respectively for training the ANNs. The first signal is a sinusoidal FM signal, given by:

$$x(n) = e^{-j\pi[\frac{5}{2}+\{0.1\sin(\frac{2\pi n}{N})\}]n} \tag{6}$$

where N refers to the number of sampling points. The spectrogram of this signal is depicted in Fig. 3(a). The respective target TFD, obtained through WVD, is depicted in Fig. 5(a).

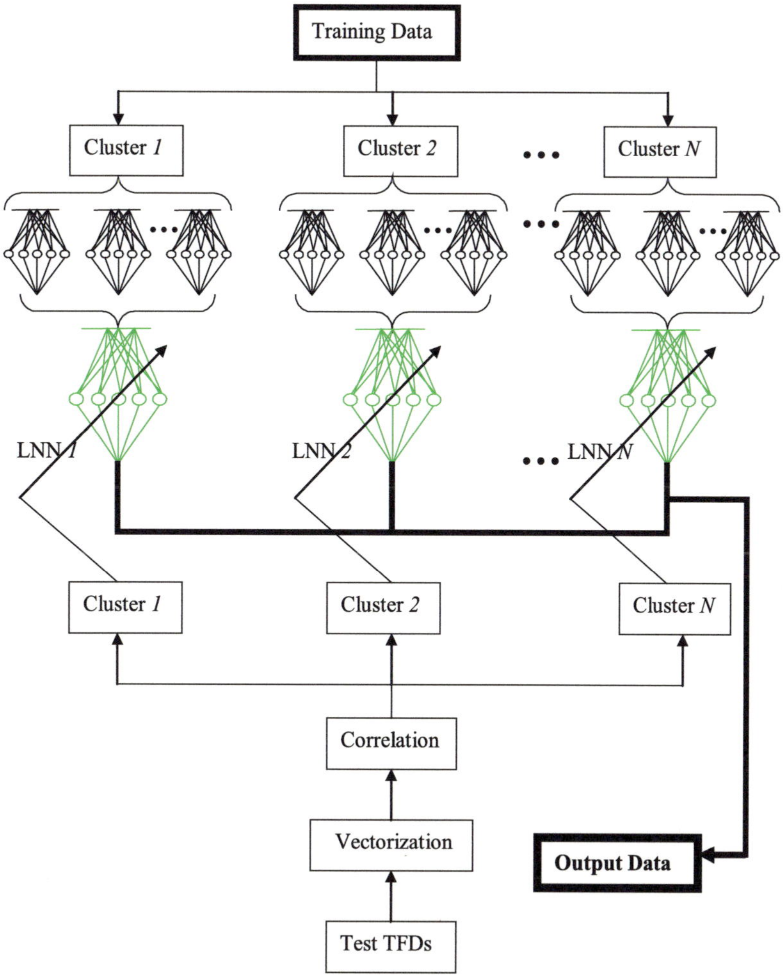

Fig. 8. Bayesian regularised neural network model

The second signal is with two parallel chirps given by:

$$y(n) = e^{j(\frac{\pi n}{4n})n} + e^{j(\frac{\pi}{3}+\frac{\pi n}{4n})n} \tag{7}$$

The spectrogram of this signal is depicted in Fig. 3(b). The respective target TFD, obtained through the WVD, is depicted in Fig. 5(a).

2.2.2 LNNs' selection

1. As mentioned above, three ANNs are being trained for each cluster and the best for each cluster is required to be selected. The "training record" is a programmed structure in which the training algorithm saves the performance of the training–set, test–set, and the validation–set, as well as the epoch number and the learning rate. By keeping track of the network error or performance, accessible via the training record, the best network is selected for each cluster. These best networks selected for various clusters are called the LNNs.

2. Using multiple networks for each cluster is found to be advantageous because the weights are initialized to random values, and when the network begins to over–fit the data, the error in the validation set typically begins to rise. If this happens for a specified number of iterations, the training is stopped, and the weights and biases at the minimum of the validation error are obtained. As a result, various networks will have different MSEs in the last training epoch. The ANN with minimum MSE is the winner and is included in the LNNs. There are three ANN trained for each of three clusters, and as recorded in Table 2 it is found that $ANN-3$ and $ANN-2$ are the best for the first and second clusters respectively, and the $ANN-1$ is found to be the best for the third cluster only. It is assumed that these selected ANN are optimally trained and will posses better generalization abilities.

2.2.3 ANN Testing

1. Test TFDs are converted to vectors (z_i) and clustered after correlating with the direction vectors, as done for the training TFDs.
2. Each test vector z_i is fed to the LNN trained for the particular type and the results are recorded.

2.3 Post–processing of the Output Data

This module is illustrated in Fig. 9. After testing phase, the resultant data is post–processed to get the resultant TFD. As we obtain one value for each vector of length three from test TFD after processing through the LNNs. There are two possibilities to fill the rest of two pixels, either (i) replicate the same value for other two places, or (ii) use zero padding around this single value to complete the number of pixels. Zero padding is optimal because it is found to reduce the blur in TF plane. Next the resultant vectors of correct length are placed at their original places from where they were correlated and clustered. These vectors are placed according to the initially stored grid positions.

3. Discussion on Experimental Results

The discussion on experimental results by the proposed approach and performance evaluation of various bilinear distributions is presented in this section. It uses objective methods of assessment to evaluate the performance of de–blurred TFDs estimated through BRNNM (henceforth the NTFDs). The objective methods allow quantifying the quality of TFDs instead of relying solely on visual inspection of their plots. Performance comparison with various other quadratic TFDs is provided too. This section is organized in two subsections. The subsection 3.1 discusses the NTFDs' performance basing on the visual results and carrying out their information quantification by measuring the entropy values only. In subsection 3.2, the concept and importance of TFDs' objective assessment is described using both real life and synthetic signals.

3.1 Visual Interpretation and Entropy Analysis

In the first phase, five synthetic signals are tested to evaluate the effectiveness of the proposed algorithm basing on visual results and their entropy analysis. They include (i) a two sets of parallel chirps signal intersecting at four places, (ii) a mono–component linear chirp signal, (iii) combined quadratic swept–frequency signals whose spectrograms are concave and convex parabolic chirps respectively, (iv) a combined crossing chirps and sinusoidal FM signal and (v) a quadratic chirp signal. The spectrograms of these signals are shown in Figs.

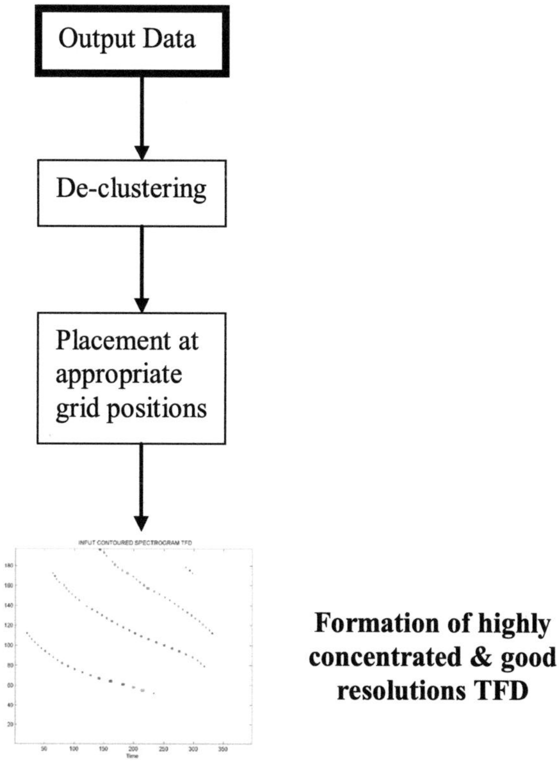

Fig. 9. Post-processing of the output data

10(a) to 10(e) respectively. Keeping in mind that estimation of the IF is rather difficult at the intersections of chirps, the first and fifth test cases are considered to check the performance of proposed algorithm at the intersection of the IFs of individual components present in the signals.

The spectrogram of the two sets of parallel chirps signals crossing each other at four points is fed as the first test signal, depicted in Fig. 10(a), is obtained by:

$$TS_1(n) = e^{j\left[\pi - \frac{\pi n}{6N}\right]n} + e^{j\left[\frac{\pi}{3} - \frac{\pi n}{6N}\right]n} + e^{j\left[\frac{\pi n}{N}\right]n} + e^{j\left[\pi + \frac{\pi n}{N}\right]n} \tag{8}$$

The second test signal is a mono–component chirp signal given by:

$$TS_2(n) = e^{j\left[\pi + \frac{\pi n}{N}\right]n} \tag{9}$$

The spectrogram of the resultant signal is depicted in Fig. 10(b).

The third test signal is obtained by point–by–point addition of two quadratic swept–frequency signals whose spectrograms are concave and convex parabolic chirps respectively. Mathematically both the signals can be obtained by manipulating different parameters of following equation:

$$TS_3(n) = \cos\left[2\pi\left(\frac{\partial}{1+\beta}\right)\left(n^{(1+\beta)}\right) + f_0 + \frac{\theta}{360}\right], \tag{10}$$

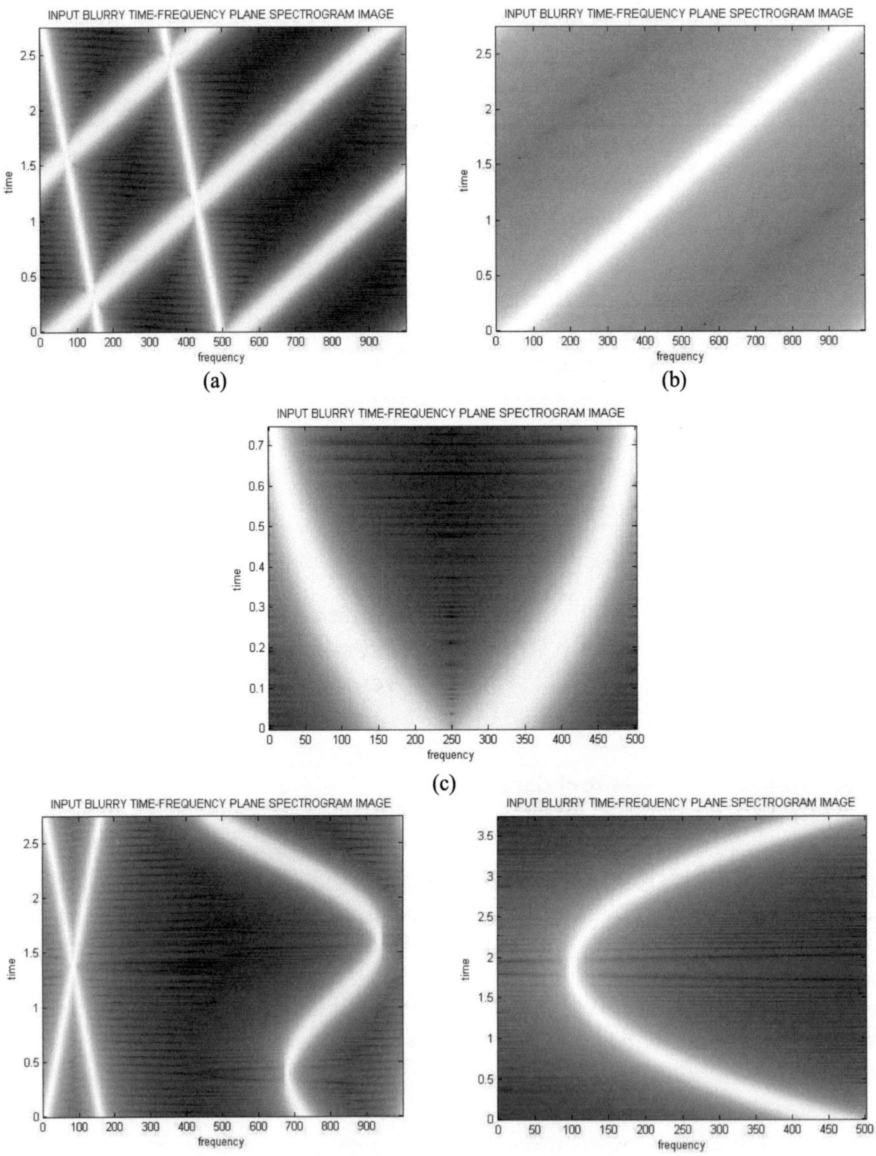

Fig. 10. Test TFDs (a) Crossing chirps (TI 1), (b) mono-component linear chirp (TI 2), (c) combined quadratic swept-frequency signals whose spectrograms are concave and convex parabolic chirps respectively (TI 3), (d) combined sinusoidal FM and crossing chirps (TI 4), and (e) quadratic chirp (TI 5)

where

$$\partial = (f_1 - f_0)\,\rho^{(-\beta)}$$

The method	Resultant E_Q (bits) for test TFDs				
	TI 1	*TI 2*	*TI 3*	*TI 4*	*TI 5*
NTFD	**8.623**	**6.629**	**5.672**	**8.175**	**6.948**
WVD	21.562	10.334	18.511	20.637	18.134
Spectrogram	28.231	18.987	27.743	28.785	23.774

Table 3. Entropy values for various techniques

here β, f_0, f_1, θ and ρ are defined as the matching string constant, start frequency, frequency after one second, initial phase of signal and sample rate respectively. The spectrogram of the first quadratic swept–frequency signal is concave parabolic chirp which starts at 250 Hz and go down to 0 Hz at a 1 kHz sample rate; whereas spectrogram of the second quadratic swept–frequency signal is a convex parabolic chirp starting at 250 Hz and going up to 500 Hz at a 1 kHz sample rate. These aspects are evident in the combined spectrogram depicted in Fig. 10(c).
Another test signal is obtained by combining crossing chirps and sinusoidal FM signal as:

$$TS_4(n) = e^{j[\frac{\pi n}{N}]n} + e^{j[\pi + \frac{\pi n}{N}]n} + e^{j\pi[\frac{1}{2} - (0.1\sin(\frac{2\pi n}{N}))n]} \quad (11)$$

The spectrogram of the signal is depicted in Fig. 10(d).
Yet another test signal is a quadratic chirp which starts at 100 Hz and cross es 200 Hz at 1 *second* with a 1 kHz sample rate. It is obtained from Eqn. (10) after necessary adjustment of different parameters. The spectrogram of this signal is depicted in Fig. 10(e).

3.1.1 Resultant NTFDs – Experimental Results

The five synthetic test signals are: a combined parallel chirps signal crossing at four points, a mono–component linear chirp signal, combined quadratic swept–frequency signals whose spectrograms are concave and convex parabolic chirps respectively, combined crossing chirps and sinusoidal FM signals without any intersection and a quadratic chirp signal. The spectrograms of these signals constitute test image 1 (TI 1), test image 2 (TI 2), test image 3 (TI 3), test image 4 (TI 4), and test image 5 (TI 5). They are depicted in Figs. 10(a–e) respectively. The entropy expression given by $E_Q = -\sum_{n=0}^{N-1} Q(n,\omega)\log_2 Q(n,\omega)\,d\omega \geq 0$ is used to quantify the TFDs' information, which has an inverse relation with the information (25, Gray 1990). The entropy values for different TFDs have been recorded in Table 3, which are the lowest for the NTFDs than other technique like WVD and the spectrogram. TI 1 and TI 5 are taken into account to check the performance of the proposed algorithm for estimation of the IFs at the intersections along the individual components in the signals. Even though estimation of IF is considered rather difficult at intersections, the algorithm performs well as depicted in Figs. 11(a) and (d). The test images including TI 2, TI 3 and TI 5 present the ideal cases to check the performance of the proposed algorithm trained with signals of different nature. The resultant TFD images are highly concentrated along the IF of individual components present in the signal as shown in Figs. 11(b), (c) and (e).

3.2 Objective Assessment

In this subsection, the objective measures are used to analyze the NTFDs' performance in comparison to other TFDs. The aim has been to find, based on these measures, the highly

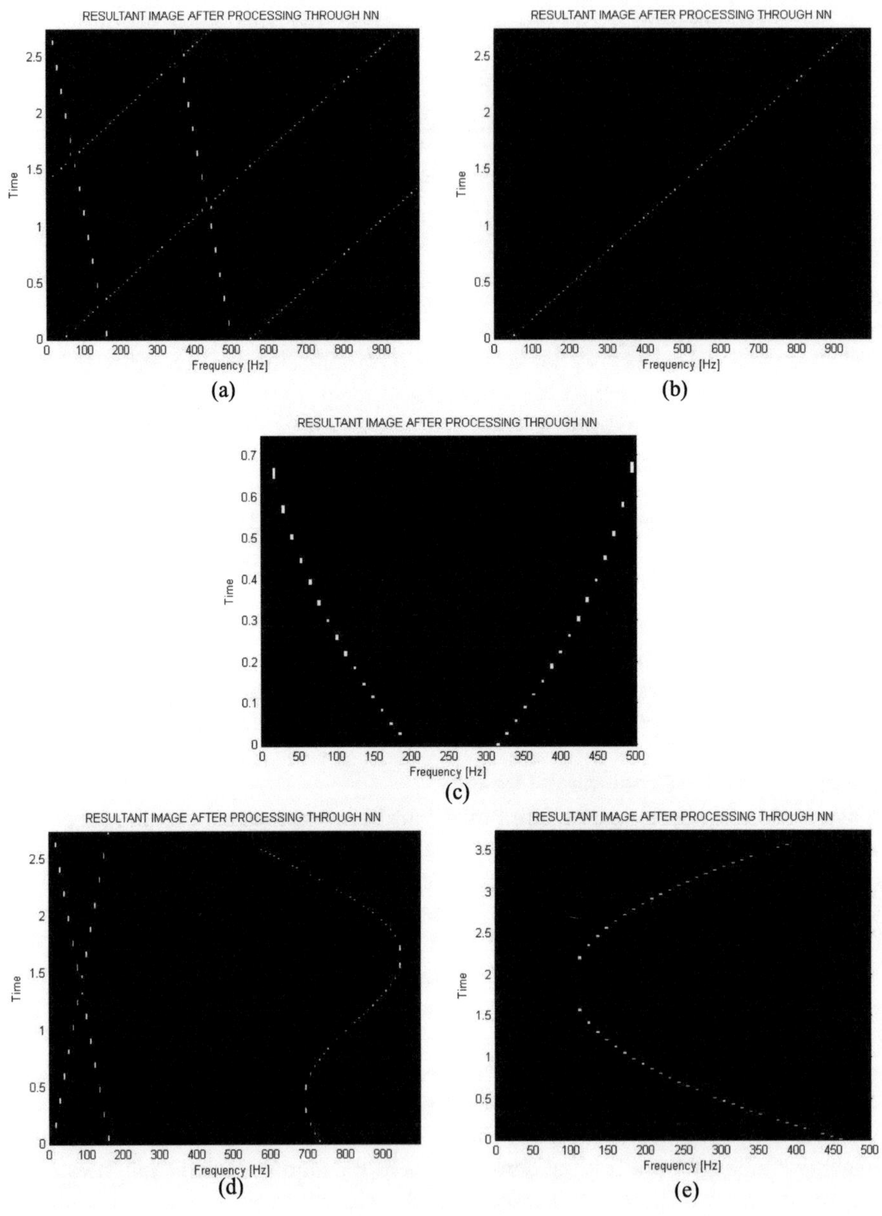

Fig. 11. Resultant TFDs after processing through correlation vectored taxonomy algorithm with LNNs for (a) Crossing chirps (TI 1), (b) mono-component linear chirp (TI 2), (c) combined quadratic swept-frequency signals whose spectrograms are concave and convex parabolic chirps respectively (TI 3), (d) combined sinusoidal FM and crossing chirps (TI 4), and (e) quadratic chirp (TI 5)

informative TFDs having the best concentration and the highest resolution. Five new examples, including both real life and synthetic multicomponent signals, are being considered. The signals include (i) a multicomponent bat echolocation chirp signal, (ii) a two–component intersecting sinusoidal FM signal, (iii) a two sets of nonparallel, nonintersecting chirps' signal, and (iv) a closely spaced three–component signal containing a sinusoidal FM component intersecting the crossing chirps. The respective spectrograms, termed as test image A (TI A), test image B (TI B), test image C (TI C), and test image D (TI D), are shown in Figs. 12(a), 14(a)–16(a) respectively. As an illustration of the evaluation of the NTFDs' performance through Boashash concentration and resolution measures in (26, Boashash & Sucic 2003), we have further considered a closely spaced multicomponent signal containing two significantly close parallel chirps. The spectrogram of this signal, termed as test image E (TI E), is depicted in Fig. 17(a). The resultant NTFDs for the test signals are shown in Fig. 12(b) & Figs. 14(b)– 17(b) respectively. The visual results are indicative of NTFDs' high resolution and concentration along the IF of the individual component present in the signals.

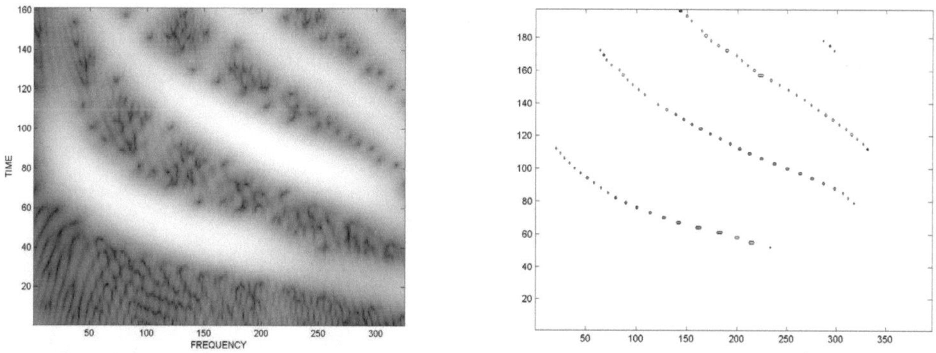

(a) (b)

Fig. 12. Test TFDs for bat chirps signal, (a) the spectrogram TFD, and (b) the resultant TFD after processing through proposed framework.

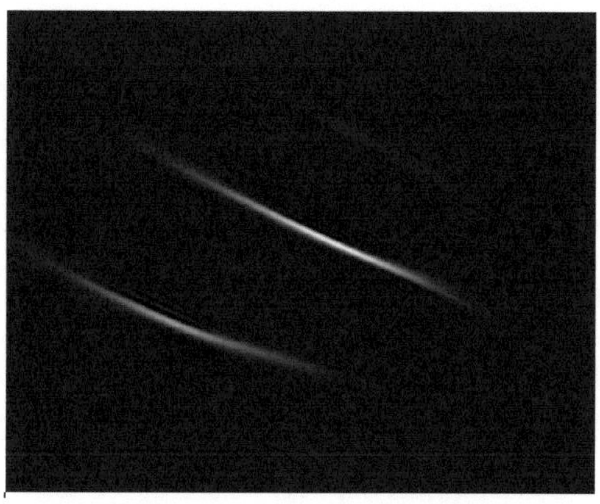

Fig. 13. Resultant TFD obtained by the method of (28, Baraniuk & Jones 1993).

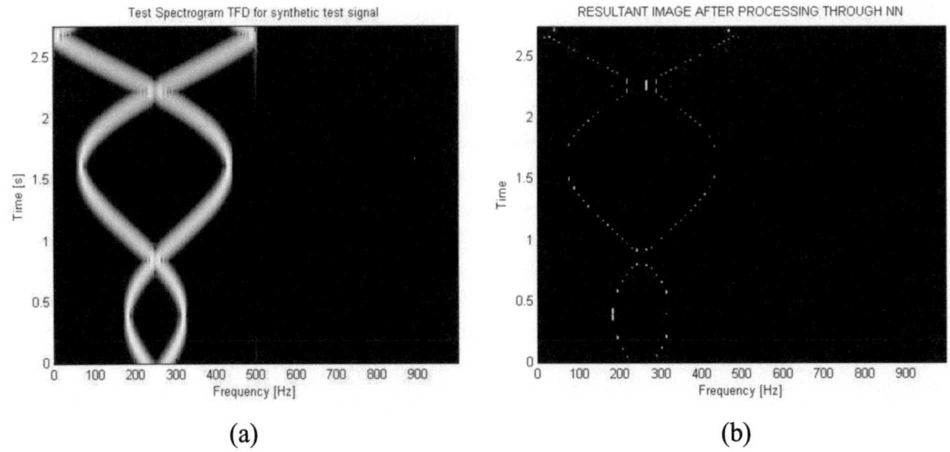

Fig. 14. (a) The test spectrogram (TI 2) $[Hamm, L = 90]$, and (b) The NTFD of a synthetic signal consisting of two sinusoidal FM components intersecting each other.

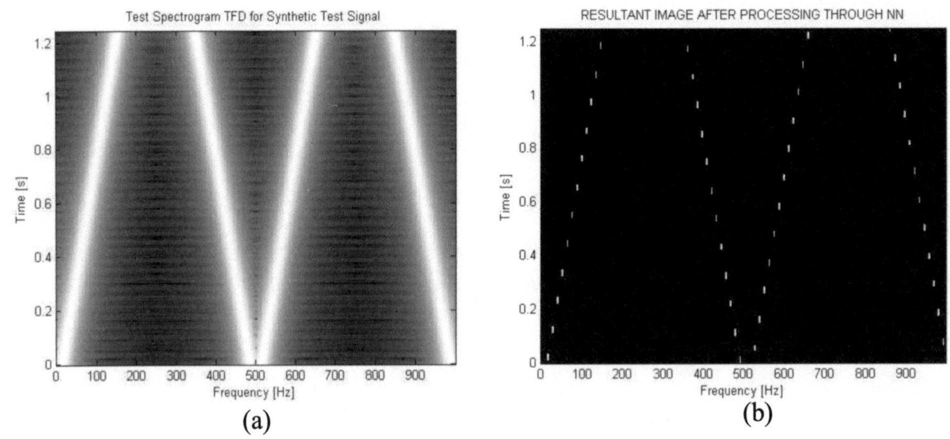

Fig. 15. (a) The test spectrogram (TI 3) $[Hamm, L = 90]$, and (b) The NTFD of a synthetic signal consisting of two-sets of non-parallel, non-intersecting chirps.

3.2.1 Real Life Test Case

Real life data for bat echolocation chirp sound (adopted from (27)) provides an excellent multicomponent test case. The nonstationary nature of the signal is only obvious from its TFD. The spectrogram of this signal is shown in Fig. 12(a), and the resultant NTFD is depicted in Fig. 12(b). The result for the same test case TFD is computed using an existing optimal kernel method (OKM) (28, Baraniuk & Jones 1993) and is plotted in Fig. 13. The OKM proposes a signal–dependent kernel that changes shape for each signal to offer improved TF representation for a large class of signals based on quantitative optimization criteria. On close monitoring the OKM's output depicted in Fig. 13, it is revealed that this TFD does not fully recover all the components, thus losing some useful information about the signal. Whereas the NTFD is

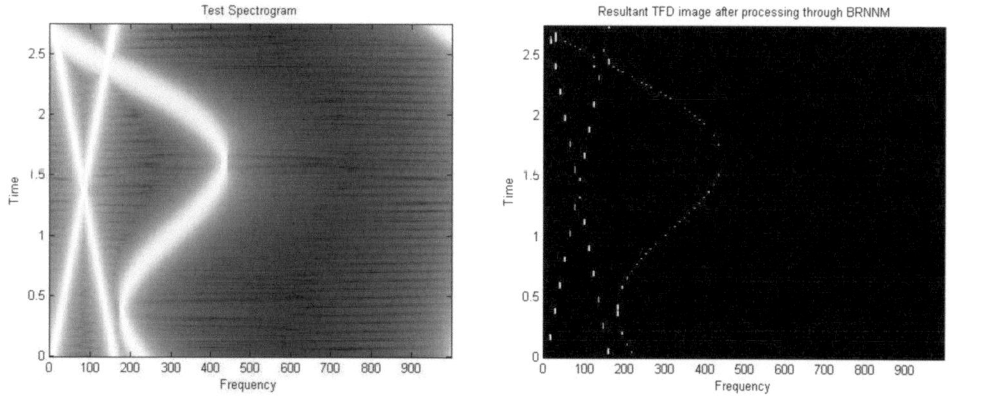

(a) (b)

Fig. 16. (a) The test spectrogram (TI 4) $[Hamm, L = 90]$, and (b) The NTFD of a synthetic signal consisting of crossing chirps and a sinusoidal FM component.

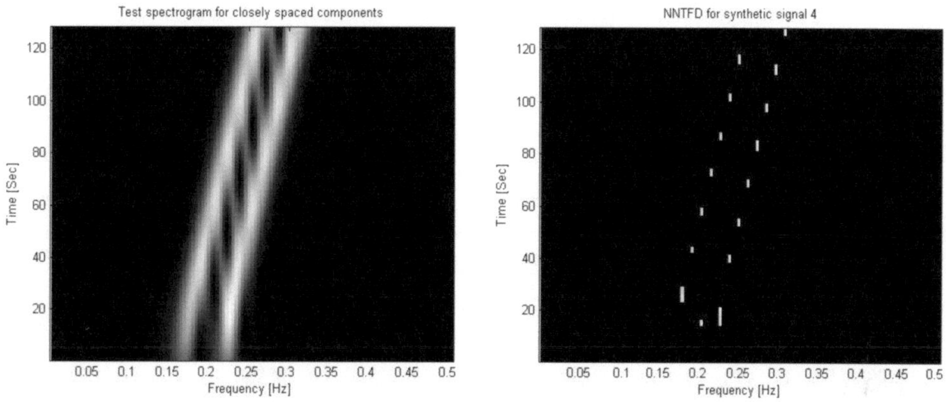

(a) (b)

Fig. 17. (a) The test spectrogram (TI 5), and (b) the NTFD of test case E.

not only highly concentrated along the IF of the individual components present in the signal but also more informative showing all the components.

For further analysis, slices of the test and resultant NTFDs are taken at the time instants $n = 150$ and $n = 310$ (recall that $n = 1, 2, \ldots, 400$) and the normalized amplitudes of these slices are plotted in Fig. 18. These instants are chosen because three chirps are visible (see Fig. 12(b)) at these time instants. Fig. 18 confirm the peaky appearance of three different frequencies at these time instants. It is worth mentioning that the NTFD not only recovers the fourth component (the weakest) but it has the best resolution i.e. narrower main lobe and no side lobes.

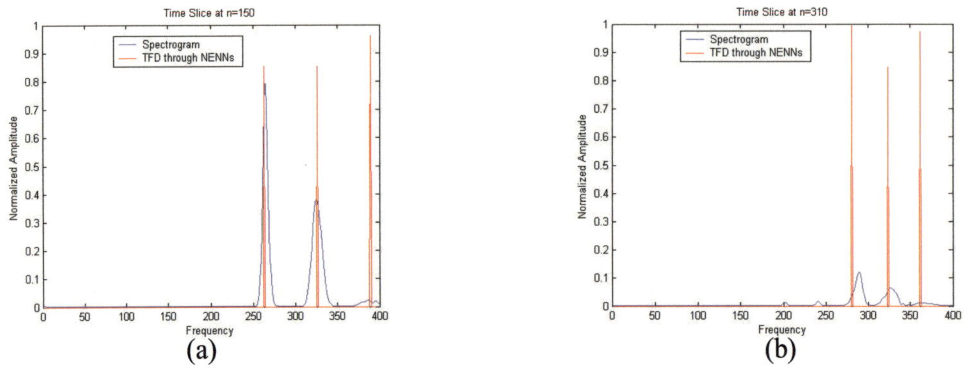

Fig. 18. The time slices for the spectrogram (blue) and the NTFD (red) for the bat echolocation chirps' signal, at n=150 (left) and n=310 (right)

3.2.2 Synthetic Test Cases

Further four specially synthesized signals of different nature are fed to the model to check its performance at the intersection of the IFs and closely spaced components, keeping in mind that estimation of the IF is rather difficult in these situations. The test cases are described as under:

3.2.2.1 Test case 1.

The first one is the synthetic signal consisting of two intersecting sinusoidal FM components, given as:

$$SynTS_1(n) = e^{-i\pi\left(\frac{5}{2}-0.1\sin(2\pi n/N)\right)n} + e^{i\pi\left(\frac{5}{2}-0.1\sin(2\pi n/N)\right)n} \tag{12}$$

The spectrogram of the signal is shown in Fig. 14(a).

3.2.2.2 Test case 2.

The second synthetic signal contains two sets of nonparallel, nonintersecting chirps once plotted on the TF plane. Mathematically it can be written as:

$$SynTS_2(n) = e^{i\pi\left(\frac{n}{6N}\right)n} + e^{i\pi\left(1+\frac{n}{6N}\right)n} + e^{-i\pi\left(\frac{n}{6N}\right)n} + e^{-i\pi\left(1+\frac{n}{6N}\right)n} \tag{13}$$

The spectrogram of the signal is shown in Fig. 15(a).

3.2.2.3 Test case 3.

It is a three–component signal containing a sinusoidal FM component intersecting two crossing chirps. It is expressed as:

$$SynTS_3(n) = e^{i\pi\left(\frac{5}{2}-0.1\sin(2\pi n/N)\right)n} + e^{i\pi\left(\frac{n}{6N}\right)n} + e^{i\pi\left(\frac{1}{3}-\frac{n}{6N}\right)n} \tag{14}$$

The spectrogram of the signal is shown in Fig. 16(a). The two components (sinusoidal FM and chirp components) are very close in between $150 - 200\ Hz$ near 0.5 *sec*. This is to confirm the model's effectiveness in de–blurring closely spaced components.

3.2.2.4 Test case 4.

This particular test case is adopted from Boashash (26, Boashash & Sucic 2003) to compare the TFDs' concentration and resolution performance at the middle of the signal duration interval by Boashash performance measures. The signal consists of two linear frequency modulated signals whose frequencies increase from 0.15 to 0.25 Hz and from 0.2 to 0.3 Hz, respectively, over the time interval $t\forall[1,128]$. The sampling frequency is $f_s = 1\ Hz$.The authors in (26, Boashash & Sucic 2003) have found the modified B distribution ($\beta = 0.01$) as the best performing TFD for this particular signal at the middle after measuring the signal components' parameters needed in Boashash resolution measure (see Table 5). The signal is defined as;

$$SynTS_4(n) = \cos\left(2\pi\left(0.15t + 0.0004t^2\right)\right) + \cos\left(2\pi\left(0.2t + 0.0004t^2\right)\right) \quad (15)$$

The spectrogram of the signal is shown in Fig. 17(a).

The above mentioned test cases are processed through the BRNNM and the resultant NTFDs are shown in Figs. 14(b)–17(b). High resolution and concentration along the IF of individual components is obvious once inspecting these plots visually.

3.2.3 Performance Evaluation

To evaluate the performance, numerical computations by the methods like the *ratio of norms based measures, Shannon & Rényi entropy measures, normalized Rényi entropy measure* and *Stankovic measure* are recorded in Table 4. The entropy measures including Shannon & Rényi entropies with or without normalization make excellent measures of the information extraction performance of TFDs. By the probabilistic analogy, minimizing the complexity or information in a particular TFD is equivalent to maximizing its concentration, peakiness, and, therefore, resolution (29, Jones & Parks 1992). To obtain the optimum distribution for a given signal, the value of ratio of norms based and Boashash resolution measures should be the maximum (30, Jones & Parks 1990), whereas TFDs' yielding the smallest values for Stankovic and Boashash concentration measures are considered as the best performing TFD in terms of concentration and resolution (26; 31, Boashash & Sucic 2003, Stankovic 2001)

The values in Table 4 refer to the NTFDs as the best TFDs by various criteria. This can be better observed by plotting these measures separately for various TI's (i.e. TI A–TI D) shown in Fig. 19. Few singularities are mainly attributable to inherent shortcomings and derivations' assumptions, e.g. simple Rényi entropies, being unable to detect zero mean CTs, indicate ZAMD as the best concentrated TFD. However the more often used volume normalized Rényi entropies are the minimum for the NTFDs[1].

[1] Here the abbreviations for different methods include the spectrogram (spec), Wigner–Ville distribution (WVD), Choi–Williams distribution (CWD), Zhao–Atlas–Marks distribution (ZAMD), neural network based TFD (NTFD), Margenau–Hill distribution (MHD), Born–Jordan distribution (BJD), Simple Neural network based method (SNN) without clustering the data and the optimal radially Gaussian kernel TFD method (OKM).

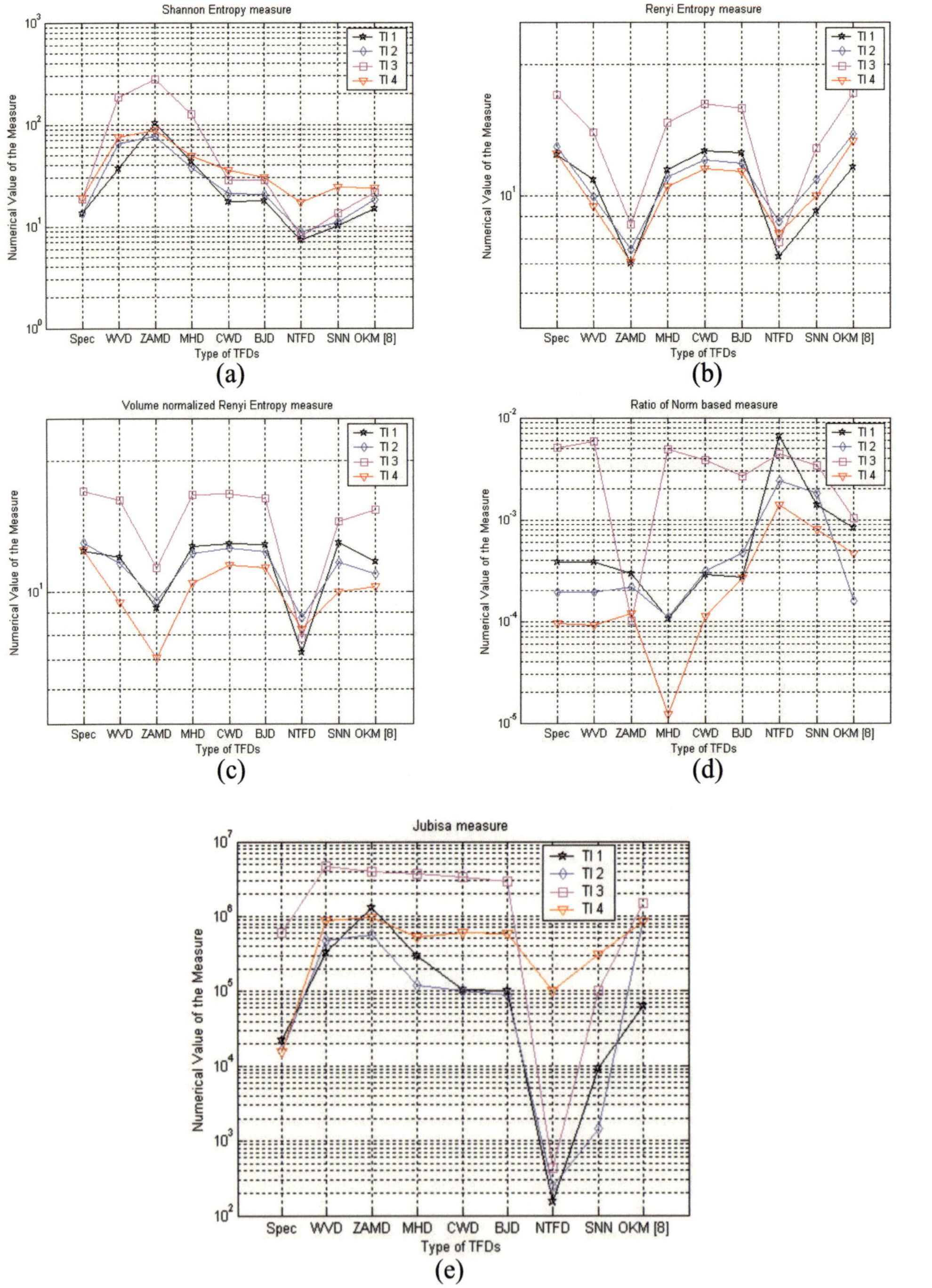

Fig. 19. Comparison plots, criteria vs TFDs, for the test images A–D, (a) The Shannon entropy measure, (b) Rényi entropy measure, (c) Volume normalized Rényi entropy measure,(d) Ratio of norm based measure, and (e) Stankovic measure.

Description	*Test TFD*	*Spec*	*WVD*	*ZAMD*	*MHD*	*CWD*	*BJD*	*NTFD*	*SNN*	*OKM*
Shannon entropy measure	TI A	13.46	36.81	102.23	42.98	17.27	17.73	7.27	10.18	14.68
	TI B	13.45	64.33	76.81	37.74	20.82	20.43	8.75	10.88	18.08
	TI C	18.66	185.49	274.73	126.02	28.08	28.05	7.87	13.45	21.42
	TI D	18.94	74.82	87.30	49.24	35.31	29.92	17.25	24.23	23.57
Ratio of Norm based measure ($\times 10^{-4}$)	TI A	3.81	3.84	2.94	1.05	2.89	2.73	66	13.88	8.32
	TI B	1.94	1.91	2.18	1.10	3.10	4.67	24	18.12	1.59
	TI C	51.23	58.0	1.02	48.71	38.53	26.37	44	33.90	10.26
	TI D	0.95	0.92	1.19	0.12	1.11	2.68	14	8	4.60
Rényi entropy measure	TI A	12.45	10.90	7	11.47	12.67	12.54	7.26	9.25	11.65
	TI B	12.98	9.95	7.56	11.03	12.06	11.85	8.74	10.89	13.82
	TI C	17.07	14.01	8.62	14.74	16.24	15.84	7.85	12.82	17.22
	TI D	12.47	9.48	7.06	10.50	11.54	11.34	8.23	10.03	13.31
Energy Normalized Rényi entropy measure	TI A	12.45	10.90	7	11.47	12.67	12.54	7.26	9.25	11.65
	TI B	12.98	9.95	7.56	11.03	12.06	11.85	8.74	10.89	13.82
	TI C	17.07	14.01	8.62	14.74	16.24	15.84	7.85	12.82	17.22
	TI D	12.47	9.48	7.06	10.50	11.54	11.34	8.23	10.03	13.31
Volume Normalized Rényi entropy measure	TI A	12.45	12.02	9.18	12.75	12.93	12.85	7.26	12.97	11.77
	TI B	12.98	11.62	9.54	12.26	12.60	12.38	8.74	11.68	10.98
	TI C	17.07	16.28	11.35	16.70	16.77	16.41	7.85	14.49	15.43
	TI D	12.47	9.48	7.06	10.50	11.54	11.34	8.23	10.03	10.31
Stankovic measure ($\times 10^{5}$)	TI A	0.2219	3.30	13.14	2.9200	1.06	1.01	0.0015	0.0912	0.6300
	TI B	0.1600	4.68	5.6266	1.1861	1.0123	0.8946	0.0024	0.0145	8.6564
	TI C	6.03	47.05	39.64	36.47	33.08	29.39	0.0043	0.9973	14.73
	TI D	0.1553	8.67	9.6253	5.1848	6.0110	5.8933	1.0030	3.0223	8.5551

Table 4. Performance Measures Comparison for Various TFDs

Boashash performance measures for concentration and resolution are computationaly expensive because they require calculations at various time instants. To limit the scope, these measures are computed at the middle of the synthetic signal defined in Eqn. (15) and the results are compared with the one reported in (26, Boashash & Sucic 2003). A slice is taken at $t = 64$ and the signal components' parameters $A_{M_1}(64)$, $A_{M_2}(64)$, $A_M(64)$, $A_{S_1}(64)$, $A_{S_2}(64)$, $A_S(64)$, $V_{i_1}(64)$, $V_{i_2}(64)$, $V_i(64)$, $f_{i_1}(64)$, $f_{i_2}(64)$and $\Delta f_i(64)$, as well as the CTs' magnitude $A_X(64)$ are measured. These are then used to calculate the TFDs' normalized instantaneous resolution and modified concentration performance measures $\mathbb{R}_i(t)$ and $\mathbb{C}_n(t)$. The measurement results are recorded in Table 5 and Table 6 seperately for $\mathbb{R}_i(64)$ and $\mathbb{C}_n(64)$. The slice of the signal's NTFD at $t = 64$ is shown in Fig. 20(f).

A TFD that, at a given time instant, has the largest positive value (close to 1) of the measure $\mathbb{R}_i$ is the TFD with the best resolution performance at that time instant for the signal under consideration. From Table 5, the NTFD of synthetic signal given by Eqn. (15) gives the largest value of $\mathbb{R}_i$ at time $t = 64$ and hence is selected as the best performing TFD of this signal at $t = 64$. On similar lines, the TFDs' concentration performance is compared at the middle of signal duration interval. A TFD is considered to have the best energy concentration for a given multicomponent signal if for each signal component, it yields the smallest

TFD (optimal parameter)	$A_M(64)$	$A_S(64)$	$A_X(64)$	$V_i(64)$	$\triangle f_i(64)$	D(64)	$\mathbb{R}(64)$
Spectrogram $(Hann, L = 35)$	0.9119	0.0087	0.5527	0.0266	0.0501	0.4691	0.7188
WVD	0.9153	0.3365	1	0.0130	0.0574	0.7735	0.6199
ZAMD $(a = 2)$	0.9146	0.4847	0.4796	0.0214	0.0420	0.4905	0.5661
CWD $(\sigma = 2)$	0.9355	0.0178	0.4415	0.0238	0.0493	0.5172	0.7541
BJD	0.9320	0.1222	0.3798	0.0219	0.0488	0.5512	0.7388
Modified B $(\beta = 0.01)$	0.9676	0.0099	0.0983	0.0185	0.0526	0.5957	0.8449
NTFD	0.9013	0	0	0.0110	0.0550	0.800	0.9333

Table 5. Parameters and the Normalized Instantaneous Resolution Performance Measure of TFDs for the Time Instant t=64

TFD (optimal parameters)	$A_{S_1}(64)$	$A_{S_2}(64)$	$A_{M_1}(64)$	$A_{M_2}(64)$	$V_{i_1}(64)$	$V_{i_2}(64)$	$f_{i_1}(64)$	$f_{i_2}(64)$	$\mathbb{C}_1(64)$	$\mathbb{C}_2(64)$
Spectrogram $(Hann, L = 35)$	0.0087	0.0087	1	0.8238	0.03200	0.0200	0.1990	0.2500	0.1695	0.0905
WVD	0.3365	0.3365	0.9153	0.9153	0.0130	0.013	0.1980	0.2554	0.4333	0.4185
ZAMD$(a = 2)$	0.4848	0.4900	1	0.8292	0.0224	0.0204	0.2075	0.2495	0.5927	0.6727
CWD$(\sigma = 2)$	0.0176	0.0179	1	0.8710	0.0300	0.0176	0.205	0.2543	0.1639	0.0898
BJD	0.1240	0.1204	1	0.8640	0.0270	0.0168	0.2042	0.2530	0.2562	0.2058
Modified B $(\beta = 0.01)$	0.0100	0.0098	1	0.9352	0.0190	0.0180	0.200	0.2526	0.1050	0.0817
NTFD	0	0	0.8846	0.9180	0.0110	0.0110	0.2035	0.2585	0.0541	0.0425

Table 6. Parameters and the Modified Instantaneous Concentration Performance Measure of TFDs for the Time Instant t=64

1. Instantaneous bandwidth relative to component IF $(V_i(t)/f_i(t))$ and,
2. Sidelobe magnitude relative to mainlobe magnitude $(A_S(t)/A_M(t))$.

The measured results are recorded in Table 6, which indicate that the NTFD of signal given by Eqn. (15) yield the smallest values of $\mathbb{C}_{1,2}(t)$ at $t = 64$ and hence is selected as the best concentrated TFD at $t = 64$. To draw a better comparison, the values of $\mathbb{R}_i$ and $\mathbb{C}_{1,2}$ computed for different TFDs are plotted in Fig. 21.

4. Conclusions

The attempt to clearly understand what a time–varying spectrum is, and to represent the properties of a signal simultaneously in time and frequency without any ambiguity, is one of the most fundamental and challenging aspects of signal analysis. A large pubished scientific literature highlights the significance of TF processing with regard to improved concentration and resolution. However as this task is achieved by many different types of TF techniques, it is important to search for the one that is most pertinent to the application. Although the WVD and the spectrogram QTFDs are often the easiest to use, they do not always provide an accurate characterization of the real data. The spectrogram results in a blurred version and the use of the WVD in practical applications has been limited by the presence of CTs and inability to produce ideal concentration for non–linear IF variations. The spectrogram, for example,

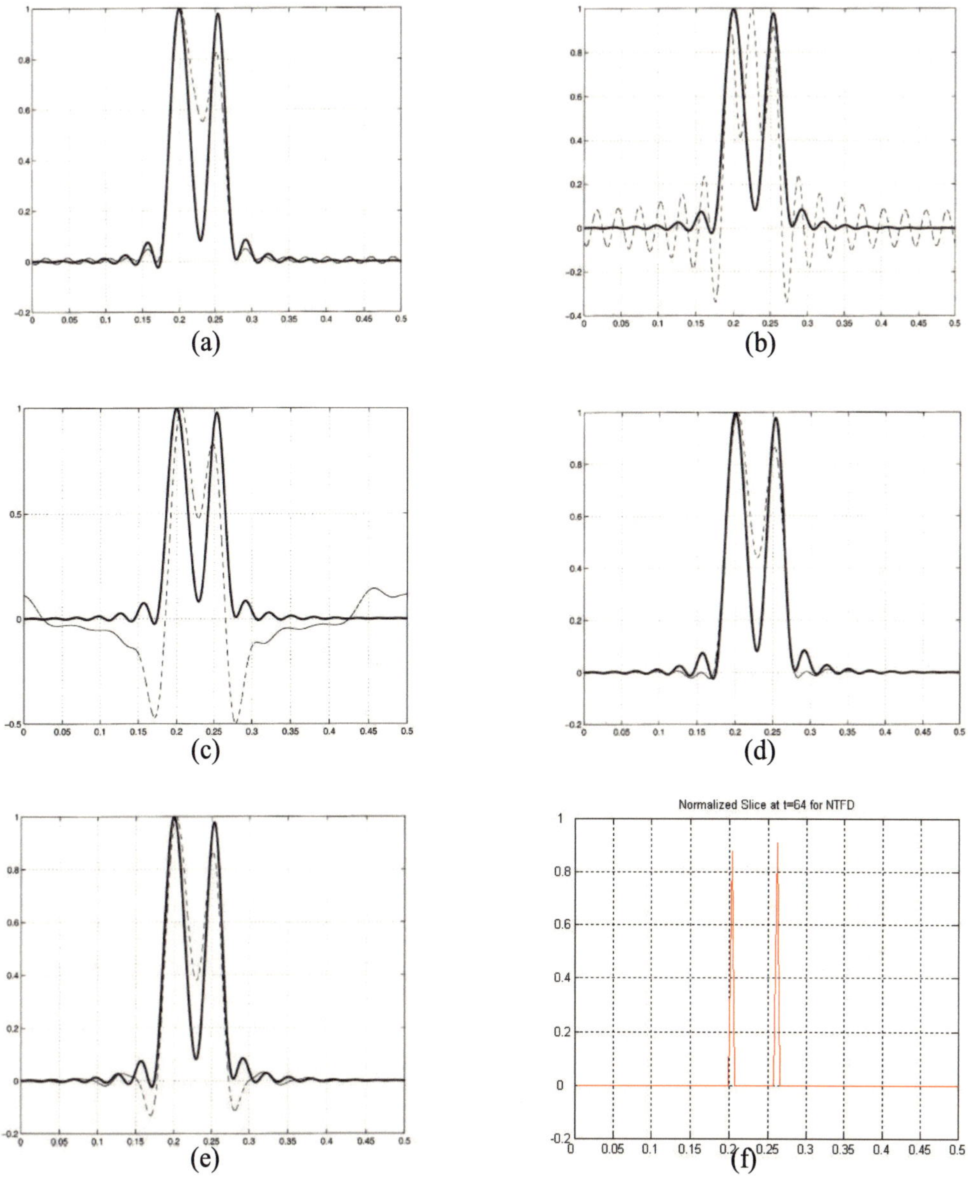

Fig. 20. The normalized slices at $t = 64$ of TFDs, (a) The spectrogram, (b) WVD, (c) ZAMD, (d) CWD, (e) BJD, (f) NTFD. First five TFDs (dashed) are compared against the modified B distribution (solid), adopted from Boashash (26, Boashash & Sucic 2003).

could be used to obtain an overall characterization of the non–stationary signals' structure, and then the information could be used to invest in another QTFD that is well matched to the data for further processing that requires information that is not provided by the spectrogram, the idea conceived and implemented in (32, Shafi et al. 2007).

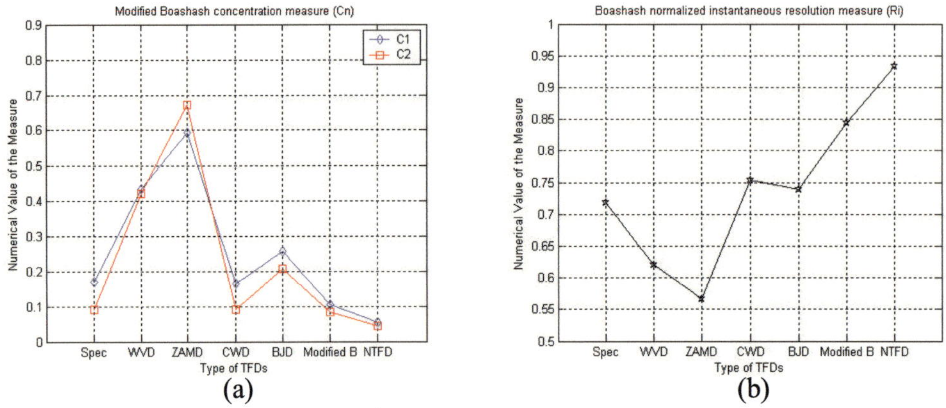

Fig. 21. Comparasion plots for Boashash TFDs' performance measures vs TFDs, (a) The modified concentration measure, and (b) Boashash normalized instantaneous resolution measure

A novel ANN based approach incorporating Bayesian regularization is implemented and evaluated of computing informative, non–blurred and high resolution TFDs. The resulting TFDs do not have the CTs that appear in case of multicomponent signals in some distributions such as WVDs, thus providing visual way to determine the IF of non–stationary signals. The technique explores that the mixture of localized neural networks focused on a specific task deliver a TFD that is highly concentrated along the IF with no CTs as compared to training the ANN which does not receive the selected input. Experimental results presented in section 3 demonstrate the effectiveness of the approach.

For the completeness of proposed framework, the NTFD's performance is further assessed by the information theoretic criteria. These quantitative measures of goodness are used instead of relying solely on the visual measure of goodness of TFDs' plots. The mathematical framework to quantify the TFDs' information is found effective in ascertaining the superiority of the results obtained by the ANN based multiprocesses technique, using both synthetic and real life examples. The NTFD is compared to some popular distributions known for their CTs' suppression and high energy concentration in the TF domain. It is shown that the NTFD exhibits high resolution, no interference terms between the signal components and is highly concentrated. Also it is found to be better at detecting the number of components in a given signal compared to the conventional distributions.

5. References

[1] Boashash, B. (2003). *Time–Frequency Signal Analysis and Processing*, B. Boashash, Ed. Englewood Cliffs, NJ: Prentice–Hall.

[2] Claasen, T. A. C. M. & Mecklenbrauker, W. F. G. (1980). The Wigner distribution–a tool for time–frequency signal analysis; part I: continuous–time signals; part II: discrete time signals; part III: relations with other time–frequency signal transformations. *Philips Journal of Research*, Vol. 35, pp. 217–250, 276–300 and 372–389.

[3] Janse,C. P. and Kaizer,J. M. (1983). Time–frequency distributions of loudspeakers: the application of the Wigner distribution. *Journal of Audio Engg. Soc.*, Vol. 31, pp.198–223.

[4] Boashash, B. (1978). Representation temps–frequence. *Soc. Nat. ELF Aquitaine*, Pau, France, Publ. Recherches, no. 373–378.
[5] Cohen, L. (1995). *Time Frequency Analysis*, Prentice–Hall, NJ.
[6] Flandrin, P. and Escudie, B. (1980). Time and frequency representation of finite energy signals: a physical property as a result of a Hilbertian condition. *Signal Processing*, Vol. 2, pp. 93–100.
[7] Wigner, E. P. (1932). On the quantum correction for thermodynamic equilibrium. *PHYS. Rev.*, Vol. 40, pp. 749–759.
[8] Ville, J. (1946). Theorie et applications de la notion de signal analytique. *cables et Transmission*, Vol. 2, No. 1, pp. 61–74.
[9] Choi, H. and Williams, W.J. (1989). Improved time–frequency representation of multicomponent signals using exponential kernels. *IEEE Trans. Acoust., Speech, Signal Process.*, Vol. 37, No. 6, pp. 862–871.
[10] Shafi, I., Ahmad, J., Shah, S.I., Kashif, F.M. (2009). Techniques to obtain good resolution and concentrated time–frequency distributions–a review. *EURASIP Journal on Advances in Signal Processing*, Vol. 2009 (2009), Article ID 673539, 43 pages.
[11] Margenau, H. and Hill, R. N. (1961). Correlation between measurements in quantum theory. *Prog. Theor. Phys.*, Vol. 26. pp. 772–738.
[12] Hippenstiel, R. D. and Oliveira, P. M. de. (1990). Time varying spectral estimation using the instantaneous power spectrum (IPS). *IEEE Trans. Acoust., Speech, Signal Process.*, Vol. 38, pp. 1752–1759.
[13] Jeong, J. and William, W.J. (1992). Alias–free generalized discrete–time time–frequency distributions. *IEEE Trans. Signal Process.*, Vol. 40, pp. 2757–2765.
[14] Daubechies, I. (1990). The wavelet transform, time–frequency localization, and signal analysis. *IEEE Trans. Inform. Theory*, Vol. 36, pp. 961–1005.
[15] Rioul, O. and Flandrin, P. (1992), Time–scale energy distributions: A general class extending wavelet transforms. *IEEE Trans. Signal Process.*, Vol. 40, pp. 1746–1757.
[16] Bertrand, J. and Bertrand, P. (1988). Time–frequency representations of broadband signals. *Proc. IEEE Intl. Conf on Acoustics, Speech, and Signal Processing (IEEE ICASSP)*, pp. 2196–2199.
[17] Hagan, M.T. Demuth, H.B. & Beale, M. (1996). *Neural Network Design*, Thomson Learning USA.
[18] http://en.wikipedia.org/wiki/Data_clustering.
[19] Gonzalez, R.C. &Wintz, P. (1987). *Digital Image Processing*, 2nd Ed., Addison–Wesley.
[20] Shafi, I., Ahmad, J., Shah, S.I., & Kashif, F.M. (2006). Impact of varying Neurons and Hidden layers in Neural Network Architecture for a Time Frequency Application. *Proc. 10th IEEE Intl. Multi topic Conf., INMIC 2006*, pp. 188-193, Pakistan.
[21] Ahmad, J., Shafi, I., Shah, S.I., & Kashif, F.M. (2006). Analysis and Comparison of Neural Network Training Algorithms for the Joint Time–Frequency Analysis. *Proc. IASTED Intl. Conf on Artificial Intelligence and application*, pp. 193–198, Austria.
[22] Shah, S.I., Shafi, I., Ahmad, J., Kashif, F.M. (2007). Multiple Neural Networks over Clustered Data (MNCD) to Obtain Instantaneous Frequencies (IFs). *Proc. IEEE Intl. Conf. on Information and Emerging Technologies*, pp. 1–6, Pakistan.
[23] Shafi, I., Ahmad, J., Shah, S.I., Kashif, F.M. (2008). Computing De–blurred Time Frequency Distributions using Artificial Neural Networks. *Circuits, Systems, and Signal Processing, Birkhäuser Boston, Springer Verlag*, Vol. 27, No. 3, pp. 277–294.

[24] MacKay, D.J.C. (1992). A Practical Bayesian Framework for Back propagation Network. *Neural Computation*, Vol. 4, No. 3, pp. 448–472.
[25] Gray, R.M. (1990). *Entropy and Information Theory*, New York Springer–Verlag.
[26] Boashash, B. and Sucic, V. (2003). Resolution Measure Criteria for the Objective Assessment of the Performance of Quadratic Time–Frequency Distributions. *IEEE Trans. Signal Process.*, Vol. 51, No. 5, pp. 1253–1263.
[27] http://www–dsp.rice.edu.
[28] Baraniuk, R. G. and Jones, D. L. (1993). Signal–Dependent Time–Frequency Analysis Using a Radially Gaussian Kernel. *Signal Processing*, Vol. 32, No. 3, pp. 263–284.
[29] Jones, D. L., Parks, T. W. (1992). A Resolution Comparison of Several Time–Frequency Representations. *IEEE Trans. Signal Process.*, Vol. 40, No. 2.
[30] Jones, D. L. and Parks, T. W. (1990). A high resolution data–adaptive time–frequency representation. *IEEE Trans. Acoust., Speech, Signal Process.*, Vol. 38, pp. 2127–2135.
[31] Stankovic, LJ. (2001). A Measure of Some Time–Frequency Distributions Concentration. *Signal Processing*, Vol. 81, No. 3, pp. 212–223.
[32] Shafi, I., Ahmad, J., Shah, S.I., Kashif, F.M. (2007). Evolutionary time–frequency distributions using Bayesian regularised neural network model. *IET Signal Process.*, Vol. 1, No. 2, pp. 97–106